Metamorphic Crystallization

RALPH KRETZ

Department of Geology
The University of Ottawa,
Ontario, Canada

JOHN WILEY & SONS
Chichester • New York • Brisbane • Toronto • Singapore

Baffins Lane, Chichester,
West Sussex PO19 1UD, England
National Chichester (0243) 779777
International +44 243 779777

Other Wiley Editorial Offices

John Wiley & Sons, Inc., 605 Third Avenue,
New York, NY 10158-0012, USA

Jacaranda Wiley Ltd, 33 Park Road, Milton,
Queensland 4064, Australia

John Wiley & Sons (Canada) Ltd, 22 Worcester Road,
Rexdale, Ontario M9W 1L1, Canada

John Wiley & Sons (SEA) Pte Ltd, 37 Jalan Pemimpin #05-04,
Block B, Union Industrial Building, Singapore 2057

Library of Congress Cataloging-in-Publication Data

Kretz, Ralph, 1930-
Metamorphic crystallization / Ralph Kretz.
p. cm.
Includes bibliographical references and index.
ISBN 0-471-94214-6
1. Metamorphism (Geology) I. Title.
QE475.A2K734 1994
552′.4 — dc20 94-4637
CIP

British Library Cataloguing in Publication Data

A catalogue record for this book is available from the British Library

ISBN 0-471-94214-6

Typeset in 10/12pt Times by Laser Words, Madras, India
Printed and bound in Great Britain by Biddles, Guildford, Surrey

Contents

Preface

'Metamorphism' refers to solid-state changes that have occurred within the Earth during the long history of the planet. Our understanding of these changes has advanced rapidly, resulting mainly from continued field studies, improved analytical methods, and new experimental investigations. This book is an introduction to the science of metamorphism, prepared for senior undergraduate and graduate students in geology and geochemistry.

It may be helpful to begin with a brief review of the traditional methods and common materials of metamorphic studies, and this is the purpose of Chapter 1. Proceeding to the more detailed study of natural crystals and crystal assemblies, it is reasonable to examine them first from the viewpoint of thermodynamic equilibrium, referring where possible to experimental results. This is the theme of Chapter 2, which is a review of thermodynamics as applied to mineral systems, and Chapter 3, where several kinds of mineral reactions are examined in terms of chemical equilibrium. The next step is a search for understanding of the kinetics of the various metamorphic processes that can be identified. This is attempted in Chapter 4, which is a review of mineral kinetics, drawing frequently on results obtained in Physical Chemistry and Metallurgy, and Chapter 5, where several kinds of granular microstructure are examined for the information they convey regarding metamorphic processes.

A very large literature of natural and experimental information is now available, and it was necessary to select certain data to illustrate the different kinds of change that are of interest. The writer may (or may not) be excused for choosing, as examples, many of his own research results, and for omitting others that every reader will no doubt feel should have been included.

It would have been appropriate to deal here with the problem of the orientation of crystals (now referred to as the 'texture') which is, admittedly, an important aspect of granular microstructure. The subject, which has grown rapidly, is well covered in a book edited by H.-R. Wenk (1985) and in several earlier texts.

A word or two on units and symbols (see Appendices A1 to A4). In general, pressure is expressed as kilobars (kbar), which is easily converted to megapascals ($\times 100$). The preferred unit of energy is the joule (J), but data from the literature, reported in calories, were not always converted to joules. The principal units of time are the second (s) and year (y). Three or four different units are used for concentration. Thus SI units are not used throughout but this should not create a difficulty for the reader.

Over the years I have derived much inspiration and benefit from former teachers, colleagues, and students, and from the writings of others, and all of these people have contributed importantly to the preparation of this book. Comments on the manuscript by J.-M. Sempels were helpful, and the assistance of Michèle Clermont, Sylvie Downing, J.-M. Sempels, and E.W. Hearn in the preparation of the manuscript is deeply appreciated. I wish also to thank the staff of John Wiley & Sons for their careful work in transforming an imperfect manuscript to this final product.

Acknowledgements

Directors and editors of the following publishing firms and societies have granted permission to reprint the following figures and tables.

FIGURES

American Ceramic Society (1.13)
American Geophysical Union (1.6)
American Institute of Physics (3.20)
American Journal of Science (1.28, 2.10, 2.22, 3.28, 3.32, 4.7, 4.8, 4.28 inset, 5.55)
American Physical Society (1.12)
Blackwell Scientific Publications, Inc. (2.9, 3.40, 3.41, 3.45, 3.46, 3.47, 5.3 to 5.12, 5.56)
Chapman & Hall (1.19, 5.65)
Elsevier Science Publishers (1.14, 5.49, 5.50)
W.H. Freeman and Company (1.29, 1.31, 3.11)
Gauthier-Villars (4.25)
Hemisphere Publishing Corp. (1.23)
IOP Publishing Limited (4.14)
John Wiley & Sons, Inc. (1.16)
Kluwer Academic Publishers (3.21 inset)
Marcel Dekker Inc. (1.4, 1.5)
McGraw-Hill, Inc. (4.1)
Mineralogical Society of America (1.1, 1.7, 1.25b, 2.2, 2.8, 2.15, 2.20, 3.2, 3.10, 3.12, 3.24, 3.31c, 3.65, 4.19, 5.26, 5.27, 5.30)
The Mineralogical Society, London (3.22, 5.39a)
Minister of Supply and Services, Canada (1.17, 5.31, 5.70)
National Research Council of Canada (1.9b, 5.15, 5.16, 5.17, 5.19, 5.37c, 5.45, 5.53, 5.59, 5.61, 5.62, 5.64 to 5.66)
R. Oldenbourgh Verlag (3.21)
Oxford University Press (1.26, 3.29, 3.62, 3.63, 4.18, 5.42 to 5.44, 5.68)
Oxford University Press, Inc. (3.6)
Pergamon Press Ltd. (1.2, 1.3, 2.14, 2.17, 2.21, 4.15, 4.17a, b, 4.32a, 5.21b, c)
Plenum Publishing Corp. (4.21)
The Royal Society of Chemistry (4.4, 4.5, 5.46)
E. Schweizerbart'sche Verlag (3.35, 3.36)

Springer-Verlag (1.10b, 1.10c, 1.20, 1.27, 2.19, 3.1, 3.3, 3.7, 3.17, 3.30, 3.39, 4.17c, 5.47, 5.48, 5.71 to 5.74)
Stäubli Verlag (3.16)
University of California Press (5.21a)
The University of Chicago Press (3.15, 3.56, 3.59, 3.60, 3.61)
J. Williams Book Co. (4.10, 4.13, 4.23)

TABLES

American Journal of Science (2.3)
Blackwell Scientific Publications Inc. (3.6, 5.1)
National Research Council of Canada (5.9)
Oxford University Press (2.1)
Pergamon Press Ltd. (3.2)
Springer-Verlag (5.10 to 5.13)
J. Williams Book Co. (4.1)

Approval to reproduce certain results was obtained, with thanks, from the following colleagues: L. Aranovich, R.G. Berman, S.R. Bohlen, W.L. Brown, C.W. Burnham, N.D. Chatterjee, J. Chernosky, R. Christofferson, M. Cohen, M.K. Debe, W.G. Ernst, E.J. Essene, J.M. Ferry, G.W. Fisher, V.I. Fonarev, W.K. Fyson, J.R. Goldsmith, S. Haffner, B.S. Hemingway, L.S. Jen, W. Johannes, D.M. Kerrick, A. Koziol, W. Lamb, D.H. Lindsley, K.L. Merkle, P. Metz, M. Morioka, R.C. Newton, T. Rivers, R.A. Robie, S.K. Saxena, H.A. Seck, J.M. Sempels, G. Skippen, J.V. Smith, F. Spear, K. Suzuki, J.B. Thompson, P.H. Thompson, V. Trommsdorff, D.R. Veblen, D. Virgo, G. Voll, P.F. Williams, B. Wood, R. Yund.

1 The Geological Background

We begin our study with a definition of metamorphism, a brief historical note, a tabulation of metamorphic minerals, a review of fluids and crystal boundaries, and a brief description of the more traditional methods that are used in metamorphic petrology. These methods include field studies, the experimental investigation of simple mineral equilibria, and the interpretation of mineral assemblages in terms of mineral reactions and gradients in temperature and pressure. Chapter 1 therefore provides a background for the more detailed study of mineral equilibria and reactions of the following chapters. More abundant information on mineral assemblages and microstructures in metamorphic terranes is available in the excellent literature reviews prepared by Barker (1990), Winkler (1974), Greenwood (1976), Turner (1981), Spry (1969), Vernon (1976), Yardley (1989), Miyashiro (1973), Spear (1993) and in a volume edited by Kerrick (1991).

1.1 METAMORPHISM DEFINED

Metamorphism (*μετα μορφη*; change of form) is a term used to refer to solid-state physical-chemical changes that take place in the Earth and other planetary bodies. It is now known with certainty that mineral assemblages of the common sedimentary and volcanic rocks, when buried and subjected to high temperatures and pressures, become thermodynamically unstable, resulting in the production of new minerals, deformed crystals, and new microstructures. Primary layering is commonly preserved, and the mineral changes are therefore presumed to be essentially solid-state reactions, rather than changes resulting from melting and crystallization from the melt. Experimental data are in support of this interpretation, but they have shown, in addition, that partial melting of certain rocks is possible in the middle and lower Crust, provided water is available.

Chemical analyses indicate that, apart from the addition or removal of H_2O and CO_2, metamorphism commonly occurs with little or no addition of material; locally, however, large amounts of K, B, Si and other elements were obviously added or removed during metamorphism — a process known as 'metasomatism'.

Metamorphic rocks are extremely abundant in the Crust, and presumably also in the Mantle. They are well exposed in mountain belts such as the Appalachians and the Alps, and in Precambrian shield areas, such as the Canadian, Baltic, and African Shields. But metamorphism, when defined as a solid-state reaction, occurs also in many igneous rocks, including gabbro and granite, as they cool below solidus temperatures. Moreover, evidence for metamorphism, including shock metamorphism is found in some meteorites and lunar rocks, and metamorphism is no doubt an important process in the history of other planetary bodies.

Metamorphic rocks are also of interest because they represent remarkable demonstrations of the behavior of matter. The many chemical changes that take place when a mudstone is transformed to garnet muscovite schist, for example, all proceed in the direction of lower energy, as required by the laws of thermodynamics. The principles of chemical kinetics are displayed in the diversity that exists in the microstructure of metamorphic rocks, resulting from variation in reaction mechanisms, and in rates of deformation, crystal nucleation, and crystal growth.

1.2 HISTORICAL NOTE

Some thoughts on the nature of minerals and rocks were recorded in Ancient Greece and Rome. For example, numerous rocks including varieties of marble were described by Theophrastus (c. 370–287 B.C.), and the atomic structure of crystals was anticipated by Lucretius (c. 95–55 B.C.). But the origin of minerals and rocks was not understood. Thus, Plato (c. 427–347 B.C.) declared that 'earth thrust together by air, so as not to be soluble by water, forms stone', and Aristotle (384–322 B.C.) suggested that rays from the Sun and other stars play a role in the 'generation of stones'.

The writings of Georgius Agricola (1494–1555) show that in 16th century Europe, naturalists had obtained an extensive knowledge of minerals, mineral deposits, and rocks, but rock-forming processes remained a mystery. In 1665 the author of a book, *Concerning the Origin of Stones*, insisted that stones, like plants, reproduce, 'each after its own kind'. An important contribution to the understanding of sedimentary rocks came from Nicolaus Steno (1638–1686), originally from Denmark, who examined folded strata in northern Italy and concluded that they were laid down as sediments on the floor of an ancient sea. Although volcanoes were active in the Mediterranean landscape, the volcanic origin for basalts remained unrecognized until the 18th century. Thus, Jean Etienne Guettard (1715–1786) and Nicholas Desmarest (1725–1815) in 1750 to 1770 determined that the basaltic hills of the Auvergne district in France are ancient volcanoes. Regarding granite, James Hutton (1726–1797) reported that certain plutonic rocks in Scotland had in his opinion crystallized from a melt. The prevailing opinion at the time was that both basalt and granite had precipitated from a primeval sea.

Concerning the origin of schist, gneiss, and other metamorphic rocks, Giovanni Arduino (1713–1795) found evidence in the Italian Alps for repeated upheavals and subsidences, causing contortion of the strata, processes he referred to as 'metamorphoses'. He also commented on the alteration of limestone to marble, and in 1779 prepared one of the earliest statements of rock metamorphism:

> 'The limestone rocks were broken and ground by the volcanic fire, they were calcined, and their dark colour gave way to the most glaring white, passing from the state of ordinary reddish limestone to that of marble.'

At the same time, James Hutton, referred to above, proposed that certain schists in Scotland represent altered sedimentary strata. These thoughts were expressed in 1785 in the following (condensed) sentence:

> 'If we find a mass which evidently formed originally in the manner of stratification, but which is now distorted and displaced — which is also extremely consolidated and variously changed in composition, — which therefore has the marks of its original or marine composition extremely obliterated, we should then have reason to suppose that here were masses of matter which, though not different in their origin from those that are gradually deposited at the bottom of the ocean, have been acted upon by subterranean heat and the expanding power, that is to say, have been changed in a greater degree by the operations of the mineral kingdom.'

Thus a change from shale to schist within the Earth's Crust was viewed by Hutton as a mineralogical change brought about by an increase in temperature and pressure. At the time, this represented a remarkable insight into the behavior of matter, and must be regarded as the beginning of the study of metamorphism. The term 'metamorphism' was introduced a little later, in 1830, by Charles Lyell (1797–1875).

It soon became evident that not all metamorphic rocks experienced the same intensity of metamorphism. Thus Harry Rosenbusch (1836–1914), in 1877, working in the Vosges region, described a granite pluton and the surrounding metamorphic aureole, which consists of zones of spotted slate, spotted mica schist, and hornfels, progressively nearer the central granite. In 1893, George Barrow (1855–1932) presented his well-known report on progressive metamorphism in the Grampian Highlands of Scotland. Three zones and isograds were defined, based on the appearance of staurolite, kyanite, and sillimanite, representing a presumed increase in temperature.

Some of the advances in Chemistry and Physics during the 19th century were of great assistance to Petrology. Early in the century, following the discovery of many of the common rock-forming elements (Si, Al, Ca, K, Na), the Swedish chemist, Jons Berzelius (1779–1848) developed methods of obtaining chemical analyses of minerals, and he recognized the law of constant proportions. Both Berzelius and Eilhard Mitscherlich (1794–1863) in Germany contributed to the discovery of the important principle of isomorphism.

The improvement of the microscope in the 19th century enabled geologists to examine rocks in much greater detail. Early work on the optical properties of crystals by William Nicol (c. 1768–1851) and others enabled Henry Sorby (1826–1908) in Scotland to initiate the microscopic study of thin slices of rock. His paper 'On the microscopic structure of crystals, indicating the origin of minerals and rocks', was published in 1858.

The development of thermodynamics during the latter half of the 19th century, was of great importance and the concepts developed by the American physicist J. Willard Gibbs (1839–1903) in his 'Equilibrium of heterogeneous substances' (1878) are particularly useful.

The discovery of X-rays by Wilhelm Röntgen (1845–1923) in 1895 and the well-known experiment by Max von Laue (1879–1960) in 1912 opened the door to the valuable information we now possess regarding the structure of minerals.

The modern approach to metamorphic petrology was initiated by Victor M. Goldschmidt (1888–1947) in 1911, in his study of contact metamorphism in the Oslo district of Norway. Pelitic and calcareous rocks of the inner zone of aureoles about small plutons of syenite and granite were found to contain cordierite, andalusite, hypersthene, diopside, grossularite, wollastonite, vesuvianite, biotite, plagioclase, orthoclase, and quartz. Goldschmidt found that these minerals were not intermixed at random, but occur together only in certain combinations. For instance, andalusite and diopside were not found together, while the chemically equivalent pair enstatite and anorthite occurred frequently. Observed restrictions of this kind indicated to Goldschmidt that chemical equilibrium was established within the contact rocks. This was in agreement with the theory of Gibbs which showed that the laws of thermodynamics impose restrictions on the kind and number of phases or minerals that may occur together at equilibrium.

Further historical information may be found in the writings of Geike (1905), Adams (1938), Loewinson-Lessing (1954), Read (1957), Turner (1958), and Burke (1966) from which the above notes were obtained.

During the past four decades, metamorphic petrology has advanced rapidly. The following areas of study may be identified.

(1) Field studies continue to lead to new discoveries, new problems, and new interpretations.
(2) The electron microprobe, which was introduced by R. Castaing in 1951, has provided abundant information on the composition of minerals, and is particularly useful for the analysis of zoned crystals and crystal intergrowths.
(3) Several new analytical instruments (e.g., the ion and proton microprobes) are now providing microanalyses of trace elements and isotopes.
(4) The study of radioactive and stable isotopes is providing information on the timing of metamorphism, and on the mobility of H_2O and other elements in contact aureoles and metamorphic terranes.
(5) Scanning and transmission electron microscopy and other techniques are revealing an astonishing diversity of crystal topography and microstructure, ranging in scale from 1 μm down to 0.1 nm.
(6) Urgently needed data on the heat of formation and heat capacity and other properties of minerals and mineral solutions are becoming available.
(7) In addition to phase equilibrium studies, the experimental approach is now being applied to investigate the distribution of trace elements and isotopes between minerals, rates of mineral reactions, and rates of diffusion.
(8) Computer-assisted calculations of mineral equilibria, and the computer simulation of various metamorphic processes are progressing rapidly.
(9) The thermal–tectonic modelling of large segments of the Crust, based on the principles of plate tectonics, heat-flow calculations, and temperature–pressure data from metamorphic rocks, has attracted much attention in recent years.

1.3 METAMORPHIC MINERALS

Bodies of metamorphic rock consist of the crystals of one or more minerals. In places it is possible to recognize relict minerals (crystals that survived the most recent metamorphism intact), and secondary minerals (crystals, usually fine-grained, that formed during cooling) in addition to the primary minerals which crystallized or achieved their present composition during the peak of metamorphism. All of the primary minerals whose crystals are intermingled in a sample taken from the rock body are referred to as a 'mineral assemblage'. In general, all minerals of an assemblage are of equal importance, regardless of abundance, and the term 'accessory mineral' should be avoided.

Most minerals of metamorphic rocks are complex solid solutions, containing numerous elements. Nevertheless it is usually possible to define simplified formulas which exclude most elements present in very low concentrations, and ignore minor compositional variation in some major elements. Mineral reactions may then be defined by use of these formulas. Table 1.1 is a list of most of the minerals found in metamorphic rock; many of these are also important constituents of the igneous and sedimentary rocks.

In the table, the minerals are grouped as silicates and non-silicates, and the silicates are divided according to structure (islands, rings, chains, sheets, frameworks) in the usual manner. General features of the silicate structures are well known and will not be reviewed here.

For some minerals the simplified formula as written in Table 1.1 expresses fairly well the composition as determined by chemical analysis. In garnet, for example, the (Mg,Fe,Ca,Mn):(Al,Fe^{3+}):(Si) ratios are commonly found to be very close to 3:2:3 as required by the ideal crystal structure. In biotite, on the other hand, the composition of natural crystals may depart considerably from the simplified formula, $K(Mg,Fe)_3AlSi_3O_{10}(OH)_2$. Thus a comparison of analyses of biotite from common gneisses (Table 1.2) with this formula leads to the following departures: (a) the ratio Al:(Si+Al) in the tetrahedral sites of the analyzed biotite exceeds 0.25, and some Al is present in the octahedral sites as well, (b) notable amounts of Ti and Fe^{3+} are present in the octahedral sites and some of these sites are vacant, (c) minor Na and Ca are present, presumably in the K site, and some of these sites are also vacant, and (d) some F and Cl are present in the (OH) sites. More complex formulas can of course be written, when necessary, for biotite and other minerals.

In addition, all minerals contain low concentrations (less than about 1000 ppm or 0.1 weight %) of various elements of the periodic table; examples of the presence of such 'trace elements' in biotite are given in Table 1.2. Commonly the positions that these atoms occupy in the crystal structures are not known with confidence. Variations also exist in the isotopic composition of oxygen, carbon, and other elements. Although trace elements and isotopes do not appear in Table 1.1, their behavior within the Earth's Crust is a subject of much interest.

Table 1.1 Table of mineral and component names, symbols, and simplified formulas

SiO_4 islands		
olivine	Ol	$(Mg,Fe)_2SiO_4$
forsterite	Fo	Mg_2SiO_4
fayalite	Fa	Fe_2SiO_4
zircon		$ZrSiO_4$
humite group		$(Mg,Ti)(OH,F,O)_2 \cdot n(Mg_2SiO_4)$
sphene (titanite)		$CaTiSiO_4(O,OH,F)$
garnet	Grt	$(Fe,Mg,Mn,Ca)_3(Al,Fe^{3+})_2Si_3O_{12}$
almandine	Alm	$Fe_3Al_2Si_3O_{12}$
pyrope	Prp	$Mg_3Al_2Si_3O_{12}$
spessartine	Sps	$Mn_3Al_2Si_3O_{12}$
grossular	Grs	$Ca_3Al_2Si_3O_{12}$
andradite	Adr	$Ca_3Fe^{3+}{}_2Si_3O_{12}$
sillimanite	Sil	Al_2SiO_5
kyanite	Ky	Al_2SiO_5
andalusite	And	Al_2SiO_5
staurolite	St	$(Fe,Mg)_4(Al,Fe^{3+})_{18}(Si,Al)_8O_{46}(OH)_2$
yoderite		$Mg_2Al_6Si_4O_{18}(OH)_2$
chloritoid	Cld	$(Fe,Mg)_2(Al,Fe^{3+})Al_3Si_2O_{10}(OH)_4$
pumpellyite	Pmp	$Ca_2Al_2(Mg,Fe,Al)SiO_4Si_2O_6(O,OH)_4$
monticellite		$CaMgSiO_4$
larnite		Ca_2SiO_4
merwinite		$Ca_3MgSi_2O_8$
spurrite		$Ca_5Si_2O_8(CO_3)$
kornerupine		$Mg_4Al_6(Si,B)_5O_{21}(OH)$
dumortierite		$Al_7Si_3O_{12}(BO_3)O_3$
grandidierite		$(Mg,Fe)Al_2(SiO_4)(BO_3)O_2$
ellenbergerite		$(Mg,Ti,Zr,\square)_2Mg_6(Al,Mg)_6(Si,P)_2$ $Si_6O_{28}(OH)_{10}$
SiO_4 and Si_2O_7 islands		
zoisite	Zo	$Ca_2Al_3Si_3O_{12}(OH)$
clinozoisite	Czo	$Ca_2Al_3Si_3O_{12}(OH)$
epidote	Ep	$Ca_2(Al,Fe^{3+})_3Si_3O_{12}(OH)$
vesuvianite		$Ca_{10}(Mg,Fe)_2Al_4(Si_2O_7)_2(SiO_4)_5(OH,F)_4$
Si_2O_7 islands		
rankinite		$Ca_3Si_2O_7$
tilleyite		$Ca_5Si_2O_7(CO_3)_2$
gehlenite		$Ca_2Al_2SiO_7$
akermanite		$Ca_2MgSi_2O_7$
lawsonite		$CaAl_2Si_2O_7(OH)_2 \cdot H_2O$
Si_6O_{18} rings		
cordierite	Crd	$(Mg,Fe)_2Al_3(Si_5Al)O_{18} \cdot n(H_2O,CO_2)$
osumalite		$(K)(Fe,Mg)_2Al_3(Si,Al)_{12}O_{30}$
tourmaline		$Na(Mg,Fe,Fe^{3+})_3Al_6Si_6O_{18}(BO_3)_3(OH,F)_4$
SiO_3 single chains		
orthopyroxene	Opx	$(Mg,Fe)SiO_3$
enstatite	En	$MgSiO_3$
ferrosilite	Fs	$FeSiO_3$
pigeonite	Pgt	$(Mg,Fe,Ca)SiO_3$

Table 1.1 *(continued)*

SiO_3 single chains (continued)		
Ca pyroxene	Cpx	$(Ca,Fe,Mg)(Mg,Fe,Fe^{3+},Al)(Si,Al)_2O_6$
diopside	Di	$CaMgSi_2O_6$
hedenbergite	Hd	$CaFeSi_2O_6$
omphacite	Om	$(Ca,Na)(Mg,Fe,Fe^{3+},Al)Si_2O_6$
jadeite	Jd	$NaAlSi_2O_6$
aegirine-augite		$(Na,Ca)(Fe^{3+},Fe,Mg,Al)Si_2O_6$
wollastonite	Wo	$CaSiO_3$
sapphirine	Spr	$(Mg,Fe,Fe^{3+},Al)_8O_2(Al,Si)_6O_{18}$
carpholite	Crp	$(Mg,Fe,Mn)_2Al_2Si_2O_6(OH)_4$
Ca-tschermak component		$CaAl(AlSi)O_6$
Si_4O_{11} double chains		
anthophyllite	At	$(Mg,Fe)_7Si_8O_{22}(OH)_2$
Mg anthophyllite	MgAt	$Mg_7Si_8O_{22}(OH)_2$
Fe anthophyllite	FeAt	$Fe_7Si_8O_{22}(OH)_2$
gedrite	Ged	$(Mg,Fe,Al)_7(Si,Al)_8O_{22}(OH)_2$
cummingtonite	Cm	$(Mg,Fe)_7Si_8O_{22}(OH)_2$
Mg cummingtonite	MgCm	$Mg_7Si_8O_{22}(OH)_2$
grunerite	Grn	$Fe_7Si_8O_{22}(OH)_2$
actinolite	Act	$Ca_2(Mg,Fe)_5Si_8O_{22}(OH)_2$
tremolite	Tr	$Ca_2Mg_5\ Si_8O_{22}(OH)_2$
Fe actinolite	FeAct	$Ca_2Fe_5Si_8O_{22}(OH)_2$
hornblende	Hbl	$(Ca,Na)_{2-3}(Mg,Fe,Fe^{3+},Al)_5(Si,Al)_8O_{22}(OH)_2$
Na amphibole		$Na_2(Mg,Fe^{3+})_3(Al,Fe^{3+})_2Si_8O_{22}(OH)_2$
glaucophane		$Na_2Mg_3Al_2Si_8O_{22}(OH)_2$
Fe glaucophane		$Na_2Fe_3Al_2Si_8O_{22}(OH)_2$
Mg riebeckite		$Na_2Mg_3Fe^{3+}{}_2Si_8O_{22}(OH)_2$
Fe riebeckite		$Na_2Fe_3Fe^{3+}{}_2Si_8O_{22}(OH)_2$
Unusual structures		
deerite		$(Fe,Mn)_6(Fe^{3+},Al)_3Si_6O_{20}(OH)_5$
howieite		$Na(Fe,Mn,Mg,Fe^{3+},Al)_{12}Si_{12}(O,OH)_{44}$
zussmanite		$K(Fe,Mn,Mg)_{13}(Si,Al)_{18}O_{42}(OH)_{14}$
$(Si,Al)_2O_5$ sheets		
muscovite	Ms	$KAl_2(Si_3Al)O_{10}(OH)_2$
paragonite	Pg	$NaAl_2(Si_3Al)O_{10}(OH)_2$
phengite	Phn	$K(Al,Mg,Fe)_2(Si,Al)_4O_{10}(OH)_2$
celadonite		$K(Mg,Fe)AlSi_4O_{10}(OH)_2$
biotite	Bt	$K(Mg,Fe,Fe^{3+},Ti,Al,\square)_3(Si,Al)_4O_{10}(OH)_2$
phlogopite	Ph	$KMg_3(Si_3Al)O_{10}(OH)_2$
annite	Ann	$KFe_3(Si_3Al)O_{10}(OH)_2$
eastonite		$K_2Mg_5Al(Si_5Al_3)O_{20}(OH)_4$
siderophyllite		$K_2Fe_5Al(Si_5Al_3)O_{20}(OH)_4$
margarite	Mrg	$CaAl_2(Si_2Al_2)O_{10}(OH)_2$
stilpnomelane	Stp	$(K,Na,Ca)(Fe,Mg,Al)_{12}(Si,Al)_{16}(O,OH)_{54}\cdot nH_2O$
pyrophyllite		$Al_2Si_4O_{10}(OH)_2$
talc	Tc	$Mg_3Si_4O_{10}(OH)_2$
chlorite	Chl	$(Mg,Fe,Al)_6(Si,Al)_4O_{10}(OH)_8$
serpentine	Srp	$Mg_6Si_4O_{10}(OH)_8$

continued overleaf

Table 1.1 (*continued*)

(Si, Al)$_2$O$_5$ sheets (continued)		
prehnite		$Ca_2Al(Si_3Al)O_{10}(OH)_2$
kaolinite		$Al_4Si_4O_{10}(OH)_8$
illite		$K_{1-1.5}Al_4(Si_{7-6.5}Al_{1-1.5})O_{20}(OH)_4$
montmorillonite		$(Ca,Na)_{0.7}(Al,Mg,Fe)_4(Si,Al)_8O_{20}(OH)_4 \cdot nH_2O$
(Si, Al)O$_2$ frameworks		
K-feldspar	Kf	$(K,Na)AlSi_3O_8$
microcline		$KAlSi_3O_8$
orthoclase		$KAlSi_3O_8$
sanidine		$KAlSi_3O_8$
albite		$NaAlSi_3O_8$
plagioclase	Pl	$(NaAlSi_3O_8),(CaAl_2Si_2O_8)$
albite	Ab	$NaAlSi_3O_8$
anorthite	An	$CaAl_2Si_2O_8$
quartz	Qtz	SiO_2
coesite		SiO_2
stishovite		SiO_2
nepheline	Ne	$(Na,K)AlSiO_4$
scapolite	Scp	$[Na_4(Al_3Si_9O_{24})Cl],[Ca_4(Al_6Si_6O_{24})CO_3]$
marialite	Mar	$Na_4(Al_3Si_9O_{24})Cl$
meionite	Mei	$Ca_4(Al_6Si_6O_{24})CO_3$
laumontite		$CaAl_2Si_4O_{12} \cdot 4H_2O$
natrolite		$Na_2Al_2Si_3O_{10} \cdot 2H_2O$
heulandite		$(Ca,Na_2)Al_2Si_7O_{18} \cdot 6\ H_2O$
Non-silicates		
graphite		C
diamond		C
periclase	Per	MgO
corundum		Al_2O_3
hematite	Hem	$Fe^{3+}{}_2O_3$
ilmenite	Ilm	$FeTiO_3$
rutile		TiO_2
perovskite		$CaTiO_3$
spinel	Spl	$(Mg,Fe,Zn)(Al,Fe^{3+},Cr)_2O_4$
MgAl spinel		$MgAl_2O_4$
hercynite		$FeAl_2O_4$
gahnite		$ZnAl_2O_4$
magnetite	Mt	$FeFe^{3+}{}_2O_4$
ulvöspinel	Usp	Fe_2TiO_4
chromite	Chr	$FeCr_2O_4$
pseudobrookite		$Fe^{3+}{}_2\ TiO_5$
brucite		$Mg(OH)_2$
högbomite		$Fe_5Al_{16}TiO_{30}(OH)_2$
pyrite		FeS_2
pyrrhotite		$Fe_{1-x}S$
calcite	Cal	$(Ca,Mg)CO_3$
magnesite	Mgs	$MgCO_3$
siderite	Sid	$FeCO_3$
dolomite	Dol	$CaMg(CO_3)_2$
ankerite	Ank	$Ca(Mg,Fe,Mn)(CO_3)_2$

Table 1.1 *(continued)*

Non-silicates (continued)		
aragonite	Arg	$CaCO_3$
apatite	Ap	$Ca_5(PO_4)_3(OH,F,Cl)$
fluorite		CaF_2
halite		$NaCl$

□ denotes vacant site.

Fe denotes ferrous iron, Fe^{3+} ferric iron.

Symbols for components are not capitalized, e.g., ab is the symbol for $NaAlSi_3O_8$ component in plagioclase.

Notes Mineral formulas from Deer, Howie and Zussman (1963, 1978), Povarennykh (1972) and Nickel and Nichols (1991). Additional structural and compositional information for the following minerals was obtained from the sources in brackets: humite group (Robinson, Gibbs and Ribbe 1973), staurolite (Griffin, Gosney and Phillips 1982, Holdaway et al. 1991), yoderite (McKie 1959, Fleet and Megaw 1962), kornerupine (Moore, Sen Gupta and Schlemper 1989), grandidierite (Lonker 1988), ellenbergerite (Chopin et al. 1986), osumilite (Armbruster and Oberhänsli 1988), carpholite (Schreyer 1988a), deerite, howieite, and zussmanite (Agrell, Bown and McKie 1965, Wenk 1974; Fleet 1977, Wood 1979, Dudek and Kienast 1989), pseudobrookite (Smith 1965), högbomite (Petersen et al. 1989).

Table 1.2 Analyses[1] of biotite from garnet biotite gneiss

					Formulas[2]		
	1	2	3		1	2	3
SiO_2/wt%	35.0	34.9	36.2	Si	2.68	2.67	2.70
Al_2O_3	18.4	17.7	15.5	Al^{iv}	1.32	1.33	1.30
TiO_2	3.86	4.51	5.40	Σ	4.00	4.00	4.00
Fe_2O_3	3.10	3.30	2.00	Al^{vi}	0.39	0.25	0.06
FeO	18.7	18.1	17.2	Ti	0.22	0.26	0.30
MgO	6.77	7.31	11.30	Fe^{3+}	0.16	0.19	0.11
MnO	0.02	—	—	Fe^{2+}	1.20	1.16	1.07
K_2O	9.16	9.42	9.91	Mg	0.77	0.83	1.25
Na_2O	0.06	0.10	—	Mn	0.001	—	—
CaO	0.01	0.09	—	Σ	2.74	2.70	2.80
H_2O	3.50	3.40	3.10	K	0.90	0.92	0.94
F	0.47	0.36	0.56	Na	0.01	0.01	—
Cl	0.08	0.07	1.09	Ca	0.001	0.01	—
				Σ	0.91	0.94	0.94
V/ppm	220	400	690	F	0.11	0.09	0.13
Cr	82	150	230	Cl	0.01	0.01	0.14
Zn	720	1000	630	OH	1.80	1.70	1.50
Rb	770	760	640	$Mg/(Mg+Fe^{2+})$	0.39	0.42	0.54
Sr	<10	<10	<10				
Y	13	17	12				
Zr	120	170	33				
Nb	110	77	42				
Ba	540	620	980				

[1]Cations by microprobe, FeO by titration, F and Cl by microprobe and ion chromatography, H_2O by infra-red absorption and Penfield methods. Trace elements by X-ray fluorescence.

[2]Formulas based on 11 oxygen atoms.

From Kretz (1990), with addition of trace-element data.

1.4 MICROSTRUCTURE

The term 'microstructure' was used by Harker (1932) and is here applied to various physical attributes of single crystals and crystal aggregates. It includes fine microstructures (crystal defects) which are normally examined by transmission electron microscopy, and granular microstructures (crystal shape, size, orientation, and distribution) which are normally examined by optical microscopy. This agrees in general with use of the term in metallurgy (Martin and Doherty 1976), where the term 'texture' refers to the orientation of crystals.

The development of the electron microscope during the past three decades has greatly increased our knowledge of fine-scale microstructures in crystals, including several mineral species. The remarkable capability of transmission electron microscopy (TEM) is demonstrated in Figure 1.1 where the six-member ring structure of cordierite is clearly visible.

Transmission electron microscopy combined with other methods has shown that virtually all crystals contain defects, which are now known to play an important role in crystal nucleation, growth, deformation, and other processes. A helpful introduction to this study is presented by Hull and Bacon (1984) and a more advanced treatment is presented by Hirth and Lothe (1982). Defects in crystals may be grouped as follows:

- point defects
  - vacancies
  - interstitials
- line defects
  - edge dislocations
  - screw dislocations
- plane defects
  - stacking faults
  - twin planes
  - anti-phase boundaries

In a perfect crystal of NaCl, every structural site is occupied by an atom. A missing atom is referred to as a vacancy; a missing Na atom and a nearby missing Cl atom (to maintain neutrality of charge) is referred to as a Shottky defect. An interstitial atom is one which resides at a place not normally occupied. A Na atom displaced from its normal position (to create a vacancy) to an interstitial position is referred to as a Frenkel defect.

Only under special conditions and by use of field ion microscopy can point defects be observed directly (Berger, Seidman and Balluffi 1973), but several indirect methods are available, as reviewed by Agullo-Lopez, Catlow and Townsend (1988). The most compelling evidence for the presence of vacancies in all crystals, however, is a theoretical argument. The presence of vacancies increases the entropy of a crystal, and every crystal is expected to possess an equilibrium concentration of vacancies which increases with increasing temperature (Frenkel 1946, Azároff 1960).

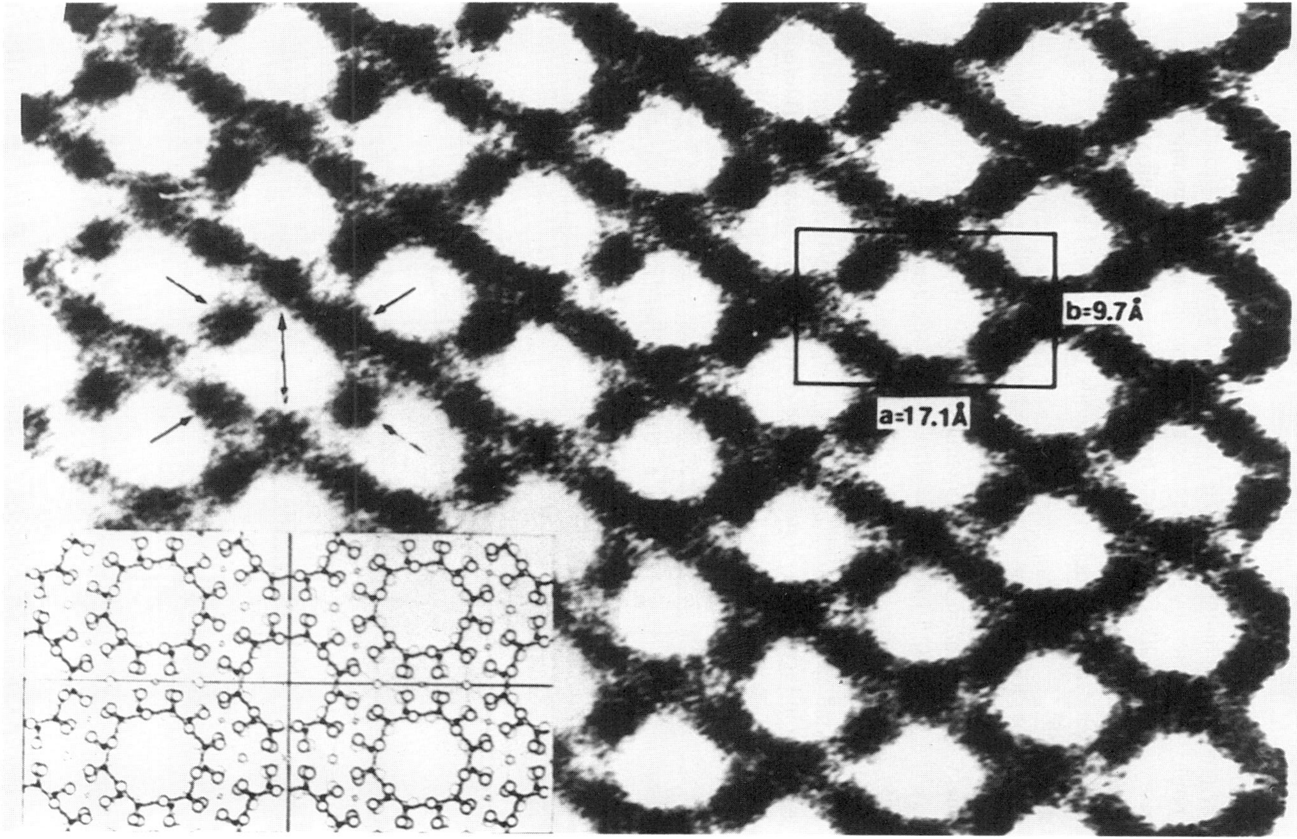

Figure 1.1 High-resolution TEM image of cordierite; the unit cell is outline. Inset shows six-member (Si_6O_{18}) rings. From Buseck and Iijima (1974)

With regard to minerals, the concept of a vacancy takes on a somewhat expanded meaning. High concentrations of Fe vacancies are known to exist in pyrrhotite. In muscovite, one-third of the octahedral sites are vacant. In biotite, as noted above (Sec. 1.3), vacancies are evidently present in some crystals to the extent of 5–10%. Thus other controls, including chemical availability during crystal growth as well as temperature, may determine vacancy concentrations in many minerals. In view of the theory it seems likely that some vacancies are present in all sites of all crystal of all minerals, and that various interstitials also occur commonly. An advanced discussion of point defects in minerals is presented by Lasaga (1981) and Kröger (1985).

An edge dislocation is a line defect created by the termination of a plane of atoms, as shown in Figure 1.2a, where the dislocation is defined by a Burgers vector. The vector is obtained by following the atom-to-atom path MNOPQ in Figure 1.2a (i.e., 3, 4, 3, 5 interatomic distances in segments MN, NO, OP, PQ, respectively), and then repeating the traverse in a corresponding perfect crystal (Figure 1.2b). The closure failure, QM, is the Burgers vector.

The second kind of line defect is a screw dislocation, which is a fault in a crystal, as shown in Figure 1.3a. Here, also, the dislocation is defined by a Burgers vector, obtained by comparing path MNOPQ in Figure 1.3a with path MNOPQ in

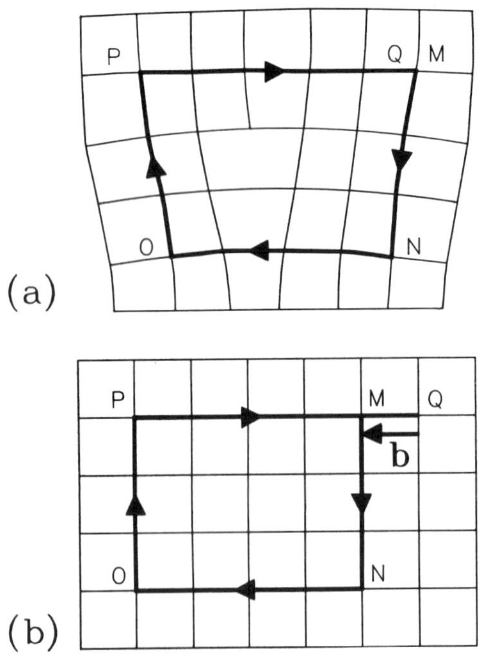

Figure 1.2 (a) Edge dislocation and Burgers circuit, and (b) same circuit in a perfect crystal; closure failure is the Burgers vector. From Hull and Bacon (1984). Reproduced by permission of Pergamon Press Ltd.

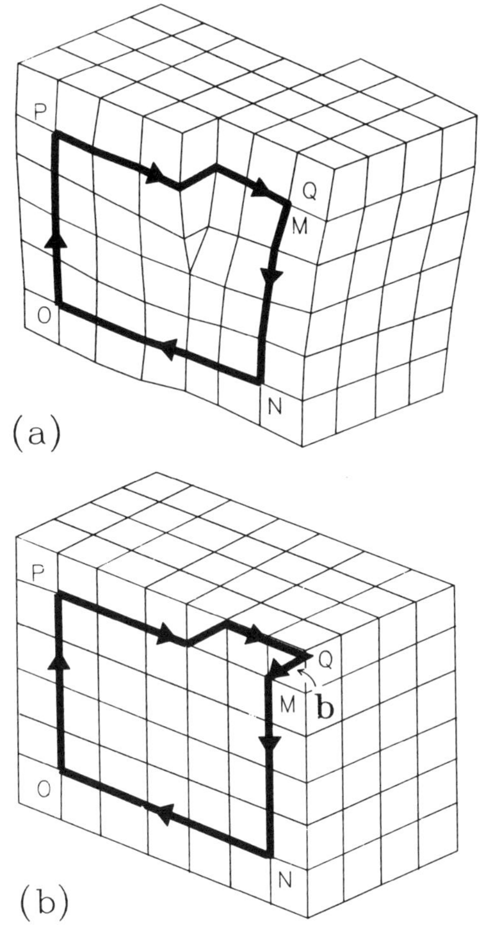

Figure 1.3 (a) Screw dislocation and Burgers circuit, and (b) same circuit in a perfect crystal; closure failure is the Burgers vector. From Hull and Bacon (1984). Reproduced by permission of Pergamon Press Ltd.

Figure 1.3b. In an edge dislocation the vector lies normal to the line defect; in a screw dislocation it lies parallel to the line defect.

A screw dislocation which emerges at a crystal face forms a step on that face, and if growth takes place at this step, a spiral step pattern will form. Patterns of this kind were predicted by Frank (1955) and were first observed on the prismatic face of a beryl crystal. Growth spirals were subsequently found on several minerals, including phlogopite (Sunagawa 1964) and garnet (Endo and Sunagawa 1968).

Initially, dislocations were detected by etching crystal faces; enhanced dissolution produced etch pits which are visible with an optical microscope. Dislocations are now examined mainly by use of transmission electron microscopy and they

Figure 1.4 TEM image of dislocation ribbons in graphite; scale bar is 1 μm

have been found in great abundance in many minerals. Dislocations in graphite, for example, as revealed by TEM, are shown in Figure 1.4. In normal (hexagonal) graphite the carbon atoms occur in alternating sheets, as shown in the left margin of Figure 1.5; the sequence is abab.... The ribbons which appear in TEM images have been interpreted by Amelinckx, Delavignette and Heerschap (1965) as lamellae in which sequences of carbon planes abc appear (rhombohedral graphite), combined with edge dislocations, as shown in the central part of Figure 1.5. No C–C bonds are broken. The Burgers vector, **b**, consists of two partials, $\mathbf{b}_1$, and $\mathbf{b}_2$. The dislocation structure here is somewhat more complex than that shown in Figures 1.2 and 1.3.

Among the planar defects, stacking faults have been known for many years to exist in sheet silicates and graphite. The imperfection in graphite described above is in part a stacking fault, i.e., abc (rhombohedral graphite, which is unstable) in place of abab.... Stacking faults may form during crystal growth, during deformation, or during a phase transformation.

An intercalation is a planar defect resulting from the presence of a slab of material of slightly different structure in an otherwise normal structure. These have been detected by high-resolution TEM, especially in crystals of pyroxene and amphibole, as reviewed by Veblen (1985a). Examples of intercalations in anthophyllite are shown in Figure 1.6. In normal anthophyllite, double chains of tetrahedra are arranged parallel to the *c* axis, and these chains viewed 'end-on' are visible in the left and right margins of Figure 1.6. This structure is interrupted by lamellae of a triple-chain structure and one lamella of a quintuple-chain structure; the latter

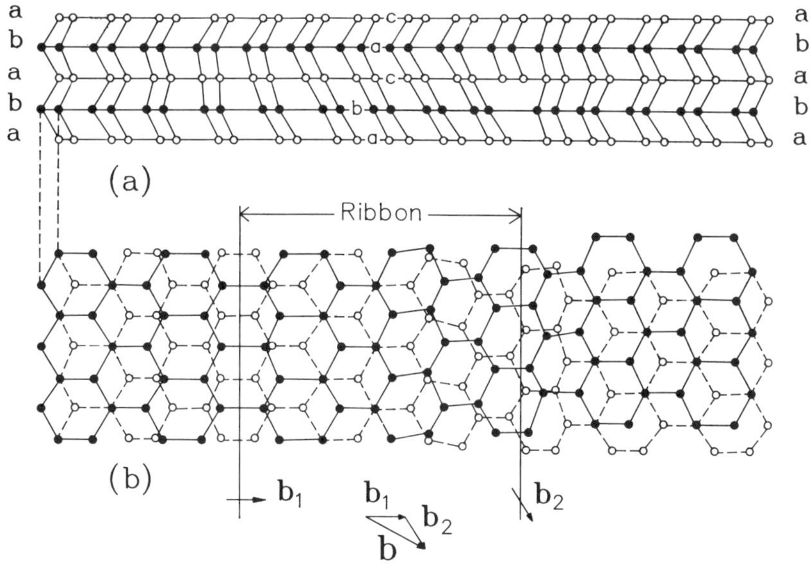

Figure 1.5 Interpretation of dislocation ribbons shown in Figure 1.4; each dot represents a carbon atom. Carbon sheets are viewed in section (a) and plan (b). The total Burgers vector (bottom) is **b**; $\mathbf{b}_1$ and $\mathbf{b}_2$ are partials. Figures 1.4 and 1.5 are reprinted from Amelinckx, Delavignette and Heerschap (1965), by courtesy of Marcel Dekker Inc.

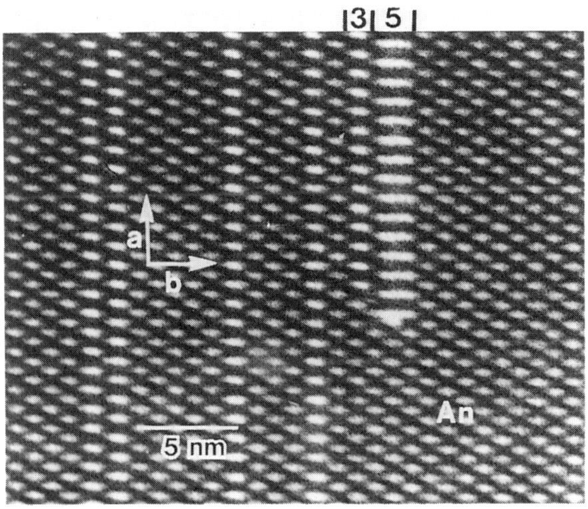

Figure 1.6 TEM image of defect structures in anthophyllite. Ideal double chains (right and left margins) are interrupted by slabs of triple chains (3) and one of five chains (5) which terminates at the bright spot in the lower-right quadrant. From Veblen (1985b)

terminates at the bright spot in the bottom-right quadrant, to form a channel. A second example of an intercalate defect is shown in Figure 1.7 which is from Eggleton and Banfield (1985). This is a drawing of an image of biotite, viewed parallel to the sheets of linked tetrahedra, and it shows imperfections, which give rise to slabs of brucite and chlorite within the structure. The intercalate defects in anthophyllite, described above, presumably formed during crystal growth, while that in biotite evidently formed by post-crystallization alteration.

Plagioclase, calcite and other minerals are commonly twinned, and the twin planes represent a third type of planar defect. These form during growth or deformation.

Anti-phase boundaries are planes that separate two domains in a crystal which are 'out of step'. For example, in a crystal of K-feldspar in which the Al and Si atoms are ordered, one could find, along a traverse, Al Si Si Si Al Si Si Si followed by Si Si Si Al. ... These domains and the boundary separating them evidently form during an ordering process.

The term 'granular microstructure' is here used to refer to properties of crystal aggregates, notably the shape, size, orientation, and distribution of crystals. The three-dimensional shape of crystals can be defined by use of descriptive terms, and crystal forms, where present, can be specified. The size of crystals can be measured and recorded, bearing in mind that the true mean size is somewhat larger than the apparent mean size, as viewed in sections. Crystal foliation and/or crystal lineation are commonly present, and are defined by the arrangement of platy or elongate crystals, as shown in Figure 1.8. Less obvious patterns of preferred orientation may be present with regard to the axes of crystals of quartz, calcite, and other minerals. In rocks composed of two or more minerals, the crystals of each mineral may be arranged at random or there may be some departure from randomness in the direction of clustering or in the direction of a more regular arrangement. The clustering of crystals of biotite or hornblende to form lenses and discontinuous layers is common and is referred to as gneissic microstructure.

Much information on granular microstructure may be obtained by examining rock thin-sections with an optical microscope, but the information so obtained is normally incomplete, and rock volumes should be examined, if possible, in three dimensions.

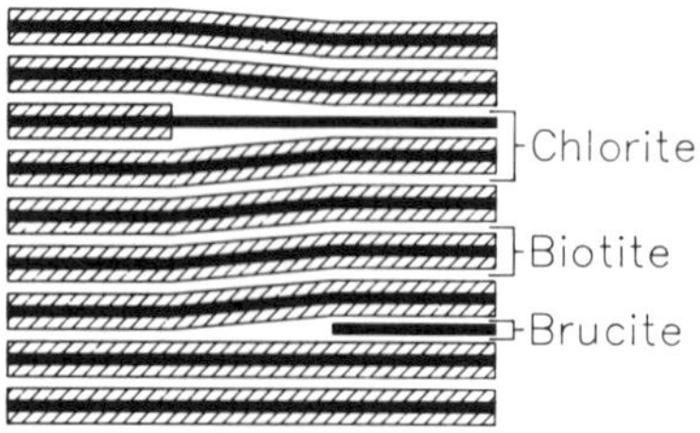

Figure 1.7 Defect structure in biotite, resulting in discontinuous layers of brucite and chlorite. Drawing of a structure illustrated by Eggleton and Banfield (1985)

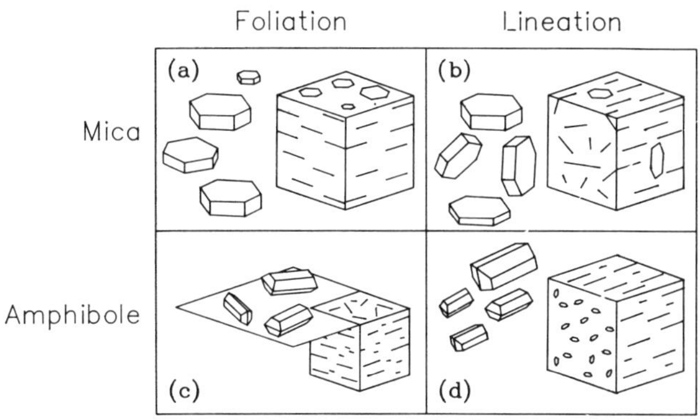

Figure 1.8 Crystal foliation (a, c) and crystal lineation (b, d) defined by alinement of platy mica crystals and elongate amphibole crystals

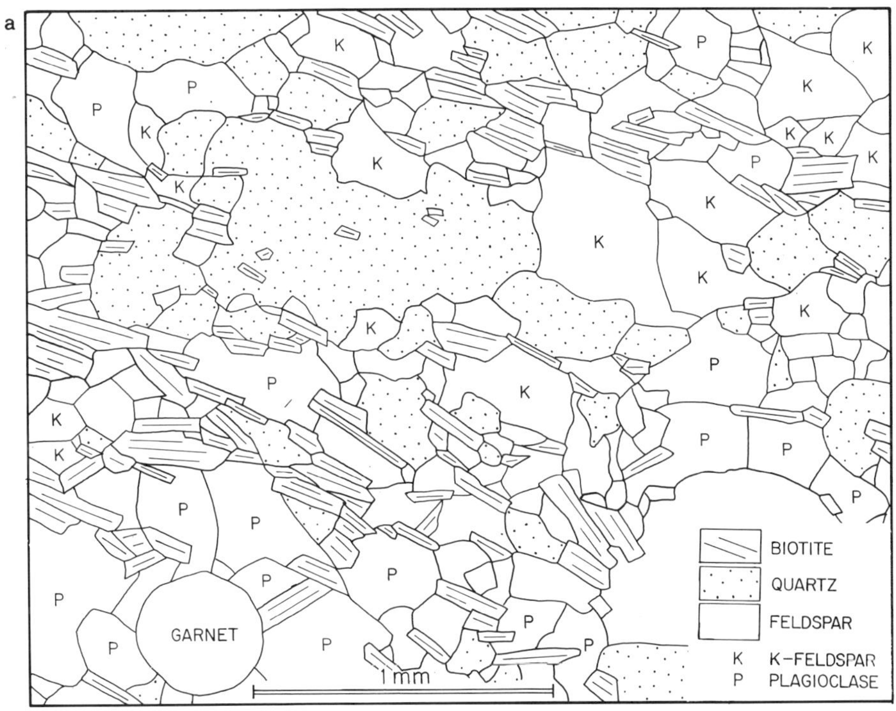

Figure 1.9 Examples of granular microstructure, (a) in biotite garnet gneiss, and (b) (*see overleaf*) in mafic metagabbro, to illustrate crystal shape, size, orientation, and distribution. In (b), Cpx(p) and Cpx(m) are interpreted as relict igneous and recrystallized Ca pyroxene respectively. Other minerals are hornblende (Hbl), plagioclase (Pl), garnet (Grt), and ilmenite (black)

b

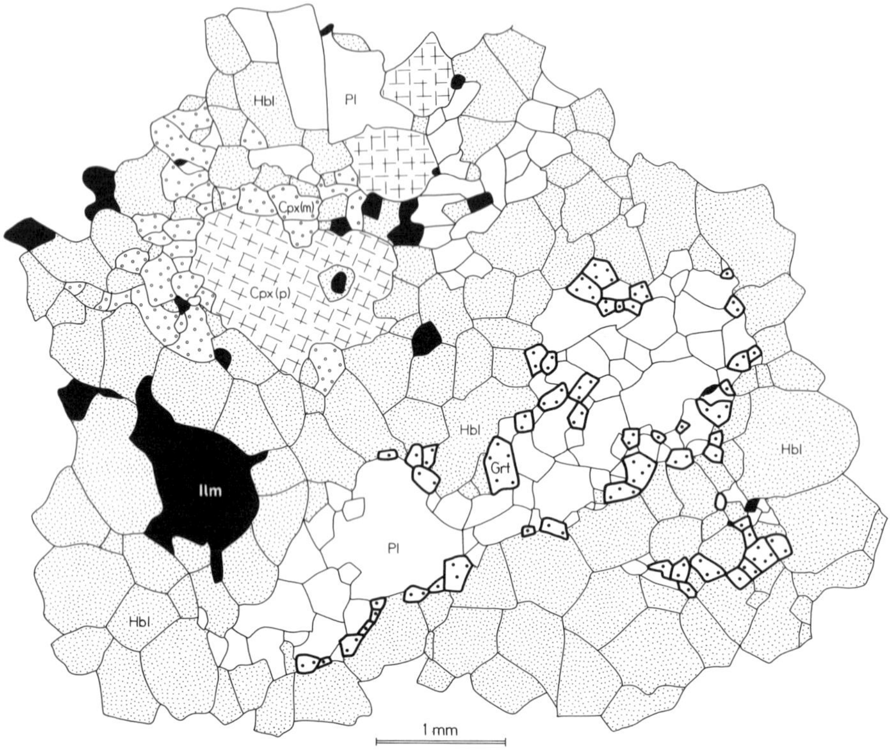

Figure 1.9 (*continued*)

Numerous examples of metamorphic microstructures were recently compiled by Yardley, MacKenzie and Guilford (1990). Additional illustrations are here shown in Figure 1.9; it is instructive to examine these in relation to crystal shape, size, orientation, and distribution. Certain granular microstructures will be examined in detail in Chapter 5.

1.5 VOLATILE COMPONENTS AND CRYSTAL BOUNDARIES

1.5.1 Volatile components

Many of the crystals of metamorphic terranes contain hydrogen and carbon as $(OH)^-$ and $(CO_3)^{2-}$ (Table 1.1), and with increasing temperature these elements may be released as H_2O and CO_2, which then migrate from the reaction sites. Simple examples of such reactions are

$$Mg^{2+}\ (OH)_2{}^- \rightarrow MgO + H_2O,$$

$$Ca^{2+}\ (CO_3)^{2-} \rightarrow CaO + CO_2.$$

Other reactions consume H_2O or CO_2, for example the production of amphibole in basaltic rocks that are initially devoid of hydrous minerals. Thus H, C, and other elements (B, N, F, S, Cl) which are capable of forming relatively volatile molecules, deserve special attention.

Many and possibly all crystals of nominally anhydrous minerals contain small amounts of hydrogen. Wilkins and Sabine (1973) detected the presence of 10 to 69 ppm H (0.009 to 0.062 weight % H_2O) in kyanite (19 specimens), and similar concentrations were measured in sillimanite by Beran, Rossman and Grew (1989), in olivine by Miller, Rossman and Harlow (1987), and in pyroxene by Skogby, Bell and Rossman (1990). Evidently H in these crystals occurs as $(OH)^-$, replacing O^{2-} (Freund 1987); charge deficiencies created by substitutions of this kind could be balanced by cation vacancies.

In brucite, muscovite, hornblende, and other minerals, H occurs abundantly in hydroxyl ions, $(OH)^-$, which form an integral part of the crystal structure. In laumontite and other zeolites, H occurs as H_2O molecules which are positioned in structural cavities, and these are readily removed or added without destroying the structure. Vry, Brown and Valley (1990) and others have shown that H_2O and CO_2 molecules reside in the channels of cordierite (Figure 1.1). In all of these examples, $(OH)^-$ and H_2O occupy structural sites in crystals.

In dehydration and decarbonation reactions of the kind written above, anions are transformed to molecules, and under laboratory conditions, a fluid phase will form. The term 'fluid' or 'vapor' is here used for a discrete and free fluid phase, e.g., supercritical H_2O or a gaseous mixture of H_2O and CO_2. Under natural conditions, additional possibilities exist for the released molecules; for example, they could become adsorbed on crystal boundaries, without forming a fluid phase.

The importance of $(OH)^-$ (hydroxyl ion; *not* water) and H_2O (water) in metamorphic processes was emphasized by Yoder (1955), Fyfe, Price and Thompson (1978), Walther and Orville (1982), and Wood and Walther (1986). During weathering, H_2O molecules are removed from the hydrosphere and are converted to $(OH)^-$ ions, which enter kaolinite, illite, chlorite, and other crystals. These crystals, together with detrital fragments and large proportions of pore water (mainly saline water), form the shale sediments of sedimentary basins. During burial and compaction, pore water is driven off, and at a depth of 3 km this water is reduced to about 1% by volume. From measurements in wells, it is known that at depths to 3 km, the pore water is under hydrostatic pressure, and at depths from 3 to 6 km the fluid pressure approaches the lithostatic pressure. Thus at depths > 6 km, convective fluid flow becomes increasingly difficult (Wood and Walther 1986).

Moving now to deeper levels of 10 to 30 km (3 to 10 kbar) where many metamorphic changes take place, several reactions occur within meta-shale to release $(OH)^-$ as H_2O molecules. Walther and Orville (1982) have calculated that between 400 and 600 °C, approximately 2.5 g of each of H_2O and CO_2 are produced in 100 g of pelitic rock. It is frequently assumed that these molecules form a fluid phase which escapes from reaction sites by fluid flow, but Ramberg (1952a, p. 179) and Walther and Orville (1982) have suggested that where reactions proceed slowly,

H_2O and CO_2 molecules could escape by diffusion. Walther and Orville have also emphasized that the H_2O produced at one locality in a metamorphic terrane must in general pass through adjoining volumes of rock, en route to fractures, where flow is concentrated. Some of this water conceivably enters bodies of mafic rock where H_2O is needed for the crystallization of amphibole. Locally, in subduction zones, hydrous minerals and H_2O may be transported to the Mantle.

It is not possible to examine here the numerous data and interpretations that exist regarding 'volatiles' in the Crust, but some information on fluid inclusions, on the activity of CO_2 and H_2O in metamorphic terranes, and on oxygen isotopes as indicators of H_2O mobility, will be reviewed briefly.

Fluid inclusions are small cavities in crystals which contain one or more fluid phases (mainly H_2O and CO_2), and in addition, one or more crystalline phases may be present (mainly NaCl). These have been detected in many minerals and appear to be almost universally present within the terrestrial Crust and upper Mantle. Attempts have been made to distinguish primary inclusions, representing volumes of fluid enclosed during crystal growth, from secondary inclusions, which are younger and represent fluid occurring along fractures, now healed (Figure 1.10). Fluid inclusions have been studied in detail, in an attempt to identify and analyze the components present, and to recover the temperature and pressure of fluid entrapment; the subject has been treated in detail by Roedder (1984, 1990) and Crawford and Hollister (1986).

In a regional metamorphic terrane in southern Norway, Touret (1971) noted that most inclusions in medium-grade rocks (amphibolite facies) are rich in H_2O, while those in high-grade rocks (granulite facies) are rich in CO_2. Similar results were reported from Finland (Schreurs 1984), where, in addition, widespread inclusions composed of N_2 and CH_4 occur. In the central Alps, Frey et al. (1980) found as dominant species, (a) higher hydrocarbons, (b) CH_4, (c) H_2O, and (d) CO_2, where (a) to (d) denotes a change from unmetamorphosed sedimentary rock to rocks of high metamorphic grade.

In some terranes, the temperature and pressure estimates obtained from fluid inclusion studies agree closely with those obtained from mineral equilibrium considerations, and the fluids are viewed as samples of a fluid phase that existed in the rock at the time of metamorphic crystallization. But Lamb, Valley and Brown (1987) have shown that locally in the Adirondack Mountains (USA), the CO_2-rich

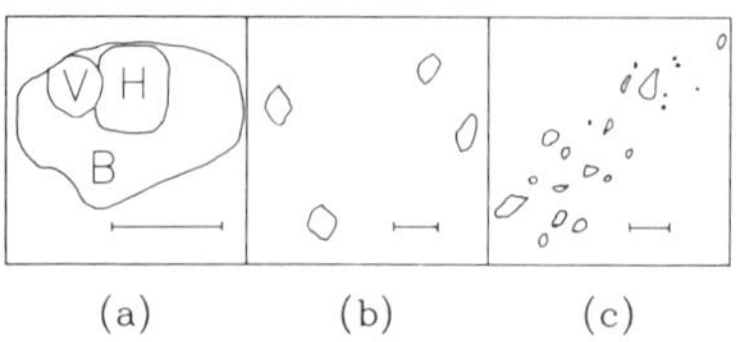

Figure 1.10 Fluid inclusions in quartz. (a) An inclusion in schist from Vermont, composed of a halite crystal (H), a vapor bubble (V), and brine (B); from Rich (1979). (b) 'Primary' and (c) 'secondary' inclusions in granulite from the Adirondack region; from Lamb, Valley and Brown (1987). Reproduced by permission of Springer-Verlag. Bar scale is 20 μm

fluid now present in the inclusions, cannot have been in equilibrium with calcite, quartz, and wollastonite at the time of wollastonite crystallization. The CO_2 was therefore introduced at a later time; this interpretation questions the criteria that are used to identify primary inclusions (Figure 1.10). Under favorable conditions, fluid inclusions can provide estimates of pressure–temperature points on the cooling path of a metamorphic terrane.

With regard to the activity of CO_2 and H_2O during metamorphism, some occurrences of medium-grade carbonate rocks have been described in which the minerals now present obviously crystallized in an environment of CO_2 activity much less than that represented by pure CO_2. An example is the crystallization of wollastonite from calcite and quartz, which at medium-grade temperatures requires a very low CO_2 activity. The results are presented by Ferry (1986) who has concluded that large volumes of H_2O have entered and flowed through the particular rock units, thus diluting CO_2 by the amount needed to reduce the activity of this component. This conclusion is built on the assumption of the presence of a fluid phase during metamorphism. Calculated volumes of H_2O which are needed, relative to 100 cm^3 of rock (the fluid–rock ratio) are in places as high as 4 or 5. Possible mechanisms for achieving a low CO_2 activity, other than dilution with H_2O, were not considered. One may note in passing that in other marble terranes of medium-to-high grade (e.g. the Grenville terrane in Quebec) calcite and quartz are commonly present, but not wollastonite (Kretz 1980). Some uncertainty obviously remains as to whether or not the 'pervasive flow along grain boundaries' envisioned by Ferry (1986) and also by Brunton and Helgeson (1983), Etheridge, Wall and Vernon (1983) and others is a local or a regional and general phenomenon.

A similar problem arises in granulite terranes where an activity of H_2O less than 1.0 is needed to (a) initiate the reaction, tremolite $\rightarrow$ enstatite + diopside + quartz + H_2O, which had occurred locally in Adirondack granulites (Lamb and Valley 1988) and (b) to prevent large-scale melting. The problem is discussed by Newton (1986) who listed three possible causes for a reduced H_2O activity — dilution with CO_2, partial melting with absorption of H_2O by the melt, and the removal of H_2O prior to granulite-facies metamorphism. The last proposal implies the absence of a fluid phase (fluid-absent metamorphism) which was considered in detail by Thompson (1983). Fluid-absent metamorphism does not mean a complete absence of H_2O molecules; a model for a solid H_2O-bearing crystal-boundary phase is presented below.

Data on oxygen isotopes have provided evidence that surface waters have locally penetrated to depths of several kilometers. For example, Wickham and Taylor (1985), working in the Trois Seigneurs Massif of France, found uniform $\delta^{18}O$ values of +11 to +13 in the high-grade rocks, and values of +15 in the lower-grade rocks, which crystallized at depths less than 12 km ($\delta^{18}O$ is defined as $\delta^{18}O = 1000 (R_{sample} - R_{smow})/R_{smow}$, where $R = {}^{18}O/{}^{16}O$, and smow refers to standard mean ocean water). The authors proposed that the increase in δ was caused by surface waters which reached depths of 12 km. On the other hand, studies by Valley and O'Neil (1984) in the high-grade Adirondack terrane, have led to the conclusion that present ^{16}O:^{18}O ratios were determined prior to metamorphism, and that evidence for deep and penetrative flow of CO_2–H_2O fluids does not exist. Similar results for a granulite terrane in India were obtained by Jiang, Clayton and Newton (1988).

The above remarks are intended only to identify some of the problems and uncertainties that exist regarding relatively volatile components in metamorphic terranes. We have focused on H_2O and CO_2, but others are also important. The cause of our lack of understanding of the behavior of these components in common metamorphic rocks is in large part the result of their inherent mobility, which (unlike crystals of feldspar and biotite) enables them to move away from reaction sites. Therefore, in dealing with mineral equilibria and reactions involving H_2O and CO_2, assumptions regarding the nature of these components at the time of metamorphism must be clearly stated.

1.5.2 Crystal boundaries

A large fraction of the vast interior of the Earth consists of crystals, and with few exceptions, a sample that is chosen for study is found to consist of very many crystals, each in contact with its neighbors. In metallurgy, these contact planes are referred to as 'grain' and 'interface' boundaries, depending on whether they separate like or unlike crystals, but in geology, the term 'grain boundary' is commonly used for both. Valiev, Mulyukov and Ovchinnikov (1990) speak of an 'intercrystalline boundary' between like or unlike crystals, and this term, shortened to *crystal boundary*, is used in the present discussion. The structure and composition of natural crystal boundaries is an important and largely unsolved problem in geology and geophysics.

Consider a boundary between two quartz crystals in metamorphic quartzite. The crystals are not easily separated from each other, and strong cohesive forces obviously exist across the boundary. Given our knowledge of the atomic structure of quartz, one can visualize the boundary as a 'surface of discontinuity' (Gibbs 1878) where interatomic distances are somewhat larger than within the adjoining crystals, and atomic bonds extend across the discontinuity. The structure of crystal boundaries was first studied by metallurgists, who recognize low-angle boundaries, which can be viewed as a number of edge dislocations (Figure 1.11) and high-angle boundaries ($> 15°$) of more complex structure (Gleiter and Chalmers, 1972).

It is now possible, by use of TEM, to obtain direct images of crystal boundaries (Merkle, 1991). At the NiO–NiO boundary, for example (Figure 1.12), a dilation of atoms is confirmed, and in addition, a periodicity (implying a structure) is discernible. The boundary width in this case is 0.5 nm ($\sim$5 atomic diameters).

Different procedures for obtaining a description of the atomic structure of crystal-boundary volumes were reviewed by Fischmeister (1985) and Duffy (1986). One of these employs the concept of the coincident-site lattice, which is formed by superimposing the lattices of two adjoining crystals, and marking all points that coincide. The boundary plane is then drawn to pass through one of these points, and all 'foreign' atoms on both sides of the plane are discarded. This procedure, applied

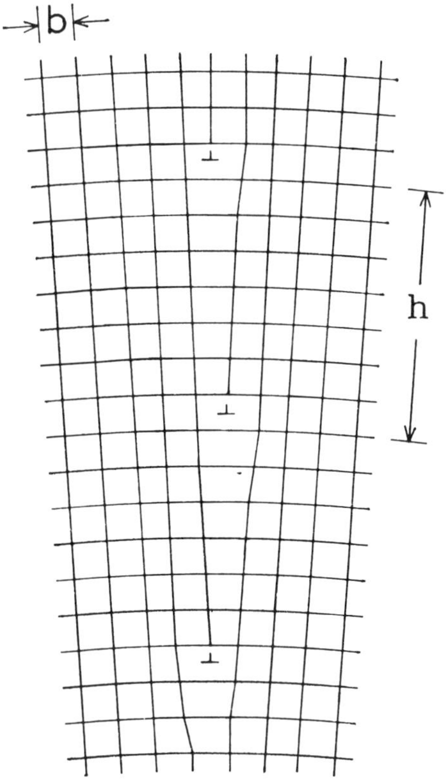

Figure 1.11 Low-angle tilt boundary, defined by edge dislocations; the tilt angle (θ) is related to the Burgers vector (**b**) and dislocation spacing (h) by $\sin\theta = \mathbf{b}/h$, or $\theta \approx \mathbf{b}/h$. From Moffatt, Pearsall and Wulff (1964). Reproduced by permission of John Wiley & Sons, Inc.

to a boundary between two Al_2O_3 crystals, produces patterns of the kind shown in Figure 1.13; note that the boundary possesses a repeated structure, analogous to the unit cell within a crystal.

It is apparent, therefore, that crystal boundaries must be viewed as volumes of matter, and the term 'crystal-boundary phase' is appropriate. Indeed, this is the view adopted by Gibbs (1878) who assigned thermodynamic properties to interfaces. Possibilities for changes in the structure of crystal boundaries with changes in temperature or pressure (grain-boundary phase transformations) were considered by Hart (1972) and Cahn (1982).

White and White (1981) have used TEM to examine crystal boundaries in highly deformed quartzite. They found narrow films of distorted structure, isolated voids (~5 to 1000 nm in diameter), and larger, tubular openings at grain edges (lines where three grains meet). The presence of foreign atoms in the boundaries could not be detected. Chlorite–chlorite and chlorite–muscovite boundaries in slate were similarly investigated by Bons et al. (1990), who found many boundaries to be less

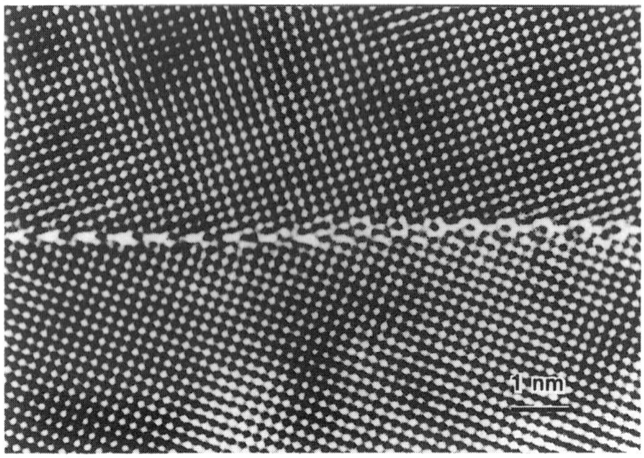

Figure 1.12 Atomic resolution micrograph of a ⟨001⟩ symmetrical NiO–NiO tilt boundary; note crystal-boundary structure. From Merkle and Smith (1987); courtesy of K.L. Merkle

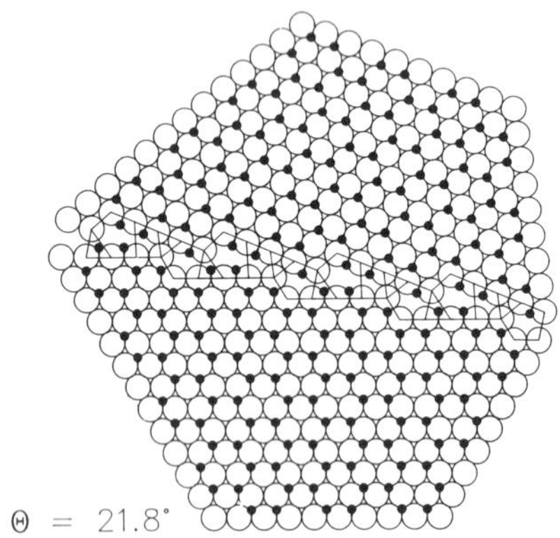

Figure 1.13 Model of a [0001] tilt boundary between two crystals of Al_2O_3 (tilt angle = 21.8 °); open circles represent O^{2-} and closed circles, Al^{3+}. Note the repeated pattern of polyhedra of sides 3, 5, 4, 3, 5, ... in the boundary phase. From Shackelford (1983), reprinted by permission of the American Ceramic Society

than 1 nm wide, while others of about 10 nm appear to possess an atomic structure. Earlier, Sella and Deicha (1962) and Brace et al. (1972), by use of scanning electron microscopy (SEM), found openings less than 1 μm wide at boundaries between quartz and feldspar crystals in granitic rocks. In another study, the writer found that considerable force was needed to break apart phlogopite doublets which were separated from marble, i.e., to initiate a fracture on the phlogopite–phlogopite boundary. But one must bear in mind that boundaries as they exist in surface samples may differ from those that exist during metamorphism.

The study of crystal–vacuum and crystal–vapor surfaces has advanced rapidly in recent years, and some of the results may apply to crystal interfaces. The subject is covered by Zangwill (1988) and reviewed by Somorjai (1978, 1985). Some of the findings are the following: distances between the first and second layers of atoms may be shortened (relaxation); surface atoms may form a new structure (reconstruction); high-index surfaces may become stepped; impurity atoms may move to the surface and cover it; certain gas molecules may adsorb more strongly on one face than on another; small adsorbed atoms may move underneath the first layer of atoms; certain adsorbed molecules may attract one component of the crystal to the surface; oxidation states of surface atoms may be altered. The surface structures that form during reconstruction may be very different from those that exist within crystals; a notable example is that which forms at clean (111) surfaces of silicon (Robinson 1986). More simple reconstruction occurs at (001) surfaces of tungsten, as shown in Figure 1.14. Similar reconstruction may possibly occur at grain boundaries.

While natural dehydration reactions are in progress, concentrations of H_2O molecules on crystal boundaries near reaction sites are expected to be high. These molecules will be in contact with crystals on either side of the boundary, and may become adsorbed on those crystals. In view of the largely ionic nature of atomic bonds in silicate minerals, chemical bonds (chemisorption) between vapor and solid atoms will likely dominate over the weaker van der Waals bonds (physisorption). Adsorbed water in general may form one or several layers (Drost-Hansen 1971), and it is well known that the density and viscosity of adsorbed 'phases' are high compared with the associated vapor, while entropy and molecular mobility are low. Some dissociation of H_2O may occur; the evidence for both H_2O and $(OH)^{-1}$ on surfaces of quartz to 400 °C was presented by Whalen (1961). Where adsorption occurs in a restricted space, the H_2O vapor that occupies the remaining space

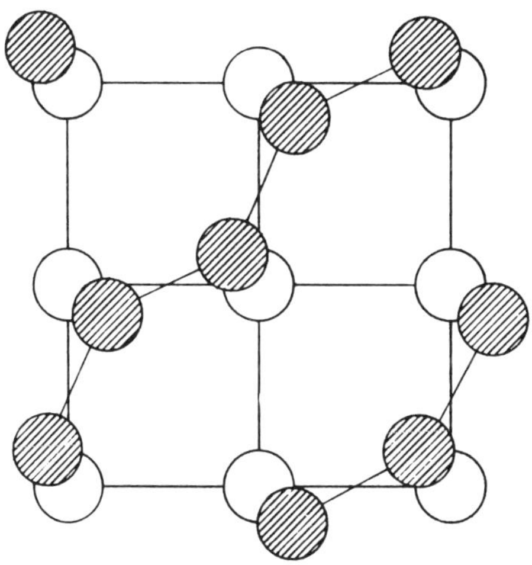

Figure 1.14 Surface structure of tungsten (001); shaded circles represent surface atoms. From Debe and King (1979)

may itself possess an enhanced density and viscosity, assuming that the characteristics of 'anomalous water' (Drost-Hansen 1971) persist to higher temperatures and pressures. That this may be so is indicated by the calculations of Belonoshko (1989) which show for example, that at 5 kbar and 600 °C, the $H_2O/(H_2O + CO_2)$ composition of vapor in 0.8 nm pores is considerably larger than in the bulk phase. Crystal–vapor interactions were also detected in the experiments of Hay and Evans (1988), who found that excess (non-adsorbed) H_2O and CO_2 vapor in a calcite aggregate occurs as pores on 2-, 3-, and 4-grain junctions, and not as a continuous film.

From the above considerations, it is conceivable that several possibilities exist for the structure of crystal-boundary phases at different depths and at different times, e.g., while reactions are in progress, and afterwards. Any single model for natural crystal boundaries must therefore be sufficiently flexible to allow for varying concentrations of H_2O (also CO_2 and other molecules and atoms) on the boundary, and for conditions that are capable of enhancing and retarding molecular and atomic mobility. As a point of departure, consider the two-dimensional biotite–biotite boundary shown in Figure 1.15, which was constructed by use of the coincident-lattice-site procedure described above. Only oxygen atoms are shown. The predicted boundary unit in the plane of oxygen atoms is a 12-member ring, but a minor relaxation is assumed (atom R) to produce 10-member rings. The result invites an analogy with the structure of zeolite crystals (Kretz 1980), e.g., chabazite which contains 6- and 8-member rings of oxygen ions (Dent and Smith 1958). In three dimensions, the cells of Figure 1.15 become tubes or cages, capable of accommodating various ions and molecules, including H_2O; one can therefore regard water in the crystal-boundary phase as being 'dissolved', a term used by Rubie (1986). From the data compilation of Freer (1981), it is clear that H_2O and cations can move through open silicate structures with ease.

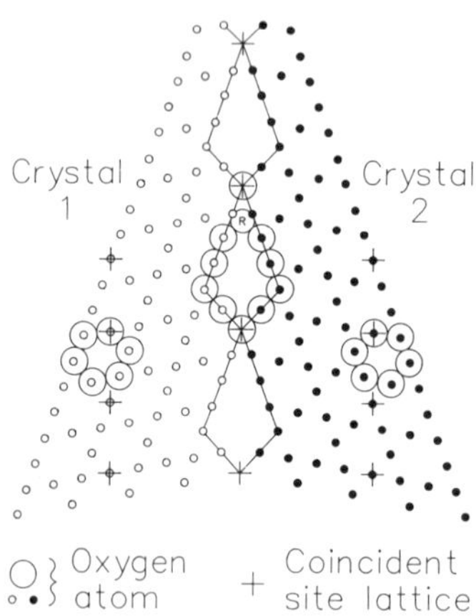

Figure 1.15 Model for the structure of the boundary between two biotite crystals (oxygen sheets are coplanar), based on the coincident-site-lattice concept. A slight relaxation (atom R) produces 10-member rings

The model can be modified to provide more intergranular space. For example, the three cells of Figure 1.15 can expand to form a single, larger cell. Processes analogous to restructuring may occur on the walls of these larger cells. Between the cells, atomic bonds continue to bind the adjacent crystals together. The model allows for much variation in the concentration and therefore the activity of H_2O in the boundary phase. If a discrete fluid phase is also present, e.g., pure H_2O, then the chemical potential of H_2O in the fluid and crystal-boundary phase must be equal, but in the absence of a fluid, the activity of H_2O in the boundary phase can decrease to zero. Thus a reduction in H_2O activity is possible without calling on an introduction of CO_2 or on melting to cause this reduction. Moreover, the boundary phase, as here conceived, is solid and capable of transmitting stress, which Gresens (1966) has noted is favorable, if not necessary, for the operation of diffusional creep and other stress-induced atomic and molecular migrations.

1.6 METAMORPHIC ROCKS

Many metamorphic rocks can be distinguished from igneous and sedimentary rocks by their microstructure, notably the parallel orientation of platy crystals and the segregation of crystals to form lenses and layers. Hence some rock names emphasize mineral content while others emphasize microstructure. Following is a list of names that are commonly used for metamorphic rocks. The terms ‘very fine’, ‘fine’, ‘medium’, and ‘coarse’ refer to $< \frac{1}{8}$, $\frac{1}{8}$–1, 1–4, 4–16 mm respectively.

Slate. A very fine-grained rock of pelitic composition with an alinement of crystals to produce slaty cleavage.

Phyllite. A very fine-grained to fine-grained rock of pelitic composition with an alinement of mica crystal, visible as a ‘silky sheen’ on cleavage surfaces.

Schist. A fine- to medium-grained rock, rich in platy or elongate crystals (e.g., mica, amphibole) that are alined to define a foliation or lineation.

Hornfels. A fine-grained rock of pelitic composition, commonly containing larger crystals of biotite, andalusite, cordierite, or other minerals; foliation and lineation are obscure or absent.

Gneiss. A medium-grained feldspar-rich rock in which mafic crystals (e.g., biotite, hornblende) occur as parallel planar or elongate aggregates, to define a foliation or lineation.

Granulite. A medium-grained feldspar-rich rock in which the mafic crystals (e.g., orthopyroxene, Ca pyroxene, garnet, biotite) do not define a conspicuous foliation or lineation, but tabular or elongate grains of quartz may do so.

Amphibolite. A fine- to coarse-grained rock composed predominantly of hornblende and plagioclase in near-equal proportions; hornblende crystals are commonly alined to define a foliation or lineation.

Marble. A layered rock composed predominantly or entirely of ‘recrystallized’ carbonate minerals.

Quartzite. A layered rock composed predominantly of ‘recrystallized’ quartz.

Greenstone. A layered, fine-grained rock composed of chlorite, amphibole, epidote, and albite.

Serpentinite. A rock composed predominantly of serpentine.

Eclogite. A rock composed principally of jadeite-rich pyroxene and Mg-rich garnet.

Skarn. A medium- to coarse-grained rock rich in Ca- and Mg-bearing minerals (diopside, phlogopite, amphibole, Ca garnet, wollastonite, and others); plagioclase is typically absent but scapolite may be present. Some varieties are rich in magnetite. Skarn may occur in a contact or regional setting.

The rocks, as defined above, are not necessarily restricted to a particular metamorphic facies. For example, amphibolite may occur in the lower granulite facies and granulite in the upper amphibolite facies.

Further information concerning a metamorphic rock may be conveyed by use of prefixes such as: foliated (biotite) quartzite, fine-grained muscovite schist, garnet biotite gneiss, forsterite marble.

Terms reserved for igneous rocks (rocks known or inferred to have crystallized from a melt) may be used, with the prefix 'meta'. Examples are: metagabbro, metaanorthosite, metapyroxenite. These names should be used only when the premetamorphic rock can be identified with confidence.

1.7 ISOGRADS, MINERAL ZONES, AND METAMORPHIC FACIES

It is locally possible to trace out in the field a change from little-altered shale to moderately altered shale (schist) to highly altered shale (gneiss or hornfels). The gradation may occur from the outer margins of mountain belts to the central, formerly deeper core, or from the outer rim of contact aureoles to the formerly hotter zones adjacent to the central pluton. This variation in metamorphic intensity, now known to be a reflection of variation in temperature or pressure or both, is referred to as 'metamorphic grade'. Little, moderately, and highly altered rocks are referred to as rocks of low, medium, and high metamorphic grade.

In the field, as one moves in a direction of increasing grade, it is often possible to locate points on the ground marking the first appearance of a metamorphic mineral, and a line joining these points is referred to as an 'isograd'. Figure 1.16, for example, shows the biotite, garnet, staurolite, and sillimanite isograds in a portion of the Appalachian Mountains. Zones between isograds are referred to as 'mineral zones'; for example, the zone between the biotite and garnet isograds is referred to as the 'biotite zone'. The minerals that appear in a metamorphic terrane obviously depend also on rock composition; for example, in mafic rocks, epidote, actinolite, and hornblende occur in abundance. The concept of *metamorphic facies* was introduced by Eskola (1915) to define metamorphic zones that are independent of rock composition. Thus all rocks, regardless of chemical composition, which correspond to the chlorite and biotite zones are assigned to the greenschist facies, and all rocks corresponding to the garnet, staurolite, and sillimanite zones are assigned to the amphibolite facies. Higher-grade rocks of the orthopyroxene zone, where present, are assigned to the granulite facies.

Figure 1.16 Isograds in metasedimentary rocks of the Appalachian Mountains (Vermont). From Albee (1968), in Studies of Appalachian geology, edited by E. An Zen, W.H. White, J.B. Hadley, and J.B. Thompson Jr. © 1968. Reprinted by permission of John Wiley & Sons, Inc.

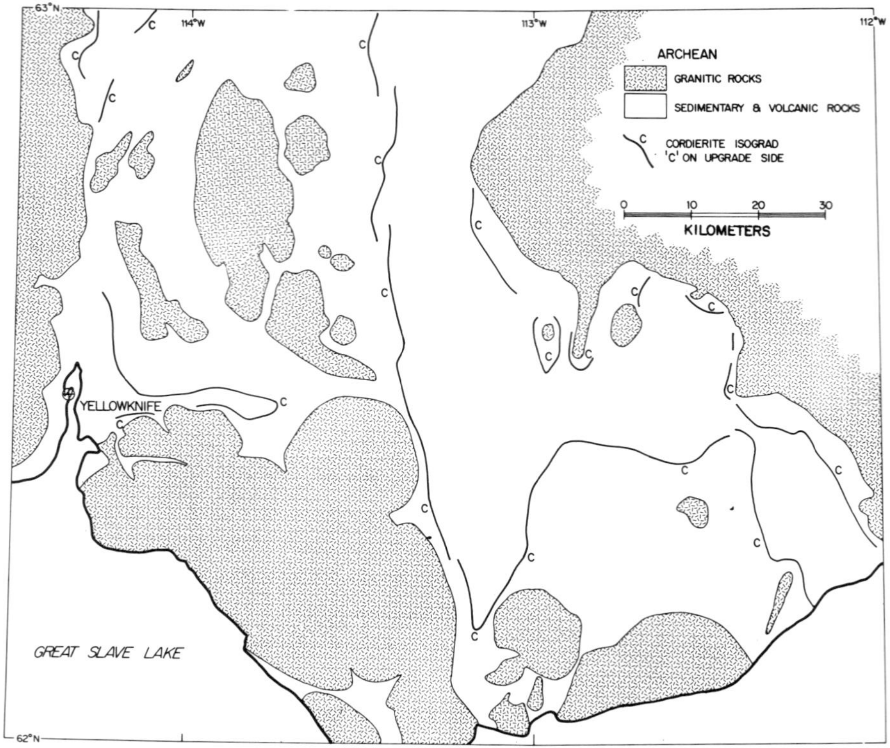

Figure 1.17 Cordierite isograd in metasedimentary rock of the southern Slave Province of the Canadian Shield. Biotite and sillimanite isograds have been mapped locally, and lie down-grade from the cordierite isograd (for biotite) and near the granite plutons (for sillimanite). From Henderson (1985)

In the Slave Province of the Canadian Precambrian Shield, in north-central North America, pelitic rocks similar in composition to those of the Appalachians, show an increase in metamorphic grade in relation to nearness to a swarm of granite plutons, as shown in Figure 1.17. Here, however, the low-density phases cordierite and andalusite appear in place of the higher-density phases garnet and kyanite. Although the increase in grade at both localities was evidently produced by an increase in temperature, the pressure increase was not as great in the Slave Province as in the Appalachians, i.e., the geothermal gradients at the two localities were not alike. The hornblende-hornfels facies was defined to accommodate cordierite-andalusite rocks which evidently crystallized at temperatures corresponding roughly to those of the amphibolite facies, but at lower pressures.

The concept of *facies series* was developed by Miyashiro (1961) to define temperature-pressure paths in different metamorphic terranes. Thus metamorphism in the Appalachians (Figure 1.16) is now referred to as 'medium-pressure metamorphism', while that in the Slave Province (Figure 1.17) is referred to as 'low-pressure metamorphism'.

Locally, for example in Japan and California, certain minerals of high density (lawsonite, aragonite, jadeite) are found, and these rocks were evidently subjected to high pressures at relatively low temperatures. This is referred to as 'high-pressure metamorphism'.

Different metamorphic facies that have been defined are shown in Figure 1.18. Experimental data, to be presented below, provide valuable information on temperature-pressure conditions for the different facies, and from these results, the axes in Figure 1.18 have been scaled approximately.

Each metamorphic facies represents a certain temperature-pressure domain (Figure 1.18) and is characterized by certain mineral assemblages which are presumably stable within that domain. Although the common metamorphic minerals are few in number, a great diversity exists in the mineral assemblages of each facies, resulting mainly from a diversity in the chemical composition of

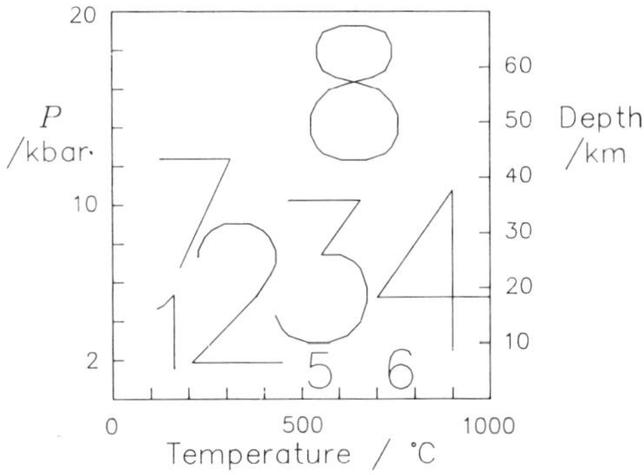

Figure 1.18 Approximate temperature-pressure domains for the metamorphic facies: 1 very low grade, 2 greenschist, 3 amphibolite, 4 granulite, 5 hornblende hornfels, 6 pyroxene hornfels, 7 blueschist, and 8 eclogite facies. Mainly from Turner (1981)

the ordinary igneous and sedimentary rocks. At the same time a remarkable uniformity exists in the sense that certain mineral assemblages are found worldwide, providing compelling evidence for the attainment of chemical equilibrium during metamorphism.

The practical value of the facies concept is that a large amount of information concerning mineral assemblages can be conveyed very briefly, by simply referring to the corresponding facies. Thus the term 'amphibolite facies' conveys to the experienced geologist a large amount of information concerning mineral assemblages within a variety of associated rocks (pelitic, calcareous, mafic) even though some uncertainty and controversy exist concerning the boundaries of the facies and the actual temperature–pressure conditions at which the rocks crystallized.

Following is a brief survey of the mineral assemblages formed under conditions corresponding to the different facies shown in Figure 1.18. This information is taken mainly from a book by Turner (1981) which may be consulted for details.

Facies of very low grade (1)

The first changes to be observed locally in certain sedimentary and volcanic rocks (graywacke, basalt, volcanic glass) resulting from burial to only a few kilometers is the crystallization of zeolites, particularly laumontite, together with other low-temperature minerals, to produce the dominant assemblage:

chlorite–quartz–albite–laumontite

Other minerals that may be present are montmorillonite, celadonite, heulandite, prehnite, pumpellyite, analcite, and sphene. These rocks are assigned to the zeolite facies.

With increasing grade, laumontite disappears, and prelmite and pumpellyite become widespread. Thus, in meta-graywacke, the dominant association is

pumpellyite–muscovite–chlorite–prehnite–quartz–albite

Epidote and sphene may be present. These rocks are assigned to the prehnite–pumpellyite facies.

With further increase in grade, prehnite disappears and actinolite appears, to produce the characteristic association:

pumpellyite–actinolite

which locally defines the pumpellyite–actinolite facies.

Greenschist facies (2)

Rocks of the greenschist facies are very abundant, occurring in Phanerozoic mountain belts, in Precambrian shield areas, and in the ocean floor. In many places they represent the first recognizable stage of metamorphism, indicating that facies of very low grade (1) have been bypassed. Characteristic minerals are albite, epidote, chlorite, phengite, and actinolite.

Where preceded by facies of very low grade, the lower boundary of the greenschist facies is marked by the elimination of pumpellyite in favor of epidote, and where preceded by the blueschist facies (to be described below), by the disappearance of glaucophane and appearance of actinolite. An important marker within the facies is the appearance of biotite and locally, Mn-rich garnet. The upper boundary is marked by the appearance of Ca-bearing plagioclase (mole fraction anorthite > 0.20) to accompany or replace nearly pure albite, and the appearance of an Al-bearing amphibole (hornblende) in place of actinolite.

Common mineral associations in the Appalachians (Vermont (Figure 1.16) and New Hampshire) and elsewhere consist of chlorite, muscovite (or phengite) and quartz, together with

chloritoid–paragonite
chloritoid–epidote
stilpnomelane–K-feldspar–albite
clinozoisite–biotite–albite
epidote–actinolite–albite

Amphibolite facies (3)

At numerous localities within mountain belts and Precambrian shields, the greenschist facies gives way to the amphibolite facies, recording the former presence of a moderate geothermal gradient (Figure 1.18). Rocks of the amphibolite facies are characterized by the widespread occurrence of almandine-rich garnet, hornblende, biotite, and plagioclase (mole fraction anorthite ≥ 0.20); kyanite and staurolite are diagnostic but are stable only in the middle part of the facies. In the upper reaches of the amphibolite facies, muscovite in the presence of quartz becomes unstable to produce K-feldspar and sillimanite. Some common mineral associations, as found in the Appalachian Mountains and in the Grenville Province (Canadian Precambrian Shield) are as follows:

garnet–kyanite–staurolite–biotite–muscovite–plagioclase–quartz
garnet–sillimanite–staurolite–muscovite–quartz
garnet–sillimanite–biotite–K-feldspar–plagioclase–quartz
garnet–clinozoisite–hornblende–biotite–plagioclase–quartz
garnet–Ca-pyroxene–hornblende–biotite–plagioclase
diopside–tremolite–phlogopite–calcite–dolomite

Granulite facies (4)

Rocks of the granulite facies occur principally in Precambrian shield areas, and represent the highest temperature–pressure conditions normally encountered in medium-pressure metamorphism. The lower boundary of the facies is marked by the appearance of orthopyroxene in pelitic and mafic rocks, which coincides approximately with the appearance of forsterite in magnesian marble. In the middle of the granulite facies, hornblende and biotite disappear, leaving the rock completely dehydrated. Alkali feldspar in granulite-facies rocks commonly consists of a lamellar intergrowth of albite and K-feldspar in equal proportions. Plagioclase typically takes on a gray-green colour. Some characteristic mineral associations are as follows:

garnet–orthopyroxene–biotite–alkali feldspar–plagioclase–quartz
garnet–sillimanite–cordierite–biotite–plagioclase–quartz
orthopyroxene–Ca-pyroxene–hornblende–plagioclase
orthopyroxene–Ca-pyroxene–plagioclase
forsterite–diopside–calcite

Within the granulite facies quartz and calcite react to produce wollastonite.

Hornblende hornfels facies (5)

The hornblende hornfels facies is distinguished from the amphibolite facies by the appearance of andalusite in place of kyanite and by the crystallization of abundant cordierite, with only sporadic garnet. Otherwise the two facies are similar. Pressures for a given temperature were obviously lower than in the amphibolite facies (Figure 1.18).

Rocks of the hornblende hornfels facies occur as zones about plutons which range in composition from gabbro to granite. In places, rocks of this facies extend to the edge of the pluton. The width of the zone varies greatly; in a typical contact aureole it may be less than 1 km wide; elsewhere, as shown in Figure 1.17 it is much wider. Where cordierite–andalusite rocks occur on a regional scale, Turner (1981) has assigned them to the amphibolite facies (low-pressure amphibolite facies).

The transition from the greenschist to the hornblende hornfels facies is marked by the appearance of Ca-bearing plagioclase and hornblende, the same minerals that define the beginning of the amphibolite facies. In places (Figure 1.17) the base of the hornblende hornfels facies may be chosen as the appearance of cordierite and disappearance of chlorite. Characteristic mineral associations of the facies are as follows:

cordierite–andalusite–biotite–muscovite–plagioclase–quartz
andalusite–sillimanite–biotite–muscovite–plagioclase–quartz
cordierite–sillimanite–biotite–plagioclase–K-feldspar–quartz
cordierite–anthophyllite
Ca-pyroxene–hornblende–plagioclase

Garnet, rich in almandine, and staurolite may appear but are not as common as in the amphibolite facies.

Pyroxene hornfels facies (6)

The pyroxene hornfels facies, which forms the inner zone of contact aureoles, is the low-pressure equivalent of the granulite facies (Figure 1.18); in both facies, the appearance of orthopyroxene defines the lower boundary. Typical mineral associations of the pyroxene hornfels facies are:

cordierite–orthopyroxene–K-feldspar–plagioclase–quartz

orthopyroxene–Ca-pyroxene–plagioclase–quartz

In addition to forsterite and wollastonite (which also appear in the granulite facies) marble of the pyroxene hornfels facies commonly contains a variety of high-temperature Ca–Mg silicates, including monticellite, merwinite, spurrite, and larnite.

Where fragments of pelitic or other rock are incorporated in mafic melts and heated to near 1000 °C, cordierite, sanidine, tridymite, mullite and other minerals crystallize, and the rock may be partially melted. These unusual rocks, which are normally assigned to the sanidinite facies, are here included with the pyroxene hornfels facies.

Blueschist facies (7)

Certain metamorphic rocks of young ($\leq$ 300 My) mobile zones (e.g., New Caledonia, California, Japan) contain various combinations of lawsonite, aragonite, jadeite-bearing pyroxene, and glaucophane, together with some minerals of the greenschist facies, and these have been assigned to the blueschist facies and the lower-grade lawsonite–albite–chlorite facies. The latter facies, which is free of glaucophane, is here placed in the blueschist facies. The presence of the high-density phases, lawsonite and aragonite, indicates that metamorphic pressures were high, while chlorite suggests that temperature was low (Figure 1.18). Examples of mineral associations are:

lawsonite–muscovite–chlorite–albite–quartz–aragonite

lawsonite–glaucophane–muscovite–paragonite–chlorite–albite–quartz

lawsonite–pumpellyite–actinolite–glaucophane–chlorite–albite

garnet–epidote–omphacite–actinolite–glaucophane–muscovite–chlorite–albite

Here omphacite refers to jadeite-bearing pyroxene, and muscovite normally contains some Mg and Fe. Stilpnomelane may be present, but not biotite.

Eclogite facies (8)

Certain rocks of basaltic composition are found to consist almost entirely of jadeite-bearing pyroxene and pyrope-rich garnet, and these are evidently metamorphic rocks that have crystallized at high temperature and pressure, corresponding to the domain of the eclogite facies (Figure 1.18). Kyanite and quartz may also be present. The rocks occur as fragments in kimberlite pipes, and as lenses in lower-grade terranes. The former were presumably derived from the Mantle, and the latter may be tectonically transported slices originating in the lower Crust.

Also assigned to the eclogite facies are certain rocks which contain coesite, for example the assemblage,

talc–kyanite–pyrope–coesite

These have experienced temperatures of about 800 °C and pressures in the vicinity of 30 kbar (Schreyer 1988a,b), and they evidently represent crustal rocks that were transported to the upper Mantle.

Shock metamorphism

A meteorite impact can produce transient pressures up to 1000 kbar and temperatures to about 5000 °C, and the changes produced in the impact rocks are referred to as 'shock metamorphism'. Engelhardt and Stöffler (1968) have proposed the following four stages or facies of increasing metamorphic intensity, based on studies of the Ries Crater in Germany.

Stage I (~ 100 to 270 kbar; ~ 100 to 250 °C): fracturing of crystals; sets of planar features (slip bands) in quartz and plagioclase; well-developed kink bands in biotite.

Stage II (~ 270 to 520 kbar; ~ 250 to 1300 °C): quartz becomes amorphous; coesite and stishovite appear; plagioclase crystals become isotropic (maskelynite), beginning in patches and twin lamellae.

Stage III (~ 520 to 620 kbar; ~ 1300 to 2500 °C): melting of quartz and feldspar, which cool to glass with vesicles and flow structures; coesite and stishovite occur in glass; decomposition of biotite.

Stage IV (~ 620 to 1000 kbar; ~ 2500 to 5000 °C): melting of all main rock-forming minerals to produce, on cooling, heterogeneous, vesicle-laden glass, partly devitrified. At the highest pressures and temperatures, vaporization occurs.

A similar scheme was proposed by Chao (1968).

1.8 EXPERIMENTAL APPROACH TO MINERAL REACTIONS

The examination of metamorphic rocks in terranes of increasing grade frequently enables one to propose particular reactions to account for the appearance or disappearance of minerals in rock of uniform composition. For example, the disappearance of muscovite in quartz-bearing pelitic hornfels in the upper part of the hornblende hornfels facies allows one to propose the reaction:

$$\underset{\text{muscovite}}{KAl_3Si_3O_{10}(OH)_2} + \underset{\text{quartz}}{SiO_2} = \underset{\text{K-feldspar}}{KAlSi_3O_8} + \underset{\text{andalusite}}{Al_2SiO_5} + H_2O. \qquad (1)$$

Several reactions, including the above, have now been investigated experimentally, providing valuable information concerning temperature–pressure conditions under which metamorphic reactions occur.

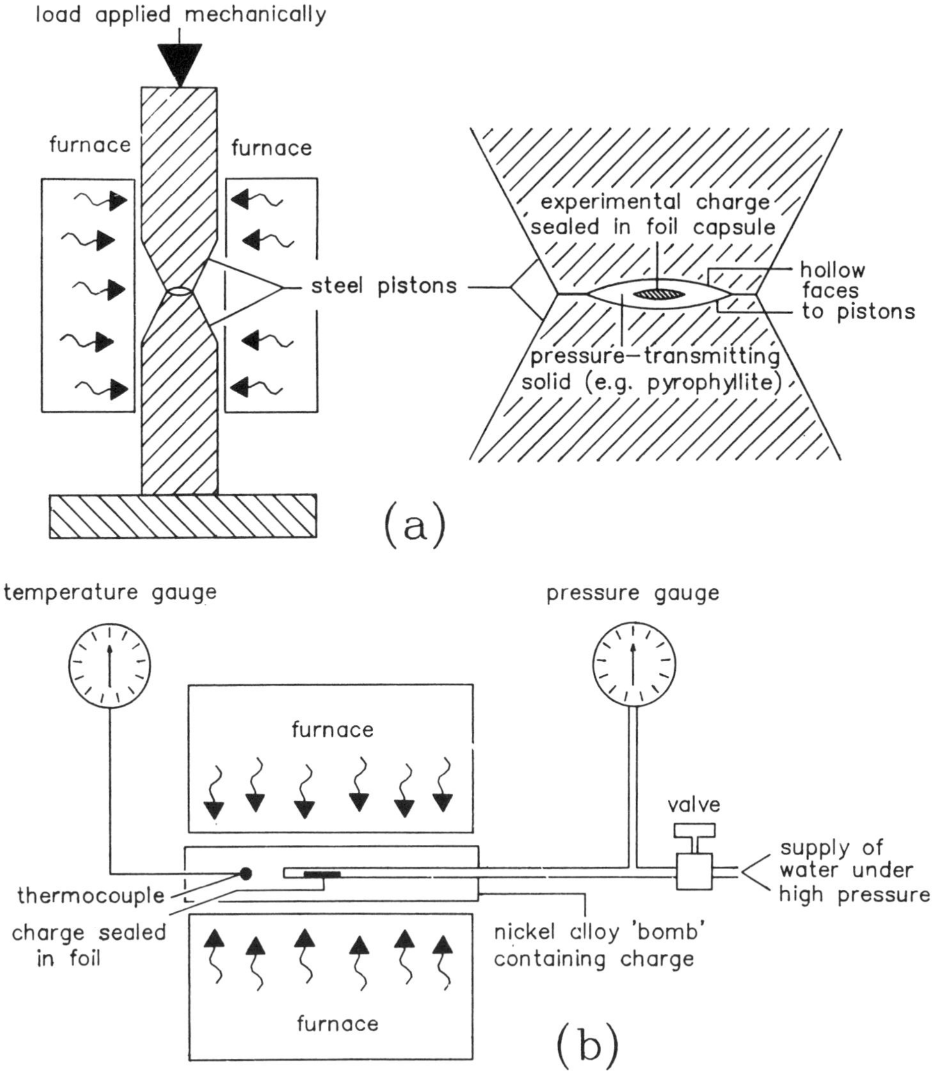

Figure 1.19 (a) A 'simple-squeezer' apparatus, and (b) a hydrothermal apparatus, used for the experimental study of mineral equilibria. From Mason (1978). Reproduced by permission of Chapman & Hall

The experiments are carried out by use of either natural or synthetically prepared minerals. Mineral powders are placed in a capsule, which is then subjected to an elevated temperature and pressure (Figure 1.19) for periods of several hours to several weeks. Often both reactants and products are combined, and the direction of the reaction is determined by observing, upon cooling, changes in amount of minerals present. When all phases involved are nearly pure substances (rather than solutions), as in the above reaction, the stability fields of the reactants and products in a temperature–pressure phase diagram, will be separated by a line (the boundary curve) on which all minerals are stable together. The boundary curve for reaction (1) as determined experimentally by Chatterjee and Johannes (1974) is shown in Figure 1.20.

Although all five phases of reaction (1) are stable only on the boundary curve, a question arises concerning combinations of minerals that are stable immediately to the left and immediately to the right of the boundary curve. This kind of information is best conveyed by use of another phase diagram in which phase changes are recorded within compositional space, defined in terms of certain components.

With regard to reaction (1), these components could be defined as the elements Si, Al, K, H, O, but Figure 1.21a shows that the composition of all of the minerals may be expressed in terms of only four units, SiO_2, Al_2O_3, K_2O, and H_2O, and these could therefore be chosen as the components of the system. The latter are preferred because at equilibrium (i.e., at conditions represented by a point on the boundary curve of Figure 1.20), each of these components may be added or removed from the system (within limits) without causing a change in the mineral assemblage. They are here referred to as *system components*. Figure 1.21a shows that reaction (1) may be perceived as a simple rearrangement of the four components.

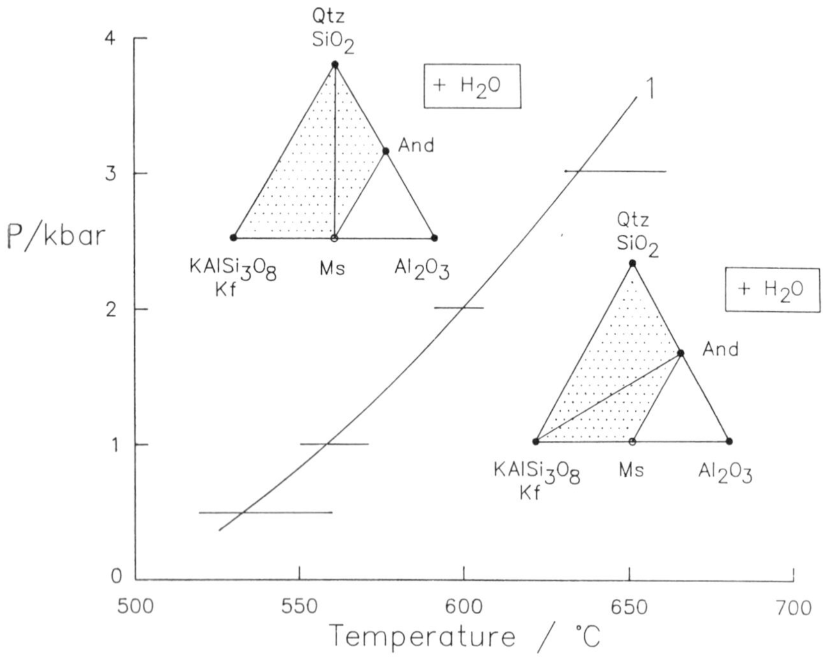

Figure 1.20 Boundary curve for the equilibrium, muscovite (Ms) + quartz (Qtz) = K-feldspar (Kf) + andalusite (And) + H_2O. Mineral associations stable to the left and right of the curve are indicated by the compositional phase diagrams. From Chatterjee and Johannes (1974). Reproduced by permission of Springer-Verlag

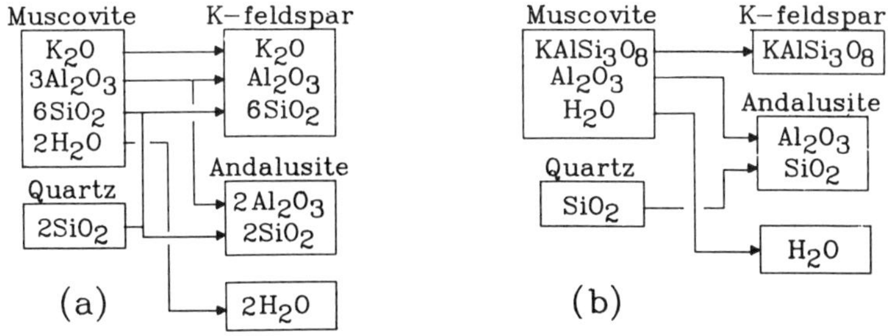

Figure 1.21 Reaction, muscovite + quartz → K-feldspar + andalusite + H_2O, expressed (a) as a rearrangement of components K_2O, Al_2O_3, SiO_2, H_2O, and (b) as a rearrangement of components $KAlSi_3O_8$, Al_2O_3, SiO_2, H_2O

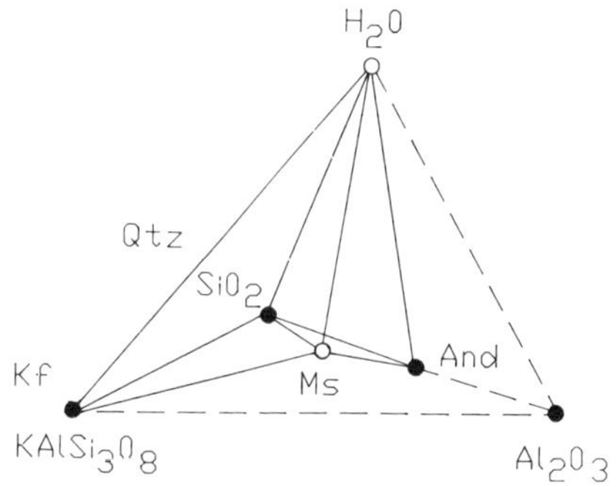

Figure 1.22 Composition tetrahedron, showing two internal tetrahedra that define mineral associations stable to the left of the boundary curve in Figure 1.20. The position of muscovite (Ms), which lies in the front triangle, has been displaced slightly for clarity

An alternative choice of system components is possible. Figure. 1.21b shows that the muscovite formula may be expressed in terms of one K-feldspar formula plus Al_2O_3 and H_2O. Hence the system may be viewed as one composed of $KAlSi_3O_8$, Al_2O_3, SiO_2, and H_2O, and reaction (1) may be perceived as a rearrangement of these components, as shown in Figure 1.21b. Chemical variation in the system will now be defined in terms of these four components.

The $KAlSi_3O_8$–Al_2O_3–SiO_2–H_2O composition tetrahedron is shown in Figure 1.22. All minerals of reaction (1) must lie on or in this tetrahedron. Observe, for example, the distorted Kf–Ms–Qtz–H_2O tetrahedron lying within the main tetrahedron (Figure 1.22); the composition of a system containing these four minerals may be represented by a point lying within the sub-tetrahedron. This point and point muscovite may for convenience be projected onto the base along straight lines originating at apex, H_2O, as shown in Figure 1.20 (left of the boundary curve). There is no loss of information in doing so, provided we acknowledge that water vapor is present in addition to the three phases that appear at the corners of the sub-triangles. The amount of H_2O present in the system can be specified, but in general the amount is unimportant.

We are not interested at present in all possible proportions of the four components, i.e., all possible points in the main tetrahedron, only in those lying within the two sub-tetrahedra that are shown in Figure 1.22, with an apex at H_2O. The projection provides an expression of mineral associations that may occur within the system to the left of the boundary curve (Figure 1.20); i.e., the following are permissible:

muscovite–quartz–K-feldspar–H_2O

muscovite–quartz–andalusite–H_2O

but not

andalusite–K-feldspar–H_2O

Similarly, the projection lying to the right of the boundary curve may be obtained. (The construction of a three-dimensional model is helpful in visualizing the composition tetrahedron.) The reaction itself is graphically represented by the disappearance of tie line muscovite–quartz and the simultaneous appearance of tie line K-feldspar–andalusite (Figure 1.20). Immediately to the right of the boundary curve, the following are permissible:

K-feldspar–andalusite–quartz–H_2O

K-feldspar–andalusite–muscovite–H_2O

but not

muscovite–quartz–H_2O

Thus the reactant minerals are stable to the right of the boundary curve, provided one is absent, and the product minerals are stable to the left of the boundary curve, provided one is absent.

In siliceous carbonate rocks, the sequential appearance of diopside, forsterite, and wollastonite has been reported at several places, and the following reactions may be proposed (among others) to produce these minerals:

$$\underset{\text{dolomite}}{CaMg(CO_3)_2} + \underset{\text{quartz}}{2SiO_2} = \underset{\text{diopside}}{CaMgSi_2O_6} + 2CO_2, \tag{2}$$

$$\underset{\text{diopside}}{CaMgSi_2O_6} + \underset{\text{dolomite}}{3CaMg(CO_3)_2} = \underset{\text{calcite}}{4CaCO_3} + \underset{\text{forsterite}}{2Mg_2SiO_4} + 2CO_2, \tag{3}$$

$$\underset{\text{calcite}}{CaCO_3} + \underset{\text{quartz}}{SiO_2} = \underset{\text{wollastonite}}{CaSiO_3} + CO_2. \tag{4}$$

The boundary curves for the first two of these reactions were located from thermochemical data, i.e., from experimentally obtained data on the heat of formation, the entropy, and the volume of the minerals, by a procedure to be described in Chapter 2, while the third was located by the direct experimental approach described above. The diopside-forming reaction is useful for defining the lower boundary of the amphibolite facies, and the forsterite-forming reaction for the lower boundary of the granulite facies. The curves are plotted in Figure 1.23.

The three reactions occur in the compositional system $CaO-MgO-SiO_2-CO_2$, which may be represented by a tetrahedron with CO_2 at the apex. Here, as in the system described above, all points representing minerals may be projected onto the $CaO-MgO-SiO_2$ base, provided the sub-tetrahedra of interest have one corner at the CO_2 apex; carbon dioxide is then acknowledged to be present in addition to the minerals that appear in the projection. At present we are interested only in the left-hand side of the projection (Figure 1.23), in systems with Ca/(Ca+Mg) ≥ 0.5, as represented by common siliceous dolomitic marble. In the composition triangles of Figure 1.23, as in those of Figure 1.20, a reaction is represented by the loss or appearance of tie lines and much information is readily conveyed regarding permissible mineral associations within the temperature-pressure domains lying between boundary curves. For example, in the domain bounded by curves 3 and 4, the following mineral associations are stable:

calcite-diopside-quartz-CO_2
calcite-diopside-forsterite-CO_2
calcite-dolomite-forsterite-CO_2

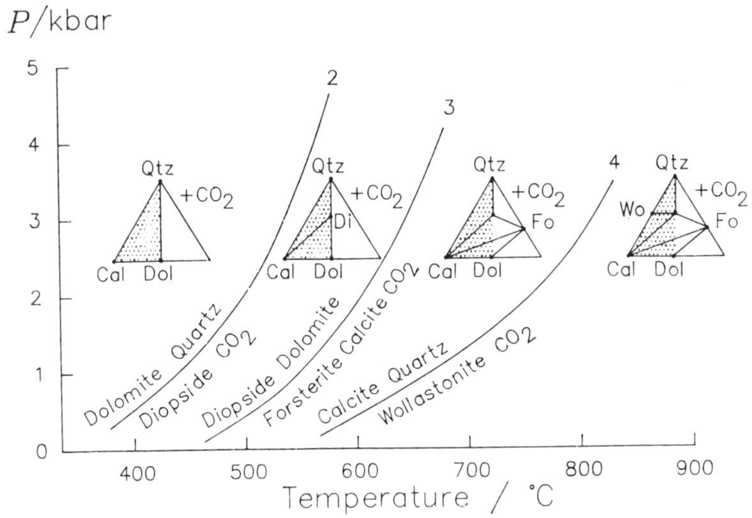

Figure 1.23 Boundary curves for three equilibria in the system $CaO-MgO-SiO_2-CO_2$. Curve 4 was located experimentally by Harker and Tuttle (1956), and curves 2 and 3 by Turner (1981) from thermochemical data. Mineral phases are calcite (Cal), dolomite (Dol), quartz (Qtz), diopside (Di), forsterite (Fo), and wollastonite (Wo); CO_2 is assumed to be present. From Metamorphic Petrology, by F.J. Turner (1981, p. 164), reproduced by permission of Hemisphere Publishing Corp., Washington D.C., all rights reserved

whereas

dolomite–quartz
dolomite–diopside

are unstable.

Reactions (1) to (4) are representative of relatively simple metamorphic reactions; they and some more complex reactions will be examined in some detail in Chapter 3.

1.9 AN EXAMPLE OF PROGRESSIVE METAMORPHISM: SHALE TO GRANULITE

Many metamorphic reactions have been proposed in the past to explain the 'appearance' and 'disappearance' of minerals in regions of increasing metamorphic grade, and numerous reactions have now been investigated experimentally. The following brief discussion of the progressive metamorphism of shale will serve to illustrate the level of uncertainty that presently exists regarding changes in mineral assemblages with changes in metamorphic grade.

The metamorphism of shale produces slate, mica schist, and in places hornfels, gneiss, and granulite. Following diagenesis, a typical shale consists of kaolinite and illite (Table 1.3) together with chlorite, quartz, and feldspar. During low-grade metamorphism, at about 200–300 °C, various reactions occur which produce combinations of pyrophyllite, muscovite, phengite, paragonite, chlorite, stilpnomelane, albite, and quartz.

Pyrophyllite presumably forms from kaolinite, as follows,

$$\underset{\text{kaolinite}}{Al_2Si_2O_5(OH)_4} + \underset{\text{quartz}}{2SiO_2} = \underset{\text{pyrophyllite}}{Al_2Si_4O_{10}(OH)_2} + H_2O$$

The boundary curve for this reaction was experimentally located by Thompson (1970), as shown in Figure 1.24 (curve 1). According to Frey (1970) pyrophyllite in the Alps appeared at about 200 °C (2 kbar), the lower temperature resulting possibly from a reduced activity of H_2O. Pyrophyllite has a small stability field,

Table 1.3 Comparison of some sheet-silicate formulas

	Site coordination			
	12	6	4	
kaolinite	—	Al_4	Si_4	$O_{10}(OH)_8$
illite	K_{1-x}	Al_2	$Al_{1-x}Si_{3+x}$	$O_{10}(OH)_2$
muscovite	K_1	Al_2	Al_1Si_3	$O_{10}(OH)_2$
paragonite	Na_1	Al_2	Al_1Si_3	$O_{10}(OH)_2$
phengite	K_1	$Al_{2-x}(Mg,Fe)_x$	$Al_{1-x}Si_{3+x}$	$O_{10}(OH)_2$
chlorite	—	$(Mg,Fe,Al)_6$	$(Al,Si)_4$	$O_{10}(OH)_8$
biotite	K_1	$(Mg,Fe,Al,Ti)_3$	$(Al,Si)_4$	$O_{10}(OH)_2$

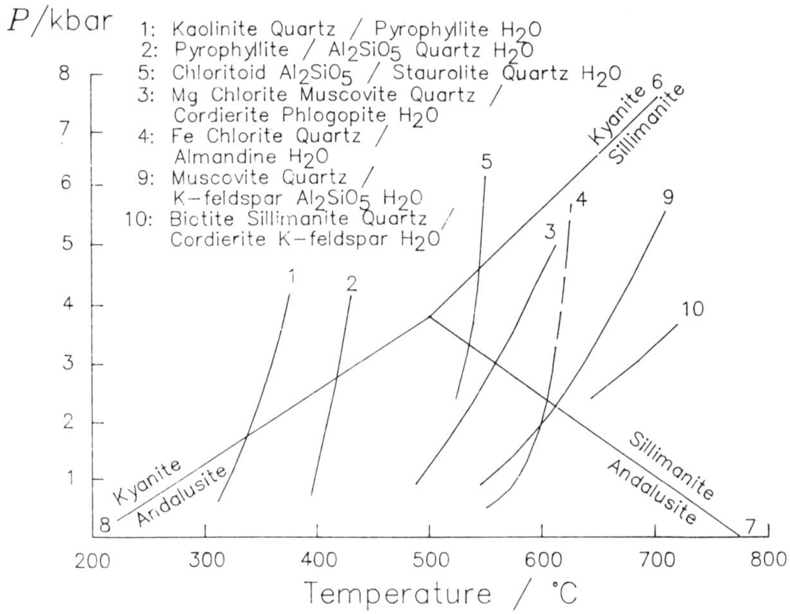

Figure 1.24 Some experimentally determined mineral equilibria relevant to the metamorphism of shale; sources of data are given in text

and with increasing temperature may break down to produce andalusite or kyanite,

$$Al_2Si_4O_{10}(OH)_2 = Al_2SiO_5 + 3SiO_2 + H_2O,$$

as shown by curve 2 (Figure 1.24), which was located by Kerrick (1968). This reaction may account for the first appearance of andalusite and kyanite in some low-grade pelitic rocks.

Chloritoid, a fairly common metamorphic mineral, appears to be confined to low-grade rocks that are rich in Fe relative to Mg. Where present, it has possibly formed by the breakdown of pyrophyllite and chlorite (Winkler 1974, p. 205).

Biotite appears near the middle of the greenschist facies. The potassium necessary to build biotite presumably comes from muscovite, phengite, or K-feldspar, while the Mg and Fe are from chlorite, although some may be derived from phengite (Table 1.3). Biotite commonly contains some Ti, which may be obtained from ilmenite. Reactions have been proposed by Mather (1970), Brown (1971), Ramsay (1973), and Wang, Banno and Takeuchi (1986); two of these are

$$\text{phengite} + \text{chlorite} = \text{biotite} + \text{muscovite} + H_2O,$$

$$\text{K-feldspar} + \text{chlorite} = \text{biotite} + \text{muscovite} + \text{quartz} + H_2O.$$

Entering the region of medium-grade metamorphism, an important reaction takes place, at relatively low pressures, to produce cordierite (see Figure 1.17).

Commonly chlorite diminishes greatly in amount as cordierite appears, and it presumably provides the necessary Mg and Fe. One possible reaction is

$$\underset{\text{chlorite}}{(Mg,Fe)_5Al_2Si_3O_{10}(OH)_8} + \underset{\text{muscovite}}{KAl_3Si_3O_{10}(OH)_2} + \underset{\text{quartz}}{2SiO_2}$$

$$= \underset{\text{cordierite}}{(Mg,Fe)_2Al_4Si_5O_{18}} + \underset{\text{biotite}}{K(Mg,Fe)_3AlSi_3O_{10}(OH)_2} + 4H_2O.$$

The boundary curve for this reaction in the Fe-free system was experimentally located by Seifert (1970) and appears as curve 3 in Figure 1.24. Meanwhile, at higher pressures, almandine-rich garnet is formed, possibly by the breakdown of an Al-rich chlorite,

$$\underset{\text{chlorite}}{Fe_9Al_6Si_5O_{20}(OH)_{16}} + \underset{\text{quartz}}{4SiO_2} = \underset{\text{almandine}}{3Fe_3Al_2Si_3O_{12}} + 8\ H_2O.$$

The boundary curve for this reaction was located by Hsu (1968) and is shown as curve 4 in Figure 1.24. A similar curve for the Mn end-members lies 200 °C to the left, in agreement with the observation that garnet, when found in the greenschist facies, is almost invariably rich in Mn.

Staurolite, like chloritoid, appears to be restricted to rocks rich in Fe relative to Mg, and, provided one of the Al_2SiO_5 polymorphs is present, it may form by the reaction

$$8\text{ chloritoid} + 10\ Al_2SiO_5 = 2\text{ staurolite} + 3\text{ quartz} + 4H_2O,$$

which, for the Fe end-members, has been located by Richardson (1968; curve 5 in Figure 1.24). In the absence of Al_2SiO_5, staurolite may form by the reaction

$$\text{chlorite} + 3\text{ muscovite} = \text{staurolite} + 3\text{ biotite} + 7\text{ quartz} + 14H_2O,$$

which occurs at approximately the same temperatures (Hoschek 1969). Staurolite has a small stability field and with increasing temperature breaks down, possibly as follows,

$$\text{staurolite} + \text{muscovite} + \text{quartz} = Al_2SiO_5 + \text{biotite} + H_2O$$

(Winkler 1974, p. 217), or

$$\text{staurolite} + \text{muscovite} + \text{quartz} = Al_2SiO_5 + \text{garnet} + \text{biotite} + H_2O$$

(Thompson and Norton 1968, Carmichael 1970).

In the upper half of the amphibolite facies, the reaction

$$\text{kyanite} = \text{sillimanite}$$

occurs, while in the hornblende hornfels facies the reaction is

$$\text{andalusite} = \text{sillimanite}.$$

The boundary curves for these two important reactions, as well as for

$$\text{kyanite} = \text{andalusite},$$

are shown in Figure 1.24 as curves 6, 7, and 8 respectively, taken from Holdaway (1971).

Near the top of the amphibolite and hornblende hornfels facies, muscovite breaks down according to the model reaction,

$$\underset{\text{muscovite}}{KAl_3Si_3O_{10}(OH)_2} + \underset{\text{quartz}}{SiO_2} = \underset{\text{K-feldspar}}{KAlSi_3O_8} + \underset{\substack{\text{sillimanite}\\\text{or andalusite}}}{Al_2SiO_5} + H_2O,$$

shown in Figure 1.20 and also in Figure 1.24 (curve 9). With decreasing activity of H_2O this curve shifts to the left; for example, if the decrease is brought about by H_2O mixing with CO_2 in an associated fluid phase, and if the mole fraction of H_2O is 0.7, the shift is about 50 °C (Kerrick 1972). Because the alkali feldspar produced by the reaction can accommodate some Na, plagioclase may also be involved in the reaction (Evans and Guidotti 1966).

Near the boundary of the granulite facies, biotite and sillimanite evidently begin to react to produce garnet (Ramberg 1952a),

$$\underset{\text{biotite}}{K(Mg,Fe)_3AlSi_3O_{10}(OH)_2} + \underset{\text{sillimanite}}{Al_2SiO_5} + \underset{\text{quartz}}{2SiO_2}$$

$$= \underset{\text{garnet}}{(Mg,Fe)_3Al_2Si_3O_{12}} + \underset{\text{K-feldspar}}{KAlSi_3O_8} + H_2O.$$

or cordierite,

$$\underset{\text{biotite}}{2K(Mg,Fe)_3AlSi_3O_{10}(OH)_2} + \underset{\text{sillimanite}}{6Al_2SiO_5} + \underset{\text{quartz}}{9SiO_2}$$

$$= \underset{\text{cordierite}}{3(Mg,Fe)_2Al_4Si_5O_{18}} + \underset{\text{K-feldspar}}{2KAlSi_3O_8} + 2H_2O.$$

The first of these is a continuous reaction, occurring over a range of temperature. The second, which is nearly univariant, has been investigated by Hoffer (1976), using natural biotite and cordierite of intermediate composition, and was observed to occur at about 700 °C, as shown by curve 10 in Figure 1.24.

As pelitic rocks are subjected to pressure–temperature conditions represented by the granulite and pyroxene hornfels facies, orthopyroxene appears and biotite eventually disappears, possibly by way of the following reaction:

$$\underset{\text{biotite}}{K(Mg,Fe)_3AlSi_3O_{10}(OH)_2} + \underset{\text{quartz}}{3SiO_2} = \underset{\text{orthopyroxene}}{3(Mg,Fe)SiO_3} + \underset{\text{K-feldspar}}{KAlSi_3O_8} + H_2O.$$

Notice that many of the above reactions release H_2O. Thus, through a number of dehydration steps, shale rich in hydroxyl-bearing minerals, is transformed to gneiss, granulite, or hornfels, containing combinations of the anhydrous minerals, plagioclase, alkali feldspar, quartz, garnet, cordierite, sillimanite, and orthopyroxene.

Concerning the graphical representation of mineral associations, it is noteworthy that although common shales contain numerous elements (Si, Al, Ti, Mg, Fe, Mn, Ca, Na, K, ...) many of the mineral reactions that occur in these rocks may be viewed as occurring in the six-component system $SiO_2-K_2O-Al_2O_3-FeO-MgO-H_2O$. If quartz is present and if H_2O is assumed to be present as a phase, the chemical activity of SiO_2 and H_2O become fixed at a given temperature and pressure, and many of the common minerals may be plotted in the composition tetrahedron $K_2O-Al_2O_3-FeO-MgO$. This represents a simplification inasmuch as Na, Ti, Mn, Fe^{3+} and other elements which participate to some extent in reactions involving biotite, garnet, alkali feldspar, etc., are disregarded.

Many of the minerals that appear in pelitic rocks contain little or no K, and can be plotted on the $Al_2O_3-FeO-MgO$ side of the composition tetrahedron, while muscovite plots on the $Al_2O_3-K_2O$ edge (Figure 1.25a). Thompson (1957) devised a procedure for examining phase relations within the six-component system by projecting phases that appear within the tetrahedron (notably biotite) onto the $Al_2O_3-FeO-MgO$ side, to give Figure 1.25b. This plot may be viewed as a phase diagram provided quartz, H_2O and muscovite are present. The figure shows, for example, permissible mineral pairs and triads at a temperature and a pressure in the middle region of the amphibolite facies. Some of these are:

1. sillimanite–staurolite–garnet
2. staurolite–garnet
3. staurolite–garnet–biotite
4. staurolite–biotite
5. sillimanite–staurolite–biotite

The Thompson diagram was used by Carmichael (1970) to illustrate changes that occurred at isograds which he located in the Whetstone Lake area of the Grenville Province (Canadian Shield). The isograds are defined by the following reactions:

1. staurolite–biotite isograd:
   chlorite + garnet + muscovite = staurolite + biotite + quartz + H_2O
2. kyanite–biotite isograd:
   chlorite + staurolite + muscovite + quartz = kyanite + biotite + H_2O
3. sillimanite isograd:
   kyanite = sillimanite

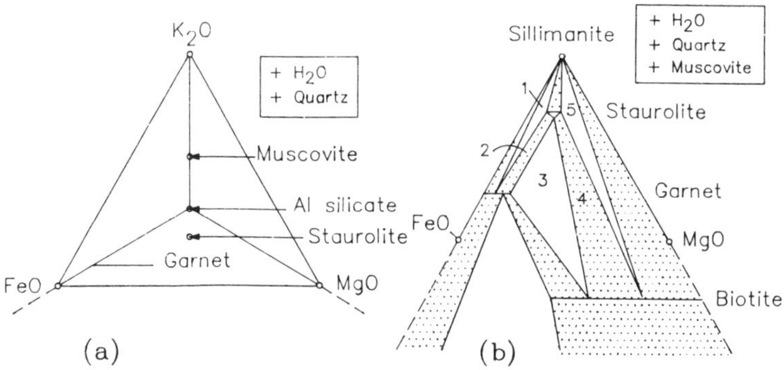

Figure 1.25 (a) The K_2O-FeO-MgO-Al_2O_3 composition tetrahedron, viewed from above. Staurolite and garnet plot on the front (Al_2O_3-FeO-MgO) side of the tetrahedron, while biotite plots as a line *within* the tetrahedron; biotite, when projected from point muscovite onto the downward extension of the Al_2O_3-FeO-MgO plane, appears as in (b), which shows various mineral associations (e.g., 1 to 5) in the staurolite zone of the amphibolite facies. (b) is the Thompson (1957) diagram

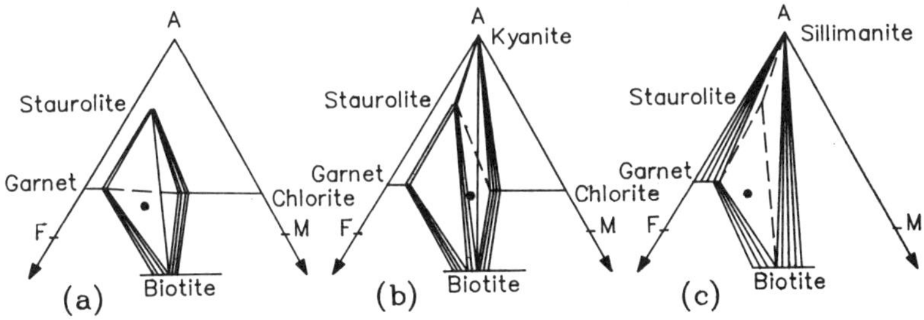

Figure 1.26 Use of the Thompson diagram (Figure 1.25b) to illustrate changes at three isograds, (a) chlorite + garnet (+ muscovite) = staurolite + biotite (+ quartz + H_2O), (b) chlorite + staurolite (+ muscovite + quartz) = kyanite + biotite (+ H_2O), (c) staurolite (+ muscovite + quartz) = sillimanite + garnet + biotite (+ H_2O). From Carmichael (1970), by permission of Oxford University Press

4. sillimanite-garnet-biotite isograd

   staurolite + muscovite + quartz = sillimanite + garnet + biotite + H_2O

The changes that occurred at isograds 1, 2, and 4 can be effectively illustrated by use of the Thompson diagram, as shown in Figure 1.26.

1.10 METAMORPHISM, STRUCTURE, AND TECTONICS

The alteration of shale to hornfels at the contact of a gabbro dike occurs in response to an influx of heat from the dike, and metamorphic crystallization occurs with little or no deformation. By contrast, the metamorphism of shale in mountain belts, in which the strata have been complexly folded, produces foliated rocks (phyllite, schist, gneiss), and it is generally understood that regional metamorphism and crustal deformation are 'broadly synchronous' (Turner and Verhoogen 1960).

The detailed timing of metamorphic crystallization *relative* to the timing of deformation has been studied in numerous terranes, and microstructural criteria for the recognition of pre-, syn-, and post-deformational crystallization have been proposed by Zwart (1960) and Spry (1969). Crystals that have been deformed (e.g., biotite with kink bands) provide examples of pre-deformational crystallization; garnet crystals with a sigmoidal arrangement of inclusions have been taken as evidence of syn-deformational crystallization (another interpretation was presented by Bell (1985)); and the random orientation of prismatic anthophyllite crystals in a foliated (biotite) schist could be viewed as evidence for the post-deformational crystallization of amphibole.

In the Scottish Highlands, Johnson (1963) recognized three major periods of folding (F_1, F_2, F_3) and by use of criteria of the kind mentioned above, he identified four periods of crystallization, (a) conversion of shale to fine-grained schist during F_1, (b) growth of garnet between F_1 and F_2, (c) growth of garnet and biotite during F_2, and (d) growth of garnet, kyanite and staurolite, followed by sillimanite between F_2 and F_3. In studies of this kind, great care is needed in the interpretation of microstructures (Jamieson 1988), and attempts should be made to distinguish between crystallization (growth) resulting from a metamorphic reaction, and growth resulting from recrystallization, e.g., the redistribution of biotite to produce a new foliation.

Related to the structural and metamorphic evolution of fold belts is the question of the disposition of isograd surfaces at the time of metamorphism. This problem was considered by P.H. Thompson (1976) who mapped the present-day three-dimensional position of the staurolite and sillimanite isograds in the Val Mesolcina area of the Swiss Alps. The spacing and dip of the isograd surfaces when viewed in relation to the spacing and slope of the experimentally located boundary curves for the isograd reactions,

$$\text{chlorite} + \text{muscovite} = \text{staurolite} + \text{biotite} + H_2O,$$
$$\text{kyanite} = \text{sillimanite},$$

produced the interpretation shown in Figure 1.27. These results emphasize that although isobaric surfaces are expected to remain nearly horizontal, isothermal surfaces in fold belts could in fact be inclined considerably, to form, for example, a thermal dome. In this case, all points on an isograd surface do not represent the same combination of temperature and pressure, i.e., the same point on a boundary curve.

The above considerations lead to questions of the source of heat and heat flow during metamorphism. With regard to metamorphism adjacent to a tabular sheet of silicate melt (which is the heat source), solutions to the heat-flow problem were presented by Lovering (1935) and Jaeger (1959). Given certain data and

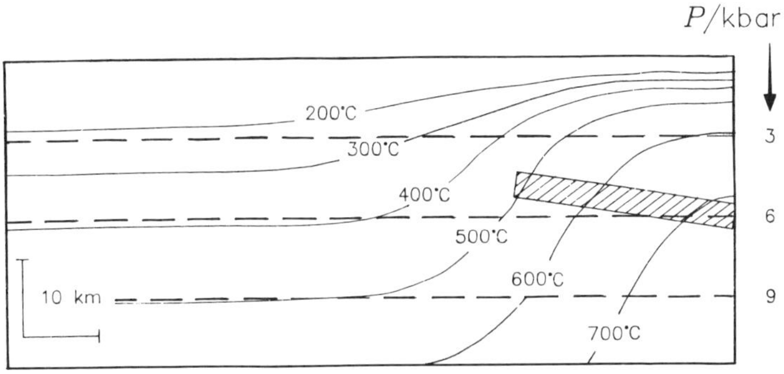

Figure 1.27 Reconstructed position of the Val Mesolcina segment (diagonal pattern) of the Swiss Alps, as it existed 35 million years ago. From P.H. Thompson (1976). Reproduced by permission of Springer-Verlag

assumptions (initial-melt and country-rock temperature, latent heat of crystallization, the density, specific heat and conductivity of the country rocks, instantaneous emplacement, transport of heat only by conduction), Jaeger (1957, 1959) calculated the country-rock temperature as a function of dike thickness, distance from the dike, and time. These results produce the temperature–time relations shown in Figure 1.28. Should the rise in temperature initiate a metamorphic reaction, the heat absorbed by the reaction would cause a change in profile slope. Note that at a given distance from the dike, the increase in temperature is nearly linear (within limits) and the maximum on the curve (representing the peak metamorphic temperature), is realized earlier or later, depending on distance from the dike. In general, heat-flow calculations of this kind confirm that rocks are notably resistant to heat conduction, even on a time scale of a million years.

On a regional and global scale, a heat flux of approximately 1.6 microcalories cm^{-2} escapes from the Earth each second; most of this is radiogenic heat, produced by isotopes of U, Th, and K (Schubert and Anderson 1974). The flux is greater over Mesozoic–Cenozoic mountain belts (~ 1.8 unit) than over Precambrian Shield areas (~ 1.0 unit). These and other data can be used to calculate approximate geothermal gradients (Fowler, Nisbet, and Pandit 1988); four such gradients, related to four tectonic settings, are shown in Figure 1.29.

Metasedimentary rocks presently exposed within fold belts have obviously experienced a journey which has taken them to depth, and then, aided by uplift and erosion, back to the surface. England and Thompson (1984) have constructed models in which the downward journey is caused by large-scale thrusting, which results in crustal thickening. In view of the thermal resistivity of rocks, as noted above, and the rapid displacement of tectonic blocks, as demanded by modern tectonic theory, pressure in a descending block is expected to increase more rapidly than temperature, to produce pressure–temperature (P–T) paths that are clockwise. Figure 1.30 shows such paths for small volumes of rock from the biotite zone (a), the garnet zone (b) and the sillimanite zone (c) of a hypothetical fold belt. It is

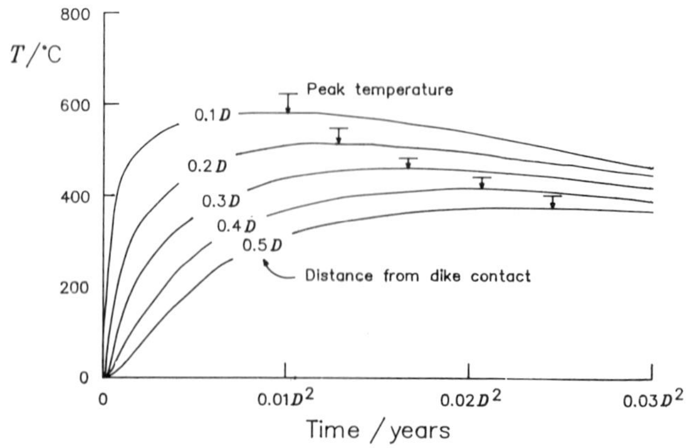

Figure 1.28 Calculated changes of temperature with time at five points of increasing distance from the contact of a gabbro dike of width D (in any length units), intruded at 1100 °C into country rock that initially was at 0 °C. From Jaeger (1957, p. 313). Reprinted by permission of *American Journal of Science*

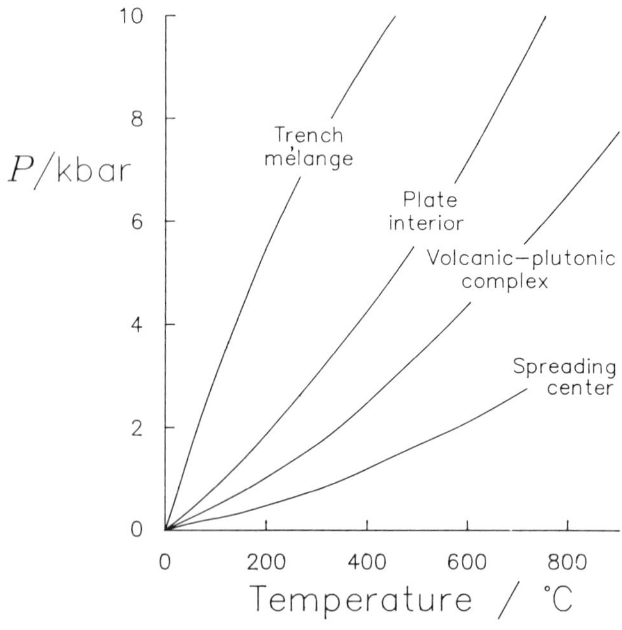

Figure 1.29 Calculated geotherms for different tectonic environments. From *Petrologic Phase Equilibria* by W.G. Ernst. © 1976 by W.H. Freeman and Company. Reprinted with permission

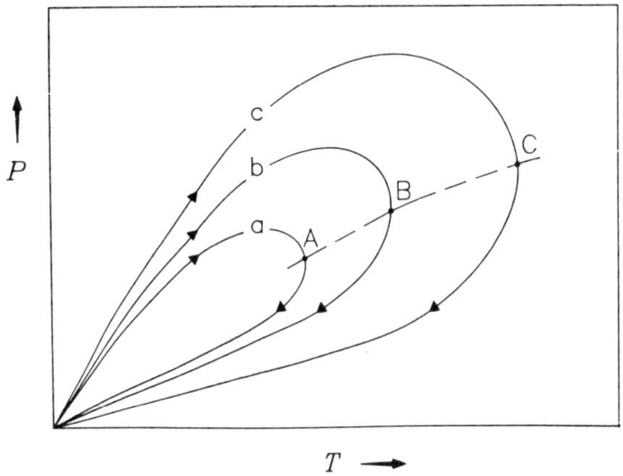

Figure 1.30 Possible temperature–pressure paths for volumes of rock, a, b, c, of increasing grade, in a metamorphic terrane. The line joining points A, B, C, of maximum temperature, is the field gradient. From Spear et al. (1984)

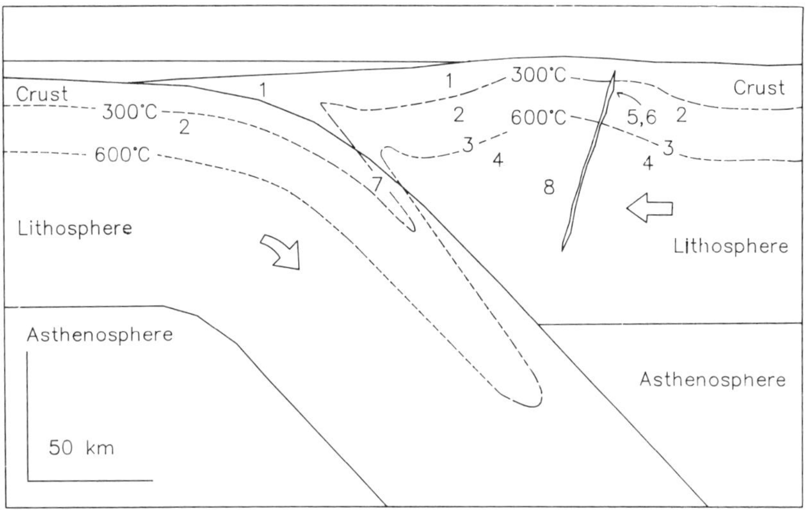

Figure 1.31 A proposal for the position of 300 °C and 600 °C isotherms, and distribution of metamorphic facies at a convergent plate margin; facies are: 1 very low grade, 2 greenschist, 3 amphibolite, 4 granulite, 5 hornblende hornfels, 6 pyroxene hornfels, 7 blueschist, 8 eclogite. From *Petrologic Phase Equilibria* by W.G. Ernst. © 1976 by W.H. Freeman and Company. Reprinted with permission

apparent from the England–Thompson models and from Figure 1.30, that the peak pressure for each volume precedes the peak temperature, and that peak temperatures for the three volumes do not necessarily occur at the same time (cf. Figure 1.28). The peak temperature is presumably that 'reported' by the mineral assemblage, and a line joining these temperatures (A–B–C in Figure 1.30), which in general is not a geothermal gradient, is referred to as the 'field gradient' (Spear et al. 1984). The use of microstructural and micro-chemical data to locate points on $P-T$ paths is now an area of active research (Spear, 1993, St.-Onge 1987), and the results are obviously important in tectonic studies.

Considerable progress has been made with regard to high-pressure metamorphism, which is almost certainly confined to subduction zones, as proposed by Ernst (1971). Here the downward-moving slab carries sediments to great depths (high pressures) while low thermal conductivity prevents a comparable rise in temperature, as shown in Figure 1.31. The cause of the relatively rapid return of the metasediments to the surface, and the absence of blueschists in Precambrian terranes are not, however, fully understood.

The sources of heat in very large terranes of medium-pressure and low-pressure metamorphism, as found in Precambrian shields remain largely unknown. Attoh and Vander Meulen (1984) have shown that granite plutons, which are commonly present, cannot, in general, provide the necessary quantity of thermal energy.

The reader is encouraged to consult Fowler, Nisbet and Pandit (1988), Ghent, Stout and Parrish (1988), Jamieson and Beaumont (1988), Condie (1989), Yardley (1989), Philpotts (1990) and references cited therein, for reviews, details, and further discussions of pressure–temperature paths, and the relation between metamorphism and tectonics.

2 Mineral Thermodynamics

Observation and experimentation have given rise to a large body of knowledge regarding mineral reactions within the solid Earth, but a satisfactory understanding and appreciation of these reactions can only be obtained through thermodynamics. The first law of thermodynamics is an expression of the conservation of energy, and the second law states that all physical-chemical processes, including mineral reactions, proceed with a local decrease in chemical energy, and an overall increase in entropy. Hence, thermodynamics deals with physical-chemical processes in terms of energy and entropy.

Thermodynamics, when applied to mineral reactions, identifies those properties of minerals that determine their behavior. In addition to chemical composition, these fundamental properties are the heat of formation (i.e., the heat energy released when crystals are constructed from the elements), the heat capacity (i.e., the capacity to absorb heat as temperature increases), the entropy (which depends on the arrangement of atoms and molecules), and the volume or density. These are the properties that determine, for example, the position and slope of the boundary curve for the reaction, aragonite = calcite, and hence determine that calcite is stable at the Earth's surface and at most places within the Crust. From this viewpoint, heat of formation, heat capacity, entropy, and volume are among the most important properties of minerals.

With regard to minerals of variable composition, thermodynamics identifies the important property referred to as 'chemical potential'. The relation between the chemical potential of a component and its concentration, together with the properties mentioned above, determine the behavior of these minerals, e.g., the presence or absence of a miscibility gap, the continuous nature of reactions involving these minerals, and the distribution of major and trace elements between associated minerals.

Thermodynamics is concerned with the properties and behavior of large aggregates of atoms or molecules. For example, it is concerned with relations between temperature, pressure and volume of a gas, with no regard for the behavior of individual molecules. It is concerned with heat and entropy changes in solid-state reactions, with no regard for the atomic structure of the crystals. Given the large amount of information we now possess on the structure of minerals, the reader may wonder why we employ a procedure that does not make use of this information. The answer is that thermodynamics enables us to obtain an understanding of the behavior of minerals which cannot be obtained only from information on atomic bonds and atomic arrangements in crystals. Frequently, however, such information can be interwoven with thermodynamic reasoning, to great advantage.

Thermodynamics is concerned with the properties of matter before and after a physical-chemical change has taken place, not with the change itself. For example, thermodynamic data inform us that at the Earth's surface, aragonite ought to change to calcite, with the release of a specified amount of energy, but the data do not tell us how fast the reaction will proceed, nor do they tell us anything about the mechanism of the reaction, e.g., the nucleation and growth of crystals of calcite. Concerns with reaction rates and mechanisms will be dealt with in Chapters 4 and 5.

2.1 REVIEW OF SOME THERMODYNAMIC CONCEPTS

Following is a brief survey of thermodynamic terms and equations useful to metamorphic petrology. These have been taken principally from books by Guggenheim (1977), Darken and Gurry (1953), and Swalin (1962) which may be referred to for detailed information.

Thermodynamic system

A particular volume of matter that is selected for study is referred to as a system. In the laboratory, system boundaries are well defined, but in Nature, the boundaries are usually imaginary. In thinking about mineral reactions within the solid Earth, we frequently focus on small volumes of rock which contain several thousands of crystals, and these volumes (about 3 cm in diameter) may be regarded as thermodynamic systems. When reactions within crystals are considered, a single crystal forms the system. Occasionally, a large volume of rock may be regarded as a system, but we shall see below that internal equilibrium demands that the composition of every mineral is uniform throughout, and this condition is rarely observed in large volumes of rock.

If the content of a system remains unchanged, the system is said to be closed; if, during an interval of time, some matter moves into or out of the system, then during that interval of time, the system is said to be 'open'. Evidently many terrestrial systems are open with respect to H_2O or CO_2 during intervals of time when H_2O or CO_2 is released or consumed by reaction. At many places evidence is found for rock systems having been open to other elements, e.g., Si and Fe to produce skarn.

Phase

A system normally contains one or more homogeneous substances referred to as phases. Although crystals of the common minerals are not perfectly homogeneous, they may nevertheless be regarded as phases. Thus all crystals of biotite in a small volume of rock constitute one phase, biotite.

In general, crystalline rocks consist of a few minerals each with a distinctive composition and atomic structure, and the phases are well defined. Problems of definition arise when crystals are compositionally zoned or when sub-microscopic intergrowths are present.

System components

Much of the solid Earth consists of oxygen, together with a few elements present in major concentrations and many additional elements present in minor or trace concentrations. The major elements tend to combine with oxygen in certain simple proportions, in a manner similar to the formation of molecules in a gaseous system. Hence the component parts of rock systems may frequently be regarded as oxides rather than elements. For example, a system composed of calcite, quartz, wollastonite, and CO_2 may be said to consist of three components, CaO, SiO_2, and CO_2, for the composition of all phases may be expressed in terms of only these three substances. Moreover, each may be added to or removed from the system (within limits), without disturbing the four-phase equilibrium:

$$CaCO_3 + SiO_2 = CaSiO_3 + CO_2,$$

$$\text{calcite} + \text{quartz} = \text{wollastonite} + CO_2.$$

Another choice of components is $CaCO_3$, SiO_2, CO_2; now the composition of $CaSiO_3$ is formulated as $CaCO_3 + SiO_2 - CO_2$.

Minor elements are always present in natural systems, and whether or not these should be viewed as system components often depends on the particular process of interest. Thus, in the system described above, if Mn is present in concentrations too small to cause a noticeable displacement of the calcite–quartz–wollastonite–CO_2 equilibrium, Mn could be disregarded. But if the behavior of Mn is of interest, or if concentrations are sufficiently high to create, in effect, a continuous (bivariant) equilibrium, then MnO or $MnCO_3$ must be counted as an additional component.

Thermodynamic state

Consider an isolated system composed of a gas of known composition and mass, for example 1 mole of CO_2. We know from experience that any change that is observed to take place in this system (e.g., changes leading to a uniformity of T and P, or dissociation reactions) will sooner or later cease and the system will come to rest.

The thermodynamic state of the system may now be defined by recording certain properties of the system, for example, the temperature and pressure. Again, from experience we know that commonly the state of a system is fully described by reporting only a few properties. For example, if the system consists of one mole of a perfect gas for which the equation of state is

$$PV = RT,$$

(where R is the gas constant) the thermodynamic state of the system is fully described by specifying, in addition to composition and mass, only two of the three properties, temperature (T), pressure (P), and volume (V). These three variables are referred to as 'thermodynamic variables', or 'state functions'.

Crystals of minerals also have equations of state, but information in addition to P, V, and T is needed to characterize these crystals fully, in particular, information on the number and kind of imperfections present (Darken and Gurry 1953). When possible and appropriate, atom-scale properties of crystals should be taken into consideration when discussing mineral transformations and other thermodynamic processes.

Pressure and work

In classical thermodynamics, the pressure acting on a system is assumed to be uniformly distributed, i.e., it is hydrostatic pressure. In examining rock bodies brought to the Earth's surface from depth, abundant evidence is found for the rocks having experienced ductile deformation, which indicates that (1) the rock bodies were subjected to directed (non-hydrostatic) pressure, and (2) that directed pressures were largely eliminated by flow. We shall therefore assume, unless stated otherwise, that metamorphic pressures within the Earth are near-hydrostatic, and that the equations of thermodynamics are generally applicable under these conditions. The range of pressure of interest extends from 1 kbar (100 MPa) at a depth of 3 km, to 10 kbar (1000 MPa) at a depth of 34 km, near the base of the continental crust, to 30 kbar (3000 MPa) at a depth of 100 km, near the base of the lithosphere.

An increase in the pressure acting on a system represents a quantity of mechanical work done *on* the system (Figure 2.1). Recall,

$$\text{work} = (\text{force})\ (\text{distance}),$$

$$w = F\,\mathrm{d}x,$$

where w is a differential of work done on the system and $\mathrm{d}x$ is a differential of displacement. If A is the cross-sectional area of a piston (Figure 2.1) then the

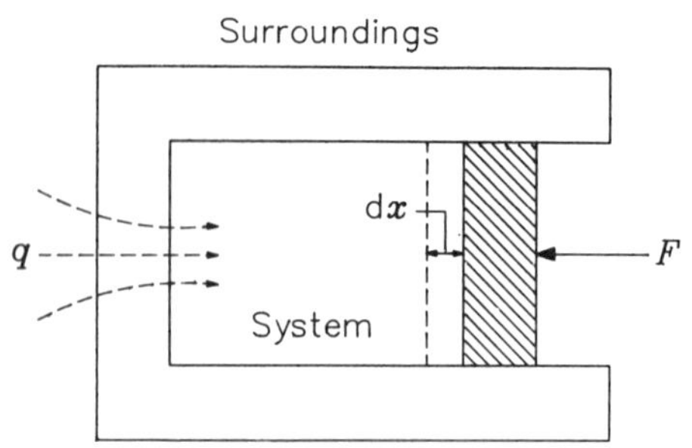

Figure 2.1 Illustration of a system and the surroundings. The work applied to the system is $w = F\,\mathrm{d}x$ (taken as negative) and the heat supplied is q (taken as positive)

differential of volume, i.e., the change in volume is

$$dV = -A\,dx.$$

But,

$$w = (F/A)(A\,dx).$$

Hence,

$$w = -P\,dV,$$

where P is pressure. If the process is carried out reversibly, i.e., the pressure on the piston from inside is equal to P, then the work done *by* the system is

$$w = +P\,dV.$$

Other kinds of work, in addition to mechanical work, may be applied, including work resulting from magnetic, electric, or gravitational forces, but these will not be taken into consideration. Work resulting from a change in interfacial area will be examined in Chapter 5.

Temperature

We know from experience that hot and cold objects, when brought into contact will eventually possess the same degree of hotness or coldness, a condition we refer to as thermal equilibrium. This leads to the postulate that when two systems are each in thermal equilibrium with a third (reference) system, they will be in thermal equilibrium with each other. If this statement is true, then there exists a particular thermodynamic variable which may be assigned to all systems that are in thermal equilibrium with a reference system; this variable is referred to as temperature.

Temperature is measured by recording in a reference system (the thermometer) a change in volume, pressure or another property which varies continuously with degree of hotness. One possible choice for a thermometer is one mole of a low-density gas. For many such gases, experiment has shown that their behavior may be described by the equation

$$P = \frac{R}{V}(273.15 + t),$$

where t is temperature on an arbitrary scale which places zero at the triple point of water and 100 at the boiling point (the Celsius scale), and R is a constant. In view of this equation, another temperature scale is now defined (the absolute scale) by placing 273.15 units at the triple point of water, i.e.,

$$T \text{ (absolute scale)} = 273.15 + t \text{ (Celsius scale)},$$

thereby transforming the gas equation to $PV = RT$. The unit on both scales is one degree or one kelvin, and symbols °C and K are used for departures from zero on the Celsius and absolute scales respectively. Thus the absolute scale of temperature and the concept of an absolute zero of temperature have their origin in the behavior of a perfect gas.

Rock metamorphism occurs at temperatures ranging from about 200 °C (473 K) to about 900 °C (1173 K). Temperatures within the Earth generally increase with depth but locally high temperatures are obtained at shallow depths (at igneous contacts) or low temperatures at great depths (in subduction zones).

Process

If a change has occurred in the macroscopic properties of a system, the change is referred to as a thermodynamic process. The changes we are principally concerned with here are chemical changes but these are almost always accompanied by physical changes, for example changes in volume, and the term physical-chemical process is generally appropriate. When no change is observed in the macroscopic properties of a system, then from a thermodynamic point of view, no change has occurred, although on the microscopic (atomic) scale, change is constantly in progress.

An infinitesimal process is one that produces a very small change in any of the macroscopic properties of a system. Such a change would presumably occur only near a condition of equilibrium, or rest. Hence we may distinguish between reversible processes occurring in a system which is at equilibrium (internal equilibrium and equilibrium with the surroundings) and an irreversible process, moving toward a condition of equilibrium.

The distinction between these two kinds of processes is very important in relation to changes in entropy, to be discussed below. Examples of near-reversible processes are the idealized increase in the volume of a gas by a very slight decrease in external pressure (Figure 2.1), and the melting of ice at a temperature very slightly above 0 °C. A rapid release of pressure, and melting at 20 °C, represent irreversible processes. All processes that occur about us are irreversible; it is possible, however, that some processes that occur very slowly within the Earth are very nearly reversible.

When a process occurs at uniform temperature it is referred to as an isothermal process; when it occurs with no flow of heat, it is referred to as an adiabatic process. The melting of ice at 0 °C with the slow absorption of heat from the surroundings is an example of an isothermal process. The rapid rise and cooling of a mass of air is an example of an adiabatic process.

The First Law

The First Law of Thermodynamics is an expression of the conservation of energy. Although blacksmiths must have known since antiquity that metal becomes hot when struck with a hammer, the notion that mechanical energy can be transformed to heat was not accepted among scientists until it was proposed by James Joule in about 1845. Joule calculated, for example, that water at the foot of a waterfall should be slightly warmer than at the top (Bent 1965). With work and heat both expressible in terms of energy, experience soon led to the concept of energy conservation.

Consider a simple system, i.e., a closed system containing only one phase, composed of a single substance (not a mixture), for example quartz. This system may be said to possess a quantity of energy, referred to as internal energy, denoted U. Now both heat and work are forms of energy, and therefore the energy of the system may be changed by adding heat, extracting heat, doing work on the system by applying pressure, or by allowing the system to do work on the surroundings, but under no circumstances can additional energy be created or destroyed. This statement is best expressed as follows (Guggenheim 1977):

$$\mathrm{d}U = q - w,$$

where $\mathrm{d}U$ is a differential change in internal energy of the system (U), q is a differential of heat added, and w is a differential of work extracted, i.e., done on the surroundings (Figure 2.1).

It is important to recognize that although we may refer to a small amount of heat energy added (q) and a small amount of work accomplished (w), we may not suppose that the system possesses a certain amount of heat energy (Q) and a certain amount of work energy (W), i.e., it is not possible to separate and identify these two forms of energy within the system. To emphasize this point, symbols q and w are here adopted (following Guggenheim 1977) in place of the more commonly used symbols, $\mathrm{d}Q$ and $\mathrm{d}W$.

Students may be reluctant to accept that a mineral such as quartz possesses an amount of energy which may be altered by adding heat or by applying pressure. The familiar crystal models in which static wooden balls depict atoms, although useful to illustrate the principal aspects of crystal structure, are somewhat deceptive. In fact, the atoms are constantly in a state of agitation, principally a vibration about their equilibrium positions, but they also occasionally jump about within the crystal. When this view of a crystal is adopted, the concept of internal energy is more readily accepted.

The absolute amount of internal energy possessed by a mineral or mineral system is not known, nor is there any need for this information. We are concerned only with changes in energy, or with the amount of energy a system possesses relative to a standard state.

The First Law implies that $\mathrm{d}U$ is an exact differential and that U is therefore a state function, i.e.,

$$\oint \mathrm{d}U = 0.$$

Experimentalists have shown that in taking a gaseous system in a cyclic path, the heat and work terms (q and w) may vary greatly, depending on the path taken, but there is no net loss or gain in energy (U). Thus, although q and w are not exact differentials, their difference ($q - w$) is an exact differential, and a state function.

The importance of U as a thermodynamic variable is brought out by the following considerations. With regard to one gram-molecular weight (1 mole) of a perfect gas, with an equation of state $PV = RT$, the state of the gas may be described completely by specifying only two of the three variables, pressure (P), volume (V), and temperature (T), i.e.,

$$f(P, V, T) = 0.$$

Hence the internal energy (U) of 1 mole of the gas is a function of these three variables only, i.e., no additional variables must be specified:

$$U = f(P, V, T).$$

But one of the three variables P, V, T may be eliminated by use of the equation of state, i.e., only two of the three variables are independent variables. For example, if V is eliminated, we have,

$$U = f(P, T).$$

Now because U is a state function and dU is an exact differential, the following useful equation applies:

$$\mathrm{d}U = \left(\frac{\partial U}{\partial P}\right)_T \mathrm{d}P + \left(\frac{\partial U}{\partial T}\right)_P \mathrm{d}T,$$

where $\partial U/\partial P$ and $\partial U/\partial T$ are partial derivatives of U with respect to P and T.

The equation $U = f(P, T)$ applies (with reservation) to crystals of specified composition. Thus for quartz, disregarding crystal imperfections,

$$U^{\mathrm{Qtz}} = f(P^{\mathrm{Qtz}}, T^{\mathrm{Qtz}}),$$

i.e., the variance, or degrees of freedom of the crystal is two, and no more than two, for only two of the variables U, T, P must be specified to define the state of the crystal.

The Second Law

Changes that occur in matter are frequently observed to proceed only in certain directions. For example, heat always flows from hotter to colder objects, and when one gas is released in another, the two will mix spontaneously. Changes in the opposite direction, i.e., up-temperature heat flow, and unmixing in gases, although not in violation of the First Law, are never observed. Hence another statement is needed regarding direction of change.

Recall the concept of a reversible change (an idealized change occurring very slowly and very near equilibrium) and the concept of an irreversible change (occurring spontaneously). Recall also that a cyclic path refers to a change (reversible or irreversible) which takes a system from an initial state (e.g., at P_1, T_1) back to the initial state.

The Second Law may be expressed in the form of two proposals (Fast 1962):

(1) In every cycle in which all changes occur reversibly:

$$\oint \frac{q}{T} = 0.$$

(2) In every cycle in which changes occur irreversibly:

$$\oint \frac{q}{T} < 0,$$

i.e., for an irreversible change,

$$\frac{q}{T}\text{ (initial)} - \frac{q}{T}\text{ (final)} < 0.$$

Here q stands for an infinitesimal amount of heat transferred to the system, and T is absolute temperature. The flow of heat from a system at temperature T_2 to the surroundings at a lower temperature, T_1, satisfies the second proposal,

$$\frac{q}{T_2} - \frac{q}{T_1} < 0. \qquad (T_2 > T_1)$$

A spontaneous flow in the opposite direction would be in violation of this proposal.

The Carnot experiment (Guggenheim 1977) confirmed that when a system is taken through a reversible cycle, there is indeed no net change in q/T, as expressed in the first proposal. This means that q/T is a state function, or a thermodynamic variable; it is referred to as entropy (S). Thus the Second Law may be stated as follows:

For a reversible process, occurring in a system which receives a differential of heat (q),

$$\mathrm{d}S = \frac{q}{T},$$

i.e., the entropy of the system has increased by q/T, while that of the surroundings has decreased by the same amount.

For an irreversible process occurring in a system which receives a differential of heat (q),

$$\mathrm{d}S > \frac{q}{T}.$$

If the heat, q, is removed from the surroundings reversibly, the change in entropy of the surroundings is $\mathrm{d}S$ (surroundings) $= -q/T$, and there is a net increase in entropy (system and surroundings).

Consider a reversible process occurring in an isolated system, i.e., the system is closed, and $q = 0$, $w = 0$. With $q = 0$, the entropy of the system remains unchanged. For an irreversible process occurring in this system, according to the Second Law, $\mathrm{d}S > 0$, i.e., the entropy of the system must increase as the process progresses. Thus every spontaneous process occurring in an isolated system results in an increase (never a decrease) in entropy, and at equilibrium the entropy approaches a maximum.

The meaning of entropy in terms of the arrangement of atoms in a crystal, liquid, or gas, has been clarified by Boltzmann, who derived the equation,

$$S = k \ln \Omega,$$

where k is the Boltzmann constant and Ω stands for the number of possible atomic or molecular arrangements. The number of arrangements in a gas or liquid is obviously greater than in a crystal, and at the triple point of H_2O (0.0061 bar, 0.01 °C), S (vapor) > S (water) > S (ice). Also the number of possible arrangements of atoms in a disordered crystal exceeds that in an ordered crystal, and a change from ordered (Al, Si) microcline to disordered sanidine is accompanied by an increase in entropy. This view of entropy, although not incorrect, is incomplete because it does not explain why the entropy of a volume of gas or a crystal of quartz increases with increasing temperature. It is better to consider the levels of energy that molecules or atoms may assume, or the number of ways energy can be stored, which increases with increasing temperature.

According to Boltzmann, the Second Law does not hold rigorously, i.e., it is not absolutely certain but only highly probable that entropy will increase in the course of an irreversible process (Fast 1962). Problems arise in very small systems. For example, in a volume of gas consisting of six H_2O and six CO_2 molecules, randomly distributed, a finite probability exists that at one point in time, all H_2O molecules will be found in one half of the volume, and all CO_2 molecules in the other half, i.e., the entropy of the system will have spontaneously decreased.

Before the Second Law was discovered, a criterion for the direction of change in chemical reactions could not be found. Some reactions are exothermic and some are endothermic, and changes in heat content could not be relied upon. The Second Law provides such a criterion, for as we have seen above, all processes (except reversible processes) are expected to proceed in the direction of increasing entropy (system and surroundings, combined). This is demonstrated by considering the reaction

$$\text{calcite} + \text{quartz} = \text{wollastonite} + CO_2,$$

which when displaced to the right produces an increase in the entropy of the system, especially through the production of the high-entropy gas phase, CO_2. In addition, the reaction absorbs heat from the surroundings, causing a decrease in entropy there. A calculation shows that when the reaction takes place at 800 K, 1 bar (irreversibly) and when the heat is extracted from the surroundings reversibly, the reaction produces an overall entropy of 47 $J{\cdot}K^{-1}$ per mole of CO_2, and the entropy of the universe (system + surroundings) will have increased by this amount.

There are of course reactions which are accompanied by a decrease of entropy (within the system): for example, the freezing of water or the crystallization of a silicate melt. All of these reactions release heat to the surroundings, causing an increase in entropy there, which exceeds that consumed in the system. Again, the net entropy change is positive. A calculation to demonstrate this is provided by Swalin (1962, p. 19). Reactions producing a net decrease in entropy have never been observed.

The Third Law

The study of the behavior of gases, and experimental studies at very low temperatures have led to the concept of an absolute zero of temperature. Statistical thermodynamics has provided the expectation that the entropy of simple crystals approaches zero as temperature approaches zero; this statement is referred to as the Third Law of thermodynamics. Experiments have indeed confirmed that the entropy of nearly all elements and many compounds is zero at $T = 0$ (Guggenheim 1977).

There are exceptions. For example, Pauling (1935) calculated that in ice, an orientational randomness at 0 K gives rise to a residual entropy of $R \ln(3/2) =$ 0.805 entropy units. This agrees well with experimental results which give 0.87 entropy units.

With regard to more complex compounds, including solid-solution metals and minerals, Darken and Gurry (1953), Ulbrich and Waldbaum (1976) and others have pointed out that for energetic or physical reasons, perfect ordering in many crystals will not be possible, and a small-magnitude configurational entropy will exist at absolute zero.

Heat capacity

The heat capacity (C) of a substance is a measure of its ability to absorb heat. It is one of the most important properties of minerals. Heat capacity is defined as follows:

$$C = \frac{q}{\mathrm{d}T},$$

$$q = C\,\mathrm{d}T.$$

Thus the heat absorbed at constant pressure as one gram formula weight of a crystal is taken from temperature T_1 to a higher temperature T_2 is,

$$\int_{T_1}^{T_2} q = \int_{T_1}^{T_2} C_P\,\mathrm{d}T \qquad \text{(constant pressure)}$$

The entropy of a mineral is determined experimentally by measuring C_P as temperature is increased by many small steps, i.e., reversibly. Recall, that for a reversible change,

$$\mathrm{d}S = \frac{q}{T}.$$

Now, the equation

$$q = C\,\mathrm{d}T,$$

when multiplied by $1/T$, gives

$$\frac{q}{T} = \left(\frac{C_P}{T}\right) \mathrm{d}T,$$

and by substitution of $\mathrm{d}S$ for q/T,

$$\mathrm{d}S = \left(\frac{C_P}{T}\right) \mathrm{d}T.$$

Integrating from 0 K to T,

$$\int_0^T \mathrm{d}S = \int_0^T \left(\frac{C_P}{T}\right) \mathrm{d}T = S_T.$$

Heat-capacity data are now available for many minerals, and when obtained at low temperatures, provide estimates of entropy. For example, low-temperature heat-capacity data for forsterite (Mg_2SiO_4) as obtained by Kelley (1943) and Robie, Hemingway and Takei (1982) are shown in Figure 2.2. From these data, the entropy of forsterite (Fo) at 298 K (25 °C) is obtained as follows,

$$S_{298}^{\mathrm{Fo}} = \int_0^{298} \mathrm{d}S^{\mathrm{Fo}} = \int_0^{298} \left(\frac{C_P^{\mathrm{Fo}}}{T}\right) \mathrm{d}T.$$

To illustrate this point, the expression for C_P^{Fo} as a function of T for $0 < T < 300$ (Figure 2.2) is very approximately

$$C_P^{\mathrm{Fo}} \approx 0.4T,$$

which yields

$$S_{298}^{\mathrm{Fo}} = \int_0^{298} \left(\frac{C_P^{\mathrm{Fo}}}{T}\right) \mathrm{d}T \approx \int_0^{298} \left(\frac{0.4T}{T}\right) \mathrm{d}T \approx 0.4(298) \approx 100 \text{ J/(gfw·K)}.$$

The exact value is 94.11 ± 0.10 J/gfw·K where gfw stands for 1 gram-formula weight, as defined in Appendix 1.

At higher temperatures (298–1800 K) the relation between heat capacity and temperature, as reported by Orr (1953), is

$$C_P^{Fo} = 2.280 \times 10^2 + 3.414 \times 10^{-3}T - 1.745 \times 10^3 T^{-0.5} - 8.940 \times 10^5 T^{-2}.$$

The entropy of forsterite at any temperature between 298 and 1800 K may be obtained by integration of this equation. For example, at 1000 K,

$$S_{1000}^{\mathrm{Fo}} = S_{298}^{\mathrm{Fo}} + \int_{298}^{1000} \left(\frac{C_P^{\mathrm{Fo}}}{T}\right) \mathrm{d}T,$$

$$S_{1000}^{\mathrm{Fo}} = 277.2 \text{ J/gfw} \cdot \text{K}.$$

This is referred to as Third Law entropy.

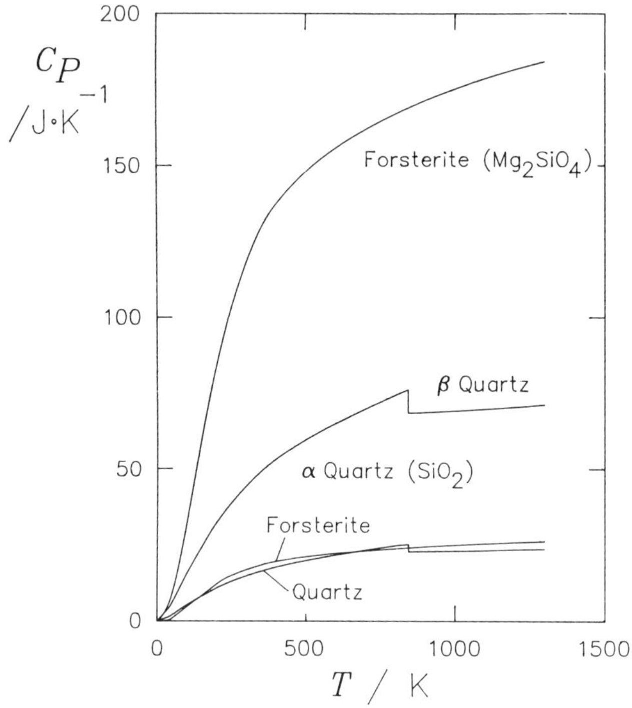

Figure 2.2 Heat capacity (C_P) of forsterite and quartz as a function of temperature (T). Data from Robie, Hemingway and Takei (1982), Orr (1953), Hemingway (1987) and Gurevich and Khlyustov (1979). C_P per 'gram atomic weight' (gfw ÷ 3 for quartz, and gwf ÷ 7 for forsterite) is plotted as the two, near-coincident lower curves

Different equations may be used to express the experimentally obtained relation between C_P and T from 298 K to higher temperatures. That used above is of the form

$$C_P = a + bT + cT^{-0.5} + \mathrm{d}T^{-2},$$

where a, b, c, d are constants which are characteristic for each mineral. Berman and Brown (1985) proposed the equation

$$C_P = k_0 + k_1 T^{-0.5} + k_2 T^{-2} + k_3 T^{-3},$$

where k_0, k_1, k_2, k_3 are other constants, with k_1, $k_2 \leq 0$. Entropy values (at 298 K) for several metamorphic minerals and C_P coefficients for the Berman–Brown equation are listed in Table 2.1. These data may be used to calculate the entropy of minerals at metamorphic temperatures.

Figure 2.2 shows two examples (forsterite and quartz) of relations between C_P and T. The high-temperature data for quartz, obtained by Hemingway (1987) reveal

Table 2.1 Thermodynamic properties

Mineral	Formula	gfw	$\Delta_f H$ (298,1) (elements) kJ gfw^{-1}	$S_{298.1}$ J K^{-1} gfw^{-1}	$C_p = k_0$ k_0	$+k_1 T^{-0.5}$ k_1 ($\times 10^{-2}$)	$+k_2 T^{-2}$ k_2 ($\times 10^{-5}$)	$+k_3 T^{-3}$ k_3 ($\times 10^{-7}$)	$V_{298.1}$ cm^3 gfw^{-1}	V_{TP} v_1 ($\times 10^6$)	v_2 ($\times 10^{12}$)	v_3 ($\times 10^6$)	v_4 ($\times 10^{10}$)
H_2O (gas)	H_2O	18.015	−241.8	188.72	115.45	−38.00	−28.71	—	24789.2	—	—	—	—
CO_2 (gas)	CO_2	44.01	−393.5	213.68	93.00	−13.41	1.238	—	24789.2	—	—	—	—
periclase	MgO	40.30	−601.5	26.95	61.11	−2.962	−6.212	0.584	11.25	−0.622	1.511	37.5	3.56
corundum	Al_2O_3	101.96	−1675.7	50.82	155.0	−8.284	−38.61	40.91	25.58	−0.385	0.375	21.3	47.2
α-quartz	SiO_2	60.09	−910.7	41.46	80.01	−2.403	−35.47	49.16	22.69	−2.43	10.14	23.9	0.0
coesite	SiO_2	60.09	−907.6	39.42	94.91	−7.673	−5.279	2.627	20.64	−1.04	3.00	7.40	43.6
spinel	$MgAl_2O_4$	142.27	−2300.3	84.54	235.9	−17.67	−17.10	4.062	39.77	−0.489	0.0	21.7	50.5
brucite	$Mg(OH)_2$	58.32	−925.9	63.06	136.8	−5.371	−43.62	55.27	24.68	−2.02	6.73	32.9	10.9
calcite	$CaCO_3$	100.09	−1206.8	91.725	178.2	−16.58	−4.827	16.66	36.90	−1.40	0.0	8.91	227.0
aragonite	$CaCO_3$	100.09	−1207.4	87.99	166.6	−14.99	0.0	5.449	34.15	—	—	—	—
magnesite	$MgCO_3$	84.31	−1113.6	65.21	162.3	−11.09	−48.83	87.47	28.03	−0.890	2.21	18.4	416.0
dolomite	$CaMg(CO_3)_2$	184.40	−2325.2	154.89	328.5	−25.54	−46.89	79.04	64.32	−1.07	0.0	14.2	364.0
kyanite	Al_2SiO_5	162.05	−2594.2	82.43	262.7	−20.01	−20.00	−6.318	44.12	−0.646	0.0	24.0	0.0
andalusite	Al_2SiO_5	162.05	−2590.0	91.43	236.5	−11.03	−75.27	93.64	51.47	−0.770	1.92	23.4	7.19
sillimanite	Al_2SiO_5	162.05	−2586.1	95.93	256.7	−18.87	−29.77	25.10	49.83	−0.753	0.0	13.4	0.0
forsterite	Mg_2SiO_4	140.69	−2174.4	94.01	238.6	−20.01	0.0	−11.62	43.66	−0.791	1.35	29.5	88.6
fayalite	Fe_2SiO_4	203.78	−1479.4	150.93	248.9	−19.24	0.0	−13.91	46.30	−0.730	0.0	26.5	79.5
grossular	$Ca_3Al_2Si_3O_{12}$	450.46	−6632.9	255.15	573.4	−20.39	−188.9	231.9	125.38	−0.654	1.635	19.0	79.7
pyrope	$Mg_3Al_2Si_3O_{12}$	403.13	−6286.5	266.36	640.7	−45.42	−47.02	0.0	113.16	−0.576	0.442	22.5	37.0
almandine	$Fe_3Al_2Si_3O_{12}$	497.78	−5265.5	339.93	574.0	−14.83	−292.9	502.2	115.11	−0.558	0.321	18.6	74.5
zoisite (ortho)	$Ca_2Al_3Si_3O_{12}(OH)$	454.37	−6889.5	297.58	749.2	−65.09	−23.81	12.49	135.88	−0.515	1.29	34.7	0.0
cordierite	$Mg_2Al_4Si_5O_{18}$	584.96	−9158.7	417.97	954.4	−79.62	−23.17	−37.02	233.11	−1.158	0.0	3.00	18.0
enstatite	$MgSiO_3$	100.39	−1545.6	66.17	166.6	−12.01	−22.71	27.92	31.33	−0.749	0.447	24.7	74.7
diopside	$CaMgSi_2O_6$	216.55	−3200.6	142.50	305.4	−16.05	−71.66	92.18	66.20	−0.872	1.707	27.8	83.1
jadeite	$NaAlSi_2O_6$	202.14	−3025.1	133.57	311.3	−20.05	−53.50	66.26	60.34	−0.860	2.15	23.1	25.8
wollastonite	$CaSiO_3$	116.16	−1631.5	81.81	149.1	−6.903	−36.59	48.44	39.83	−1.245	3.11	28.2	0.0
tremolite	$Ca_2Mg_5Si_8O_{22}(OH)_2$	812.37	−12305.6	551.15	1229.4	−64.02	−320.9	420.9	272.68	−1.392	3.48	24.4	98.3

anthophyl-lite	$Mg_7Si_8O_{22}(OH)_2$	780.84	−12069.0	535.20	1219.3	−57.66	−347.7	440.1	265.60	−1.259	0.0	27.1	31.3
muscovite	$KAl_3Si_3O_{10}(OH)_2$	398.31	−5976.7	293.16	651.5	−38.73	−185.2	274.2	140.87	−1.717	4.30	33.5	0.0
phlogopite	$KMg_3AlSi_3O_{10}(OH)_2$	417.26	−6207.3	334.16	610.4	−20.84	−215.3	284.1	149.77	−1.697	0.0	34.4	0.0
talc	$Mg_3Si_4O_{10}(OH)_2$	379.27	−5897.4	261.24	664.1	−51.87	−21.47	−32.74	136.10	−1.699	5.67	29.4	0.0
anorthite	$CaAl_2Si_2O_8$	278.21	−4228.7	200.19	439.4	−37.34	0.0	−31.70	100.75	−1.272	3.18	10.9	42.0
albite (low)	$NaAlSi_3O_8$	262.23	−3935.1	207.44	393.6	−24.15	−78.93	107.1	100.43	−1.945	4.86	26.3	32.4
microcline	$KAlSi_3O_8$	278.33	−3970.8	214.14	381.4	−19.41	−120.37	183.6	108.69	−1.805	5.11	15.1	54.8
nepheline	$NaAlSiO_4$	142.06	−2092.1	124.35	—	—	—	—	54.16	—	—	—	—

Notes: Gram-formula weight (gfw) from Robie, Hemingway and Fisher (1979). Mole volume ($V_{298.1}$) from Berman (1988). Heat of formation ($\Delta_f H_{298.1}$) and entropy ($S_{298.1}$) are from Berman (1988), and are based on internal consistency with regard to phase equilibrium. Heat capacity (C_P) coefficients are from the Berman (1988) compilation; these are experimental results expressed by use of the Berman and Brown (1985) equation, and estimates, where data are lacking. Compressibility (v_1, v_2) and expansion (v_3, v_4) coefficients from measurements and estimates compiled by Berman (1988); gfw volume is obtained by use of the equation

$$V_{T,P} = V_{298.1}[1 + v_1(P - 1) + v_2(P - 1)^2 + v_3(T - 298) + v_4(T - 298)^2]$$

(units: bars, K). Values of precision, expressed as % relative standard deviation, are approximately as follows: $\Delta_f H_{298.1}$ 0.05; $S_{298.1}$ 0.1 to 0.5; $V_{298.1}$ 0.01 to 0.14; for details see Robie, Hemingway and Fisher (1979). For information on other sets of thermodynamic data, see Engi (1992).

a discontinuity in C_P, at 844 ± 5 K, which results from an absorption of 7.5 J/gfw during the $\alpha \rightarrow \beta$ transformation.

For many metals, C_P at high temperatures approaches 25 J/K for one gram atomic weight of metal, i.e., about $3R$, where R is the gas constant (Swalin 1962). A body of theory now exists which relates heat capacity, and hence heat itself, to atomic vibrations. The $C_P - T$ functions for minerals, as for metals, are similar, provided the molar C_P values are divided by the number of atoms in the standard formula (by 3 for quartz and 7 for forsterite); this is shown by the two lower curves in Figure 2.2. A detailed discussion of this subject is given by Kieffer (1985).

Characteristic thermodynamic functions

Consider again a simple system, i.e., one containing a single phase composed of a single substance. This system may undergo infinitesimal changes,

$$q = T\,\mathrm{d}S, \qquad \text{(Second Law, reversible change)}$$

and

$$w = P\,\mathrm{d}V.$$

Recall,

$$\mathrm{d}U = q - w. \qquad \text{(First Law)}$$

Therefore, by substitution,

$$\mathrm{d}U = T\,\mathrm{d}S - P\,\mathrm{d}V,$$

which relates changes of internal energy to changes in entropy and volume at constant temperature and pressure. Entropy and volume are not the most convenient variables, and other energy functions were obtained by the addition or subtraction of terms, all of which have units of energy. One of these, referred to as the enthalpy or heat function (H), is obtained as follows:

$$H = U + PV.$$

By differentiation,

$$\mathrm{d}H = \mathrm{d}U + P\,\mathrm{d}V + V\,\mathrm{d}P,$$

and by substitution,

$$\mathrm{d}H = T\,\mathrm{d}S + V\,\mathrm{d}P.$$

At constant pressure,

$$\mathrm{d}H = T\,\mathrm{d}S.$$

But

$$T\,\mathrm{d}S = q. \qquad \text{(reversible change)}$$

Thus

$$\mathrm{d}H = q. \qquad \text{(constant pressure; reversible change)}$$

Two other functions are defined as follows:

$$F = U - TS,$$
$$G = U - TS + PV$$

Differentiating, as above, yields

$$\mathrm{d}F = -S\,\mathrm{d}T - P\,\mathrm{d}V,$$
$$\mathrm{d}G = -S\,\mathrm{d}T + V\,\mathrm{d}P.$$

F is referred to as the Helmholtz function, and G as the Gibbs function, or Gibbs (free) energy of a phase or system. In summary,

$$\mathrm{d}U = T\,\mathrm{d}S - P\,\mathrm{d}V,$$
$$\mathrm{d}H = T\,\mathrm{d}S + V\,\mathrm{d}P,$$
$$\mathrm{d}F = -S\,\mathrm{d}T - P\,\mathrm{d}V,$$
$$\mathrm{d}G = -S\,\mathrm{d}T + V\,\mathrm{d}P.$$

We have seen above that entropy (S) is an indicator of direction of chemical change and that changes occur naturally or spontaneously only if they result in an overall increase in entropy. We shall now prove that the Gibbs function (G) is also an indicator of direction of change, and moreover, it is a more convenient indicator, for it is necessary to consider only changes in the system, not in the surroundings (Gibbs 1878).

For a closed system,

$$\mathrm{d}U = q + w, \qquad \text{(First Law)}$$

where work done *on* the system is now taken as positive, and

$$q \leq T\,\mathrm{d}S, \qquad \text{(Second Law)}$$

where $=$ refers to a reversible change and $<$ to an irreversible change. Combining these equations,

$$\mathrm{d}U - T\,\mathrm{d}S \leq w.$$

At constant T,

$$\mathrm{d}(U - TS)_T \leq w,$$

or,

$$\mathrm{d}(U - TS)_T - w \leq 0.$$

Suppose P as well as T is to be held constant. It is still possible for the system, through expansion ($\mathrm{d}V$) to supply work to the surroundings. This work is $-P\,\mathrm{d}V$ where P is pressure on the system; the negative sign results from the above sign convention. Thus,

$$\mathrm{d}(U - TS + PV)_{T,P} \leq 0.$$

But by definition, $U - TS + PV = G$. Therefore at constant T and P,

$$\mathrm{d}G = 0, \qquad \text{(reversible process)}$$

$$\mathrm{d}G < 0, \qquad \text{(irreversible process)}$$

and any spontaneous change that is possible in the system will proceed with a reduction in Gibbs energy, until at equilibrium, G is the lowest value possible.

Similarly, one may show that for a given T and V, F is a minimum, and for a given H and P, S is a maximum (Guggenheim 1977).

Changes in Gibbs energy may be viewed as consisting of two parts, an enthalpy change and an entropy change. Recall that by definition,

$$G = U - TS + PV,$$

$$H = U + PV.$$

Thus,

$$G = H - TS,$$

and at constant temperature,

$$\Delta G = \Delta H - T\Delta S,$$

where Δ refers to a change resulting from a mineral reaction. Of greatest interest in the study of metamorphism are prograde reactions, i.e., those that occur in response to an increase in temperature. For these reactions, heat is absorbed and ΔH is positive. Consequently, if ΔG is to be negative (signalling a decrease in the Gibbs energy of the system), the entropy change must be positive, and moreover it must be positive by an amount that, when multiplied by temperature (T), more than compensates for the increase in enthalpy. Many of these reactions release H_2O or CO_2 from crystals, which is the principal source of entropy production.

Changes in the Gibbs energy of a chemical system may be compared with changes in the potential energy of a mechanical system, as shown in Figure 2.3. In both, a condition of rest or equilibrium is obtained when the energy is minimized. Metastable equilibrium in mineral systems is obtained when a mineral has crystallized outside of its stability field (for example, the crystallization of aragonite at low pressure), or when a mineral acquires an unstable composition (for example, the crystallization of high-Mg calcite at low temperature). When a mineral crystallizes within its stability field and persists outside of it, the term 'metastable persistence' is frequently used. The essential feature of metastable equilibrium is an energy barrier, but in dealing with metamorphic minerals, it is often difficult to identify this barrier. Consequently the term 'frozen equilibrium' will here be used for equilibria that are preserved at lower temperatures and pressures, i.e., far from equilibrium. Metamorphic rocks as found today at the Earth's surface form examples of frozen equilibrium.

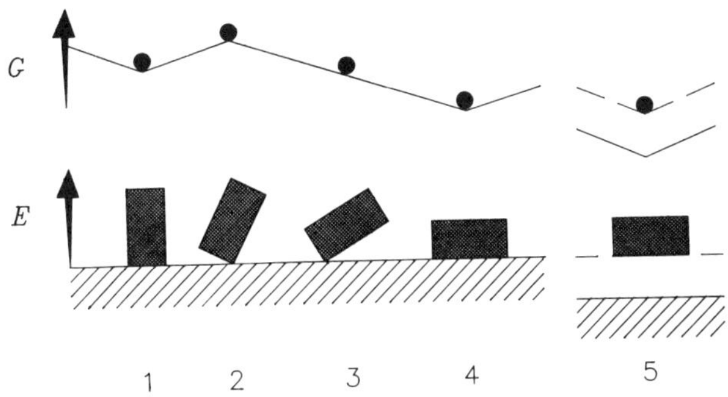

Figure 2.3 Illustrative comparison of the Gibbs energy (G) of a chemical system and the potential energy (E) of a mechanical system: 1 metastable equilibrium, 2 unstable equilibrium, 3 non-equilibrium, 4 stable equilibrium, and 5 frozen equilibrium

Gibbs–Helmholtz relations

The following two equations are useful in dealing with mineral equilibria:

$$\frac{\partial(G/T)}{\partial T} = -\frac{H}{T^2},$$

$$\frac{\partial(G/T)}{\partial(1/T)} = H.$$

There are different derivations for these equations; the following is from Darken and Gurry (1953). Recall,

$$\mathrm{d}G = -S\,\mathrm{d}T + V\,\mathrm{d}P.$$

At constant pressure,

$$\left.\frac{\partial G}{\partial T}\right)_P = -S.$$

Recall,

$$G = H - TS,$$

or,

$$-S = \frac{G-H}{T},$$

and by substitution,

$$\left.\frac{\partial G}{\partial T}\right)_P = \frac{G-H}{T}$$

$$\mathrm{d}G = \frac{G-H}{T}\,\mathrm{d}T,$$

$$T\,\mathrm{d}G = (G-H)\,\mathrm{d}T,$$

$$T\,dG - G\,dT = -H\,dT.$$

Dividing by T^2,

$$\frac{T\,dG - G\,dT}{T^2} = \frac{-H\,dT}{T^2}.$$

The left side is,

$$d(G/T).$$

Thus,

$$d(G/T) = -(H/T^2)\,dT,$$

$$\left.\frac{\partial(G/T)}{\partial T}\right]_P = \frac{-H}{T^2}.$$

Also,

$$\left.\frac{\partial(\Delta G/T)}{\partial T}\right]_P = -\frac{\Delta H}{T^2}.$$

and,

$$\left.\frac{\partial(\Delta G/T)}{\partial(1/T)}\right]_P = \Delta H,$$

where ΔG and ΔH represent the changes in Gibbs energy and enthalpy resulting from a mineral reaction, at the point of equilibrium.

Thermal expansion and isothermal compressibility

Crystalline solids, as well as liquids and gases, expand when heated and contract when uniform pressure is applied, but the resulting changes in volume (V) are small. For a crystal of specified composition,

$$dV = \left(\frac{\partial V}{\partial T}\right)_P dT + \left(\frac{\partial V}{\partial P}\right)_T dP.$$

The coefficient of thermal expansion is defined by

$$\alpha = \frac{1}{V}\left(\frac{\partial V}{\partial T}\right)_P,$$

and the coefficient of isothermal compressibility by

$$\beta = -\frac{1}{V}\left(\frac{\partial V}{\partial P}\right)_T.$$

α and β are small positive numbers with units of K^{-1} and bar^{-1} respectively. By substitution,

$$dV = \alpha V\,dT - \beta V\,dP,$$

$$d\ln V = \alpha\,dT - \beta\,dP.$$

The coefficients α and β are normally measured by subjecting crystals to high temperatures and pressures and measuring cell dimensions and hence cell volumes by X-ray diffraction. The methods were described by Hazen (1976, 1985) who also presented interpretations of expansion and compressibility in terms of crystal chemistry. In the range of metamorphic temperature (200–800 °C), α for many common minerals lies in the range of 17×10^{-6} to 44×10^{-6} K^{-1} (Skinner 1966). For example, at 400 °C, representative α values are, cordierite 18×10^{-6}, almandine 25×10^{-6}, and forsterite 38×10^{-6} K^{-1}, while the value for quartz is exceptionally large at 69×10^{-6} K^{-1}. The increase in volume implied by these numbers is more readily understood in terms of the ratio V/V_0 where V is the molar volume of a crystal at a specified temperature, and V_0 is molar volume at 20 °C. Values of this ratio for some common metamorphic minerals in the interval of 400 to 800 °C lie in the range of 1.01 to 1.03 (Table 2.2), i.e., the crystals experience a 1 to 3 percent increase in volume on heating. More detailed information for forsterite, as obtained by Hazen (1976), is shown in Figure 2.4.

Some representative data on the isothermal compressibility of metamorphic minerals, also expressed as a ratio V/V_0 (where V_0 is molar volume at 1 bar), are listed in Table 2.2. At 10 kbar the decrease in volume resulting from pressure amounts to $\frac{1}{2}$ to 2%.

From the above data, the molar volume of some minerals may be estimated at different metamorphic conditions or at different depths. For example, under amphibolite-facies conditions of 600 °C and 5 kbar, the effect of temperature is to increase the volume of almandine by 1.3%, and the effect of pressure is to decrease the volume by 0.3%, relative to Earth-surface conditions, and the volume change (ΔV) is $1.3 - 0.3 = +1.0\%$. Corresponding ΔV values for some other minerals are:

microline	$1.0 - 0.9$	=	$+0.10\%$
plagioclase	$1.0 - 0.85$	=	$+0.15\%$
hornblende	$1.6 - 0.5$	=	$+1.1\%$

Table 2.2 Thermal expansion and isothermal compressibility of some minerals expressed as ratio of unit-cell volume at the indicated temperature and pressure (V) to volume at ~ 20 °C and 1 bar (V_0). Data from Birch (1966) and Skinner (1966)

			V/V_0		
T/°C:	400	600	800	20	20
P/bar:	1	1	1	5000	10000
plagioclase	1.006	1.010	1.015	0.992	0.984
microcline	1.006	1.010	1.014	0.991	0.982
almandine	1.008	1.013	1.019	0.997	0.995
hornblende	1.010	1.016	1.022	0.995	0.990
diopside	1.010	1.015	1.022	0.996	0.991
forsterite	1.012	1.021	1.029	0.996	0.992
quartz	1.019			0.987	0.976

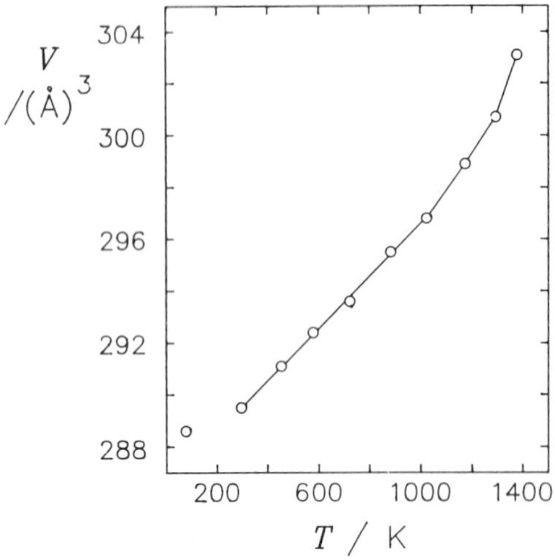

Figure 2.4 Unit cell volume of forsterite (V) as a function of temperature (T). Data from Hazen (1976, Table 1A)

diopside	1.5 − 0.4	=	+1.1%
forsterite	2.1 − 0.4	=	+1.7%

These calculations assume that α is independent of pressure and β is independent of temperature.

Berman (1988) has proposed the following equation to describe existing experimental data on the expansion and compressibility of minerals:

$$V_{T,P} = V_{298,1}[1 + v_1(P - 1) + v_2(P - 1)^2 + v_3(T - 298) + v_4(T - 298)^2].$$

Molar volume data ($V_{298,1}$) and the coefficients, v_1, v_2, v_3, v_4 for several minerals are listed in Table 2.1; these may be used to calculate $V_{T,P}$ at desired values of temperature and pressure.

Although volume changes in crystals resulting from changes in temperature and pressure are small, they must nevertheless be taken into consideration when volume changes are calculated for certain solid–solid reactions. These changes may also play a role in the course of exsolution and other reactions that occur during cooling and decompression.

2.2 THE BOUNDARY CURVE

Thermodynamics provides us with an understanding of mineral reactions in terms of fundamental properties of the minerals; thus the nature of the boundary curves presented in Chapter 1 may be more fully understood. We proceed now to obtain a general equation for a boundary curve for reactions involving minerals of fixed composition.

Recall that, by definition,

$$H = U + PV, G = U - TS + PV.$$

Therefore,

$$G = H - TS. \tag{1}$$

Now for any mineral reaction,

$$\alpha = \beta,$$

where each of α and β represents a single mineral but could in principle represent two or more minerals,

$$\Delta G = \Delta H - T\Delta S, \tag{2}$$

where

$$\Delta G = G^{\beta} - G^{\alpha},$$

$$\Delta H = H^{\beta} - H^{\alpha},$$

$$\Delta S = S^{\beta} - S^{\alpha},$$

and G, H, S refer to molar quantities, i.e., energy, enthalpy, and entropy per gram molecular weight (mol) or gram formula weight (gfw) of mineral.

At equilibrium,

$$G^{\alpha} = G^{\beta}, \tag{3}$$

i.e.,

$$\Delta G = 0. \tag{4}$$

(1) The $\alpha = \beta$ reaction at 25 °C (= 298.15 K), 1 bar, i.e., at standard temperature and pressure (STP).

By convention, the enthalpy of the elements at STP in their most stable form is taken as zero. The heat released when elements combine to form a mineral is the heat of formation, $\Delta_f H_{298.1}$, of that mineral (Table 2.1). For example, the heat of formation of calcite at STP from calcium, graphite, and oxygen is -1207 kJ/gfw. The negative sign signifies a release of heat. By the Third Law, the entropy of simple crystals at 0 K approaches zero and at STP is obtained by heat capacity measurements. Thus the entropy of calcite, for example, at STP is 91.7 J/gfw·K. Now by use of equation (2), the Gibbs energy change for the $\alpha = \beta$ reaction at STP is

$$\Delta G_{298.1} = \Delta H_{298.1} - 298\Delta S_{298.1}. \tag{5}$$

(2) The $\alpha = \beta$ reaction at $T > 298$ K, P = 1 bar.

ΔG for a reaction at any temperature $T (> 298$ K) and at 1 bar pressure is obtained as follows. Recall, at constant pressure,

$$\mathrm{d}H = q,$$

$$dS = q/T, \qquad \text{(reversible)}$$

$$q = C_P \, dT.$$

Therefore,

$$dH = C_P \, dT, \tag{6}$$

$$d\Delta H = \Delta C_P \, dT, \tag{7}$$

$$dS = \left(\frac{C_p}{T}\right) dT, \tag{8}$$

$$d\Delta S = \left(\frac{\Delta C_p}{T}\right) dT. \tag{9}$$

Thus, for the reaction $\alpha = \beta$,

$$\Delta H_{T,1} = \Delta H_{298,1} + \int_{298}^{T} \Delta C_P \, dT, \tag{10}$$

where $\Delta C_P = C_P^{\beta} - C_P^{\alpha}$, and

$$\Delta S_{T,1} = \Delta S_{298,1} + \int_{298}^{T} \left(\frac{\Delta C_P}{T}\right) dT. \tag{11}$$

By combining (10), (11), and (2), the Gibbs energy change for the reaction $\alpha = \beta$ at $T > 298$ K, $P = 1$ bar is

$$\Delta G_{T,1} = \Delta H_{T,1} - T\Delta S_{T,1},$$

$$\Delta G_{T,1} = \Delta H_{298,1} + \int_{298}^{T} \Delta C_P \, dT - T\Delta S_{298,1} - T\int_{298}^{T} \left(\frac{\Delta C_P}{T}\right) dT. \tag{12}$$

(3) The $\alpha = \beta$ reaction at $T > 298$ K, $P > 1$ bar.

The final step in obtaining an equation for the boundary curve is to allow for the effect of pressure on ΔG and then set ΔG equal to zero. Recall, for a phase consisting of a single substance,

$$dG = -S \, dT + V \, dP.$$

At constant T,

$$\left(\frac{\partial G}{\partial P}\right)_T = V,$$

$$\left(\frac{\partial \Delta G}{\partial P}\right)_T = \Delta V,$$

and

$$d(\Delta G) = (\Delta V) \, dP, \tag{13}$$

where V is the molar volume of a phase in units of J bar^{-1}, and ΔV is the volume change for the reaction,

$$\Delta V = V^{\beta} - V^{\alpha}.$$

Thus at $T > 298$ K, $P > 1$ bar, the equation for the boundary curve is

$$\Delta G_{T,P} = 0 = \Delta G_{T,1} + \int_{1}^{P} \Delta V_T \, \mathrm{d}P, \tag{14}$$

and by substitution of $\Delta G_{T,1}$ from equation (12),

$$0 = \Delta H_{298,1} + \int_{298}^{T} \Delta C_P \, \mathrm{d}T - T\Delta S_{298,1} - T \int_{298}^{T} \left(\frac{\Delta C_P}{T}\right) \mathrm{d}T + \int_{1}^{P} \Delta V_T \, \mathrm{d}P. \tag{15}$$

The boundary equation is a rather complex function, which for a particular mineral reaction (involving only phases of fixed composition), contains two variables, temperature and pressure. The equilibrium coexistence of all phases of the reaction is possible only at a unique temperature for a given pressure, or a unique pressure for a given temperature, i.e., the equilibrium is univariant.

It is instructive to view the boundary curve for an equilibrium, $\alpha = \beta$, as the line of intersection of two energy surfaces. We have seen above that for a mineral, α, consisting of a single substance,

$$\mathrm{d}G = -S \, \mathrm{d}T + V \, \mathrm{d}P,$$

i.e.,

$$f(G, T, P) = 0,$$

and the Gibbs energy is a function of temperature and pressure, as is shown schematically in Figure 2.5. The variance of this phase is two. Another surface may be drawn for β. The line of intersection of these two curved planes is a locus of points at which

$$G^{\alpha} = G^{\beta}, \Delta G = 0,$$

as required at equilibrium. The boundary curve as normally shown is the projection of this line of intersection onto the P–T base.

Provided α is the low-temperature phase, then in the G–T section, at the point of intersection,

$$-\partial G^{\beta}/\partial T > -\partial G^{\alpha}/\partial T,$$

but

$$\partial G/\partial T = -S,$$

therefore,

$$S^{\beta} > S^{\alpha},$$

and the high-temperature phase is invariably the high-entropy phase.

The slope of the boundary curve at any temperature may be obtained by differentiating the boundary equation, as follows. At equilibrium, at $P \gg 1$ bar,

$$\Delta G_{T,P} = 0 = \Delta H_{T,P} - T\Delta S_{T,P} + \Delta V_{T,P} P.$$

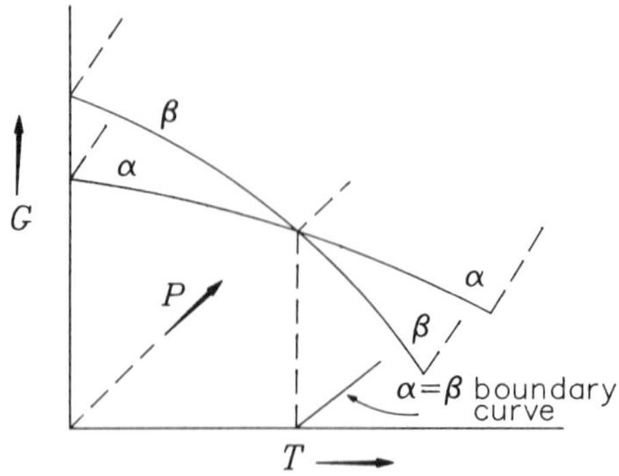

Figure 2.5 Illustration of Gibbs energy (G)–temperature (T)–pressure (P) surfaces for phases α and β, that are related by the reaction $\alpha = \beta$. A projection of the line of intersection onto the P–T plane gives the $\alpha = \beta$ boundary curve

Rearranging terms,

$$P = \frac{-\Delta H_{T,P}}{\Delta V_{T,P}} + \frac{\Delta S_{T,P}(T)}{\Delta V_{T,P}}.$$

By differentiation,

$$\frac{dP}{dT} = \frac{\Delta S_{T,P}}{\Delta V_{T,P}}, \qquad \text{(Clapeyron equation)}$$

assuming negligible changes in ΔH, ΔS, and ΔV for small changes in T. The Clapeyron equation may also be obtained as follows. For the reaction $\alpha = \beta$ at an equilibrium temperature (T) and pressure (P),

$$G^{\alpha}_{T,P} = G^{\beta}_{T,P},$$

$$dG^{\alpha}_{T,P} = dG^{\beta}_{T,P},$$

But,

$$dG = -S\,dT + V\,dP.$$

Thus,

$$-S^{\alpha}_{T,P}\,dT + V^{\alpha}_{T,P}\,dP = -S^{\beta}_{T,P}\,dT + V^{\beta}_{T,P}\,dP.$$

Rearranging terms,

$$\frac{dP}{dT} = \frac{(S^{\beta} - S^{\alpha})_{T,P}}{(V^{\beta} - V^{\alpha})_{T,P}} = \frac{\Delta S_{T,P}}{\Delta V_{T,P}}.$$

Calculation of the slope in units of bars/K is carried out by use of the conversion factor

$$1\ \text{J} = 10.00\ \text{cm}^3\ \text{bar}$$

Thus the slope at any point is determined by $\Delta S/\Delta V$. When α is the low-temperature phase in the equation $\alpha = \beta$, then ΔS is always positive (as noted above), and a positive slope means that ΔV is positive, as it usually is. When H_2O or CO_2 are released, ΔV decreases sharply with increasing pressure, and the slope of the boundary curve increases with increasing pressure.

The boundary-curve equation will be used below in discussions of certain mineral reactions. The intention at present is to demonstrate that an equation for the boundary curve can be derived from purely thermodynamic considerations, and to draw attention to the terms, in addition to temperature and pressure, that appear in the equation. These quantities — the heat of formation, entropy, heat capacity, and volume — of the participating minerals determine the location and shape of the boundary curve.

2.3 THE IMPERFECT GAS

Numerous metamorphic reactions involve CO_2 and H_2O, and in the direct experimental investigation of these reactions, a discrete H_2O and CO_2 gas phase of relatively large volume is present together with the solid mineral phases. It is important therefore to give some attention to the properties of H_2O and CO_2 at elevated temperature and pressure, and to introduce and define the concept of fugacity which has been helpful in dealing with these and other real (imperfect) gases. The following treatment is mainly from Denbigh (1961).

Consider one mole of a single-component perfect gas at any temperature (T) and pressure (P):

$$V = RT/P.$$

At constant temperature T',

$$dG = V\,dP,$$

$$dG = (RT'/P)\,dP.$$

Integrating from a chosen standard pressure of one bar to a pressure of interest, P',

$$\int_1^{P'} dG = RT' \int_1^{P'} d\ln P,$$

$$G_{T'P'} = G_{T'1} + RT' \ln(P'/1)$$

(Reference state : $P = 1$ bar, $T = T'$).

For a single-component imperfect gas at temperature T' and pressure P', we write an analogous equation,

$$G_{T'P'} = G_{T'1} + RT' \ln(f/1),$$

(Reference state: $f = 1$ bar, $T = T'$)

noting that,

$$(f/P) \to 1 \text{ as } P \to 0.$$

We have simply replaced P' by f, which is referred to as the fugacity of an imperfect gas, without changing the units (bar). $G_{T'1}$ in the perfect gas equation refers to a pressure at one bar, and in the imperfect gas equation, to a fugacity of 1 bar, but most gases at one bar show near-perfect behavior, and these two terms are very nearly equal.

For an imperfect gas, the fugacity is greater or less than the pressure acting on it by a factor ϕ referred to as the fugacity coefficient:

$$f = \phi P,$$

$$\phi = f/P.$$

Most gases including H_2O and CO_2 approach perfect behavior at low pressures, but at pressures of a few kilobars and temperatures of a few hundred kelvin they are far from perfect. Several equations have been proposed to describe experimentally obtained relations between P, V, and T for these gases. The modified Redlich–Kwong equation which was proposed by Kerrick and Jacobs (1981) is

$$P = \frac{RT}{V - b} - \frac{a}{(\sqrt{T})V(V + b)},$$

where V is molar volume, b is a constant, and a is a function of temperature and pressure. Note that as a and b approach zero, the equation approaches $PV = RT$; terms a and b therefore describe departure from perfect behavior. By use of the above equation and existing PVT data on H_2O and CO_2, Kerrick and Jacobs derived fugacity coefficients (ϕ) for these gases over a wide range of T and P; representative values are listed in Table 2.3.

The numbers in Table 2.3 may be used to evaluate the last term in the boundary-curve equation,

$$\int_1^P \Delta V \, dP.$$

If the mineral reaction, as written, releases 1 mole of a perfect gas, and the comparatively small volume change resulting from the solids is disregarded, ΔV may be replaced by V(gas), and this by RT/P,

$$\int_1^P \Delta V \, dP = \int_1^P (RT/P) \, dP = RT \ln P.$$

To accommodate the expected imperfect behavior of the gas, P is replaced by f or by ϕP,

$$\int_1^P \Delta V \, dP = RT \ln(\phi P).$$

Table 2.3 Fugacity coefficients for H_2O and CO_2

H_2O				
		T/°C		
P/kbar	400	600	800	1000
0.5	0.40	0.78	0.91	
2.0	0.20	0.52	0.79	0.94
4.0	0.21	0.54	0.84	1.03
6.0	0.28	0.67	1.01	1.22
8.0	0.40	0.89	1.27	1.49
CO_2				
		T/°C		
P/kbar	377	577	777	977
0.5	1.02	1.10	1.12	1.12
2.0	1.79	1.86	1.82	1.75
4.0	4.91	4.18	3.63	3.22
6.0	13.85	9.48	7.20	5.83
8.0	38.73	21.33	14.15	10.44

From Kerrick and Jacobs (1981), reprinted by permission of *American Journal of Science.*

Discrete volumes of H_2O and CO_2 at metamorphic pressures possess thermodynamic properties which do not differ greatly from those of liquids. For example, the volume of H_2O at high temperature and pressure, as listed in Table 2.4, may be compared with a specific volume of 1.00 for water at 25 °C, 1 bar.

2.4 PHASES OF VARIABLE COMPOSITION

Most minerals are solid solutions composed of two or more components. Olivine, for example, may be regarded as a solid solution of components Mg_2SiO_4 (fo) and Fe_2SiO_4 (fa). These are here referred to as '*solution components*', to distinguish them from the system components defined in Sec. 2.1. The concentration of a

Table 2.4 Specific volume (cm^3/g) of H_2O at elevated temperature and pressure

			P/bar		
T/°C	100	1000	3000	5000	7000
200	1.15	1.08	1.00	0.95	0.92
300	1.40	1.21	1.08	1.01	0.96
400	26.40	1.45	1.18	1.07	1.01
500	32.80	1.89	1.30	1.15	1.07
600	38.30	2.67	1.45	1.24	1.14
700	43.50	3.54	1.63	1.34	1.21
800	48.60	4.34	1.82	1.44	1.28

From Burnham, Holloway and Davis (1969).

component in a solution may be expressed as a mole fraction, for example the mole fraction (X) of fa in olivine (Ol) is defined as

$$X_{\text{fa}} = n_{\text{fa}}/(n_{\text{fa}} + n_{\text{fo}}), \tag{16}$$

where n denotes the number of moles or gram-formula-weights of a component in the olivine crystal. Note that provided olivine contains only fa and fo components,

$$X_{\text{fa}} + X_{\text{fo}} = 1.0 \tag{17}$$

Many minerals are capable of change in the concentration of certain ions on a particular structural site or group of similar sites, while the remaining sites remain virtually unchanged. Such variation in site composition may be expressed by use of an atomic ratio. For example, the atomic ratio of Fe on the six coordinated sites of olivine (M1 and M2 sites combined) may be defined as

$$X_{\text{Fe}} = n_{\text{Fe}}/(n_{\text{Fe}} + n_{\text{Mg}}), \tag{18}$$

where n refers to the number of atoms in a crystal. The right-hand side is often written Fe/(Fe + Mg). Note that $X_{\text{fa}} = X_{\text{Fe}}$.

Consider now a number of crystals (A, B) similar to olivine, composed of X_{A} mol of component A and X_{B} mol of component B, where $X_{\text{A}} + X_{\text{B}} = 1.0$. The mass, M, of the crystals may be described by the equation

$$M = X_{\text{A}}M^{\text{A}} + X_{\text{B}}M^{\text{B}}, \tag{19}$$

where M^{A} and M^{B} are the molar masses of the individual components in their pure form. The equation states that there is no loss or gain of mass as mixing occurs. We may not, however, write a general equation of this kind with respect to volume, i.e.,

$$V \neq X_{\text{A}}V^{\text{A}} + X_{\text{B}}V^{\text{B}}, \tag{20}$$

where V^{A} and V^{B} are the molar volumes of pure A and pure B. The volume of the mixture may be greater or less than the value indicated by the right-hand side of the above equation, depending on whether there is a gain or loss of volume upon mixing. Such changes in volume are commonly observed when two liquids are mixed together. It is, however, permissible to write a generally valid equation in the form

$$\mathrm{d}V = V_{\text{A}}\,\mathrm{d}X_{\text{A}} + V_{\text{B}}\,\mathrm{d}X_{\text{B}}, \tag{21}$$

where V_{A} and V_{B} are *partial* molar volumes, defined by the following partial derivatives:

$$V_{\text{A}} = \left(\frac{\partial V}{\partial X_{\text{A}}}\right)_{T,P,X_{\text{B}}}, \; V_{\text{B}} = \left(\frac{\partial V}{\partial X_{\text{B}}}\right)_{T,P,X_{\text{A}}}. \tag{22}$$

By use of Euler's theorem,

$$V = V_{\text{A}}X_{\text{A}} + V_{\text{B}}X_{\text{B}}. \tag{23}$$

Similarly, the Gibbs energy of the (A, B) mixture must be described by the equation,

$$G = \mu_A X_A + \mu_B X_B, \tag{24}$$

where,

$$\mu_A = \left(\frac{\partial G}{\partial X_A}\right)_{T,P,X_B}, \mu_B = \left(\frac{\partial G}{\partial X_B}\right)_{T,P,X_A} \tag{25}$$

These very useful partial quantities, denoted μ, are referred to as the chemical potential. Thus the molar Gibbs energy of olivine is

$$G = \mu_{fa} X_{fa} + \mu_{fo} X_{fo}. \tag{26}$$

Partial molar quantities may be obtained from molar quantities by the following procedure in which volume is chosen as the variable of interest.
Let:

V = molar volume of an (A, B) mixture,

V^A = molar volume of pure A,

V^B = molar volume of pure B,

V_A = partial molar volume of A,

V_B = partial molar volume of B,

X_A = mole fraction of A,

X_B = mole fraction of B.

Recall equations (23) and (21),

$$V = V_A X_A + V_B X_B, \tag{27}$$

$$dV = V_A \, dX_A + V_B \, dX_B \tag{28}$$

Note that

$$X_A + X_B = 1. \tag{29}$$

Therefore,

$$dX_A = -dX_B. \tag{30}$$

By combining equations (28) and (30),

$$dV = (V_B - V_A) \, dX_B. \tag{31}$$

Divide by dX_B:

$$\frac{dV}{dX_B} = V_B - V_A, \tag{32}$$

$$V_B = V_A + \frac{dV}{dX_B}. \tag{33}$$

Rearranging (27) and replacing X_A by $1 - X_B$ (from (29)),

$$V_A = \frac{V - V_B X_B}{1 - X_B}. \tag{34}$$

Place (34) in (33); then cross-multiply and simplify, to give

$$V_B = V + (1 - X_B)\frac{dV}{dX_B}. \tag{35}$$

Similarly for A,

$$V_A = V + X_B \frac{dV}{dX_B}. \tag{36}$$

Note, from equation (35),

$$\text{when } X_A = 0, X_B = 1, V_B = V = V^B,$$

and from (36),

$$\text{when } X_B = 0, X_A = 1, V_A = V = V^A.$$

V_A and V_B for a given composition, for example $X_B = 0.4$, are obtained by drawing a tangent to the V–X curve at point $X_B = 0.4$, as shown in Figure 2.6.

Crystals of Mg_2SiO_4 and Fe_2SiO_4 if brought together at a sufficiently high temperature, would presumably mix spontaneously, given enough time, resulting in a decrease in Gibbs energy, an increase in entropy, and possibly a slight increase

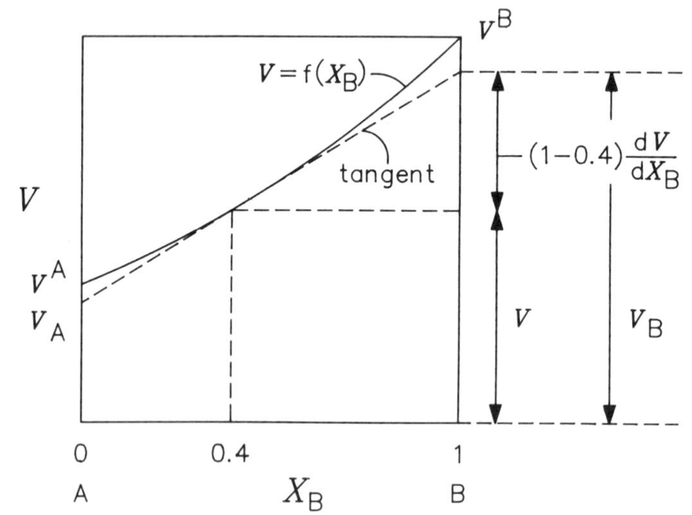

Figure 2.6 Illustration of a method for determining the partial molar volume of component A (V_A) and component B (V_B) in an (A, B) mixture of composition, X_B = B/(A + B) = 0.4. Molar volumes of pure A and pure B are V^A and V^B

or decrease in enthalpy and volume. Expressions for such changes were initially derived by considering the mixing of perfect gases, as follows. The equations for the Gibbs energy per mole of perfect gases A and B at temperature T' and pressure P' are

$$G^{\mathrm{A}}_{T'P'} = G^{\mathrm{A}}_{T'1} + RT' \ln(P'/1)^{\mathrm{A}}, \tag{37}$$

$$G^{\mathrm{B}}_{T'P'} = G^{\mathrm{B}}_{T'1} + RT' \ln(P'/1)^{\mathrm{B}}, \tag{38}$$

When gases A and B are allowed to mix, the molar Gibbs energy of the mixture is,

$$\begin{aligned} G_{T'P'} &= X_{\mathrm{A}} G^{\mathrm{A}}_{T'1} + X_{\mathrm{A}} RT' \ln X_{\mathrm{A}} (P')^{\mathrm{A}} \\ &\quad + X_{\mathrm{B}} G^{\mathrm{B}}_{T'1} + X_{\mathrm{B}} RT' \ln X_{\mathrm{B}} (P')^{\mathrm{B}}, \\ &= (X_{\mathrm{A}} G^{\mathrm{A}}_{T'1} + X_{\mathrm{A}} RT' \ln (P')^{\mathrm{A}}) + X_{\mathrm{A}} RT' \ln X_{\mathrm{A}} \\ &\quad + (X_{\mathrm{B}} G^{\mathrm{B}}_{T'1} + X_{\mathrm{B}} RT' \ln (P')^{\mathrm{B}}) + X_{\mathrm{B}} RT' \ln X_{\mathrm{B}} \\ &= X_{\mathrm{A}} G^{\mathrm{A}}_{T'P'} + X_{\mathrm{B}} G^{\mathrm{B}}_{T'P'} + X_{\mathrm{A}} RT' \ln X_{\mathrm{A}} + X_{\mathrm{B}} RT' \ln X_{\mathrm{B}}, \end{aligned} \tag{39}$$

where X_{A} and X_{B} stand for the mole fractions of A and B respectively.

The last two terms in equation (39) represent the energy loss upon mixing, i.e., the Gibbs energy of mixing. Because $X < 1.0$, $\ln X$ is negative, and the energy of mixing is negative, as expected.

Comparing equation (39) with (24), which is

$$G = \mu_{\mathrm{A}} X_{\mathrm{A}} + \mu_{\mathrm{B}} X_{\mathrm{B}}, \tag{40}$$

we see that

$$\begin{aligned} \mu_{\mathrm{A}} &= G^{\mathrm{A}} + RT \ln X_{\mathrm{A}}, \\ \mu_{\mathrm{B}} &= G^{\mathrm{B}} + RT \ln X_{\mathrm{B}}. \end{aligned} \tag{41}$$

Thus the chemical potential of a component in a mixture of two perfect gases decreases exponentially as concentration decreases linearly, as shown in Figure 2.7.

Similarly, an equation was derived for the entropy of a perfect gas mixture,

$$S = X_{\mathrm{A}} S^{\mathrm{A}} + X_{\mathrm{B}} S^{\mathrm{B}} - X_{\mathrm{A}} R \ln X_{\mathrm{A}} - X_{\mathrm{B}} R \ln X_{\mathrm{B}}, \tag{42}$$

where S is the molar entropy of the mixture, and S^{A} and S^{B} represent the molar entropy of pure gas components A and B. The last two terms in this equation define the entropy of mixing, which is always positive, as expected.

And similarly,

$$H = X_{\mathrm{A}} H^{\mathrm{A}} + X_{\mathrm{B}} H^{\mathrm{B}}, \tag{43}$$

$$V = X_{\mathrm{A}} V^{\mathrm{A}} + X_{\mathrm{B}} V^{\mathrm{B}}, \tag{44}$$

i.e., the cnthalpy and volume of mixing are zero.

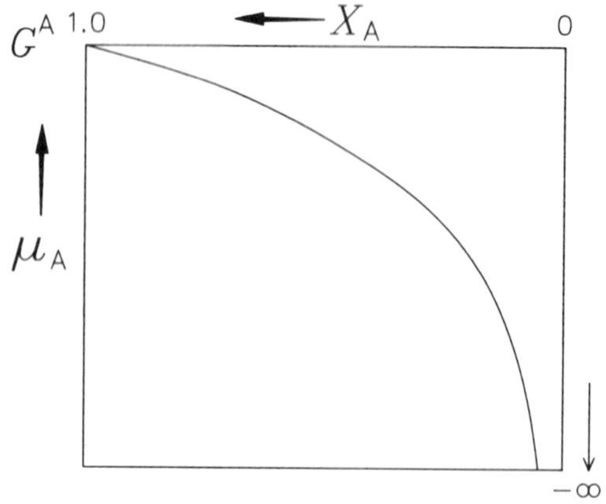

Figure 2.7 Change in the chemical potential of component A (μ_A) with concentration (X_A) in an ideal mixture, (A, B); G^A is the molar Gibbs energy of pure A. The relation is, $\mu_A = G^A + RT \ln X_A$, where R is the gas constant and T is absolute temperature

Liquid and crystal mixtures that obey these equations are referred to as ideal mixtures or ideal solutions. Non-ideal crystals may be characterized by specifying various amounts of departure from the ideal-solution model.

The absolute activity of a component is a concept that was introduced to obtain a simpler and more convenient expression of the exponential relation shown in Figure 2.7. Thus for component A in an ideal mixture, the absolute activity of A (λ_A) in the mixture is defined as follows:

$$\mu_A = RT \ln \lambda_A,$$

$$G^A = RT \ln \lambda^A,$$

$$\mu_A - G^A = RT \ln(\lambda_A/\lambda^A), \tag{45}$$

where λ^A is the absolute activity of pure A. The ratio λ_A/λ^A is referred to as the 'relative activity', or simply the activity, denoted a_A. Thus,

$$\mu_A = G^A + RT \ln a_A, \tag{46}$$

and for an ideal mixture,

$$a_A = X_A. \tag{47}$$

A departure from ideal mixing may now be expressed in terms of a departure of a from X, i.e.,

$$a = \gamma X, \tag{48}$$

where γ is the activity coefficient.

Figure 2.8 shows that olivine at 1200 °C and 1 bar exhibits slight positive departure from ideal behavior. Similar a–X relations occur in many mixtures. Note that for each component, at low concentrations (X small), γ is nearly constant; this

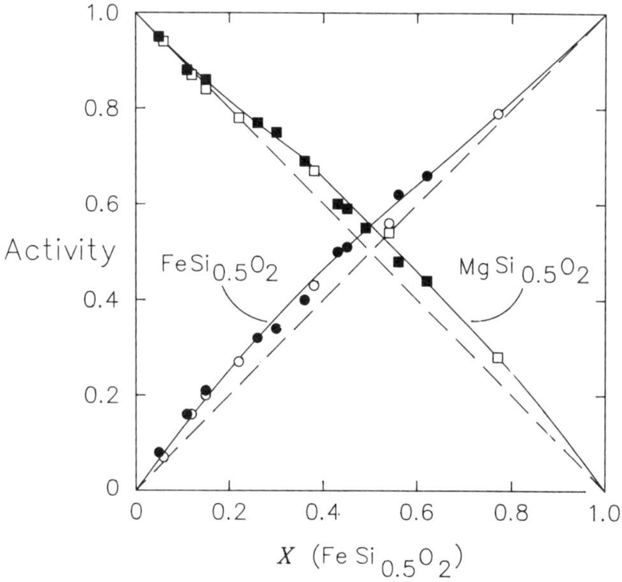

Figure 2.8 Activity (a) versus mole fraction (X) relations in olivine ($MgSi_{0.5}O_2$, $FeSi_{0.5}O_2$) at 1200 °C. Solid and open symbols refer to activity determined in experiments with orthopyroxene and with magnesiowüstite respectively. From Nafziger and Muan (1967)

is referred to as 'Henry's rule'. At high concentrations (X approaching 1.0), γ approaches unity.

An application and clarification of the concept of chemical potential is provided by the following exercise. Consider two crystals of plagioclase in a rock, namely crystal a of composition $X_{an} = 0.60$ and crystal b of composition $X_{an} = 0.40$. Plagioclase may be viewed as a solid solution of two components, ab and an, and we may define for each component a chemical potential, μ, which will be a function of concentration, X. The requirement for equilibrium between crystals a and b is

$$\mu^{a}_{ab} = \mu^{b}_{ab}, \quad X^{a}_{ab} = X^{b}_{ab},$$

$$\mu^{a}_{an} = \mu^{b}_{an}, \quad X^{a}_{an} = X^{b}_{an}.$$

Chemical equilibrium will be obtained if some anorthite migrates from a to b, and some albite from b to a, until a uniformity of composition is achieved. Hence, components will tend to migrate from phases in which their chemical potentials are high to those in which they are low, thereby levelling off energy gradients. In plagioclase, the migration could presumably occur by the simultaneous migration of NaSi in one direction and CaAl in the opposite direction, i.e., by exchange diffusion. Thus the homogenization of zoned plagioclase crystals during metamorphism, whether by diffusion or recrystallization may be understood in terms of chemical potential gradients.

2.5 REACTIONS INVOLVING MINERALS OF VARIABLE COMPOSITION

Most mineral-forming reactions involve one or more minerals of variable composition, for example,

$$\text{(Mn,Fe,Mg) chlorite} + \text{quartz} = \text{(Mn,Fe,Mg) garnet} + H_2O.$$

From the examination of metamorphic rocks, from experimentation, and from thermodynamic considerations, we know that many reactions of this kind proceed over a range of temperature for a given pressure, and they cannot therefore be expressed by use of the boundary-curve equation and a corresponding line in a P–T diagram. The importance of these reactions was emphasized by Ramberg (1952a); they are now referred to as 'continuous reactions'.

We shall consider here, by way of illustration, the thermodynamic analysis of one kind of continuous reaction, namely that expressed above but simplified by excluding Mg. The simplified reaction equation is

$$(Fe,Mn)_9Al_6Si_5O_{20}(OH)_{16} + 4SiO_2 = (Fe,Mn)_9Al_6Si_9O_{36} + 8H_2O,$$

which for the Fe and Mn end-members was experimentally investigated by Hsu (1968). Because Fe is readily oxidized, the reaction temperature for the Fe end-member is influenced by the activity of oxygen in the system. Thus the boundary curve for this reaction was located by controlling the oxygen activity, maintaining it at a level determined by the quartz–fayalite–magnetite buffer. Provided all three of these minerals are present in the system, together with Fe chlorite, Fe garnet, quartz, and H_2O, the partial pressure of oxygen is determined by the equilibrium,

$$\underset{\text{fayalite}}{3Fe_2SiO_4} + O_2 = \underset{\text{magnetite}}{2\,Fe_3O_4} + \underset{\text{quartz}}{3SiO_2},$$

and is fixed at a value corresponding to about 10^{-20} bar at 600 °C. Under these conditions, the breakdown of Fe chlorite at 2 kbar occurs at 600 °C, and the breakdown of Mn chlorite occurs at 405 °C (Hsu 1968).

The breakdown of Fe–Mn chlorite, according to the above reaction, is expected to take place at an intermediate temperature. If the Mn:Fe ratio of the garnet so produced is identical to that in the initial chlorite, this reaction would be univariant. But garnet shows a strong preference for Mn, and the reaction presumably proceeds over a range of temperature in a manner not yet investigated experimentally. We might anticipate, however, that the phase relations, as shown in a composition–temperature diagram, will appear as a loop, similar to the familiar plagioclase loop, which results from the melt phase showing a preference for the albite component. We now proceed to derive equations for the changes in composition of both chlorite and garnet as a function of rising temperature. Two reactions must be written to define the equilibrium fully:

$$(1/9)\,\underset{\text{in chlorite}}{Fe_9Al_6Si_5O_{20}(OH)_{16}} + (4/9)\,SiO_2 = (1/9)\,\underset{\text{in garnet}}{Fe_9Al_6Si_9O_{36}} + (8/9)\,H_2O, \quad \text{(f)}$$

$(1/9)\ Mn_9Al_6Si_5O_{20}(OH)_{16} + (4/9)\ SiO_2 = (1/9)\ Mn_9Al_6Si_9O_{36} + (8/9)\ H_2O$ (m)
in chlorite in garnet

The following symbols will be used:

Component	*Name*	*Symbol*
$(1/9)\ Fe_9Al_6Si_5O_{20}(OH)_{16}$	Fe chlorite	fc
$(1/9)\ Mn_9Al_6Si_5O_{20}(OH)_{16}$	Mn chlorite	mc
$(1/9)\ Fe_9Al_6Si_9O_{36}$	Fe garnet	fg
$(1/9)\ Mn_9Al_6Si_9O_{36}$	Mn garnet	mg
Phase	*Name*	*Symbol*
$(1/9)\ (Fe,Mn)_9Al_6Si_5O_{20}(OH)_{16}$	chlorite	c
$(1/9)\ (Fe,Mn)_9Al_6Si_9O_{36}$	garnet	g
H_2O	vapor	w

At equilibrium, the following two energy-balance equations must be satisfied:

$$\mu_{\text{fg}} + (8/9)G^{\text{w}} - \mu_{\text{fc}} - (4/9)G^{\text{Qtz}} = 0,$$

$$\mu_{\text{mg}} + (8/9)G^{\text{w}} - \mu_{\text{mc}} - (4/9)G^{\text{Qtz}} = 0.$$

Substituting four equations of type

$$\mu_{\text{fg}} = G^{\text{fg}} + RT \ln a_{\text{fg}},$$

gives

$$G^{\text{fg}} + RT \ln a_{\text{fg}} + (8/9)G^{\text{w}} - G^{\text{fc}} - RT \ln a_{\text{fc}} - (4/9)G^{\text{Qtz}} = 0,$$

$$G^{\text{mg}} + RT \ln a_{\text{mg}} + (8/9)G^{\text{w}} - G^{\text{mc}} - RT \ln a_{\text{mc}} - (4/9)G^{\text{Qtz}} = 0.$$

Rearranging terms,

$$\ln(a_{\text{fg}}/a_{\text{fc}}) = -\Delta G^{\text{f}}/RT,$$

$$\ln(a_{\text{mg}}/a_{\text{mc}}) = -\Delta G^{\text{m}}/RT,$$

where ΔG^{f} and ΔG^{m} are the changes in Gibbs energy for reactions (f) and (m), above, with components in their pure form. These last two equations must be differentiated to determine how the activity ratios vary with temperature (at constant pressure),

$$\frac{\partial \ln(a_{\text{fg}}/a_{\text{fc}})}{\partial T} = -\frac{1}{R}\frac{\partial(\Delta G^{\text{f}}/T)}{\partial T} = \frac{\Delta H^{\text{f}}}{RT^2},$$

$$\frac{\partial \ln(a_{\text{mg}}/a_{\text{mc}})}{\partial T} = -\frac{1}{R}\frac{\partial(\Delta G^{\text{m}}/T)}{\partial T} = \frac{\Delta H^{\text{m}}}{RT^2},$$

where ΔH^{f} and ΔH^{m} are the changes in enthalpy for reactions (f) and (m), above, with components in their pure form. Integrating now from T_{f} and from T_{m}, the equilibrium temperature for equations (f) and (m), with components in their pure form,

$$\int_{T_{\mathrm{f}}}^{T} \mathrm{d}\ln(a_{\mathrm{fg}}/a_{\mathrm{fc}}) = \int_{T_{\mathrm{f}}}^{T} (\Delta H^{\mathrm{f}}/RT^2)\,\mathrm{d}T,$$

$$\int_{T_{\mathrm{m}}}^{T} \mathrm{d}\ln(a_{\mathrm{mg}}/a_{\mathrm{mc}}) = \int_{T_{\mathrm{m}}}^{T} (\Delta H^{\mathrm{m}}/RT^2)\,\mathrm{d}T,$$

we obtain

$$\ln(a_{\mathrm{fg}}/a_{\mathrm{fc}}) = (-\Delta H^{\mathrm{f}}/R)(T^{-1} - T_{\mathrm{f}}^{-1}),$$

$$\ln(a_{\mathrm{mg}}/a_{\mathrm{mc}}) = (-\Delta H^{\mathrm{m}}/R)(T^{-1} - T_{\mathrm{m}}^{-1}).$$

Assume now that the activities of the four components are approximately equal to their concentrations, expressed as mole fractions (X),

$$(X_{\mathrm{fg}}/X_{\mathrm{fc}}) = \exp[(-\Delta H^{\mathrm{f}}/R)(T^{-1} - T_{\mathrm{f}}^{-1})],$$

$$(X_{\mathrm{mg}}/X_{\mathrm{mc}}) = \exp[(-\Delta H^{\mathrm{m}}/R)(T^{-1} - T_{\mathrm{m}}^{-1})].$$

Because

$$X_{\mathrm{fg}} + X_{\mathrm{mg}} = 1.0,$$

$$X_{\mathrm{fc}} + X_{\mathrm{mc}} = 1.0,$$

we have only three unknowns $(X_{\mathrm{fg}}, X_{\mathrm{fc}}, T)$ in two equations, provided ΔH^{f}, ΔH^{m}, T_{f}, T_{m}, can be obtained from the two univariant equilibria, (f) and (m). Thus explicit $X_{\mathrm{fg}} - T$ and $X_{\mathrm{fc}} - T$ functions may be found:

$$X_{\mathrm{fg}} = f\,(m-1)/(m-f),$$

$$X_{\mathrm{fc}} = (m-1)/(m-f),$$

where

$$f = \exp[(-\Delta H^{\mathrm{f}}/R)(T^{-1} - T_{\mathrm{f}}^{-1})],$$

$$m = \exp[(-\Delta H^{\mathrm{m}}/R)(T^{-1} - T_{\mathrm{m}}^{-1})].$$

These are the van Laar equations (van Laar (1908), Prigogine and Defay (1954)) which were used by Bowen (1928) to describe the liquidus and solidus curves in the plagioclase system.

Under the following conditions:

$$P = 2\ \mathrm{kbar}$$

$$T_f = 543\,°C = 816\ K\ \text{(iron–magnetite buffer, Hsu 1968)}$$

$$T_m = 405\,°C = 678\ K\ \text{(Hsu 1968)}$$

$$\Delta H^f = 140\ J\ gfw^{-1}\ \text{(estimate)}$$

$$\Delta H^m = 100\ J\ gfw^{-1}\ \text{(estimate)}$$

the equations plot as shown in Figure 2.9.

According to this model, chlorite of composition $X_{fc} = 0.95$ in the presence of quartz will begin to break down at 790 K to produce garnet of composition $X_{fg} \sim 0.5$, and the reaction will continue to completion at 810 K. While the reaction is in progress, both chlorite and garnet become progressively enriched in Fe, provided near-equilibrium is maintained. If the reaction proceeds rapidly relative to the time required for Fe–Mn exchange diffusion in garnet, the garnet crystals that are produced will be compositionally zoned, with Mn-rich centers, as commonly found in medium-grade metamorphic rocks.

The above model is intended only as an example of the use of chemical potentials in dealing with mineral reactions. The presence of Mg in chlorite would make the reaction considerably more complex.

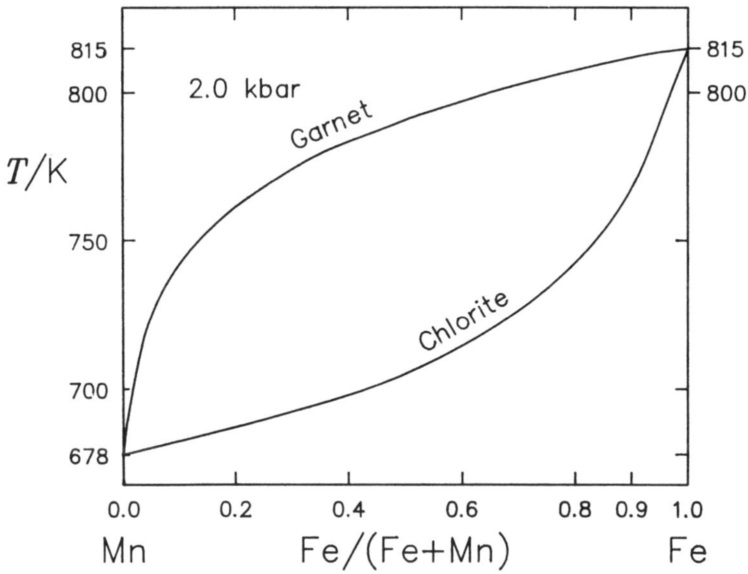

Figure 2.9 Calculated temperature (T) versus composition (Fe/(Fe + Mn)) relations for chlorite and garnet in the equilibrium, (Mn, Fe) chlorite + quartz = (Mn, Fe) garnet + H_2O. Temperatures for the end-member reactions were determined experimentally by Hsu (1968). From Kretz (1993), reprinted by permission of Blackwell Scientific Publications Inc.

2.6 THE GIBBS–DUHEM EQUATIONS

The Gibbs–Duhem equations express a relationship between temperature, pressure, and the m chemical-potential (or activity) terms of an m-component chemical mixture. Consider a mixture of components A and B. From equation (24), the molar Gibbs energy of the mixture is

$$G = X_{\mathrm{A}}\,\mu_{\mathrm{A}} + X_{\mathrm{B}}\mu_{\mathrm{B}}. \tag{49}$$

By differentiation,

$$\mathrm{d}G = X_{\mathrm{A}}\,\mathrm{d}\mu_{\mathrm{A}} + \mu_{\mathrm{A}}\,\mathrm{d}X_{\mathrm{A}} + X_{\mathrm{B}}\,\mathrm{d}\mu_{\mathrm{B}} + \mu_{\mathrm{B}}\,\mathrm{d}X_{\mathrm{B}}. \tag{50}$$

But in addition, by applying the equation for partial differentiation (Sec. 2.1), we have

$$\mathrm{d}G = \left(\frac{\partial G}{\partial X_{\mathrm{A}}}\right)\mathrm{d}X_{\mathrm{A}} + \left(\frac{\partial G}{\partial X_{\mathrm{B}}}\right)\mathrm{d}X_{\mathrm{B}}, \tag{51}$$

$$\mathrm{d}G = \mu_{\mathrm{A}}\,\mathrm{d}X_{\mathrm{A}} + \mu_{\mathrm{B}}\,\mathrm{d}X_{\mathrm{B}}, \tag{52}$$

at constant temperature and pressure, and in general,

$$\mathrm{d}G = -S\,\mathrm{d}T + V\,\mathrm{d}P + \mu_{\mathrm{A}}\,\mathrm{d}X_{\mathrm{A}} + \mu_{\mathrm{B}}\,\mathrm{d}X_{\mathrm{B}}. \tag{53}$$

From equation (50) we subtract (53), and obtain

$$S\,\mathrm{d}T - V\,\mathrm{d}P + X_{\mathrm{A}}\,\mathrm{d}\mu_{\mathrm{A}} + X_{\mathrm{B}}\,\mathrm{d}\mu_{\mathrm{B}} = 0, \tag{54}$$

and at constant temperature and pressure,

$$X_{\mathrm{A}}\mathrm{d}\mu_{\mathrm{A}} + X_{\mathrm{B}}\,\mathrm{d}\mu_{\mathrm{B}} = 0. \tag{55}$$

An analogous equation in terms of activity is obtained as follows. Recall equation (46),

$$\mu_{\mathrm{A}} = \mu^{\mathrm{A}} + RT\ln a_{\mathrm{A}}, \tag{56}$$

which gives

$$\mathrm{d}\mu_{\mathrm{A}} = RT\,\mathrm{d}\ln a_{\mathrm{A}}. \tag{57}$$

Substituting equation (57) and an analogous equation for B in (55), and dividing by RT,

$$0 = X_{\mathrm{A}}\,\mathrm{d}\ln a_{\mathrm{A}} + X_{\mathrm{B}}\,\mathrm{d}\ln a_{\mathrm{B}}, \tag{58}$$

which relates the activities of components A and B. This equation may be used to estimate the activity of one component, A, if that of the other, B, is known.

The Gibbs–Duhem equation in the form of (58) has been used to answer the following question. What is the activity–concentration relation for a major component, A, if a minor component, B, obeys the equation

$$a_{\mathrm{B}} = kX_{\mathrm{B}}, \tag{59}$$

where k is a constant? (This behavior for minor components has been observed frequently and is here referred to as Henry's rule.) From equation (59), we obtain

$$\ln a_B = \ln k + \ln X_B, \tag{60}$$

$$d \ln a_B = d \ln X_B. \tag{61}$$

By substitution in (58) above, and rearranging terms,

$$d \ln a_A = \frac{-X_B \, d \ln X_B}{X_A}. \tag{62}$$

But, by differentiation,

$$d \ln X_B = \frac{1}{X_B} \, dX_B. \tag{63}$$

Also, X_A and X_B are related by $X_A + X_B = 1$. Therefore,

$$dX_B = -dX_A. \tag{64}$$

By substitution of (63) and (64) into (62),

$$d \ln a_A = d \ln X_A, \tag{65}$$

$$a_A = I X_A, \tag{66}$$

where I is a constant of integration. But when $a_A = 1, X_A = 1$. Therefore $I = 1$, and

$$a_A = X_A, \tag{67}$$

which is the expression for Raoult's rule. Thus, in the concentration range within which the activity coefficient for B is constant, that for A is unity. This relation is shown in Figure 2.8.

2.7 THE GIBBS PHASE RULE

The phase rule of Gibbs (1878) is an expression of the variance that a multiphase system possesses, taking into consideration the variance that each phase possesses by itself, and the overall reduction in variance resulting from the restrictions imposed by equilibrium. The restrictions are, (1) all phases are at the same temperature, (2) the pressure acting on all phases is the same, and (3) for every phase interaction, a minimum of Gibbs energy is obtained. In a discussion of this subject, the following concepts and symbols are helpful.

- r number of phases
- w variability of an unrestricted system, i.e., one in which the only restrictions are those imposed by the Gibbs–Duhem equation
- v variance, or degrees of freedom
- m number of solution components

n number of system components
e number of energy-balance equations

Consider a system composed of a specified amount of a single substance, A, for example a crystal of forsterite, Mg_2SiO_4. The expression

$$dU = T\,dS - P\,dV, \qquad \text{(reversible process, } T \text{ and } P \text{ constant)}$$

which embodies the First and Second Laws, when applied to this system is also a statement regarding the variance of the system, i.e.,

$$f\,(U, S, V) = 0.$$

The system possesses two degrees of freedom (two of U, S, V) and the variance is two. Alternatively,

$$dG = -S\,dT + V\,dP, \qquad \text{(reversible process, } S \text{ and } V \text{ constant)}$$

$$f\,(G, T, P) = 0.$$

Consider now a crystalline mixture (olivine) composed of N_A moles of component A (forsterite, Mg_2SiO_4) and N_B moles of component B (fayalite, Fe_2SiO_4), and let μ_A and μ_B denote the chemical potential of A and B. The Gibbs–Duhem equation,

$$S\,dT - V\,dP + N_A\,d\mu_A + N_B\,d\mu_B = 0,$$

requires that, compared with pure A, the variance of the mixture has increased from two to three,

$$f\,(G, T, P, \mu_A) = 0,$$

and in general, the variance (v) of a chemical mixture is

$$v = 2 + (m - 1),$$

where m stands for the number of solution components present.

Consider next a system composed of olivine crystal (α) and olivine melt (β), each composed of solution components A (forsterite) and B (fayalite). We have seen above that the variance of each phase *by itself* is $2 + (m - 1)$. Then, if each of the r phases present is composed of the same m components, and the only restrictions imposed on the system are those imposed by the Gibbs–Duhem equations, the variability (w) is

$$w = r[2 + (m - 1)],$$

$$w = r[m + 1],$$

and for the olivine crystal–melt system, $w = 6$.

A number of equalities must, however, be satisfied for the system to enter a state of complete equilibrium, and each equality will reduce w by one. With regard to the olivine system there are four of these equations,

$T^\alpha = T^\beta$, and in general $(r-1)$ equations of this kind,

$P^\alpha = P^\beta$, and in general $(r-1)$ equations of this kind,

$$\left.\begin{aligned}\mu_A^\alpha &= \mu_A^\beta \\ \mu_B^\alpha &= \mu_B^\beta\end{aligned}\right\}$$ and in general, $e = m(r-1)$ equations of this kind,

for a total of $[2(r-1)+e]$ equations. Thus the variability of the system *at equilibrium*, i.e., the variance, is

$$v = w - [2(r-1)+e]. \tag{68}$$

By substitution of $w = r(m+1)$ and $e = m(r-1)$,

$$v = m + 2 - r, \tag{69}$$

and for the olivine system $(m = 2, r = 2)$, $v = 2$. Note that only $(r-1)$ equalities in each of T, P, and μ, are needed. For example, in a three-phase system, α, β, γ, when we write

$$T^\alpha = T^\beta,$$
$$T^\beta = T^\alpha,$$

it follows that $T^\gamma = T^\alpha$. In the system we have chosen to obtain equation (69), m refers to the number of *solution components* present; these reside in each of the phases and are therefore able to migrate from one phase to another.

With regard to systems composed only of phases of *fixed composition* ($v = 2$ for each phase by itself), the variance from equation (68), is

$$\begin{aligned} v &= [r(2)] - [2(r-1)+e], \\ &= 2 - e. \end{aligned} \tag{70}$$

If *no* interactions are possible, that is, $e = 0$, then $v = 2$, regardless of the number of phases present. If interactions do occur, v is reduced to one or to zero, depending on whether one or two interactions exist. For example in the system andalusite–sillimanite, the interaction is Al_2SiO_5 (andalusite) = Al_2SiO_5 (sillimanite), and the energy-balance equation is

$$G^{\text{And}} = G^{\text{Sil}},$$

where G denotes molar Gibbs energy. Thus $v = 2 - 1 = 1$. One may also arrive at this conclusion by deriving the equation for the andalusite–sillimanite boundary curve, which shows that $T = f(P), f(T,P) = 0$.

The variance of the andalusite–sillimanite system remains unchanged as other phases are added to the system (e.g., quartz, muscovite, biotite), provided these phases do not interact with andalusite or sillimanite, or with each other. In this case we could view andalusite and sillimanite as forming a subsystem.

Gibbs (1878) pointed out that the equation (69), i.e., $v = m + 2 - r$, which refers to a system of r solution phases each composed of m solution components, may be applied to any system, provided m is redefined as the 'number of independently variable components of the bodies taken as a whole', i.e., by replacing the number of solution components (m) by the number of system components (n),

$$v = n + 2 - r. \tag{71}$$

Thus for the system andalusite–sillimanite viewed *as a whole*, the number of moles of Al_2SiO_5 may be varied independently (but not Al_2O_3 or SiO_2), $n = 1$, and $v = 1 + 2 - 2 = 1$. The reasoning here is that because one component (not two) is adequate to describe the two-phase system, an interaction must exist which reduces the variance by one.

In summary,

$$v = [w] - [2(r - 1) + e], \tag{68}$$

(general expression, any system)

$$v = 2 - e, \tag{70}$$

(system of phases of fixed composition)

$$v = n + 2 - r. \tag{71}$$

(phase rule, any system)

We shall now consider six examples.

(1) For a system composed of andalusite, sillimanite, and kyanite, all of composition Al_2SiO_5, there are three interactions and three energy-balance equations,

$$G^{\text{And}} = G^{\text{Sil}},$$
$$G^{\text{Sil}} = G^{\text{Ky}},$$
$$G^{\text{Ky}} = G^{\text{And}},$$

but only two of these are *independent*, i.e., the third follows from the first two, and $e = 2$. Thus by use of equation (70), $v = 2 - e = 2 - 2 = 0$.

By use of the phase rule (equation (71)), $v = n + 2 - r$, $n = 1$, and $v = 0$.

(2) For a system composed of crystals of olivine (forsterite, fayalite) and plagioclase (albite, anorthite), there are no interactions ($e = 0$) and by use of (68), $v = [w] - [2(r - 1) + e]$, noting that $w = 6$ (3 for each phase), we find that $v = 4$.

By use of the phase rule, $n = 4$ (i.e., the four solution components), and therefore $v = 4$.

Thus all of T, P, and the composition of both phases may be varied independently.

(3) For a system composed of dolomite ($CaMgC_2O_6$), quartz (SiO_2), diopside ($CaMgSi_2O_6$), and CO_2, all phases are of fixed composition and are interrelated by

$$CaMgC_2O_6 + 2SiO_2 = CaMgSi_2O_6 + 2CO_2,$$

$$G^{\text{Dol}} + 2G^{\text{Qtz}} = G^{\text{Di}} + 2G^{\text{CO}_2},$$

i.e., $e = 1$, and by use of equation (70), $v = 2 - e = 1$.

In terms of the phase rule, $v = n + 2 - r$, if there were no interactions, n would be 4, i.e., $CaMgC_2O_6$, SiO_2, $CaMgSi_2O_6$, and CO_2. But only three independently variable components exist for the system as a whole, and these could be chosen as SiO_2, CO_2, and $CaMgO_2$. This *implies* that an interaction exists, i.e. that $e = 1$. Thus $v = 3 + 2 - 4 = 1$. Notice that CaO and MgO are not system components because their concentration within the system cannot be changed independently. This example emphasizes the care that is needed in defining system components.

(4) For a system composed of olivine (forsterite, fayalite) and orthopyroxene (enstatite, ferrosilite), the single interaction is an exchange reaction,

$$MgSi_{1/2}O_2 + FeSiO_3 = FeSi_{1/2}O_2 + MgSiO_3,$$

with energy balance,

$$\mu_{\text{fo}} + \mu_{\text{fs}} = \mu_{\text{fa}} + \mu_{\text{en}},$$

i.e., $e = 1$. Thus from equation (68), $v = [w] - [2(r - 1) + e] = 6 - [3] = 3$, and at constant T and P, the mole fraction of fayalite in olivine is a function of the mole fraction of ferrosilite in orthopyroxene.

In view of the phase rule, the solution components are not common to both phases, and the system components may be defined as $FeSiO_3$, $MgSiO_3$, SiO_2. Thus $v = 3 + 2 - 2 = 3$.

(5) For a system composed of olivine (forsterite, fayalite), orthopyroxene (enstatite, ferrosilite), and quartz, which provides an example of the coexistence of phases of fixed and variable composition, we find that $w = 3(\text{olivine}) + 3(\text{orthopyroxene}) + 2(\text{quartz}) = 8$. Two energy-balance equations must be satisfied simultaneously:

$$\mu_{\text{fo}} + G^{\text{Qtz}} = 2\mu_{\text{en}},$$

$$\mu_{\text{fa}} + G^{\text{Qtz}} = 2\mu_{\text{fs}},$$

referring to interactions

$$Mg_2SiO_4 \text{ (in olivine)} + SiO_2 = 2MgSiO_3 \text{ (in orthopyroxene)},$$

$$Fe_2SiO_4 \text{ (in olivine)} + SiO_2 = 2FeSiO_3 \text{ (in orthopyroxene)}.$$

Thus from (68), $v = [w] - [2(r - 1) + e] = 8 - 6 = 2$.

In terms of the phase rule, $n = 3$ (as in example 4), and $v = 3 + 2 - 3 = 2$.

(6) For a system composed of andalusite (Al_2SiO_5), sillimanite (Al_2SiO_5), quartz (SiO_2), muscovite ($KAl_3Si_3O_{10}(OH)_2$), K-feldspar ($KAlSi_3O_8$), and H_2O, all of fixed composition, from an inspection of the formulas we detect three interactions:

$$\text{andalusite} = \text{sillimanite},$$

$$\text{muscovite} + \text{quartz} = \text{K} - \text{feldspar} + \text{andalusite} + H_2O,$$

$$\text{muscovite} + \text{quartz} = \text{K} - \text{feldspar} + \text{sillimanite} + H_2O,$$

but only two of these are independent, i.e., $e = 2$, and $v = 2 - e = 0$. The system represents the invariant point, where the three boundary curves meet.

If we choose to apply the phase-rule equation, $v = n + 2 - r$, $r = 6$, and n is two less than six, e.g., SiO_2, $KAlSi_3O_8$, Al_2SiO_5, H_2O. Andalusite and sillimanite need only one component rather than two (implying one interaction), and andalusite, muscovite, quartz, K-feldspar, H_2O need only four components, rather than five (implying another interaction). Thus $v = 4 + 2 - 6 = 0$. In this case, variation in Al_2O_3, SiO_2, K_2O, and H_2O is possible, considering the system as a whole, and these could also have been chosen as the four system components.

Is it possible for a system to possess a variance of -1? This question was considered by Halliwell and Nyburg (1963). Suppose in example 6, above, the muscovite + quartz decomposition curve happened to pass through the Al_2SiO_5 invariant point, and kyanite appeared in the system together with the phases already present. The reader is invited to consider the consequences of this possibility.

In conclusion, the variance of a system can usually be evaluated by use of the fundamental relations which underlie the phase-rule equation, as expressed by Gibbs (1878). This approach leads to an understanding of the nature of the restrictions that are imposed on the system by the laws of thermodynamics.

2.8 PROPERTIES OF GAS MIXTURES

Abundant information is available on the properties of gas mixtures at moderate temperature and pressure. With regard to metamorphic processes, H_2O and CO_2 are of greatest interest, but other species (HF, HCl, H_2S, SO_2, CH_4) may be present. Some metamorphic reactions involve both H_2O and CO_2, for example,

$$\text{dolomite} + \text{quartz} + H_2O = \text{talc} + \text{calcite} + CO_2,$$

and in experimental investigations of such reactions the thermodynamic properties of the H_2O–CO_2 gas mixture which accompanies the crystals are important (Greenwood 1967a,b). Briefly presented here are some thermodynamic equations for gas mixtures, taken mainly from Denbigh (1961), and some information on the nature of H_2O–CO_2 mixtures at temperatures and pressures corresponding to crustal metamorphism.

Given a perfect gas mixture, i.e., a mixture of two gases, A and B, each of which by itself is a perfect gas, and the mixture behaves as a perfect gas, at a given temperature T' and pressure P',

$$\mu_{A(T'P')} = G^{A}_{T'P'} + RT' \ln X_A,$$

$$\mu_{A(T'P')} = (G^{A}_{T'1} + RT' \ln P') + RT' \ln X_A,$$

where $G^{A}_{T'1}$ is the molar Gibbs energy of pure A at temperature T' and standard pressure of 1 bar. Thus,

$$\mu_{A(T'P')} = G^{A}_{T'1} + RT' \ln P'X_{A}.$$

The partial pressure of A in the gas mixture (denoted P'_{A}) is by definition

$$P'_{A} = X_{A}P',$$

and therefore,

$$\mu_{A(T'P')} = G^{A}_{T'1} + RT' \ln P'_{A}.$$

An analogous expression exists for μ_{B}.

These equations are modified to deal with those mixtures of A and B for which the equations do not apply. Thus for component A,

$$\mu_{A(T'P')} = G^{A}_{T'1} + RT' \ln f_{A(T'P')},$$

noting that

$$\frac{f_{A}}{P_{A}} \rightarrow 1 \text{ as } P \rightarrow 0.$$

Here the partial pressure of A (P_{A}) has been replaced by the term f_{A} which is referred to as the 'fugacity' of A in the gas mixture; it should not be confused with the fugacity of pure A (f^{A}). The term f_{A} is related to P_{A} by $f_{A} = \gamma_{A}P_{A}$, where γ_{A} is the fugacity coefficient. In the above equation which contains f_{A}, $G^{A}_{T'1}$ is the molar Gibbs energy of A at temperature T' and at unit fugacity of A, which can be taken as 1 bar.

Recall that for an imperfect *pure* gas, A,

$$G^{A}_{T'P'} = G^{A}_{T'1} + RT' \ln f^{A},$$

where f^{A} is the fugacity of pure A. Now, from the above expression for the chemical potential of A in the A–B mixture, i.e.,

$$\mu_{A(T'P')} = G^{A}_{T'1} + RT' \ln f_{A},$$

where f_{A} is the fugacity of A in the mixture, we obtain by combining the above two equations,

$$\mu_{A(T'P')} = G^{A}_{T'P'} - RT' \ln f^{A} + RT \ln f_{A},$$

or

$$\mu_{A(T'P')} = G^{A}_{T'P'} + RT \ln(f_{A}/f^{A}).$$

Comparing this with the general equation for the chemical potential, i.e.,

$$\mu_{A} = G^{A} + RT \ln a_{A},$$

we see that the fugacity ratio can be equated to activity, i.e.,

$$a_A = (f_A / f^A),$$

Two imperfect gases, A and B, when mixed could produce a gas mixture which satisfies the definition of an ideal mixture, i.e.,

$$a_A = (f_A / f^A) = X_A,$$
$$a_B = (f_B / f^B) = X_B.$$

This behavior is common at low pressures and is referred to as the 'Lewis and Randall rule'.

If the equations of state for two gases are known, it is possible to calculate $a - X$ relations for the components at different combinations of temperature and total pressure. Thus Kerrick and Jacobs (1981) have shown that at 500 °C, 500 bar, mixing of H_2O and CO_2 is nearly ideal, in good agreement with results obtained earlier by Greenwood (1973). At higher pressures increasing departure from the $a = X$ relation is found, as shown in Figure 2.10. Thus if an H_2O–CO_2 gas phase is present in marble at 500 °C, 2 kbar, for example, the activity of the H_2O and

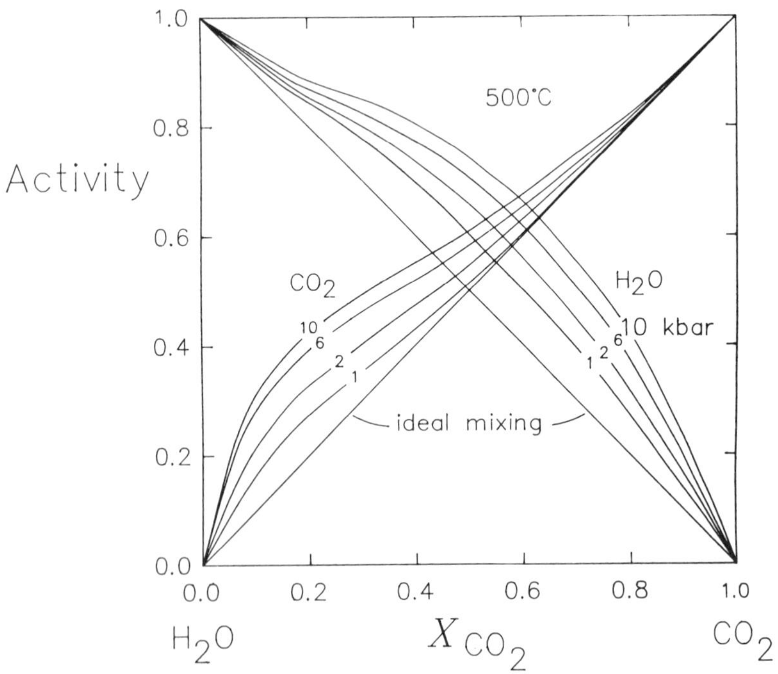

Figure 2.10 Activity of H_2O and CO_2 in H_2O–CO_2 mixtures as a function of composition ($X_{CO_2} = CO_2/(H_2O + CO_2)$) and pressure; lines of unit slope (referring to ideal mixing) are drawn for comparison. From Kerrick and Jacobs (1981). Reprinted by permission of *American Journal of Science*

CO_2 components will exceed their mole fractions by a factor that can be obtained from this figure.

For a more advanced treatment of this subject and for more information of gas mixtures at metamorphic temperatures and pressures, the reader is referred to Kerrick and Jacobs (1981), Flowers and Helgeson (1983), and Spycher and Reed (1988).

2.9 MINERAL MIXTURES: A MICROSCOPIC VIEW

The molecules of two gases, for example H_2O and CO_2, will mix spontaneously, resulting in an increase in entropy and a decrease in Gibbs energy. On a microscopic scale, the mixing process may be attributed to the inherent mobility of the molecules, and the tendency for increasing disorder, as expressed by the Second Law of Thermodynamics. Suppose now that a crystal of copper and one of zinc are brought together at 600 °C, 1 bar. Here also, given sufficient time, a mixing of atoms will occur, but now forces between atoms become important, in addition to atomic mobility. The importance of atomic bonds is indicated by the formation of a discrete phase, Cu_1Zn_1, and some more complex compounds, in addition to solid solution which occurs in the range of X_{Zn} from 0 to 0.5 (Lumsden 1952).

Consider now a crystal of Mg_2SiO_4 (forsterite) and one of Fe_2SiO_4 (fayalite) in contact at 800 °C, 1 bar. Both crystals have the same crystal structure, which may be viewed as an SiO_4 'framework' composed principally of large oxygen atoms, with cavities or sites occupied by Mg and Fe atoms. In a sense, the structure is intermediate between that of a metal alloy, where like and unlike atoms are in direct contact, and that of a gas, where the empty space that separates the molecules is analogous to the oxygen framework. Given sufficient time, a change will occur at the forsterite–fayalite interface, and this change may be viewed as mixing of a special kind. The oxygen framework, with its negative charge and clearly defined sites for Mg^{2+} and Fe^{2+} ions, requires that for every Mg^{2+} atom displaced from forsterite to fayalite, an atom of Fe^{2+} must be displaced in the opposite direction. When an Fe atom becomes located in the forsterite crystal, bonds are formed with the surrounding and contacting oxygen atoms, but interactions will also occur with the more distant Mg or Fe atoms, as shown in Figure 2.11. The latter interactions will be relatively weak, and from this point of view it will not be surprising to find rather small departures from ideal mixing in some silicate minerals.

Another important consideration is the nature of the oxygen framework, which may or may not easily accommodate cations of unusual size. This results in limited solid solution, for example, the limited amount of Ca that will enter orthopyroxene at a given temperature. The charge-balance requirement is also important, and may give rise to coupled substitutions, such as $(NaSi)^{5+}$ substituting for $(CaAl)^{5+}$ in plagioclase.

The ease of substitution that occurs in certain sites of silicates, while other sites in the same structure allow for a limited substitution or none at all, encourages the view that mixing occurs by the intermingling of atoms on sites rather than

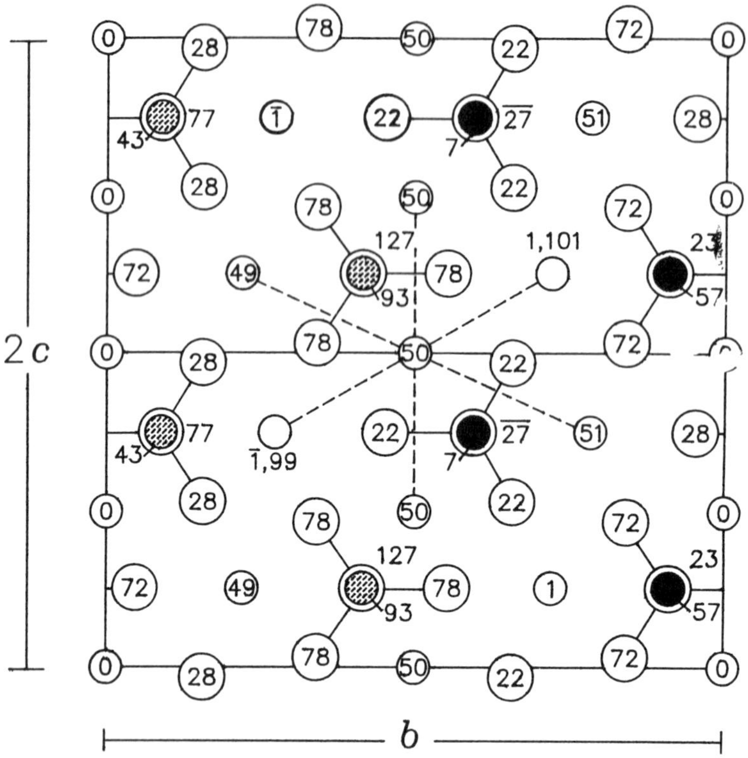

Figure 2.11 The structure of forsterite, Mg_2SiO_4, with cell dimensions, a = 4.76, b = 10.21, c = 5.98 Å, projected on (100). Broken lines extend from the central Mg atom to the eight nearest Mg neighbors. From Bragg and Claringbull (1965)

the intermingling of 'molecular' end-members. The atoms themselves cannot, however, be regarded as solution components (as they are in metals) for it is not possible in general to imagine an introduction of one kind of atom (e.g., Fe) into a silicate crystal, as an independent process. But it is possible to imagine the addition of 1 Fe atom accompanied by the removal of 1 Mg atom, and Thompson (1982) has demonstrated that $FeMg_{-1}$ forms a possible choice for a component of ferromagnesian silicates. Abstract components of this kind are particularly useful in dealing with complex solid solutions such as hornblende. For the present, however, we shall select $FeSi_{1/2}O_2$ and similar 'molecular' species as components, for many of these exist as pure substances and their thermodynamic properties can be measured.

In searching for a mineralogical analogue to the ideal gas mixture, considerations of the entropy of mixing are most helpful. Recall that for one mole of an ideal gas mixture of components A and B, the entropy of mixing is given by the expression

$$\Delta_m S = -X_A R \ln X_A - X_B R \ln X_B.$$

The statistical definition of the entropy of mixing applied to a crystal of olivine is obtained as follows.

Let the number of configurations or arrangements of Fe and Mg atoms in the crystal be Ω. Then the entropy of mixing (S^{M}) is

$$S^{\mathrm{M}} = k \ln \Omega,$$

where k is the Boltzmann constant. Suppose the olivine crystal contains N octahedral sites (N_{Mg} occupied by Mg and N_{Fe} by Fe atoms). Then the number of possible arrangements of Mg and Fe atoms is

$$\Omega = (N_{\mathrm{Mg}} + N_{\mathrm{Fe}})!/(N_{\mathrm{Mg}}!N_{\mathrm{Fe}}!).$$

N is large and hence Stirling's approximation may be used:

$$N! = (N/\mathrm{e})^N,$$

where e is the base of natural logarithms. Thus,

$$\Omega = -N_{\mathrm{Mg}} \ln(N_{\mathrm{Mg}}/N) - N_{\mathrm{Fe}} \ln(N_{\mathrm{Fe}}/N),$$

and the entropy of mixing is

$$S^{\mathrm{M}} = -kN_{\mathrm{Mg}} \ln(N_{\mathrm{Mg}}/N) - kN_{\mathrm{Fe}} \ln(N_{\mathrm{Fe}}/N).$$

Thus the entropy of mixing per mole ($\Delta_{\mathrm{m}}S$) is

$$\Delta_{\mathrm{m}}S = -X_{\mathrm{Mg}}R \ln X_{\mathrm{Mg}} - X_{\mathrm{Fe}}R \ln X_{\mathrm{Fe}},$$

which is identical to the expression (above) for ideal gas mixtures.

Hence in dealing with mixing in olivine it is appropriate to define the components as $MgSi_{1/2}O_2$ and $FeSi_{1/2}O_2$ and to consider entropy and energy changes in terms of mixing of Fe and Mg on the octahedral sites.

Several ferromagnesian minerals have two or more similar sites occupied by Fe and Mg atoms, in different proportions. For example, in orthopyroxene, both of the M sites (M1 and M2) are occupied by Fe and Mg atoms, with M2 favoring Fe, M1 favoring Mg (Ghose 1965). What kind of activity–concentration relations might be expected for these crystals?

One may deal with this problem by creating for orthopyroxene a special model of on-site mixing, or by defining and measuring the activity of $MgSiO_3$ (en) and $FeSiO_3$ (fs) in the crystal as a whole, the approach adopted by Nafziger and Muan (1967). This latter approach, combined with crystal-chemical considerations is here preferred, and the ideal model for orthopyroxene is defined as

$$a_{\mathrm{fs}} = X_{\mathrm{fs}},$$

$$a_{\mathrm{en}} = X_{\mathrm{en}},$$

where X = Fe/(Mg + Fe) in the crystal as a whole.

2.10 SOLUTION MODELS FOR MINERALS

Several models have been proposed, mainly in the fields of Chemistry, Metallurgy, and Chemical Engineering, and some of these have been used to describe mixing and unmixing in silicates, oxides, and other minerals. Three of these models are now examined briefly. In a following section (2.11) we shall see that the models can be helpful in treating and interpreting thermodynamic data on olivine, garnet, plagioclase, and other minerals.

Recall that the molar Gibbs energy of an ideal solution (G^{I}) of components A and B is expressed as

$$G^{\mathrm{I}} = X_{\mathrm{A}}G^{\mathrm{A}} + X_{\mathrm{B}}G^{\mathrm{B}} + X_{\mathrm{A}}RT\ln X_{\mathrm{A}} + X_{\mathrm{B}}RT\ln X_{\mathrm{B}},$$

where G^{A} and G^{B} stand for the molar Gibbs energy of pure A and pure B, R is the gas constant, T is thermodynamic temperature, and X denotes mole fraction. The last two terms contain the molar energy of mixing,

$$\Delta_{\mathrm{m}}G^{\mathrm{I}} = X_{\mathrm{A}}RT\ln X_{\mathrm{A}} + X_{\mathrm{B}}RT\ln X_{\mathrm{B}}.$$

The Gibbs energy of mixing for a non-ideal solution may be expressed as

$$\Delta_{\mathrm{m}}G = \Delta_{\mathrm{m}}G^{\mathrm{I}} + G^{\mathrm{E}},$$

where G^{E} stands for the excess energy of mixing.

A general expression for G^{E} was proposed by Guggenheim (1937):

$$G^{\mathrm{E}} = X_{\mathrm{A}}X_{\mathrm{B}}[A_0 + A_1(X_{\mathrm{B}} - X_{\mathrm{A}}) + A_2(X_{\mathrm{B}} - X_{\mathrm{A}})^2],$$

to which additional terms could be added if necessary. The coefficients, A_0, A_1, A_2, are independent of X_{A} and X_{B}, and in general are expected to vary with temperature and pressure.

Three models are now defined in terms of these coefficients.

Model I: All A coefficients equal to 0

This is the ideal solution model described above, with $G^{\mathrm{E}} = 0$. In addition, by definition, $S^{\mathrm{E}}, H^{\mathrm{E}}, V^{\mathrm{E}}$ are zero, and $RT\ln\gamma = 0$, i.e., the activity coefficients, γ_{A} and γ_{B}, equal 1.0.

The ideal model was considered by Ramberg and DeVore (1951) in their study of the distribution of Mg and Fe between olivine and orthopyroxene. Soon after, abundant information became available on the distribution of Mg and Fe between various ferromagnesian silicates in medium- to high-grade metamorphic rocks, and Mueller (1961a, 1972), Saxena (1973), Kretz (1978) and others pointed out that frequently the shape of the distribution curves (isotherms) was of the kind to be expected if both minerals involved in the exchange behaved as ideal or near-ideal mixtures.

Model II: $A_0 \neq 0, A_1 = 0, A_2 = 0$

Now,

$$G^E = X_A X_B A_0.$$

Let W stand for A_0; (W is referred to as the Margules parameter):

$$G^E = X_A X_B W.$$

The Gibbs energy of mixing is,

$$\Delta_m G^{II} = [\Delta_m G^I] + [G^E],$$

$$\Delta_m G^{II} = [X_A RT \ln X_A + X_B RT \ln X_B] + [X_A X_B W].$$

This equation for different values of W and an arbitrary value of RT (10 000 J/mol) is plotted in Figure 2.12, where the different curves may be compared with that describing ideal mixing (model I, $W = 0$). Negative values of W result in increased stability (lower Gibbs energy), and positive values of W, if they exceed $2RT$ J/mol, result in instability, i.e., exsolution. Thus with $W = 30\,000$ J/mol, two phases, of composition a and b, are more stable than one of intermediate composition.

Recall that departure from ideal mixing can be accommodated by use of activity coefficients (γ), i.e., by writing $a_A = \gamma_A X_A$. Thus, in general,

$$G^E = X_A RT \ln \gamma_A + X_B RT \ln \gamma_B.$$

Now for model II,

$$G^E = X_A X_B W.$$

Thus,

$$G^E = X_A X_B W = X_A (RT \ln \gamma_A) + X_B (RT \ln \gamma_B).$$

This equation has the form

$$V = X_A V_A + X_B V_B,$$

for which we determined above (Sec. 2.4), that

$$V_B = V + (1 - X_B)(dV/dX_B).$$

Thus, by analogy,

$$RT \ln \gamma_B = G^E + (1 - X_B)(dG^E/dX_B),$$

$$RT \ln \gamma_B = X_A X_B W + (1 - X_B)\frac{d}{dX_B}[(1 - X_B)(X_B)W].$$

Carrying out the differentiation, we obtain

$$RT \ln \gamma_B = X_A^2 W,$$

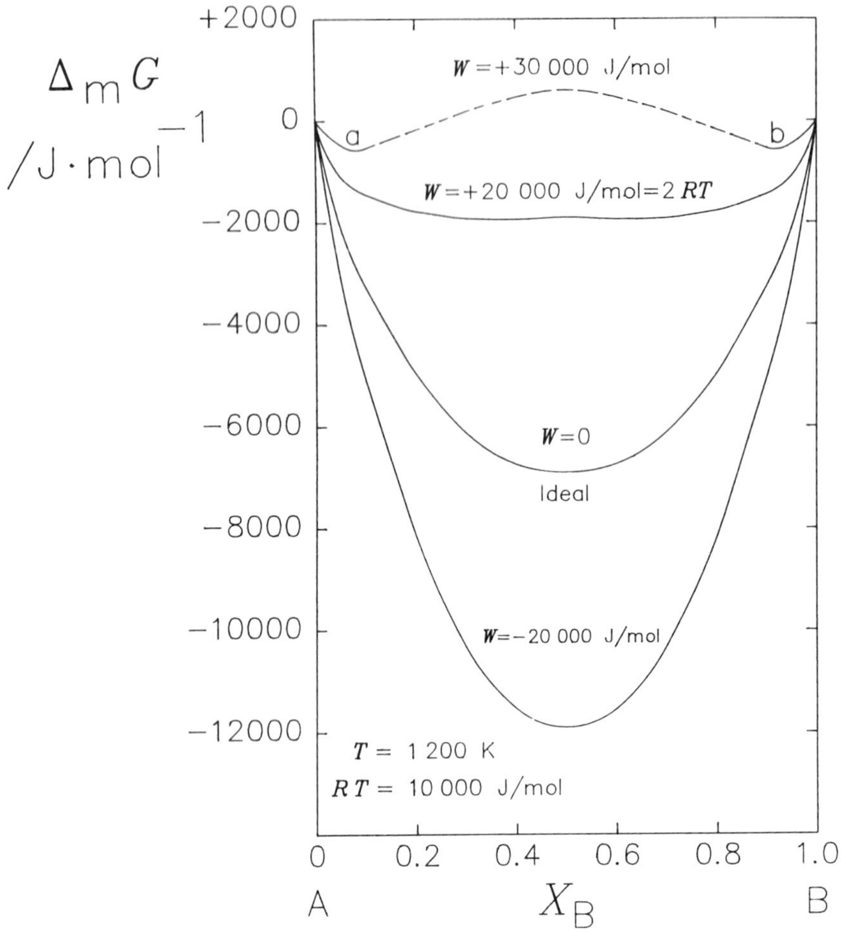

Figure 2.12 Plot of the equation for the Gibbs energy of mixing ($\Delta_m G$) for symmetrical mixtures (model II), compared with the ideal mixture. The curves are for an arbitrary value of RT = 10 000 J/mol, and for different values of Margules parameter (W), as indicated. When $W > 2RT$, two phases of different composition are stable, for example with W = 30 000 J/mol, the compositions are represented by points a and b

and similarly,

$$RT \ln \gamma_A = X_B^2 W,$$

to give the activity coefficients in terms of W.

Expressions for $\Delta_m S^{II}$, $\Delta_m H^{II}$, and $\Delta_m V^{II}$ are obtained by noting that

$$S^E = -\partial G^E / \partial T,$$

$$H^{\mathrm{E}} = G^{\mathrm{E}} + TS^{\mathrm{E}},$$

$$V^{\mathrm{E}} = \partial G^{\mathrm{E}}/\partial P,$$

to yield, for mixtures of type II,

$$\Delta_{\mathrm{m}}S^{\mathrm{II}} = [\Delta_{\mathrm{m}}S^{\mathrm{I}}] + [S^{\mathrm{E}}],$$

$$\Delta_{\mathrm{m}}S^{\mathrm{II}} = [-X_{\mathrm{A}}R\ln X_{\mathrm{A}} - X_{\mathrm{B}}R\ln X_{\mathrm{B}}] - \left[X_{\mathrm{A}}X_{\mathrm{B}}\frac{\partial W}{\partial T}\right],$$

$$\Delta_{\mathrm{m}}H^{\mathrm{II}} = [\Delta_{\mathrm{m}}H^{\mathrm{I}}] + [H^{\mathrm{E}}],$$

$$\Delta_{\mathrm{m}}H^{\mathrm{II}} = [0] + \left[X_{\mathrm{A}}X_{\mathrm{B}}\left(W - T\frac{\partial W}{\partial T}\right)\right],$$

$$\Delta_{\mathrm{m}}V^{\mathrm{II}} = [\Delta_{\mathrm{m}}V^{\mathrm{I}}] + [V^{\mathrm{E}}],$$

$$\Delta_{\mathrm{m}}V^{\mathrm{II}} = [0] + \left[X_{\mathrm{A}}X_{\mathrm{B}}\frac{\partial W}{\partial P}\right].$$

Also, for model II,

$$\mu_{\mathrm{A}} = G^{\mathrm{A}} + RT\ln X_{\mathrm{A}} + X_{\mathrm{B}}^{2}W,$$

$$\mu_{\mathrm{B}} = G^{\mathrm{B}} + RT\ln X_{\mathrm{B}} + X_{\mathrm{A}}^{2}W,$$

$$a_{\mathrm{A}} = X_{\mathrm{A}}\gamma_{\mathrm{A}} = X_{\mathrm{A}}\exp[X_{\mathrm{B}}^{2}(W/RT)],$$

$$a_{\mathrm{B}} = X_{\mathrm{B}}\gamma_{\mathrm{B}} = X_{\mathrm{B}}\exp[X_{\mathrm{A}}^{2}(W/RT)].$$

Model II is referred to as a ‘simple mixture’ by Guggenheim (1977) and as a ‘symmetric regular solution’ by Thompson (1967). It was first applied to minerals by Mueller (1962, 1964) to describe the non-ideal distribution of Mg and Fe between cummingtonite and associated minerals, and to provide a general interpretation of limited miscibility in some mineral systems.

Model III: $A_0 \neq 0, A_1 \neq 0, A_2 = 0$

Now,

$$G^{\mathrm{E}} = X_{\mathrm{A}}X_{\mathrm{B}}[A_0 + A_1(X_{\mathrm{B}} - X_{\mathrm{A}})],$$

or

$$G^{\mathrm{E}} = X_{\mathrm{A}}X_{\mathrm{B}}[A_0 + A_1X_{\mathrm{B}} - A_1X_{\mathrm{A}}].$$

Multiply A_0 by $(X_{\mathrm{A}} + X_{\mathrm{B}}) = 1.0$ and let

W_{B} stand for $(A_0 - A_1)$

W_{A} stand for $(A_0 + A_1)$

Then,

$$G^{\mathrm{E}} = X_{\mathrm{A}} X_{\mathrm{B}} (X_{\mathrm{A}} W_{\mathrm{B}} + X_{\mathrm{B}} W_{\mathrm{A}}),$$
$$G^{\mathrm{E}} = X_{\mathrm{A}}^2 X_{\mathrm{B}} W_{\mathrm{B}} + X_{\mathrm{A}} X_{\mathrm{B}}^2 W_{\mathrm{A}}.$$

And the Gibbs energy of mixing for solutions of type III is given by

$$\Delta_{\mathrm{m}} G^{\mathrm{III}} = [\Delta_{\mathrm{m}} G^{\mathrm{I}}] + [G^{\mathrm{E}}],$$
$$\Delta_{\mathrm{m}} G^{\mathrm{III}} = [X_{\mathrm{A}} RT \ln X_{\mathrm{A}} + X_{\mathrm{B}} RT \ln X_{\mathrm{B}}] + [X_{\mathrm{A}} X_{\mathrm{B}} (X_{\mathrm{A}} W_{\mathrm{B}} + X_{\mathrm{B}} W_{\mathrm{A}})].$$

A plot of this equation for different values of W_{A} and W_{B} with $W_{\mathrm{B}} > W_{\mathrm{A}}$, and an arbitrary value of RT (10 000 J/mol) is shown in Figure 2.13, on which also is plotted the curve for model I ($W_{\mathrm{A}} = W_{\mathrm{B}} = 0$). Notice that compared with model II (Figure 2.12), the energy of mixing curves is not symmetrical about $X_{\mathrm{B}} = 0.5$. In model III, as in II, increasing values of W can result in instability, shown by the uppermost curve in Figure 2.13. This curve defines a miscibility gap (*a* to *b*) for an asymmetric solvus; points *a* and *b* are located by the common tangent.

Equations for the activity coefficients in terms of W_{A} and W_{B} are obtained as above, for model II,

$$RT \ln \gamma_{\mathrm{B}} = G^{\mathrm{E}} + (1 - X_{\mathrm{B}})\, \mathrm{d}G^{\mathrm{E}}/\mathrm{d}X_{\mathrm{B}},$$
$$RT \ln \gamma_{\mathrm{B}} = X_{\mathrm{A}}^2 X_{\mathrm{B}} W_{\mathrm{B}} + X_{\mathrm{A}} X_{\mathrm{B}}^2 W_{\mathrm{A}} + X_{\mathrm{A}} \frac{\mathrm{d}}{\mathrm{d}X_{\mathrm{B}}} [(1 - X_{\mathrm{B}})^2 X_{\mathrm{B}} W_{\mathrm{B}} + (1 - X_{\mathrm{B}}) X_{\mathrm{B}}^2 W_{\mathrm{A}}].$$

Upon differentiation,

$$RT \ln \gamma_{\mathrm{B}} = W_{\mathrm{B}} X_{\mathrm{A}}^2 - 2X_{\mathrm{A}}^2 X_{\mathrm{B}} (W_{\mathrm{B}} - W_{\mathrm{A}}),$$

and similarly,

$$RT \ln \gamma_{\mathrm{A}} = W_{\mathrm{A}} X_{\mathrm{B}}^2 + 2X_{\mathrm{B}}^2 X_{\mathrm{A}} (W_{\mathrm{B}} - W_{\mathrm{A}}).$$

Expressions for $\Delta_{\mathrm{m}} S^{\mathrm{III}}$, $\Delta_{\mathrm{m}} H^{\mathrm{III}}$, and $\Delta_{\mathrm{m}} V^{\mathrm{III}}$ for model III mixtures are obtained as for model II:

$$\Delta_{\mathrm{m}} S^{\mathrm{III}} = [\Delta_{\mathrm{m}} S^{\mathrm{I}}] + [S^{\mathrm{E}}],$$
$$\Delta_{\mathrm{m}} S^{\mathrm{III}} = [-X_{\mathrm{A}} R \ln X_{\mathrm{A}} - X_{\mathrm{B}} R \ln X_{\mathrm{B}}] + \left[-X_{\mathrm{A}}^2 X_{\mathrm{B}} \left(\frac{\partial W_{\mathrm{B}}}{\partial T} \right) - X_{\mathrm{A}} X_{\mathrm{B}}^2 \left(\frac{\partial W_{\mathrm{A}}}{\partial T} \right) \right],$$
$$\Delta_{\mathrm{m}} H^{\mathrm{III}} = [\Delta_{\mathrm{m}} H^{\mathrm{I}}] + [H^{\mathrm{E}}],$$
$$\Delta_{\mathrm{m}} H^{\mathrm{III}} = [0] + \left[X_{\mathrm{A}}^2 X_{\mathrm{B}} \left(W_{\mathrm{B}} + T \frac{\partial W_{\mathrm{B}}}{\partial T} \right) + X_{\mathrm{A}} X_{\mathrm{B}}^2 \left(W_{\mathrm{A}} + T \frac{\partial W_{\mathrm{A}}}{\partial T} \right) \right],$$
$$\Delta_{\mathrm{m}} V^{\mathrm{III}} = [\Delta_{\mathrm{m}} V^{\mathrm{I}}] + [V^{\mathrm{E}}],$$
$$\Delta_{\mathrm{m}} V^{\mathrm{III}} = [0] + \left[X_{\mathrm{A}}^2 X_{\mathrm{B}} \frac{\partial W}{\partial P} + X_{\mathrm{A}} X_{\mathrm{B}}^2 \frac{\partial W}{\partial P} \right].$$

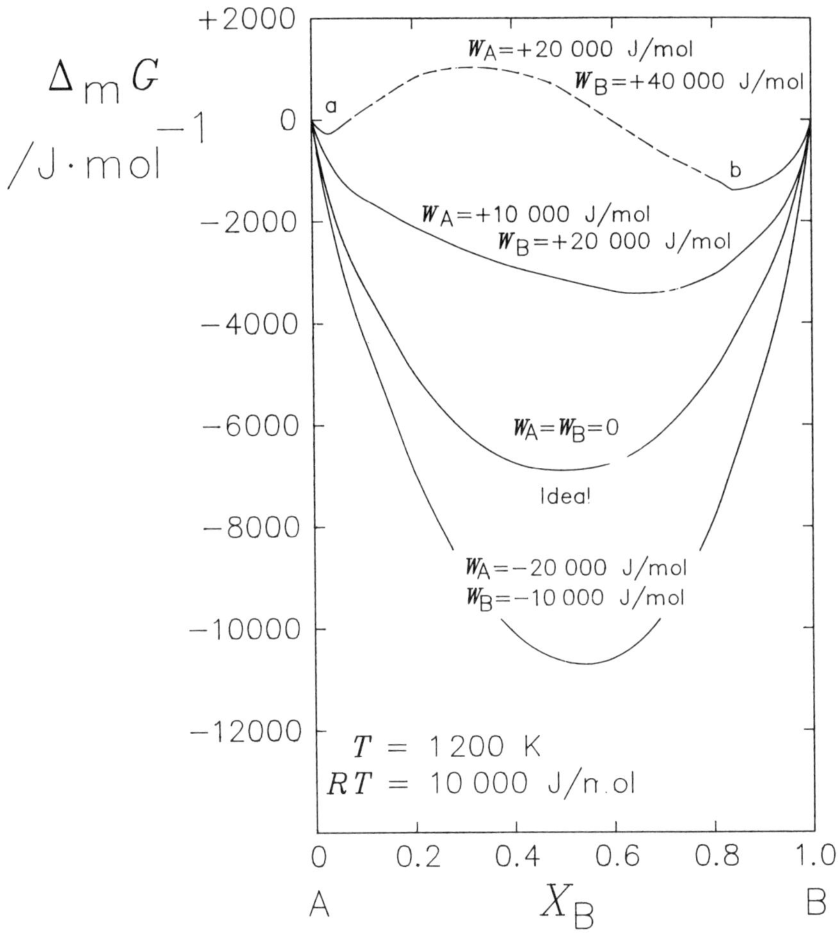

Figure 2.13 Plot of the equation for the Gibbs energy of mixing ($\Delta_m G$) for asymmetrical mixtures (model III) compared with the ideal mixture. The curves are for an arbitrary value of RT = 10 000 J/mol, and for different combinations of Margules parameters, W_A and W_B. Compare with Figure 2.12

Also, for model III,

$$\mu_A = G^A + RT \ln X_A + W_A X_B^2 + 2X_B^2 X_A (W_B - W_A),$$

$$\mu_B = G^B + RT \ln X_B + W_B X_A^2 - 2X_A^2 X_B (W_B - W_A),$$

$$a_A = X_A \gamma_A = X_A \exp\left[\frac{W_A X_B^2 + 2X_B^2 X_A (W_B - W_A)}{RT}\right],$$

$$a_B = X_B\gamma_B = X_B \exp\left[\frac{W_B X_A^2 - 2X_A^2 X_B(W_B - W_A)}{RT}\right].$$

W_A and W_B are referred to as 'Margules parameters'. W_A is commonly written W_{AB}, implying that it is the parameter associated with component A in the A–B solution. Thompson (1967) has pointed out that these parameters may be viewed as consisting of three parts, associated with excess enthalpy (W^H), excess entropy (W^S), and excess volume (W^V), i.e., for W_A (which is associated with excess Gibbs energy and is now written W_A^G),

$$W_A^G = W_A^H - TW_A^S + (P - 1)W_A^V.$$

Model III was referred to as the 'asymmetric regular solution' by Thompson (1967) but it is also known as the 'subregular' or 'Margules solution'. The terms 'Margules asymmetrical' for III, and 'Margules symmetrical' for II are proposed here.

Model III was examined in detail by Thompson (1967) and was applied by Thompson and Waldbaum (1968) in their analysis of mixing in alkali feldspar, and by Lerman (1965) in an analysis of solid solution in calcite.

Other models

Several other models exist. A mixture for which $A_0 \neq 0$, $A_1 = 0$, $A_2 \neq 0$, is referred to as a symmetrical mixture by Guggenheim (1977), and one for which $A_0 + A_1 = 0$, $A_2 = 0$, as an unsymmetrical mixture. Additional models have been evaluated in relation to mineral mixtures by Fei, Saxena, and Eriksson (1986) but the impression obtained is that models I to III as defined above are sufficiently versatile to satisfy present needs.

There is also the 'ideal-site' model which has been used frequently, and is defined as

$$a = X^n,$$

where n is the number of sites in the standard formula. Thus for olivine, written $(Mg,Fe)_2SiO_4$, the activity of fayalite is

$$a_{fa} = X_{fa}^2.$$

In isolation, this model seems unreasonable because as X_{fa} increases to 0.5, a_{fa} increases only to 0.25, and moreover the relation is not supported by the experimental data reported above (Figure 2.8, p. 87).

In terms of a particular reaction, this model becomes identical to model I. For example, in the ideal-site model, the activity of grossular in Ca–Fe garnet is written

$$a_{grs} = X_{grs}^3.$$

Now, when the reaction

$$\text{anorthite} = \text{grossular} + \text{Al silicate} + \text{quartz}$$

is written in terms of 1 mole of grossular, defined as $CaAl_{2/3}SiO_4$, and we adopt model I ($a_{grs} = X_{grs}$), we have

$$-RT \ln X_{grs} = \Delta G^\circ.$$

When we write the reaction in terms of 1 mole of $Ca_3Al_2Si_3O_{12}$ and adopt the ideal-site model, $a_{grs} = X^3_{grs}$, we have

$$-RT \ln X^3_{grs} = 3\Delta G^\circ,$$

$$-3RT \ln X_{grs} = 3\Delta G^\circ,$$

$$-RT \ln X_{grs} = \Delta G^\circ,$$

which is the same as above.

2.11 PROPERTIES OF MINERAL MIXTURES: OLIVINE, GARNET, AND PLAGIOCLASE

A reaction to produce garnet from chlorite was examined above (Sec. 2.5) as an example of reactions involving mineral solutions. The lack of information on the thermodynamic properties of Mn–Fe chlorite and Mn–Fe garnet at the temperature and pressure of interest made necessary the assumption that the solutions are ideal mixtures. This assumption points to and emphasizes the need that exists for more information on the mixing properties of common metamorphic minerals.

Such data are, however, becoming available for some minerals, and presented here is a review of data on three minerals, olivine, garnet, and plagioclase. In viewing the results it will be helpful, initially, to refer to the ideal mixture (I) as a model, and to compare the observed mineral data with those that are prescribed by this model. Model II or III may then be adopted, as necessary, to describe the experimental results.

The properties that will be considered at present are the heat and volume of mixing and relations between component activity and concentration. The initial comparison with model I can be carried out by recalling that for an ideal mixture the heat of mixing ($\Delta_m H$) and the volume of mixing ($\Delta_m V$) are zero, and the activity of each component is equal to its mole fraction ($a = X$).

Olivine

The heat of solution ($\Delta_{solution} H$) of olivine has been measured by dissociating crystals of different composition in a solvent and recording, by use of a calorimeter, the heat released or absorbed in the process. Details of the procedure are provided by Navrotsky (1977) and by Holloway and Wood (1988). Torgeson and Sahama (1948) used HF at 75 °C to dissolve seven natural olivine crystals and found a linear relation between the heat of solution and the mole fraction of fayalite present, in agreement with a zero or very small heat of mixing for olivine.

Subsequently, Thierry et al. (1981) measured the heat absorbed by dissolving three synthetic olivine crystals in a $NaBO_2$–$LiBO_2$ melt at 910 °C, with similar results. At the same time, Wood and Kleppa (1981) measured the heat of solution of six synthetic crystals in a $Pb_2B_2O_5$ melt at 700 °C and obtained the results shown in Figure 2.14. In this experiment, crystals of intermediate composition consumed slightly less heat than expected, implying a slight positive heat of mixing for olivine, at 700 °C.

Estimates of the volume of mixing in olivine were obtained by Fisher and Medaris (1969) who measured the cell dimensions of six synthetic crystals (prepared at 2000 bar); the results are shown in Figure 2.15. These data, together with data for crystals that were synthesized at 500 bar, reveal a small positive volume of mixing. Louisnathan and Smith (1968) measured the cell volume of several natural olivine crystals and found anomalously low volumes for

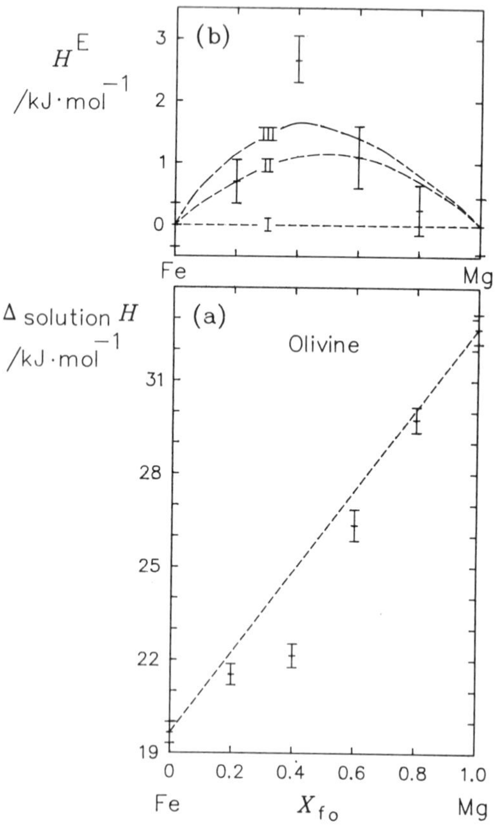

Figure 2.14 (a) (lower part) Heat of solution of olivine per one cation mole ($M_1Si_{0.5}O_2$). From Wood and Kleppa (1981), with permission from Pergamon Press Ltd. Error bars are slightly modified. Broken line refers to ideal mixing. (b) (upper part) Excess heat of mixing (H^E), from data in (a)

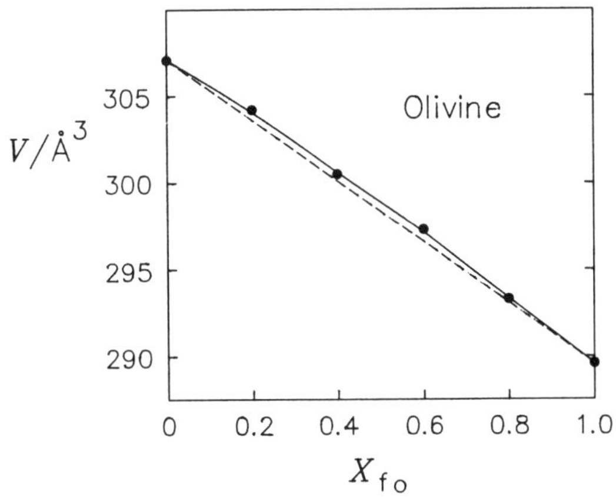

Figure 2.15 Cell volume (V) of olivine crystals of composition X_{fo} = Mg/(Mg + Fe), at 2.0 kbar; broken line represents zero volume of mixing. From Fisher and Medaris (1969)

Mg-rich crystals (Mg/(Mg+Fe) > 0.8); the results were discussed by Newton and Wood (1980).

Information on the activity of forsterite and fayalite components in olivine at 1200 °C was obtained by Nafziger and Muan (1967). First, the reaction

$$\underset{\text{fs component in orthopyroxene}}{FeSiO_3} = Fe + SiO_2 + 1/2\ O_2$$

was investigated to obtain measurements of the activity of $FeSiO_3$ (fs) in orthopyroxene. At equilibrium, energy-balance considerations produce the equation

$$(\mu_{fs})_{T,P} = G^{Fe}_{T,P} + G^{SiO_2}_{T,P} + \frac{1}{2}(G^{O_2}_{T,1} + RT \ln P_{O_2}),$$

$$\ln a_{fs} = (\Delta G^\circ/RT) + (1/2) \ln P_{O_2}.$$

Here ΔG° is the Gibbs energy change for the reaction when components and phases are in their standard states (pure ferrosilite, Fe, and SiO_2 at ambient temperature and pressure, oxygen at ambient temperature and 1 bar pressure), and P_{O_2} is the partial pressure of oxygen. Having obtained values for the activity of ferrosilite, the corresponding values for the activity of enstatite were calculated by use of the Gibbs–Duhem equation.

Experiments were then carried out to measure the Mg:Fe ratios of associated orthopyroxene (Opx) and olivine (Ol), which are determined by the following exchange reaction:

$$\underset{\text{fs in Opx}}{FeSiO_3} + \underset{\text{fo in Ol}}{MgSi_{0.5}O_2} = \underset{\text{en in Opx}}{MgSiO_3} + \underset{\text{fa in Ol}}{FeSi_{0.5}O_2}.$$

At equilibrium,

$$\mu_{\text{fs}} + \mu_{\text{fo}} = \mu_{\text{en}} + \mu_{\text{fa}},$$

$$\ln\left(\frac{a_{\text{en}}a_{\text{fa}}}{a_{\text{fs}}a_{\text{fo}}}\right) = -\Delta G^{\circ}/RT,$$

where ΔG° is the Gibbs energy change for the reaction with all components in their pure form.

Having determined a_{fs} and a_{en} from the previous experiment, the above equation enabled the authors to obtain measurements of a_{fo} and a_{fa} as a function of composition, and the results are shown as solid symbols in Figure 2.8 (p. 87). Similar experiments with magnesiowüstite, for which activity–concentration relations were determined previously, produced the results shown as open symbols in Figure 2.8. Thus at 1200 °C, activity–concentration relations in olivine depart slightly from those prescribed by the ideal solution model.

Similar experimental results were obtained at 1204 °C by Kitayama and Katsura (1968), and at 1130 °C by Wiser and Wood (1989).

The data of Figures 2.8 and 2.14 reveal that at 700 to 1200 °C, olivine is not an ideal mixture as defined above ($H^{\text{E}} = 0$, $V^{\text{E}} = 0$, $a = X$), but departures from ideal behavior are not large. Thus model II or III could be applied.

Wood and Kleppa (1981) used the following equation from model III:

$$H^{\text{E}} = X_{\text{A}}^2 X_{\text{B}} W_{\text{B}}^{\text{H}} + X_{\text{A}} X_{\text{B}}^2 W_{\text{A}}^{\text{H}}$$
$$(T = 970\ \text{K},\ W_{\text{A}}^{G} = \text{constant},\ W_{\text{B}}^{G} = \text{constant})$$

to derive

$$W_{\text{Fe}}^{H} = 4180 \text{ and } W_{\text{Mg}}^{H} = 8370 \text{ J/mol MSi}_{0.5}\text{O}_2$$

as values for W that best fit their heat-of-solution data. The plotted equation is shown in Figure 2.14b. Chatterjee (1987) proposed that olivine is a symmetrical solution (model II), with $W^{G} = 4500$ J/mol $MSi_{0.5}O_2$, assuming that $W^{H} \sim W^{G}$. Placing this value in the equation

$$H^{E} = X_A X_B W^{H},$$

produces the symmetrical curve (II) in Figure 2.14b. Line I in this figure is for model I, i.e., $H^{\text{E}} = 0$.

Provided the above values for W hold to 1200 °C, the appropriate equations for activity coefficients (models III and II) may be used to construct activity (a_{Fe}) versus mole fraction (X_{Fe}) curves for fayalite. The results are shown in Figure 2.16, where they may be compared with experimental results obtained by Nafziger and Muan (1967).

Garnet

The common garnet of metamorphic rocks can be viewed as a mixture of four components: almandine, pyrope, grossular, and spessartine. The formula is

$$(\text{Fe, Mg, Ca, Mn})\text{Al}_{0.66}\text{SiO}_4$$

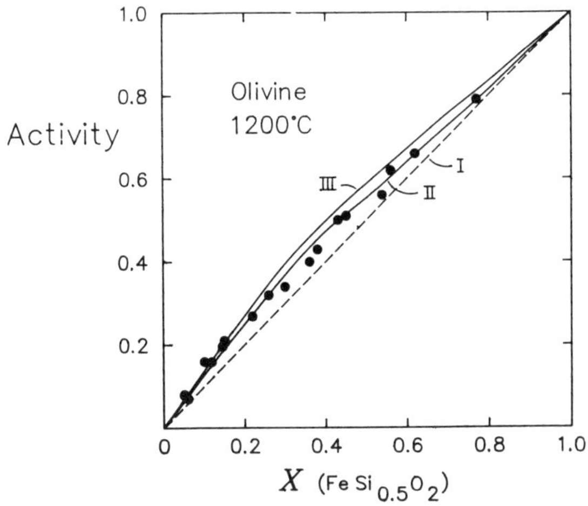

Figure 2.16 Three models for the activity (a) of fayalite ($FeSi_{0.5}O_2$) in Fe–Mg olivine, compared with experimental data points of Nafziger and Muan (Figure 2.8). The models are: I ideal mixing, II the symmetrical Margules model, with W values proposed by Chatterjee (1987), and III the asymmetrical Margules model with W values obtained from heat of mixing measurements of Wood and Kleppa (Figure 2.14)

but some of the data presented below are in terms of $M_3Al_2Si_3O_{12} = 1$ mole, where M stands for the divalent cations. Following is a summary of experimental data, principally on the three binary solutions, Fe–Mg, Fe–Ca, and Mg–Ca garnet.

Newton, Charlu and Kleppa (1977) measured the heat of solution of synthetic Mg–Ca crystals in a $2PbO{\cdot}B_2O_3$ melt at 700 °C, and Geiger, Newton and Kleppa (1987) measured the heat of solution of synthetic Fe–Mg and Fe–Ca crystals in a $(Li,Na)_2B_2O_4$ melt at 760 °C, and the results are shown in Figure 2.17. There is obviously a positive excess heat of mixing in Mg–Ca garnet over the entire range of composition, and possibly in Fe–Mg garnet over part of the composition range, but for Fe–Ca garnet the heat of mixing is very close to zero.

Molar volume data on the three binaries are shown in Figure 2.18. All refer to synthetic crystals for which volumes were calculated from cell-edge measurements. The Mg–Ca results are from different sources, compiled by Wood (1988), and those for Fe–Mg and Fe–Ca are from Geiger, Newton and Kleppa (1987). There appears to be a very small excess positive volume of mixing in Mg–Ca garnet, recently confirmed by Ganguly, Cheng and O'Neil (1993), and Fe–Ca garnet, and possibly also in Mg–Fe garnet (Figure 2.18) but the values are extremely small. A negative volume anomaly in Fe–Ca garnet near the Fe termination was reported by Cressey, Schmid, and Wood (1978), but has not been confirmed by more recent studies. A noticeable positive volume anomaly appears in Mg–Ca garnet, near the Ca end; this is not shown in Figure 2.18.

The displaced-equilibrium method, similar to that used by Nafziger and Muan (1967) was used by Cressey, Schmid, and Wood (1978), Wood (1988), and Koziol

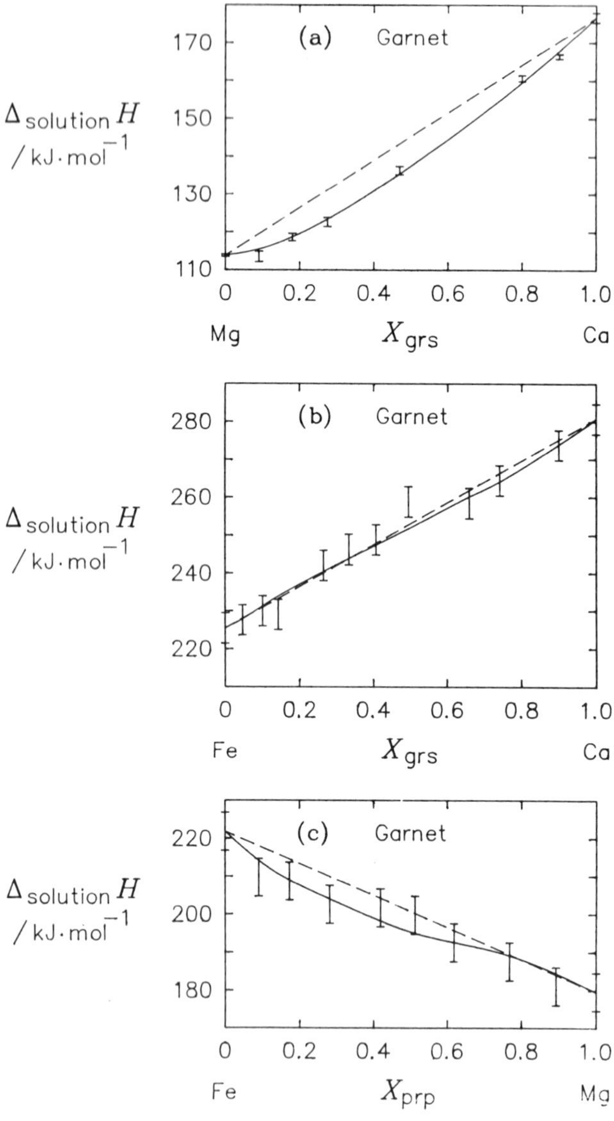

Figure 2.17 Molar heat of solution (1 mole = $M_3Al_2Si_3O_{12}$) of (a) synthetic Mg–Ca garnet at 700 °C (Newton, Charlu and Kleppa 1977), (b) Fe–Ca garnet at 760 °C, and (c) Fe–Mg garnet at 760 °C; (b) and (c) are from Geiger, Newton and Kleppa (1987). Error bars are approximate 0.95 confidence limits. Broken lines refer to zero heat of mixing. Reprinted from *Geochimica et Cosmochimica Acta,* with permission from Pergamon Press, Ltd, Oxford

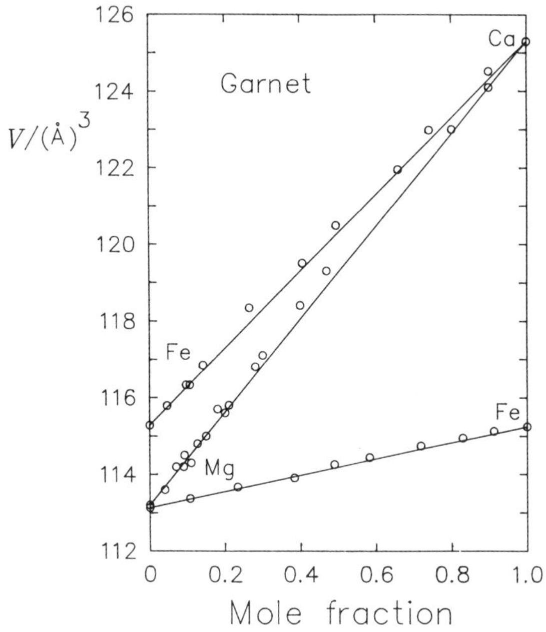

Figure 2.18 Cell volume (V) of Mg–Fe and Fe–Ca garnet, from Geiger, Newton and Kleppa (1987), and of Mg–Ca garnet, from Wood (1988) (compilation). Straight lines represent zero volume of mixing

and Newton (1989) to obtain measurements of the activity of grossular in garnet solid solutions. The mineral reaction that was selected to obtain this information is

$CaAl_2Si_2O_8$	=	$CaAl_{0.67}SiO_4$	+	2/3 Al_2SiO_5	+	1/3 SiO_2
anorthite		grossular in garnet solid solution		Al silicate (Als) kyanite or sillimanite		quartz

At a set temperature (T′), within the range 850–1100 °C, and set pressure (P′), within the range 10–23 kbar, the reaction is displaced to the right or left until equilibrium is obtained (Figure 2.19). Then,

$$G^{\text{An}}_{T'P'} = \mu_{\text{grs}(T'P')} + (2/3)G^{\text{Als}}_{T'P'} + (1/3)G^{\text{Qtz}}_{T'P'},$$

$$-RT' \ln(a_{\text{grs}})_{T'P'} = (\Delta G^{\circ})_{T'P'},$$

where $(\Delta G^{\circ})_{T'P'}$ is the Gibbs energy change for the reaction with grossular in its pure form, i.e., as pure grossular crystals, at the specified temperature and pressure. The standard state pressure (P°) is now taken as the pressure at which $\Delta G^{\circ}_{T'} = 0$,

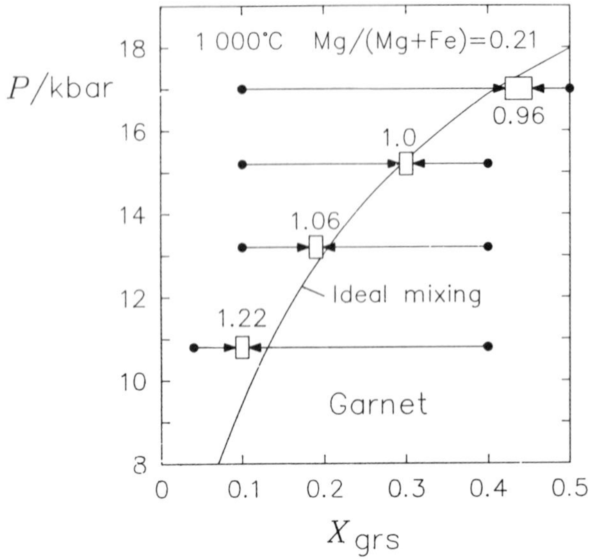

Figure 2.19 Results of displaced-equilibrium experiments which produce a change in the grossular content of garnet (arrows) in the reaction, anorthite = grossular (in Ca–Mg–Fe garnet) + Al silicate + quartz, as a function of pressure (P). Numbers beside data points are activity coefficients for grossular in Ca–Mg–Fe garnet in which Mg/(Mg+Fe) = 0.21. From Koziol and Newton (1989). Reproduced by permission of Springer-Verlag

i.e., at the point on the boundary curve where $T = T'$. Then,

$$-RT\ln(a_{\text{grs}})_{T'P'} = (\Delta G^{\circ})_{T'P^{\circ}} + \int_{P^{\circ}}^{P} (\Delta V^{\circ})_{T'P'}\,\mathrm{d}P,$$

$$-RT\ln(a_{\text{grs}})_{T'P'} = \int_{P^{\circ}}^{P} (\Delta V^{\circ})_{T'P'}\,\mathrm{d}P.$$

The integral is evaluated by taking into consideration the expansion and compressibility of all of the crystals involved in the end-member reaction.

By use of this procedure, Koziol (1990) obtained information on the activity of Ca garnet in Ca–Fe garnet, and the results are shown in Figure 2.20a. For X_{grs} greater than about 0.3, a_{grs} appears to depart, in a positive sense, from the ideal $a_{\text{grs}} = X_{\text{grs}}$ relation. This is in agreement with earlier results obtained by Cressey, Schmid and Wood (1978). The same method applied to Ca–Mn garnet produced the relation $a_{\text{grs}} = X_{\text{grs}}$ (Figure 2.20b), which confirms near-ideal mixing in Ca–Mn garnet at 1000 °C.

The displaced-equilibrium method was also used by Koziol and Bohlen (1992) to measure the activity of almandine in Fe–Mg garnet. The equilibrium that was studied is

$$\text{almandine (in Fe–Mg garnet)} + \text{rutile} = \text{ilmenite} + \text{sillimanite} + \text{quartz}.$$

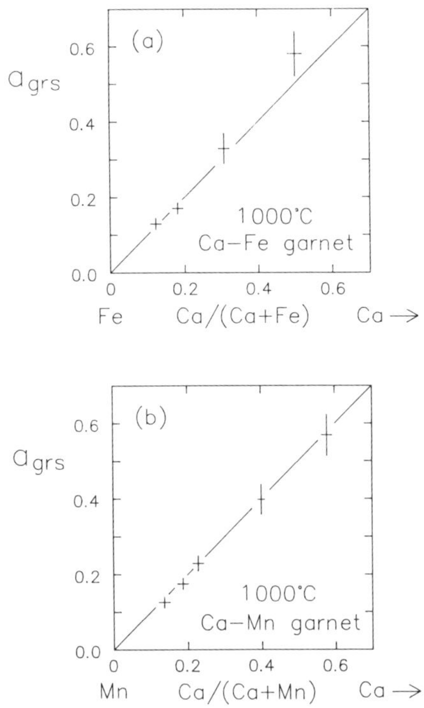

Figure 2.20 (a) Activity of grossular in Ca-Fe garnet, and (b) in Ca-Mn garnet, determined experimentally by Koziol (1990). Straight lines of unit slope refer to ideal mixing

It is clear from the results that at 900 to 1000 °C and 8 to 14 kbar, and in the concentration range X_{alm} = 0.89 to 0.61, Fe-Mg garnet is 'essentially' an ideal solution.

The heat-of-mixing data have been interpreted as follows. The excess heat of mixing in Mg-Ca garnet (Figure 2.17) is obviously asymmetrical (as in olivine), and Newton, Charlu and Kleppa (1977) used these data to obtain W_{Ca}^{H} and W_{Mg}^{H} parameters (model III) for the solution series. Haselton and Newton (1980) and Wood (1988) evaluated W^S and W^V for Mg-Ca garnet mixtures and have found these to be small relative to W^H.

With regard to Fe-Ca garnet and Fe-Mg garnet, the excess enthalpy of mixing (H^E) at 760 °C is comparatively small (Figure 2.17), and one could argue that the ideal model (I) is satisfactory, especially for Fe-Ca garnet. Geiger, Newton and Kleppa (1987) have, however, adopted the asymmetrical model for these solutions, and have derived W^H parameters which produce the best-fit curves shown in Figure 2.17. The asymmetrical model was also used by Cressey, Schmid and Wood (1978) and by Koziol and Newton (1989) to describe their experimental data on the activity of grossular in these two garnet series.

In summary, near-ideal mixing occurs at ~1000 °C in Ca–Fe, Mg–Fe, and Ca–Mn garnet, while Ca–Mg garnet is distinctly non-ideal. Berman (1990) and Berman and Koziol (1991) have presented a detailed discussion of present knowledge regarding mixing in aluminous garnet, including values for W^H, W^S, and W^V parameters. As more data become available, it seems likely that the asymmetrical model (III), expanded to accommodate three and four components, will be most useful, although for some applications to geological problems, the ideal solution model with $a = X$ may be adequate.

Plagioclase

Plagioclase is a complex solid solution of components $NaAlSi_3O_8$ (which at metamorphic temperatures displays varying degrees of order in the arrangement of Al and Si atoms) and $CaAl_2Si_2O_8$ (which is ordered). At about 1000 °C the solid-solution series is evidently continuous; but at temperatures corresponding to low-grade metamorphism, discontinuities were demonstrated by Maruyama, Liou and Suzuki (1982) and by others. In general, ideal-solution behavior at metamorphic temperatures is not expected.

The heat of solution of several plagioclase samples of different composition and from various localities was measured by Kracek and Neuvonen (1952). The samples were dissolved in HF at 75 °C and the heat of solution was found to bear a linear relation to composition, but only in the range of $X_{an} = 0$ to 0.65 (implying very little heat of mixing in this range); beyond $X_{an} = 0.65$ the heat of solution (i.e., the heat released) was less than expected by up to 6 kcal/mol, implying positive heat of mixing in this range. Newton, Charlu and Kleppa (1980) used high-temperature synthetic plagioclase crystals (high structural state) to measure the heat of solution in $Pb_2B_2O_5$ melt at 700 °C, and the results are shown in Figure 2.21. There is obviously, for these crystals, a positive heat of mixing over the entire composition range, and with a maximum of about 5000 J/mol.

The cell volume of plagioclase varies little across the series, from albite to anorthite. Kroll and Müller (1980) found (in high-temperature plagioclase) an increase from 666.8 to 670.5 Å^3, and in view of measurement uncertainty, it is not possible to determine the nature of the $V - X_{an}$ relation with confidence, but the volume of mixing does appear to be negative at the albite end. Kroll and Müller (1980) proposed that the series be divided into three segments: (1) analbite–high albite, (2) a transitional region, and (3) anorthite structure; mixing, even at high temperatures, is evidently complex.

Activity–concentration relations for plagioclase at 700 °C and 2 kbar were obtained by Orville (1972), proceeding as follows. Synthetic crystals prepared at 800 °C and 1 kbar were immersed in a $NaCl–CaCl_2$ solution at the specified temperature and in the presence of quartz. The reaction

$$2NaAlSi_3O_8 + CaCl_2 = CaAl_2Si_2O_8 + 2NaCl + 4SiO_2$$

was observed to proceed to the right or to the left, until equilibrium was obtained.

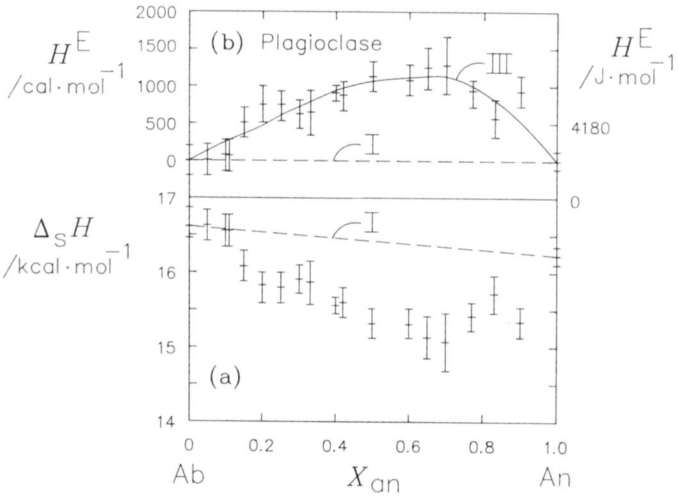

Figure 2.21 (a) Heat of solution of synthetic high-structural-state plagioclase crystals at 700 °C. Data from Newton, Charlu and Kleppa (1980), with error bars modified to 0.95 confidence limits, taking number of measurements (1 to 10 for each composition) into consideration. (b) Excess heat of mixing (H^E), data from (a). Curve III located by the above authors, using the Margules asymmetrical solution model. Broken lines I in (a) and (b) are for zero heat of mixing. Reproduced by permission of Pergamon Press Ltd

At this point, the equation

$$2\mu_{ab} + \mu_{CaCl_2} = \mu_{an} + 2\mu_{NaCl} + 4G^{Qtz}$$

was balanced. Substituting equations of the kind

$$\mu_{ab} = G^{Ab} + RT \ln a_{ab},$$

and rearranging terms, gives

$$\frac{a_{an}\, a_{NaCl}^2}{a_{ab}^2\, a_{CaCl_2}} = \exp\left(-\frac{\Delta G^\circ}{RT}\right).$$

Here ΔG° is the Gibbs energy change for the reaction, with all components in a standard state, and is a function only of temperature and pressure. Thus the entire term on the right, which is referred to as the 'equilibrium constant' (K), is a function of T and P, but not of the composition of plagioclase crystals or of chloride solution.

By replacing $a_{an}, a_{ab}, a_{NaCl}, a_{CaCl_2}$ by $\gamma_{an}X_{an}, \gamma_{ab}X_{ab}, \gamma_{NaCl}M_{NaCl}, \gamma_{CaCl_2}M_{CaCl_2}$, where γ refers to activity coefficient, X to mole fraction, and M to molar concentration, we obtain

$$\left(\frac{X_{\mathrm{an}}\,M^2_{\mathrm{NaCl}}}{X^2_{\mathrm{ab}}\,M_{\mathrm{CaCl_2}}}\right)\left(\frac{\gamma_{\mathrm{an}}}{\gamma^2_{\mathrm{ab}}}\right)\left(\frac{\gamma^2_{\mathrm{NaCl}}}{\gamma_{\mathrm{CaCl_2}}}\right) = K\,.$$

Under the given experimental set-up, the last term in brackets was found to equal one. The quotient in the first brackets on the left is referred to as the 'distribution coefficient' (K_D). Thus,

$$K_{\mathrm{D}}\left(\frac{\gamma_{\mathrm{an}}}{\gamma^2_{\mathrm{ab}}}\right) = K\,,$$

or,

$$\log K_{\mathrm{D}} = \log \gamma^2_{\mathrm{ab}} - \log \gamma_{\mathrm{an}} + \log K\,.$$

In differential form,

$$\mathrm{d}(\log K_{\mathrm{D}}) = 2\,\mathrm{d}(\log \gamma_{\mathrm{ab}}) - \mathrm{d}(\log \gamma_{\mathrm{an}})\,.$$

The Gibbs–Duhem relation must also be satisfied:

$$X_{\mathrm{an}}\,\mathrm{d}(\log \gamma_{\mathrm{an}}) + X_{\mathrm{ab}}\,\mathrm{d}(\log \gamma_{\mathrm{ab}}) = 0.$$

For each experimental run, $X_{\mathrm{an}}, X_{\mathrm{ab}}, K_{\mathrm{D}}$ were measured, and with two equations and two unknowns ($\gamma_{\mathrm{an}}, \gamma_{\mathrm{ab}}$), the activity coefficients could be calculated. It was thus possible to derive $a_{\mathrm{ab}} - X_{\mathrm{ab}}$ and $a_{\mathrm{an}} - X_{\mathrm{an}}$ relations (700 °C, 2 kbar) and the results are shown in Figure 2.22.

These data combined with results on the heat of mixing (Figure 2.21a) show that under conditions of high-grade metamorphism, an appreciable departure from ideal mixing occurs in plagioclase.

The heat-of-solution data at 700 °C (Figure 2.21a) imply an asymmetrical excess heat of mixing. Newton, Charlu and Kleppa (1980) obtained the following asymmetrical equation to fit their data:

$$H^{\mathrm{E}} = 2025X_{\mathrm{an}}X^2_{\mathrm{ab}} + 6746X_{\mathrm{ab}}X^2_{\mathrm{an}}$$

(i.e., W^H_{an} = 2025 cal/mol and W^H_{ab} = 6746 cal/mol), and the equation is plotted in Figure 2.21b.

Complex entropy changes are expected to occur in plagioclase solutions, and the W^H parameters deduced above cannot be equated to W^G, i.e., activity coefficients cannot be derived as they were for olivine.

A theoretical expression for activity as a function of concentration may, however, be obtained as follows. Kerrick and Darken (1975) used a statistical thermodynamic approach to derive equations for the 'activity' of albite and anorthite in plagioclase, taking into consideration only the arrangement of atoms (i.e., the configurational entropy), these arrangements being governed in part by the Al-avoidance

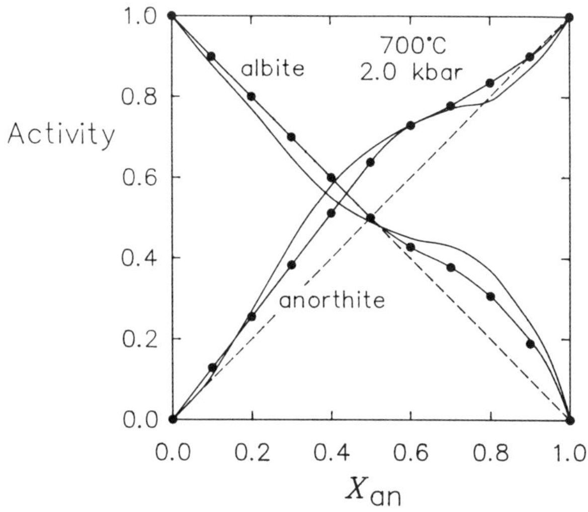

Figure 2.22 Activity-concentration (X_{an}) relations in plagioclase at 700 °C. Data points are experimental results of Orville (1972), and adjacent curved lines are plots of the Newton-Charlu-Kleppa (1980) equation. Broken lines are for ideal mixing

principle, i.e., an avoidance of Al-O-Al bonds in the structure. The equations are

$$a'_{ab} = (X_{ab})^2(2 - X_{ab}),$$

$$a'_{an} = (1/4)(X_{an})(1 + X_{an})^2.$$

These may be viewed as representing conditions ideal with respect to entropy, while disregarding enthalpy. Newton, Charlu and Kleppa (1980), by use of their W^H estimates referred to above, calculated the 700 °C enthalpic activity coefficients, employing the appropriate equation from model III,

$$RT \ln \gamma_{ab} = 6746X_{an}^2 + 2X_{an}^2X_{ab}(2025 - 6746),$$

$$RT \ln \gamma_{an} = 2025X_{ab}^2 - 2X_{ab}^2X_{an}(2025 - 6746).$$

These were then combined with the Kerrick-Darken equations (above), as follows,

$$a_{ab} = a'_{ab}\gamma_{ab},$$

$$a_{an} = a'_{an}\gamma_{an},$$

to give

$$a_{ab} = (X_{ab})^2(2 - X_{ab}) \exp\left[\frac{6746X_{an}^2 + 2X_{an}^2X_{ab}(2025 - 6746)}{RT}\right],$$

$$a_{an} = \left(\frac{1}{4}\right) X_{an}(1 + X_{an})^2 \exp\left[\frac{2025X_{ab}^2 - 2X_{ab}^2X_{an}(2025 - 6746)}{RT}\right],$$

which simplify (Newton and Haselton 1981) to

$$a_{ab} = (X_{ab})^2(2 - X_{ab}) \exp\left[\frac{(1 - X_{ab})^2(6746 - 9440X_{ab})}{RT}\right],$$

$$a_{an} = \left(\frac{1}{4}\right) X_{an}(1 + X_{an})^2 \exp\left[\frac{(1 - X_{an})^2(2025 + 9440X_{an})}{RT}\right].$$

(Note that with W^H in units of cal/mol, R = 1.987 cal/K·mol.)

These equations, which are plotted in Figure 2.22, are in good agreement with the results of Orville (1972).

The above is a brief introduction to the thermodynamic properties of minerals of variable composition. For an extended treatment of this subject, and for data on other minerals, the reader is referred to Thompson (1967), Saxena (1973), Grover (1977), Navrotsky (1987), and Ganguly and Saxena (1987).

3 Phase Equilibrium

The most striking mineral reactions that occur within the solid Earth are phase transformations which produce crystals of new minerals such as garnet, staurolite, and cordierite. There are, however, several different kinds of more subtle reactions that occur during a rise or fall of temperature, for example order–disorder transformations, which are changes in the arrangement of atoms in crystals, and exchange reactions, resulting in changes in the chemical composition of crystals. These different kinds of reactions are here defined, and for each class, one or more representative examples are examined and discussed. Information is drawn from observations of natural occurrences and from experimental results, and where possible, the thermodynamic relations presented in Chapter 2 will be called on to facilitate our analysis and enhance our understanding of the transformations.

The term 'reaction' is used above, but the focus at present is, in fact, on equilibrium relations. When a system arrives at a state of complete equilibrium, either within the Earth or in a laboratory, it presumably does so, in general, by way of a chemical reaction. Indeed, an equilibrium may be viewed as resulting from reactions on the microscopic scale, in both directions, at equal rates. Reaction mechanisms will be considered in Chapters 4 and 5.

3.1 TRANSFORMATIONS INVOLVING MINERALS OF NOMINALLY FIXED COMPOSITION

Conceptually and thermodynamically, single-component phase transformations, as represented by the transformation of graphite to diamond, calcite to aragonite, and andalusite to sillimanite, are relatively simple. They result in a conspicuous change in crystal structure and are sometimes referred to as 'first-order transformations'; a familiar analogue is the freezing of water to produce crystals of ice. Complexity arises when minor amounts of another component are present, for example, minor Fe in andalusite and sillimanite. Experimentally, phase transformations can, however, present difficulties because of small changes in entropy, resulting in small changes in Gibbs energy at temperatures rather far from the equilibrium temperature. When the boundary curves are located with confidence, they can be helpful in setting limits to temperature and pressure conditions within the Earth.

3.1.1 The aragonite–calcite equilibrium

$$CaCO_3 \text{ (aragonite)} = CaCO_3 \text{ (calcite)}$$

Calcite and aragonite occur abundantly in sediments and sedimentary rock, and until recently, calcite was assumed to be the only form of $CaCO_3$ to occur at depth. In 1962, Coleman and Lee discovered aragonite in blueschists of the Mesozoic Franciscan formation in California, where it occurs as beds, lenses, and veins associated with metamorphosed pillow basalt, chert, and shale, that now consist of glaucophane, lawsonite, muscovite, pumpellyite, clinozoisite, garnet, stilpnomelane, and other minerals. Subsequently, aragonite was discovered at other blueschist localities in California and Oregon (Coleman and Lee 1962), in Italy (Hoffmann 1970), in Turkey (Cogulu 1967), and elsewhere, and it also occurs in rocks of the prehnite–pumpellyite facies in Washington (Vance 1968). It has not been detected in rocks of the greenschist and amphibolite facies. Thus the position of the aragonite–calcite boundary curve is of interest with regard to information it may present on the pressure of metamorphism and on pressure–temperature gradients within the Earth.

Calcite I is rhombohedral R3c (Reeder 1983), with a molar volume of 36.93 cm^3, while aragonite is orthorhombic Pmcn (Villiers 1971), with a molar volume of 34.15 cm^3; consequently an abrupt first-order transformation is expected.

From the data of Table 2.1 (Sec. 2.1) one may calculate that the Gibbs energy change for the aragonite $\rightarrow$ calcite transformation at 25 °C and 1 bar is -1050 J/mol, which implies that calcite is stable at Earth-surface conditions. The entropy change is +3.74 J/mol·K, i.e., calcite is the high-entropy phase, and therefore the high-temperature phase. The volume change is +2.78 cm^3/mol, and therefore the slope of the boundary curve, $dP/dT = \Delta S/\Delta V$, is positive (+13.4 bar/K), and aragonite is the high-pressure phase. With increasing temperature, ΔS increases (Robie, Hemingway and Fisher 1979) and ΔV decreases, and the slope of the boundary is therefore expected to increase with increasing temperature.

Several investigations have been carried out to locate the boundary curve. Jamieson (1953) obtained an estimate for the transition at 4 to 5 kbar, 50 °C based on electrical conductivity measurements in solutions which were in metastable equilibrium with aragonite and calcite. In addition, many high-pressure and high-temperature experiments have been carried out in which the reaction has been reversed, thereby defining brackets to the reaction curve. These results have been compiled and discussed by Boettcher and Wyllie (1968) and by Newton and Fyfe (1976).

The results presented here (Figure 3.1) are those of Johannes and Puhan (1971) who used both a hydrothermal apparatus and a piston cylinder device to study the reaction. The charge consisted of a mixture of pure synthetic calcite and natural aragonite, nearly pure but for 0.12 wt % MgO. Progress in the reaction was monitored by use of X-rays. The boundary curve so located is shown in Figure 3.1,

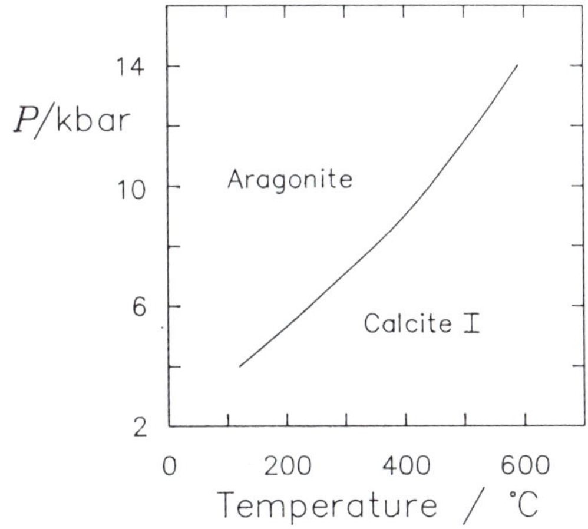

Figure 3.1 Boundary curve for the equilibrium aragonite = calcite, from Johannes and Puhan (1971). Reproduced by permission of Springer-Verlag

where limits of uncertainty extend about 20 °C on either side. This curve does not differ by more than 2 kbar from curves located in other laboratories.

Numerous problems of interpretation have arisen. For example, Vance (1968) found that the aragonite-bearing low-grade rocks from Washington could not have experienced pressures great enough to take them into the experimentally located stability field of aragonite, and he concluded that aragonite crystallized metastably, either within the Earth or within the laboratory. One possible solution to this problem is that the boundary curve is depressed by the presence of Sr in aragonite, but Froese and Winkler (1966) have determined that this is unlikely. Another possibility is that strain in calcite produces an appreciable depression of the curve, and Newton, Goldsmith and Smith (1969) have indeed produced aragonite from calcite by prolonged grinding at low pressure.

Although the above and other uncertainties remain, some broad conclusions may be drawn, assuming that the boundary in Figure 3.1 is correctly located. (1) Aragonite is not stable at Earth-surface conditions, and where it crystallizes from sea water it evidently forms metastably, with a Gibbs energy greater than that of calcite but less than that of $CaCO_3$ in supersaturated sea water. (2) Under conditions of low-pressure and medium-pressure metamorphism, calcite is stable relative to aragonite. (3) At temperatures of 200–300 °C, which are estimates for blueschist facies metamorphism, a pressure of at least 6 kbar is required to stabilize aragonite. (4) At temperatures of 500–600 °C, estimated for upper amphibolite facies metamorphism, the presence of calcite indicates that pressures could not have exceeded 12 kbar, barring the complete inversion of aragonite to calcite subsequent to metamorphism.

es fehlt Hinweis auf Experimente zur Kinetik von Carlson + Rosenfeld (1981)

3.1.2 Al silicate equilibria

$$Al_2SiO_5 \text{ (kyanite)} = Al_2SiO_5 \text{ (andalusite)}$$
$$Al_2SiO_5 \text{ (kyanite)} = Al_2SiO_5 \text{ (sillimanite)}$$
$$Al_2SiO_5 \text{ (andalusite)} = Al_2SiO_5 \text{ (sillimanite)}$$

The three forms of Al_2SiO_5 — kyanite, andalusite, and sillimanite — are widely distributed, especially in metamorphosed shale, but they also occur locally in granite, pegmatite, and quartz veins, and kyanite is found in some eclogites and whiteschists. The Al atoms in the three structures occupy different positions, as follows,

kyanite	$Al^{vi}Al^{vi}Si^{iv}O_5$
andalusite	$Al^{vi}Al^{v}Si^{iv}O_5$
sillimanite	$Al^{vi}Al^{iv}Si^{iv}O_5$

where iv, v, and vi are the coordination numbers (Kerrick 1990). Only in kyanite are all Al atoms surrounded by six oxygen ions, resulting in a density which is relatively high (3.60 g/cm^3 compared to 3.14 g/cm^3 for andalusite and 3.25 g/cm^3 for sillimanite). Thus the equilibria 'andalusite = kyanite' and 'sillimanite = kyanite' must be displaced to the right with increasing pressure. In sillimanite, half the Al atoms occupy four-coordinated sites, together with Si, and some disorder in the arrangement of these atoms might be expected at high temperatures.

In view of the above differences in crystal structure, we expect the three minerals to be related by first-order phase transformations, which in a phase diagram would appear as three boundary curves meeting at an invariant point. Before any experimental data were available, the nature of the phase diagram was correctly deduced by Miyashiro (1953) based only on the physical properties of the minerals and their occurrence. Knowledge of the exact location of the boundary curves would obviously be helpful in assigning limits to peak temperatures and pressures of terranes in which the minerals occur, and in defining possible temperature–pressure paths for these terranes.

The experimental approach used by Holdaway (1971) to locate the boundaries is here described briefly. Holdaway used a cold-seal pressure vessel to investigate two-phase equilibria at pressures below 5 kbar and at temperatures in the range of 350–900 °C. In one set of experiments, designed to locate a point on the kyanite–andalusite boundary, a crystal of natural andalusite of known weight was placed in a silver capsule, together with powdered kyanite and some water. The capsule was sealed and subjected to a known pressure and temperature for several days. The andalusite crystal was then removed and re-weighed to determine the loss or gain of weight. Repeated runs at intervals of 10 °C made possible the detection of a 'zero point' temperature at which there was no loss or gain, representing a point on or close to the kyanite–andalusite boundary. Thus the equilibrium point was approached from lower and higher temperatures. In other experiments, both minerals were present as small crystals, and the reaction direction could be determined by examining the products with a petrographic microscope. The numerous experimental results that have been reported from different laboratories (not all are in agreement) are reviewed by Kerrick (1990) and by Holdaway and Mukhopadhyay (1993).

The phase diagram that is here presented (Figure 3.2) is from Hemingway et al. (1991). It is based on thermodynamic calculations, making use of volume data of Skinner, Clark and Appleman (1961), Winter and Ghose (1979), and Hemingway et al. (1991), heat-of-reaction data of Topor et al. (1989), and new heat capacity data of Hemingway et al. (1991); it is in good agreement with the experimental results of Weill (1966), Newton (1966), Holdaway (1971), and Bohlen et al. (1989). The triple point lies at 3.9 ± 0.3 kbar and $511 \pm 20\,^{\circ}C$. A similar synthesis by Holdaway and Mukhopadhyay (1993) produced a nearly identical phase diagram, with a triple point at 3.75 ± 0.25 kbar and $504 \pm 20\,^{\circ}C$.

The occurrence of the Al silicate minerals will now be examined briefly in relation to Figure 3.2. It is not uncommon to find two of the three minerals together, and occasionally all three occur in contact. For example, Hietanen (1956) reported from the Precambrian Belt series in Idaho, the coexistence of all three minerals, and she proposed that the complex sequence of regional and contact metamorphic events in this part of the Cordillera occurred not far from the invariant point. The more common association of andalusite and sillimanite has been reported from numerous localities: for example, the Steinbach aureole in Bavaria (Okrusch 1969), the aureoles of the Donegal granite in the Irish Republic (Pitcher and Read 1963, Naggar and Atherton 1970), the Maritime Alps (Compagnoni, Lombardo and Prato 1974), the Ryoke Belt of Japan (Yokoi 1983), the Truchas Range in New Mexico (Grambling and Williams 1985), the Kiglapait aureole in Labrador (Kerrick and Speer 1988) and the Ballachulish aureole in Scotland (Pattison 1992).

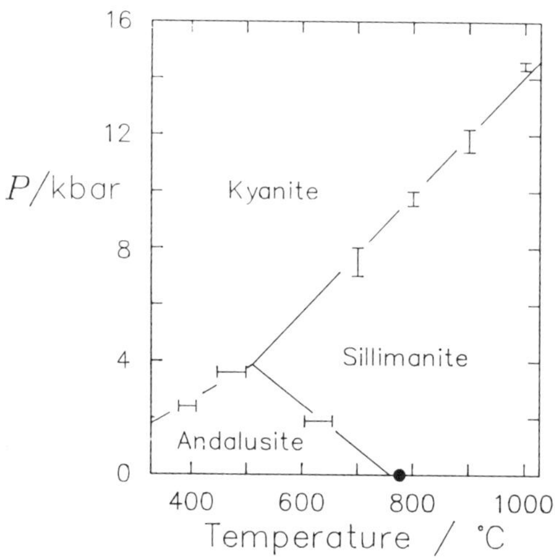

Figure 3.2 The aluminum silicate phase diagram, presented by Hemingway et al. (1991). Data points are from Weill (1966, circle), Holdaway (1971, horizontal bars), and Bohlen et al. (1989, vertical bars)

In view of Figure 3.2 one may propose that samples from these localities experienced metamorphic pressures less than 4 kbar and temperatures ~530–630 °C, i.e., peak metamorphic conditions represented by points lying not far to the right of the andalusite–sillimanite boundary curve.

With regard to the andalusite–sillimanite isograd, at some places the change is abrupt, with andalusite confined to one side and sillimanite to the other, while elsewhere it forms a zone in which both minerals occur together. An example of the second kind, from a low-pressure terrane in the Canadian Shield, is illustrated in Figure 3.3. Additional examples are described by Kerrick (1990).

Similar observations relate to kyanite and sillimanite, but these two minerals do not occur together as frequently. This may result from the larger entropy change for the kyanite = sillimanite reaction, compared with the andalusite = sillimanite reaction (Brown and Fyfe 1971), the values being 11.0 J/mol·K (at 8.5 kbar and 1023 K) for the former and 2.8 J/mol·K (at 1 bar and 1048 K) for the latter (Robie and Hemingway 1984). The driving force ($\partial \Delta G/\partial T = -\Delta S$) for the production of sillimanite from kyanite is therefore greater than for the production of sillimanite from andalusite, i.e., ΔG at some temperature (say 10 °C) past the equilibrium

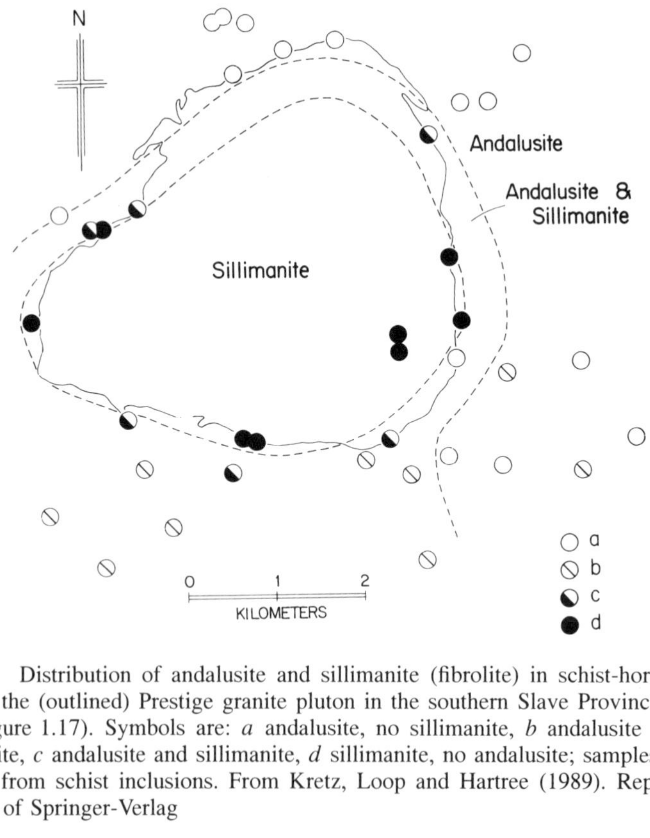

Figure 3.3 Distribution of andalusite and sillimanite (fibrolite) in schist-hornfels in the vicinity of the (outlined) Prestige granite pluton in the southern Slave Province, Canadian Shield (Figure 1.17). Symbols are: *a* andalusite, no sillimanite, *b* andalusite and a trace of sillimanite, *c* andalusite and sillimanite, *d* sillimanite, no andalusite; samples within the pluton are from schist inclusions. From Kretz, Loop and Hartree (1989). Reproduced by permission of Springer-Verlag

temperature is greater (see Figure 2.5). In addition, the activation energy for the kyanite-to-sillimanite reaction may be less, which would favor a more rapid rate of reaction for a given ΔG.

We have, up to now, viewed the Al silicate minerals as being essentially pure, but they do in fact contain other elements, which may be important in relation to phase equilibrium. In an early spectrographic study, Pearson and Shaw (1960) found kyanite, andalusite, and sillimanite to contain < 200 ppm of B, Ga, Cr, Li, V, and Mn, and < 2000 ppm of Mg and Ti, i.e., very low concentrations of these elements. Later studies, by Albee and Chodos (1969), Chinner, Smith and Knowles (1969), Okrusch and Evans (1970), and Grambling and Williams (1985), showed that apart from some unusual Mn-rich occurrences, concentrations of Cr, V, and Ti obtained by microprobe analysis are generally < 0.2 wt %, but Fe may be present to about 2 wt % Fe_2O_3, and occasionally higher. In places, andalusite was found to contain more Fe than associated sillimanite (by a factor of ~ 1.6; Okrusch and Evans 1970), and sillimanite and kyanite were found to contain equal amounts of Fe (Grambling and Williams 1985). Kyanite in eclogite may contain Cr to 7 wt % Cr_2O_3 (Delor and Leyreloup 1986).

With regard to andalusite and sillimanite in common meta-shale, the abundant data from the Kiglapait aureole, reported by Kerrick and Speer (1988) are perhaps representative. Fe in 35 andalusites was found to range from 0.03 to 2.6 wt % Fe_2O_3, and in 27 sillimanites, from 0.02 to 1.26 wt % Fe_2O_3. Commonly the Fe content of andalusite exceeds that of associated sillimanite by a small factor. For example, in a typical specimen Fe_2O_3 was determined at 1.98 wt % in andalusite and 1.22 wt % in sillimanite (2 wt % translates to four Fe atoms per 100 (Fe+Al) atoms in six-coordinated sites).

There is general agreement that concentrations of Fe and other elements in the Al silicates of common meta-shale are too low to cause a large displacement of the boundary curves, but Grambling and Williams (1985) and Kerrick and Speer (1988) have argued (following Strens 1968), that the effect may be measurable.

Consider an equilibrium between Fe-bearing andalusite and Fe-bearing sillimanite, writing the formulas as $(Fe,Al)^{vi}AlSiO_5$. Thus the two solution components in both minerals are $FeAlSiO_5$ and $AlAlSiO_5$. The equation of equilibrium is

$$(Fe,Al)\ AlSiO_5\ (andalusite) = (Fe,Al)AlSiO_5\ (sillimanite),$$

and at equilibrium,

$$\mu^{And}_{FeAlSiO_5} = \mu^{Sil}_{FeAlSiO_5},$$

$$\mu^{And}_{AlAlSiO_5} = \mu^{Sil}_{AlAlSiO_5},$$

where μ denotes chemical potential. This equilibrium is analogous to the following which was considered in Chapter 2 (Sec. 2.5),

$$(Mn,Fe)\ chlorite + quartz = (Mn,Fe)\ garnet + H_2O,$$

and the two van Laar equations which were derived there to express relations between phase composition and temperature, apply also to the andalusite–sillimanite equilibrium. In particular, at constant pressure,

$$X^{\mathrm{Sil}}_{\mathrm{FeAlSiO_5}} = f\,(a-1)/(a-f),$$

$$X^{\mathrm{And}}_{\mathrm{FeAlSiO_5}} = (a-1)/(a-f),$$

where X is the mole fraction of $FeAlSiO_5$ component in andalusite (And) and sillimanite (Sil), and

$$f = \exp[(-\Delta H^{\mathrm{Fe}}/R)(T^{-1} - T_{\mathrm{Fe}}^{-1})],$$

$$a = \exp[(-\Delta H^{\mathrm{Al}}/R)(T^{-1} - T_{\mathrm{Al}}^{-1})].$$

Here ΔH^{Al} is the enthalpy change for the reaction

$$\mathrm{AlAlSiO_5\ (andalusite) = AlAlSiO_5\ (sillimanite)}$$

at temperature T_{Al}, and ΔH^{Fe} is the enthalpy change for the imaginary reaction

$$\mathrm{FeAlSiO_5\ (andalusite) = FeAlSiO_5\ (sillimanite)}$$

at temperature T_{Fe}. Symbol R is the gas constant.

Numerous possibilities exist for relations between X and T, depending on the unknowns ΔH^{Fe} and T_{Fe} and on activity coefficients (the above equations assume ideal mixing). One cannot be predictive, but one possibility is $T_{\mathrm{Al}} = 900$ K at 2 kbar (Figure 3.2), $\Delta H^{\mathrm{Al}} = 4100$ J/mol (Table 2.1), $\Delta H^{\mathrm{Fe}} = 9500$ J/mol (arbitrary), $T_{\mathrm{Fe}} = 1600$ K (arbitrary) with the last two values chosen to give a ratio $X^{\mathrm{And}}_{\mathrm{Fe}}/X^{\mathrm{Sil}}_{\mathrm{Fe}} = 1.6$, as observed, and the resulting van Laar curves are shown in Figure 3.4. There is, in this case, a small displacement (~30 °C) of the equilibrium temperature (now a small range in temperature) and this would appear to be too small to account for the transitional nature of some andalusite–sillimanite isograds, particularly so where the two minerals contain nearly equal concentrations of Fe (Kerrick and Woodsworth 1989), and other explanations must be considered.

The status of fine-grained sillimanite (fibrolite) has attracted much attention and is not fully understood. Holdaway (1971) found fibrolite to be more reactive than sillimanite, while Kerrick (1987) found it to behave as sillimanite. Some samples of fibrolite examined by Bell and Nord (1974) consist of an intergrowth of Al_2SiO_5 and quartz. The fibrolite problem is reviewed in detail by Kerrick (1990) who noted that with regard to chemical composition, crystal structure, and thermodynamic properties, fibrolite and sillimanite are essentially indistinguishable. This conclusion is strengthened by the recently obtained heat-capacity data of Hemingway et al. (1991). Thus there is at present no compelling evidence for fibrolite differing from sillimanite except in crystal size. In this regard one must remember that sillimanite crystals at the time of nucleation were also small in size, and that large variations in crystal size can result from differences in the rate of nucleation.

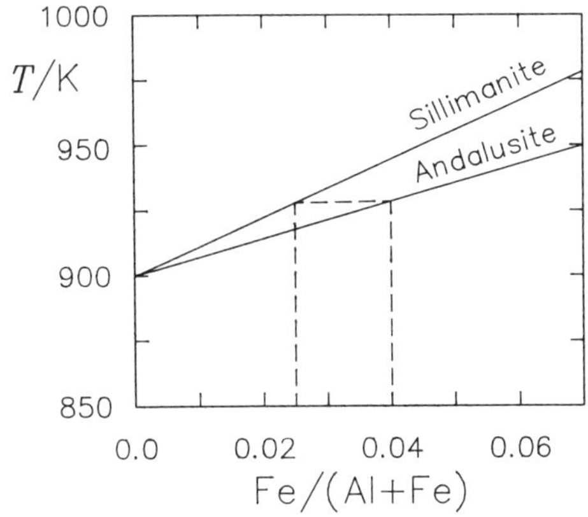

Figure 3.4 Possible influence of Fe on the andalusite = sillimanite transformation at 2.0 kbar. Parameters used to construct the van Laar lines are given in the text. Broken lines represent coexisting andalusite and sillimanite at ~925K

The question of $(AlSi)^{iv}$ disorder in sillimanite was discussed by Holdaway (1971), Greenwood (1972), and Saxena (1974). In a calorimetric study of sillimanite, Navrotsky, Newton and Kleppa (1973) detected a thermal anomaly at ~1300 °C which they attributed to a disordering event, and some disorder may possibly be present at lower temperatures. It is not yet known, however, if some of the variation that is found in the behavior of sillimanite (and fibrolite) under natural and laboratory conditions may be attributed to non-equilibrium with regard to the arrangement of four-coordinated Al and Si atoms in the crystals. This topic is reviewed in detail by Kerrick (1990) who cites recent spectroscopic results and other considerations which lead to the conclusion that most natural sillimanite and fibrolite are virtually ordered.

In summary, early uncertainties in the Al_2SiO_5 phase diagram appear now to be resolved, and the equilibrium relations as presented in Figure 3.2 can be accepted with confidence. Numerous questions remain, however, with regard to the interpretation of natural occurrences, and the answers to these questions are likely to be found in small-magnitude departures from equilibrium, and in the kinetics of natural crystallization processes.

3.1.3 Equilibria involving jadeite

$$\begin{array}{ccccc} NaAlSi_2O_6 & + & SiO_2 & = & NaAlSi_3O_8 \\ \text{jadeite} & + & \text{quartz} & = & \text{albite} \end{array}$$

$$\begin{array}{ccccc} 2NaAlSi_2O_6 & = & NaAlSi_3O_8 & + & NaAlSiO_4 \\ \text{jadeite} & = & \text{albite} & + & \text{nepheline} \end{array}$$

The above two equations provide examples of equilibria involving three minerals of fixed composition. Although pure jadeite is not common, crystals composed dominantly of this component are widespread in some blueschist terranes, and the stability of jadeite relative to the common mineral albite is therefore of some importance.

The history of the study of jadeite is interesting. Beginning at about 1930, attempts were made by several workers at the Geophysical Laboratory of the Carnegie Institution of Washington, including P. Eskola, to synthesize this mineral, but always without success (Yoder 1950). The first synthesis was achieved by L. Coes in 1948. The location of the jadeite = albite + nepheline boundary curve was then determined by Adams (1953) by use of enthalpy and volume data, and the assumption $\Delta C_P = 0$. This was one of the first thermochemical determinations of a mineral boundary curve. With improvements in high-pressure apparatus, boundary curves for both of the above equilibria were determined directly; those shown in Figure 3.5 were located by Robertson, Birch and MacDonald (1957), Birch and LeComte (1960), Newton and Smith (1967), and Holland (1980). In an inter-laboratory study involving workers in six laboratories (Johannes et al. 1971)

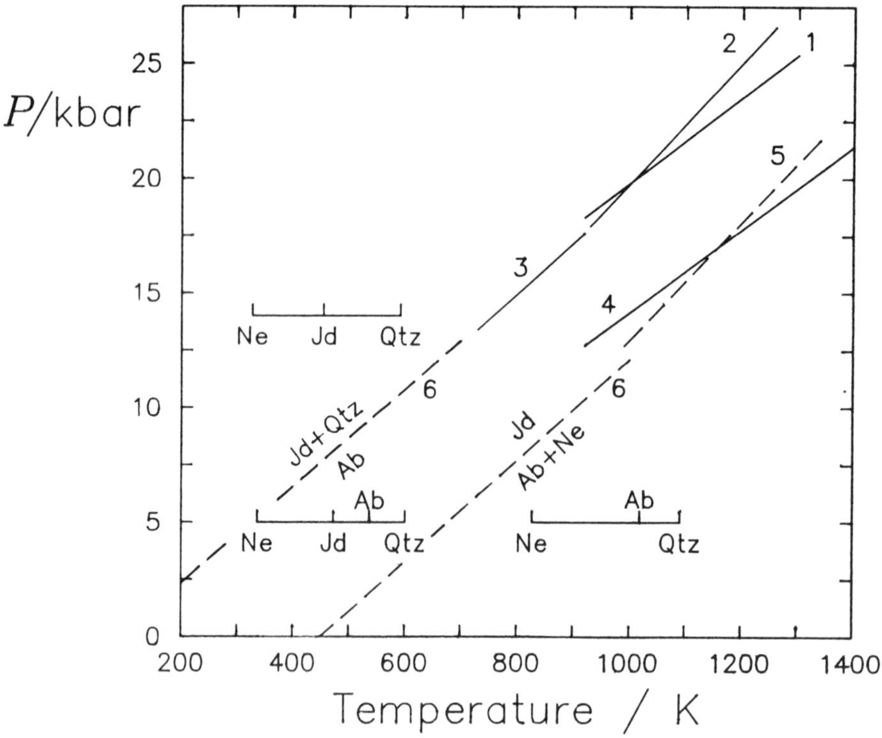

Figure 3.5 Boundary curves for albite + nepheline = jadeite, and albite = jadeite + quartz. Sources are: 1 Birch and LeComte (1960), 2 Holland (1980), 3 Newton and Smith (1967), 4 Robertson, Birch and MacDonald (1957), 5 and 6 Hlabse and Kleppa (1968), high albite and low albite respectively

the jadeite + quartz = albite equilibrium at 600 °C was located by consensus, near 16.3 kbar, and this point, assuredly, is reliable.

Early extrapolations of the high-temperature boundaries to lower temperatures created some problems. For example, the jadeite + quartz = albite curve of Birch and LeComte, when extrapolated to 300 °C indicated that at least 10 kbar of pressure are needed to stabilize jadeite + quartz in blueschist terranes, and that seemed excessive. These problems were largely solved by Hlabse and Kleppa (1968) who obtained improved enthalpy measurements for the minerals and combined them with entropy and volume data to calculate the two lower-temperature boundary curves shown in Figure 3.5. Hlabse and Kleppa also brought into their calculations the enthalpy change associated with disordering in albite (but not nepheline) near 600 °C, which amounts to about 3 kcal/mol. They suggested that with regard to jadeite = albite + nepheline, some of the disagreement between the experimental and thermochemical results at high temperatures (Figure 3.5) may be attributed to the use of low-temperature (ordered) albite in the experiments.

The occurrence of jadeite in relation to metamorphism has been reviewed by Newton and Fyfe (1976) and Turner (1981). Na pyroxene containing 70–90 mole % jadeite is widespread in the low-grade graywacke of the Franciscan blueschist terrane in California (Ernst et al. 1970) where it commonly occurs with albite and locally with albite and quartz. Figure 3.5 shows that jadeite without quartz is stable at low pressures but jadeite with quartz is stable only at pressures of 7 or 8 kbar, at the estimated blueschist temperature of 150–300 °C (Ernst 1971).

With regard to those crystals of Na pyroxene in the Franciscan which contain 20 to 30 mole % of acmite and diopside components, reactions more complex than those represented by Figure 3.5 must have occurred, and the boundary curves cannot be applied directly. Reactions to produce these more complex crystals have been proposed. Most recently, Patrick and Day (1989) have mapped a jadeite-in followed by a pumpellyite-out isograd, which together with mineral analyses provide evidence for the following continuous reaction:

$$\begin{gathered}\text{pumpellyite} + \text{chlorite} + \text{albite} + \text{hematite} + H_2O \\ = \text{jadeite} + \text{glaucophane} + \text{lawsonite} + \text{quartz}.\end{gathered}$$

Note that, if correctly stated, this would form one of the few reactions to consume H_2O with increasing grade. Further data are needed to define reaction pressures for these and less sodic pyroxenes (omphacites) of blueschists and eclogites.

Additional uncertainties are encountered when attempting to draw inferences regarding temperature–pressure paths for jadeite-bearing blueschists. Complications resulting from metasomatism, metastability, overpressure, and the participation of highly reactive (disordered) albite, have been mentioned in several reports. Ernst (1971) has, however, argued persuasively that the phase relations for jadeite as shown in Figure 3.5 are entirely consistent with his model of blueschist crystallization within subduction zones.

The boundary jadeite = albite + nepheline is of interest in relation to nepheline–albite gneisses which occur, for example, in the southern Grenville Province of the

Canadian Shield (Osborne 1930, Gummer and Burr 1946, Appleyard 1969). If the amphibolite-facies temperature here was about 630 °C, the pressure according to Figure 3.5 could not have exceeded 10 kbar.

3.1.4 High-pressure forms of SiO_2, C, and Mg_2SiO_4

Recent discoveries of *in situ* coesite and diamond in metasedimentary rock by Chopin (1984) and by Sobolev and Shatsky (1990) have focused attention on the SiO_2 and C phase relations, while phase transformations involving Mg_2SiO_4 continue to be of interest in relation to conditions in the Earth's Mantle. The occurrence of polymorphs of SiO_2 in rocks affected by shock metamorphism is also of importance. Phase relations of SiO_2, C, and Mg_2SiO_4 are therefore described briefly in this section, drawing mainly on the review by Liu and Bassett (1986).

SiO_2

The dominant form of SiO_2 in the Earth's crust is α-quartz; β-quartz, tridymite, and cristobalite crystallize in some volcanic rocks and enstatite chondrites and locally in quartzite affected by contact metamorphism. Coesite, which is more dense than quartz by a factor of 1.5, was synthesized by L. Coes in 1953 and was discovered at the Arizona Meteorite Crater by Chao, Shoemaker and Madsen (1960). Stishovite, which is denser than quartz by a factor of 1.6, was synthesized by S.M. Stishov and S.V. Popova in 1961 and this phase was also found at the Arizona Crater (Chao, Fahey and Littler 1962). Stishovite is the only known phase in which Si atoms are six-coordinated. Both high-pressure forms have been found at other crater sites and provide compelling evidence for an impact origin.

The α-quartz–coesite and coesite–stishovite boundary curves have been located experimentally and are shown in Figure 3.6a. The thermodynamics of the transformations were considered by Dorogokupets, Karpov and Lashkevich (1988).

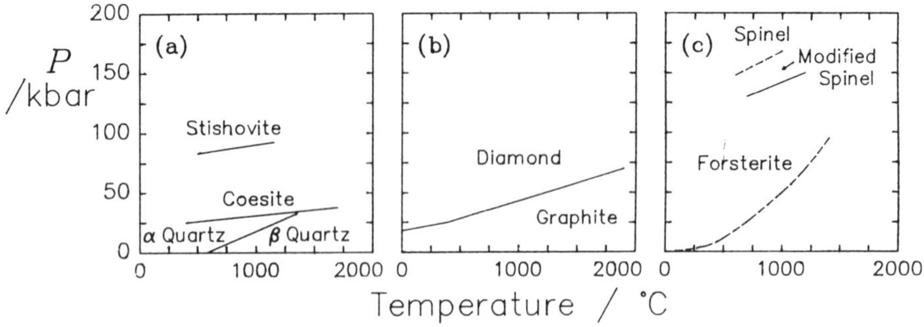

Figure 3.6 High-pressure phase equilibria in the systems (a) SiO_2 (small stability fields of cristobalite and tridymite not shown), (b) carbon, and (c) Mg_2SiO_4; lower broken line is geothermal gradient of Clark and Ringwood (1964). From Liu and Bassett (1986). Reproduced by permission of Oxford University Press, Inc.

Chopin (1984) discovered coesite in the Dora Maira Massif of the western Alps, where it occurs as inclusions in pyrope. Additional information on this important discovery is presented by Goffé and Chopin (1986) who state:

> 'The regional occurrence of coesite and other uncommon or new minerals in part of the Dora Maira massif implies that continental crust may be buried to depths of about 100 km along a rather low temperature gradient, which is best accounted for by the deep subduction of continental lithosphere. A more difficult problem is that these rocks must then reach the surface without complete overprinting.'

Based on various considerations, including the position of the quartz–coesite boundary (Figure 3.6a), metamorphic conditions for the Dora Maira Massif are estimated at 700–800 °C, 27 kbar.

Mineral reactions in the massif were considered by Schreyer (1988b) who presented evidence for an initial reaction that caused sedimentary chlorite and quartz to produce talc and kyanite, followed by the reaction

$$\underset{\text{talc}}{Mg_3Si_4O_{10}(OH)_2} + \underset{\text{kyanite}}{Al_2SiO_5} = \underset{\text{pyrope}}{Mg_3Al_2Si_3O_{12}} + \underset{\text{coesite}}{2SiO_2} + H_2O.$$

Much of the coesite reverted to quartz during subsequent decompression, but where present as inclusions in garnet this change was evidently suppressed due to the required increase in volume.

C

The change of carbonaceous matter of organic origin to graphite has been examined at numerous places (e.g., Grew 1974), and normally the change is largely completed in the chlorite or biotite zones (greenschist facies); TEM studies by Buseck and Bo-Jun (1985) reveal that the change is progressive and complex. Graphite is common in schist, gneiss, and marble of higher grade. Diamond occurs in kimberlite pipes and rarely in ultramafic rocks (Pearson et al. 1989) and is obviously stable at depth.

In the latter 19th century, several attempts were made to produce diamond in the laboratory. One of these was described by Bundy, Strong, and Wentorf (1973) as follows:

> 'In 1880 J.B. Hannay reported on an extensive series of experiments in which he sealed mixtures of hydrocarbons, bone oil, and lithium in strong wrought-iron tubes and heated them to red heat for several hours in a furnace. Most of the tubes exploded, but a few survived. The carbonaceous residue in one or two of the surviving tubes contained some crystals that had the properties of diamond. These crystals were submitted later to the British Museum, and by modern tests are indeed diamond. No one since has succeeded in growing diamonds by this process ...'

The graphite–diamond boundary curve was eventually located by the direct experimental approach and by calculation, and is shown in Figure 3.6b. Diamond is stable at a depth greater than about 120 km (40 kbar). The density of diamond exceeds that of graphite by a factor of 1.6.

Other forms of carbon occur and are evidently unstable. These include rhombohedral graphite in which the carbon layers are arranged abc... rather that abab... (Lipson and Stokes 1942, Laves and Baskin 1956), and hexagonal diamond (lonsdaleite) which occurs rarely in meteorites and eclogite (Liu and Bassett 1986).

Recently Sobolev and Shatsky (1990) discovered diamond in the Kokchetav Massif in northern Kazakhstan. Diamond and graphite occur as inclusions in crystals of garnet and zircon within garnet–biotite–plagioclase gneiss and other rocks of apparent sedimentary origin. Thus, in view of Figure 3.6b, these strata are evidently surface sediments that were transported into the Mantle where they were metamorphosed at 900–1000 °C and ≥ 40 kbar pressure. During their subsequent return to the surface, retrograde metamorphism produced biotite and other minerals and converted some of the diamond to graphite. These rocks raise some important questions regarding vertical tectonics within the Earth.

Mg_2SiO_4

Mg-rich olivine is common in ultramafic inclusions in basalt, some of which are composed almost entirely of olivine and have a composition close to Mg_2SiO_4. A large amount of experimental work, including work on analogue systems such as Co_2SiO_4, has been carried out in an attempt to identify high-pressure forms of Mg_2SiO_4. Liu and Bassett (1986) have summarized the results, and Ringwood (1975) has discussed the problem in detail.

The evidence at present, as shown in Figure 3.6c, is that with increasing pressure, forsterite changes to a spinel-like phase, resulting in a 2.3% decrease in volume, which then changes to a true spinel, with a further reduction in volume. These changes occur between 150 and 200 kbar at a temperature of ~1500 °C, i.e., well within the upper Mantle.

Some evidence exists for forms of Mg_2SiO_4, even more dense, at higher pressures. Thus vertical movements within the Mantle are expected to produce phase transformations (metamorphism) analogous to those that occur in the Crust.

3.2 TRANSFORMATIONS INVOLVING MINERALS OF NOMINALLY FIXED COMPOSITION AND ONE OR BOTH OF H_2O AND CO_2

Most minerals that are produced during weathering and diagenesis (clay minerals, chlorite, calcite, dolomite), contain $(OH)^-$ or $(CO_3)^{-2}$ ions as integral parts of the crystal structures. During burial and metamorphism, many different reactions occur which release these ions as H_2O and CO_2. In large part, the H_2O and CO_2 molecules migrate away from the reaction sites, presumably as the reactions are in progress or soon afterwards. During retrograde metamorphism the molecules may reappear to produce a reversal of some of the earlier prograde reactions.

Reactions involving H_2O or CO_2 or both are of great interest and only a few of the many that have been identified are here selected for study.

3.2.1 Muscovite + quartz = K-feldspar + Al silicate + H_2O

$$KAl_3Si_3O_{10}(OH)_2 + SiO_2 = KAlSi_3O_8 + Al_2SiO_5 + H_2O$$

Muscovite is very common in low- to medium-grade metamorphic terranes, and we have noted in Chapter 1 (Sec. 1.9) that the above reaction provides an explanation for the disappearance of this mineral in the upper reaches of the amphibolite and hornblende hornfels facies. The stability of muscovite is also of interest in relation to the occurrence of muscovite granite. The above reaction was recognized by Harker (1932) and was described briefly in Sec. 1.8; here it is examined in greater detail.

In the experimental study of dehydration equilibria, 'excess' water is added to the crystalline phases, and it is of interest to consider the properties of this water at elevated temperatures and pressures, and how it might differ from water present under natural conditions. At temperatures somewhat less than 300 °C and pressures less than 200 bar, pure water, if it exists as a discrete phase, occurs as a liquid or a vapor (gas) depending on temperature and pressure, these being separated by a well-defined boundary curve. But at a point on the boundary curve, at 374 °C and 218 bar (the critical point) the distinction between liquid and vapor is lost, and at temperatures and pressures beyond this point, the phase is referred to as a supercritical fluid. Examples of the specific volume of this fluid at some metamorphic temperatures and pressures are listed in Table 2.4, p. 81. Note that the volume of water at high temperatures and pressures does not greatly exceed that of liquid water; for example at 500 °C and 5 kbar (amphibolite facies) it is 1.15 $cm^3\ g^{-1}$. The crowding of water molecules under these conditions is also reflected in the entropy, which does not depart greatly from the range 60–80 J/mol·K, which may be compared with 200 J/mol·K at 130 °C and 1 bar.

In Chapter 1 (Sec. 1.5) we have noted that a fluid phase is not necessarily present as natural dehydration reactions are in progress (the water molecules being removed as fast as they are produced), and that the activity of H_2O could vary from place to place, and be less than that which exists in a hydrothermal experiment. Nevertheless, the appropriate procedure in the experimental study of dehydration equilibria is to include abundant water together with the interacting crystalline phases; the interpretation of the experimental results must then take into consideration the uncertainties referred to above.

The boundary curve for the equilibrium

$$\text{muscovite} + \text{quartz} = \text{K-feldspar} + \text{Al silicate} + H_2O$$

was determined experimentally by Evans (1965), Althaus et al. (1970), and Day (1973). The results presented here (Figure 3.7) are from Chatterjee and Johannes (1974).

The Chatterjee–Johannes experiments were carried out by use of a cold-seal hydrothermal apparatus. Synthetic $2M_1$ muscovite, synthetic pure sanidine, pure

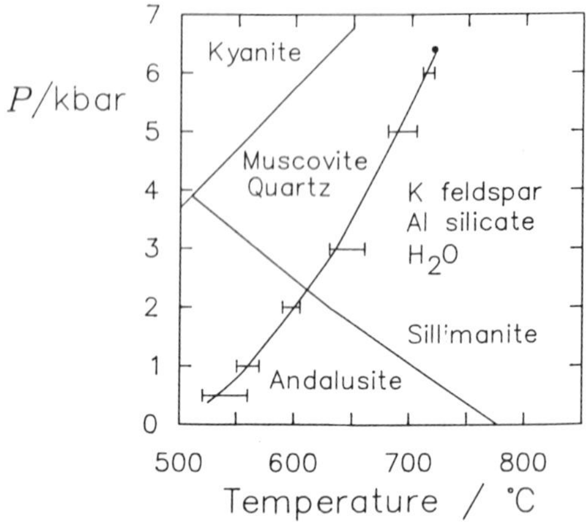

Figure 3.7 Boundary curve for the equilibrium, muscovite + quartz = K-feldspar + Al silicate + H_2O; dot marks the appearance of a melt phase. From Chatterjee and Johannes (1974). Reproduced by permission of Springer-Verlag. Al_2SiO_5 boundaries from Figure 3.2

quartz, excess water, and either natural andalusite or natural sillimanite were subjected to preset temperature and pressure for periods ranging from 18 days (at 700 °C) to 173 days (at lower temperatures). Upon quenching, the reaction direction was determined by noting (using optical and X-ray methods) if the volume of muscovite had decreased or increased. Thus the equilibrium temperature at 5 kbar, for example, could be bracketed by 680 and 705 °C. Metastable equilibria were obtained with sillimanite at low pressures and with andalusite at high pressures, but these results are not shown in Figure 3.7.

The general equation for a boundary curve, which was derived in Chapter 2 (Sec. 2.2), can be used to describe the muscovite and other univariant dehydration equilibria. At an equilibrium temperature (T') and pressure (P'), corresponding to a point on the boundary curve,

$$\Delta G_{T'P'} = \Delta H_{298,1} + \int_{298}^{T'} \Delta C_P \, \mathrm{d}T - T\Delta S_{298,1} - T\int_{298}^{T'} (\Delta C_P/T)\,\mathrm{d}T$$
$$+ \int_{1}^{P'} \Delta V_{T'} \, \mathrm{d}P = 0.$$

Note that the standard state for all phases is chosen as 298 K, 1 bar. Let $\Delta H_{T',1}$ and $\Delta S_{T',1}$ refer to the enthalpy and entropy changes at the equilibrium temperature and 1 bar. Then,

$$\Delta G_{T'P'} = \Delta H_{T',1} - T'\Delta S_{T',1} + \int_{1}^{P'} \Delta V \, \mathrm{d}P = 0.$$

The term ΔV can be divided in two parts, the volume change, considering only the solid phases (ΔV^{s}) which is now equated to $\Delta V^{\mathrm{s}}_{\mathrm{STP}}$ and taken as constant, and the volume of the gas phase (V^g) which for the muscovite breakdown, as written, is the volume of one mole of H_2O. Should this gas be a perfect gas, $V = RT/P$, and

$$\int_1^{P'} V^g \, dP = RT' \int_1^{P'} (1/P)\, dP = RT' \ln(P'/1).$$

Under the $P-T$ conditions of interest, H_2O is not a perfect gas, and this is accommodated by substituting fugacity (f) for pressure (P), specifying also that the standard state is the gas at a fugacity of 1 bar. Thus,

$$\int_1^{P'} \Delta V \, dP = \Delta V^s(P' - 1) + RT' \ln f.$$

Placing this in the above expression for ΔG, and rearranging terms,

$$\ln f = -\frac{\Delta H_{T',1}}{RT'} + \frac{T'\Delta S_{T'1}}{RT'} - \frac{\Delta V^s(P')}{RT'}.$$

Differentiating with respect to $1/T$,

$$\frac{d\ln f}{d(1/T)} = -(\Delta H_{T'1} + \Delta V^s(P'))/R.$$

This modified form of the Clapeyron equation gives the slope of a boundary curve when plotted in $\ln f - (1/T)$ coordinates. The muscovite + quartz = K-feldspar + Al silicate + H_2O boundary (Figure 3.7), replotted in this manner, is shown in Figure 3.8. Values of fugacity are from Burnham, Holloway and Davis (1969).

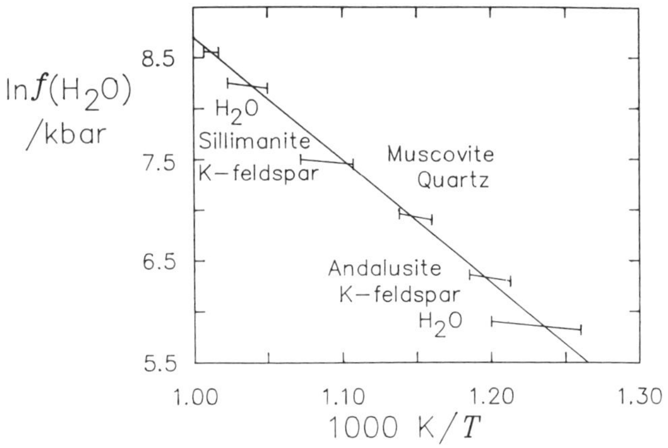

Figure 3.8 Data of Figure 3.7 replotted with ln fugacity of H_2O and $1/T$ as variables. The equation of the boundary curve in this form is $\ln f\,(H_2O) = 21 - (12\,000/T)$. The expected small change in slope at $(1000/T) = 1.13$ (the And–Sil boundary) is not shown. Data from Chatterjee and Johannes (1974)

Commonly such plots yield straight lines; this results from ΔH for the reaction being nearly constant, and from ΔV^s being negligible. These plots are helpful in extrapolating boundaries into unexplored regions and for obtaining a boundary-curve equation from experimental data.

When examining all of the experimental data that are now available on the muscovite + quartz reaction, we find only slight disagreement in results obtained in different laboratories. Skippen (1977) has shown that this disagreement can be attributed to differences in the amount of Al–Si order in the K-feldspar used in the experiments, i.e., equilibrium with regard to order–disorder in K-feldspar did not exist in all experimental runs.

A reduction in the activity of water in the above equilibrium will cause a reduction in the Gibbs energy of the products, combined, as shown in Figure 3.9, and will cause the reaction to proceed at a lower temperature, i.e., the boundary curve in Figure 3.7 will shift to the left. Kerrick (1972) investigated this problem experimentally. First the equilibrium at 2 kbar in the presence of pure water was located and was found to be $605 \pm 10\,^\circ\text{C}$, in agreement with results shown in Figure 3.7. Then the equilibrium at 2 kbar in the presence of a H_2O–CO_2 mixture ($X_{CO_2} = 0.5$) was located and was found to be $555 \pm 10\,^\circ\text{C}$, i.e., $50\,^\circ\text{C}$ lower. The experiment with a H_2O–CO_2 mixture was repeated at 3.5 kbar. From these results it is clear that the reaction, as it occurs naturally, could take place at temperatures that are considerably lower than those obtained in laboratory studies in the presence of a discrete and pure H_2O phase.

Numerous field studies have demonstrated that although the boundary curve in Figure 3.7 may be viewed as being accurately located, the reaction as it occurs within the Earth is not yet accurately understood. Initially, one might expect to find in medium- to high-grade terranes, a well-defined muscovite + quartz = K-feldspar

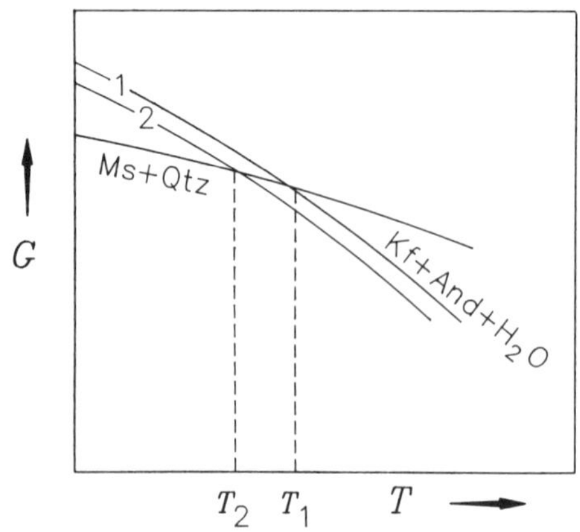

Figure 3.9 Gibbs energy (G) versus temperature (T) curves for reactants (muscovite + quartz) and products (K-feldspar + andalusite + H_2O), showing decrease in reaction temperature, T_1 to T_2, resulting from a decrease in the chemical potential of H_2O (curves 1 to 2)

+ Al silicate isograd, but at most localities, a sharp muscovite-out K-feldspar-in isograd does not exist. For example, in the Strontian area of Scotland (Tyler and Ashworth 1982) the four-phase assemblage occurs in a zone 1 km wide, and in the Appalachians of Maine (Evans and Guidotti 1966) the four-phase assemblage occurs within a zone up to 10 km wide.

With regard to the occurrence in Maine, Evans and Guidotti (1966) mapped a sillimanite–K-feldspar isograd that marks the first appearance of K-feldspar in rocks which in the down-grade direction contain muscovite, quartz, and sillimanite. The four-phase assemblage then persists in the broad zone above the isograd, with little evidence of a displacement of the reaction to the high-temperature side.

Below the isograd, muscovite contains 18 to 8.5 mole % paragonite (Na) component, and in crossing the isograd this decreases to 8.6 to 5.2%. The equilibrium

$$\text{paragonite} + \text{quartz} = \text{albite} + \text{Al silicate} + H_2O$$

lies at a lower temperature than the corresponding one with muscovite (Chatterjee 1972), and the observed decrease in Na content of muscovite evidently results from the breakdown of the paragonite component of muscovite before the muscovite (K) component of muscovite begins to react. A relation of this kind was subsequently reported from other localities, referred to by Guidotti (1984, p. 405).

Phase relations involving paragonite, muscovite, albite, and K-feldspar were explored by Evans and Guidotti (1966) and Thompson (1974); the results obtained by Chatterjee and Froese (1975), based on experimental and thermodynamic data, are here reproduced as Figure 3.10. According to these relations, muscovite containing 10% paragonite in solution (broken line in Figure 3.10) will, with increasing temperature, first produce albite, thereby becoming poorer in Na, and this agrees with data from the metamorphic terrane in Maine and elsewhere. With a further rise in temperature, K-feldspar is produced, and this point corresponds to the K-feldspar–sillimanite isograd as mapped. According to Figure 3.10 the reaction should then continue, producing more K-feldspar and sillimanite, as the muscovite becomes progressively poorer in Na, but this latter behavior has not been detected in the broad K-feldspar–sillimanite zone in Maine.

The reaction is obviously complex, and Evans and Guidotti (1966) have considered the part played by plagioclase (with about 20 mole % anorthite) which is present both above and below the K-feldspar–sillimanite isograd. From phase-rule considerations (Chapter 2, Sec. 2.7), within the system '(Na, K) plagioclase (free of anorthite), (K, Na) muscovite, quartz, (K, Na) K-feldspar, sillimanite, H_2O' there are six interactions:

$$\mu_{\text{ms}}^{\text{Ms}} + G^{\text{Qtz}} = \mu_{\text{or}}^{\text{Kf}} + G^{\text{Als}} + G^{H_2O},$$

$$\mu_{\text{pg}}^{\text{Ms}} + G^{\text{Qtz}} = \mu_{\text{ab}}^{\text{Kf}} + G^{\text{Als}} + G^{H_2O},$$

$$\mu_{\text{ms}}^{\text{Ms}} + G^{\text{Qtz}} = \mu_{\text{or}}^{\text{Pl}} + G^{\text{Als}} + G^{H_2O},$$

$$\mu_{\text{pg}}^{\text{Ms}} + G^{\text{Qtz}} = \mu_{\text{ab}}^{\text{Pl}} + G^{\text{Als}} + G^{H_2O},$$

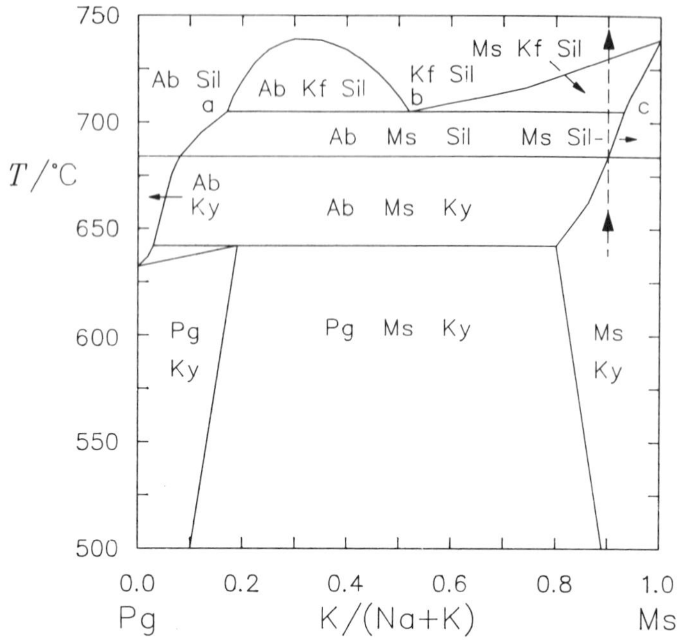

Figure 3.10 Phase relations involving muscovite (Ms), paragonite (Pg), kyanite (Ky), sillimanite (Sil), K-feldspar (Kf), and albite (Ab), in the presence of quartz and H_2O at 7.0 kbar. From Chatterjee and Froese (1975)

$$\mu_{\text{or}}^{\text{Kf}} = \mu_{\text{or}}^{\text{Pl}},$$

$$\mu_{\text{ab}}^{\text{Kf}} = \mu_{\text{ab}}^{\text{Pl}},$$

but only four are independent. Thus the variance is

$$\begin{aligned} v &= w - [2(r-1) + e], \\ &= 15 - [2(6-1) + 4], \\ &= 1, \end{aligned}$$

or from $v = n + 2 - r = 5 + 2 - 6 = 1$, and at constant pressure the system is invariant. This condition is represented in Figure 3.10 by line abc, at 710 °C, with all phases fixed in composition. With the addition of anorthite to albite, the variance increases by one, and Evans and Guidotti (1966) have pointed out that in this case, line abc becomes a field, which permits all phases to persist over a range in temperature. This may explain the widespread occurrence of all phases involved within the K-feldspar–sillimanite zone.

In brief, the reaction can be expressed as follows, with (a) occurring below the K-feldspar isograd and with all occurring at and above the isograd:

$NaAl_3Si_3O_{10}(OH)_2$ (paragonite component in muscovite) + SiO_2

$= NaAlSi_3O_8$ (albite component in plagioclase) + $Al_2SiO_5 + H_2O$, (a)

$KAl_3Si_3O_{10}(OH)_2$ (muscovite component in muscovite) + SiO_2

$= KAlSi_3O_8$ (orthoclase component in K-feldspar) + $Al_2SiO_5 + H_2O$, (b)

$NaAlSi_3O_8$ (albite component in plagioclase)

$= NaAlSi_3O_8$ (albite component in K-feldspar). (c)

Another interpretation, one that is favored by Evans and Guidotti (1966) and also by Cheney and Guidotti (1979) and Tyler and Ashworth (1982) is that in the K-feldspar–sillimanite zone, as temperature increased the chemical potential of H_2O also increased, causing the boundary curve to shift from its initial position (reduced H_2O activity) progressively to the right. This is represented in Figure 3.9 by a change from temperature T_2 to T_1.

3.2.2 Equilibria involving talc, anthophyllite, forsterite, and enstatite (system $MgO–SiO_2–H_2O$) and a review of the Schreinemakers method

In our study of andalusite, kyanite, and sillimanite, we encountered a pressure–temperature phase diagram for a one-component system (Al_2SiO_5). In this diagram (Figure 3.2) three univariant lines meet at an invariant point and divide the P–T area into three divariant sectors. Imagine now a two- or three-component system; how many univariant lines radiate from the invariant points in this system, and what rules govern the positioning of the lines relative to each other? These and other questions were considered in detail by F.A.H. Schreinemakers in the years 1915–1925, and his results were applied to mineral equilibria by Niggli (1954), Korzhinskii (1959), and Zen (1966). The subject was recently reviewed by Philpotts (1990). A brief description of the Schreinemakers method for the construction and analysis of multi-component phase diagrams will now be presented. This will be followed by the examination of a three-component system of moderate complexity, in which some relations involving talc, anthophyllite, forsterite, and enstatite have been established by use of both the Schreinemakers method and by the direct experimental approach.

The Schreinemakers method applies only to systems of phases of fixed composition. The variance of such systems is expressed by use of the phase-rule equation, $v = n + 2 - r$, where n and r stand for the number of components and phases respectively. Components are defined (Zen 1966) as 'distinct units that cannot be formed by a combination of other units'. Phases consist of different combinations of these components. Ambiguity arises when a compositional system is defined in terms of a number of components (e.g., MgO, SiO_2, H_2O; $n = 3$) and the volume of matter under consideration (i.e., the thermodynamic system) consists of fewer than n phases, e.g., two phases. Then the equation $v = n + 2 - r$ results in a variance > 2, which in general is impossible. With $r < n$, the components must be redefined.

The combinatorial formula may be used to determine the number of univariant assemblages (i.e., the number of univariant lines) in a phase diagram and the number of bivariant assemblages of a given number of phase combinations in an n-component system. The formula gives the number of ways (N) of combining a number of objects (k), taken j at a time ($j < k$):

$$N(k, j) = k!/j!(k - j)!.$$

In terms of the phase rule, $k = n + 2, j = r$ (the number of phases), and

$$N(n + 2, r) = (n + 2)!/r!(n + 2 - r)!,$$

as given by Zen (1966). For example, in a three-component system ($n = 3$), how many four-phase assemblages are possible, the variance of each being $v = 3 + 2 - 4 = 1$, i.e., how many univariant lines are present? By use of the above formula,

$$N(5, 4) = 5!/4!(3 + 2 - 4)! = 5,$$

and we expect five lines to radiate from the five-phase ($v = 0$) invariant point. As a second example, how many possible three-phase assemblages exist in this system? The answer is

$$N(5, 3) = 5!/3!(5 - 3)! = 10.$$

Further application of this formula leads to the results shown in Table 3.1. Exceptions to these statements result from degeneracy, to be described below.

Consider (following Niggli 1954), a hypothetical collection of phases, A, B, C, D, E whose composition can be expressed in terms of three components, x, y, z, ($n = 3$) as shown in Figure 3.11a. From the position of the phases, as plotted in the composition triangle, we see that phase D is chemically equivalent to B + C + E, in unspecified proportions, and therefore one possible (four-phase) univariant equilibrium is

$$\text{B} + \text{C} + \text{E} = \text{D} \qquad \text{[A]}$$

Table 3.1 Number of phases (r) in n-component non-degenerate invariant, univariant, and divariant ($v = 0, 1, 2$) systems

	$v = 0$		$v = 1$		$v = 2$	
n	r	N_0	r	N_1	r	N_2
1	3	1	2	3	1	3
2	4	1	3	4	2, 1	6, 4
3	5	1	4	5	3, 2, 1	10, 10, 5
4	6	1	5	6	4, 3, 2, 1	15, 20, 15, 6
5	7	1	6	7	5, 4, 3, 2, 1	21, 35, 35, 21, 7

N_0 number of r-phase invariant assemblages = number of invariant points
N_1 number of r-phase univariant assemblages = number of univariant lines = number of bivariant sectors
N_2 number of r-phase divariant assemblages, assuming phases consist of the n components in any combination

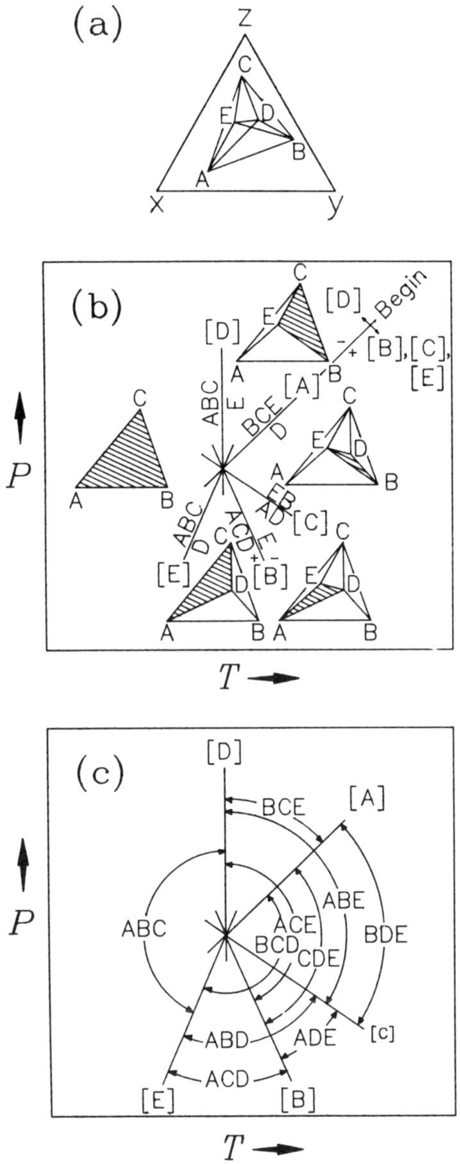

Figure 3.11 (a) Hypothetical system of three components (x, y, z) and five phases (A, B, C, D, E); (b) disposition of univariant lines about the invariant point in pressure (*P*) versus temperature (*T*) space; the unique divariant (three-phase) assemblage in each sector is indicated by the shaded triangle; (c) unique and other three-phase assemblages stable in the five sectors. From Niggli (1954) with minor changes. Reproduced by permission of W.H. Freeman and Co.

The other possibilities, obtained by inspection (Figure 3.11a) are

$$A + C + D = E \qquad [B]$$

$$A + D = E + B \qquad [C]$$

$$A + B + C = E \qquad [D]$$

$$A + B + C = D \qquad [E]$$

An algebraic method may also be used to define these equations. Note that each invariant assemblage is labelled by use of the phase that is *missing*, enclosed in square brackets.

Any two equations involve each phase at least once, and therefore only two equations are independent. Thus the variance (when all five phases are present) expressed in the form $v = 2 - e = 2 - 2 = 0$ agrees with that expressed as $v = n + 2 - r = 3 + 2 - 5 = 0$.

From the above geometric and phase-rule considerations, we anticipate that the phase diagram for our model system will consist of five (four-phase) univariant lines radiating from a (five-phase) invariant point.

Consider any one of the univariant lines, for example that formed by the last equation, A + B + C = D. The 'fundamental axiom' states that the assemblage on one side of the line (assemblage A,B,C) is more stable than that on the other side (D), and vice versa. This follows from a consideration of intersecting Gibbs energy surfaces, as shown for example in Figure 2.5 of Chapter 2 (p. 78).

By use of the above axiom one may prove that the angle between any two adjacent univariant lines cannot exceed 180°. This is the Morey–Schreinemakers rule.

We now proceed to construct a phase diagram for our model system. Normally, some information is available on a system in advance. Suppose that in our example, the following is known:

(1) The slope of [A] is positive, and D lies on the high-temperature side, i.e.,

$$B + C + E = D \qquad [A]$$

(2) The slope of [B] is negative and E lies on the high-temperature side, i.e.,

$$A + C + D = E \qquad [B]$$

Symbols (+) and (–) are now introduced to refer to the field encountered in a clockwise (+) or counterclockwise (–) direction on either side of a univariant boundary, when viewed from the invariant point.

We begin by drawing line [A] with positive slope as shown in Figure 3.11b. Assume that the stable part of [A] lies to the right of the invariant point, the metastable part to the left, as shown. Accordingly, phase D lies in the (+) half, B, C, E in the negative half.

One of the Schreinemakers rules states that when a phase, e.g., D, is stable on the (+) side of a univariant line (e.g., [A]) then line [D] must lie within the (–)

half of the $P-T$ area that is divided by line [A]. Similarly, in our example, the three lines [B], [C], [E] must lie in the (+) half. Line [B] may now be positioned approximately and labelled, taking into consideration the information in item (2), above. Applying the above rule to [B], line [A] must lie in the (−) half of the area divided by [B], as it does. Thus [A] and [B] are correctly positioned relative to each other.

The remaining lines may now be placed. Line [C] must lie on the (+) side of [A] and the (−) side of [B], line [D] must lie on the (−) side of [A] and on the (−) side of [B], and line [E] must lie on the (+) side of [A] and the (+) side of [B] (Figure 3.11b). This results in a clockwise sequence of the lines [A], [C], [B], [E], [D].

Suppose there was some doubt regarding the information provided in advance for [B], and the possibility of E lying on the low-temperature side, i.e., E = A + C + D, should also be considered. Following the same rules as before, we would now find [A] to extend down and to the left of the invariant point, and in the final construction, the same sequence of lines as above would appear, but in a counterclockwise direction.

Thus for a three-component system, independent experimental or paragenetic information is needed on two of the univariant equilibria to enable one to construct with assurance the correct topology. Further information is then needed to fix the invariant point in terms of temperature and pressure and to fix the slopes of the univariant lines.

The important question of the particular divariant phase assemblages that occur in each sector can be answered as follows. In the sector formed by [A] and [C] (Figure 3.11b), the assemblage B, D, E is unique (i.e., all phases except A and C), and in the sector formed by [C] and [B], the assemblage A, D, E is unique (i.e., all phases except C and B), and so on. These unique assemblages are confined to their respective sectors. The same method is used to identify other divariant assemblages, as shown in Figure 3.11c. For example, in the *two* sectors lying between [D] and [C] (the angle between these lines must not exceed 180°), the assemblage A, B, E is stable (omitting D and C). In addition, various two-phase assemblages are, of course, also permitted.

Composition triangles may now be placed in each sector, showing the assemblages that are unique and others (Figure 3.11b). The various non-unique assemblages can also be deduced from an inspection of these triangles.

Suppose that for our hypothetical system, there was no information whatsoever to begin with, only that the system is ternary and the phase compositions are known. It would still be possible to determine a particular line sequence which in one topology would be a clockwise sequence, and in the other possible topology, the same sequence, but in a counterclockwise sense. The slopes would remain unknown and each bundle of univariant lines could rotate like a wheel, about the invariant point.

In the above model we have assumed that four phases lie on each univariant line. It is not uncommon, in real systems, for less than four (i.e., less than $n+1$)

phases to define a line. In a three-component system, this occurs when three phases (within the composition triangle) are collinear, i.e., the composition of three phases can be expressed by use of only two components, rather than three. As a result, the number of univariant lines is reduced from five to four or from five to three. These are degenerate systems.

What happens to univariant lines as they extend outward from an invariant point? One possibility is that they simply terminate. This occurs in the H_2O system, where at the critical point the distinction between the phase on one side (liquid) and that on the other (vapor) disappears. But this behavior is unlikely to occur when crystalline phases are involved, and it seems more likely that univariant lines eventually join onto other invariant points.

Zen and Roseboom (1972) and Day (1972) have analyzed ternary systems in which six phases appear. Only five of these can occupy an invariant point. The single added phase gives rise to 6 possible invariant points and 15 possible univariant lines. These are systems referred to as 'multisystems'. Hypothetically, some lines could have sufficient curvature to produce a closed net of the kind proposed by Zen and Roseboom (1972).

We shall consider now some equilibria involving quartz (Qtz), talc (Tc), anthophyllite (At), forsterite (Fo), enstatite (En), and H_2O, which were first studied in detail by Greenwood (1963). The composition of all phases can be expressed by use of only three components, MgO, SiO_2, and H_2O, as shown in Figure 3.12 (inset). All of the solids are known to have crystallized in metamorphosed ultramafic rocks within medium- to high-grade contact aureoles (Evans 1977), and several univariant boundaries are therefore expected to lie between 500 and 800 °C and 1 to several kbar. H_2O is assumed to be present as a phase. With six phases under consideration two or more invariant points are anticipated.

It is possible, from an inspection of Figure 3.12 (inset) to define six possible (four-phase) univariant reactions. As in the illustrative example above, these are recognized wherever two tie-lines cross, where one phase lies in a triangle defined by three others, or where three phases are collinear. Forsterite is present in three of these reactions and they presumably emanate from one invariant. The other three reactions involve quartz and they presumably emanate from another invariant. The points are labelled [Qtz] and [Fo] respectively, i.e., by use of the symbol for the phase that is missing. The six reactions are:

$$\mathrm{Tc + Fo = En + H_2O} \qquad [\mathrm{At,Qtz}]$$

$$\mathrm{Tc + Fo = At + H_2O} \qquad [\mathrm{En,Qtz}]$$

$$\mathrm{At + Fo = En + H_2O} \qquad [\mathrm{Tc,Qtz}]$$

$$\mathrm{Tc = At + Qtz + H_2O} \qquad [\mathrm{En,Fo}]$$

$$\mathrm{Tc = En + Qtz + H_2O} \qquad [\mathrm{At,Fo}]$$

$$\mathrm{At = En + Qtz + H_2O} \qquad [\mathrm{Tc,Fo}]$$

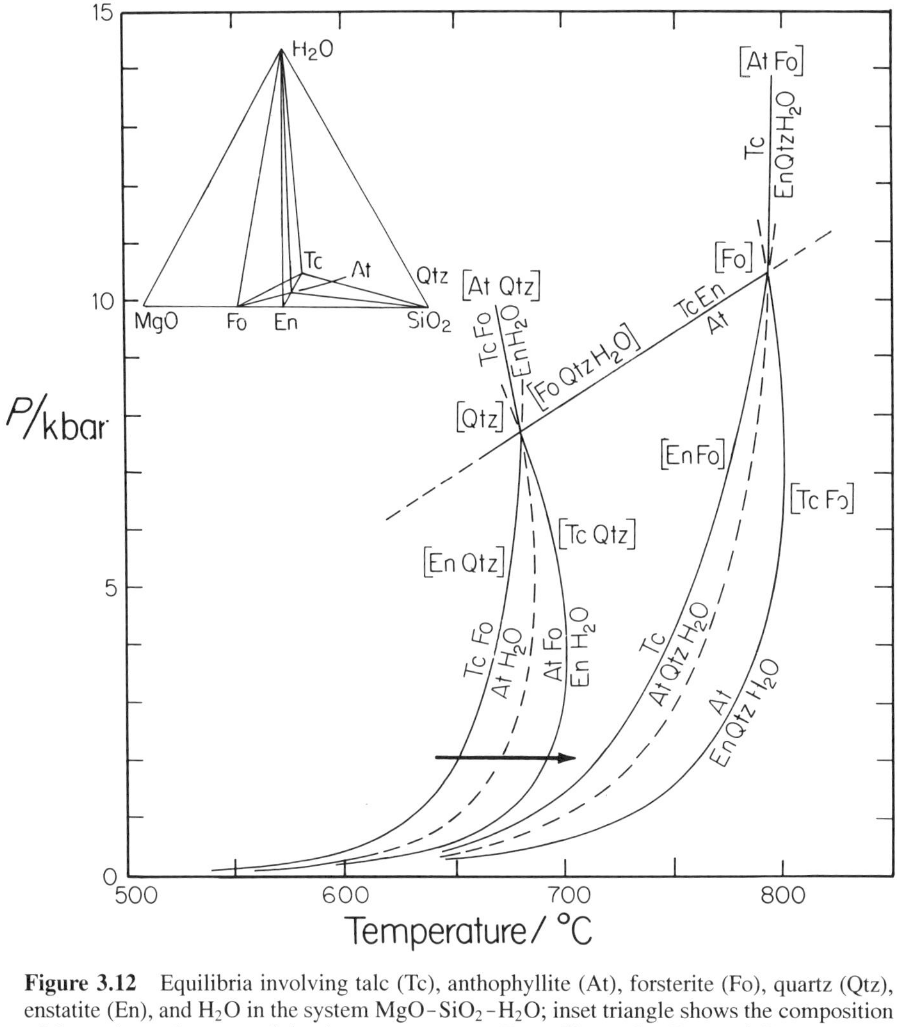

Figure 3.12 Equilibria involving talc (Tc), anthophyllite (At), forsterite (Fo), quartz (Qtz), enstatite (En), and H_2O in the system $MgO-SiO_2-H_2O$; inset triangle shows the composition of these phases in terms of the three components. From Chernosky, Day and Caruso (1985)

The multisystem can be viewed as consisting of these two systems. A final, degenerate, equilibrium,

$$Tc + En = At \qquad [Fo,Qtz,H_2O]$$

involves three phases that are common to both systems, and the two invariant points [Qtz] and [Fo] presumably lie on this line.

It is now possible, in principle, to use the Schreinemakers method to construct a number of possible topologies for this multisystem, but fortunately, some experimental data are available. These data were obtained by Greenwood (1963), Hemley et al. (1977), and Chernosky, Day and Caruso (1985). The phase relations presented by the last authors are shown in Figure 3.12. The results are in agreement with thermochemical data, as reviewed by Day, Chernosky and Kumin (1985), but minor uncertainties in the locations of the two invariant points remain. (Two additional invariants may possibly lie at low pressures.) The topology is identical to that proposed by Greenwood (1963), which was based on experimental data combined with restrictions imposed by the phase rule and by Schreinemakers rules. The reader will notice that in Figure 3.14 the various Schreinemakers expectations referred to above are fulfilled.

The phase diagram is of great interest in relation to the metamorphism of ultramafic rocks, especially those that are low in Ca and Fe. Evans (1977) has reviewed results from the Alps, where the emplacement of granitic and gabbroic rocks has produced contact aureoles two or more kilometers wide within serpentinite bodies. Ideally, following the disappearance of antigorite, the following Mg silicates appear in the Alps, with increasing grade:

forsterite talc
forsterite anthophyllite
forsterite enstatite

This agrees with expectations drawn from the phase diagram, as shown by the arrow at 2 kbar in Figure 3.12. The restricted occurrence of anthophyllite in the aureoles agrees with the experimental finding that in the presence of forsterite, anthophyllite is stable within a temperature range of less than 50 °C.

The phase relations are also of interest in relation to certain occurrences of talc together with kyanite, described from Valais, Switzerland (Chinner and Dixon 1973) and from central Zambia (Vrána and Barr 1972), as reviewed by Schreyer (1988b). Figure 3.12 shows that talc in the presence of H_2O, alone or with enstatite and quartz, is stable to pressures in the vicinity of 12 kbar. Additional experimental data indicate that talc and kyanite are stable to 30 kbar (Schreyer 1988a), where at 800 °C talc and kyanite produce the remarkable assemblage pyrope–coesite, which was described from the western Alps by Chopin (1984).

3.2.3 Equilibria involving calcite, dolomite, quartz, talc, tremolite, diopside, enstatite, forsterite, and a H_2O–CO_2 gas mixture (system CaO–MgO–SiO_2–CO_2–H_2O)

Numerous reactions to produce the above minerals occur in impure limestone during progressive metamorphism. The primary source of Ca and Mg now contained by the silicates is sedimentary calcite and dolomite, while Si was derived from sedimentary quartz or chert. At low grades of metamorphism, H_2O is evidently introduced, to produce talc and possibly tremolite, and is then removed as these give way to diopside and forsterite.

The talc-forming reaction,

$$3\text{ dolomite} + 4\text{ quartz} + H_2O = \text{talc} + 3\text{ calcite} + 3CO_2, \qquad \text{(a)}$$

(which was studied by Metz and Winkler (1963), Metz and Puhan (1970), and Gordon and Greenwood (1970)) involves H_2O *and* CO_2, as do some others, and it seems likely that in general both H_2O and CO_2 are present as the various reactions proceed. The importance of variation in CO_2:H_2O ratio during metamorphism was emphasized by Korzhinskii (1959), Greenwood (1962), Wyllie (1962), and Dobretsov (1966); a helpful review of mixed volatile metamorphic equilibria was prepared by Kerrick (1974).

The uncertainty regarding the nature of H_2O in muscovite-bearing schists and gneisses, referred to above, also exists with regard to H_2O and CO_2 in impure marble at depth, but we shall, for the present, assume that a discrete gas phase is present, as it is in all experimental systems that are designed to investigate carbonate–silicate equilibria.

Before examining these equilibria we must digress to a discussion of some principles regarding reactions in gases and the question of the species (CO_2, CO, CH_4, etc) that are present in C–O–H gases at metamorphic temperatures and pressures. To this end, the following equations for gas mixtures which were derived in Sec. 2.8 will be useful:

$$\mu_{A(T,P)} = G^{A}_{T,P} + RT \ln a_{A(T,P)}, \tag{1}$$

$$\mu_{A(T,P)} = G^{A}_{T,1} + RT \ln(f_A/1), \tag{2}$$

$$P_A = f_A/\gamma_A, \tag{3}$$

where the symbols are a_A activity, f_A fugacity, P_A partial pressure, and γ_A fugacity coefficient for component A in a gas mixture.

Consider a gas consisting of carbon, oxygen, and hydrogen in equilibrium with graphite at 2000 bar and 500 °C. According to French and Eugster (1965) the principal species within this gas are CO_2, CO, CH_4, O_2, H_2, and H_2O; others are present but are assumed to be very minor in amount. Numerous equilibria exist in this system, but only four are independent, for example,

$$C + O_2 = CO_2, \tag{b}$$

$$C + (1/2)O_2 = CO, \tag{c}$$

$$C + 2H_2 = CH_4, \tag{d}$$

$$H_2 + (1/2)O_2 = H_2O. \tag{e}$$

With six solution components present in the gas phase, and four equations, the variance of this system, from Sec. 2.7, is [7(gas) + 2(graphite)] − [2(1) + 4] = 3, or by use of $v = n + 2 - r = 3 + 2 - 2 = 3$.

We shall now obtain an expression for equilibrium (b),

$$\text{C (graphite)} + \text{O}_2 \text{ (gas)} = \text{CO}_2 \text{ (gas)}.$$

The energy-balance equation at any temperature, T, and at 1 bar pressure, is

$$\Delta G_{T,1} = 0 = \mu_{CO_2(T,1)} - \mu_{O_2(T,1)} - G^{C}_{T,1}. \tag{4}$$

From equation (1), above,

$$\mu_{CO_2(T,1)} = G^{CO_2}_{T,1} + RT \ln a_{CO_2(T,1)}, \tag{5}$$

$$\mu_{O_2(T,1)} = G^{O_2}_{T,1} + RT \ln a_{O_2(T,1)}. \tag{6}$$

By substitution and rearrangement,

$$\ln(a_{CO_2}/a_{O_2})_{T,1} = -\Delta G^{\circ}_{T,1}/RT, \tag{7}$$

$$(a_{CO_2}/a_{O_2})_{T,1} = K_{T,1}. \tag{8}$$

The activity ratio on the left is referred to as the 'equilibrium constant' (denoted K) and at a pressure of 1 bar, is a function of temperature.

Consider now the equilibrium at pressure $P\,(> 1)$, and temperature T,

$$\Delta G_{T,P} = 0 = \mu_{CO_2(T,P)} - \mu_{O_2(T,P)} - G^{C}_{T,P}. \tag{9}$$

From equation (2), above,

$$\mu_{CO_2(T,P)} = G^{CO_2}_{T,1} + RT \ln f_{CO_2(T,P)}, \tag{10}$$

$$\mu_{O_2(T,P)} = G^{O_2}_{T,1} + RT \ln f_{O_2(T,P)}. \tag{11}$$

Also,

$$G^{C}_{T,P} = G^{C}_{T,1} + V^{C}(P - 1), \tag{12}$$

where we assume that V^C, the molar volume of graphite, is constant. Thus by substitution of equations (10), (11), and (12) in (9),

$$\ln(f_{CO_2}/f_{O_2}) = K_{T,1} + V^{C}(P - 1)/RT.$$

The molar volume of graphite is 0.5298 J/bar, and with $R = 8.314$ J/K·mol, $V^C/\mathrm{R} = 0.0637$ K/bar, and

$$\ln(f_{CO_2}/f_{O_2})_{T,P} = K_{T,1} + 0.0637(P - 1)/T. \tag{13}$$

This equation, which was applied by Skippen (1971) states that in the graphite–gas system, the fugacity *ratio* (on the left) is determined at a given temperature and pressure. Similar equations hold for the three other equilibria that are written above.

There is, however, an added restriction, that the total pressure must equal the sum of the partial pressures,

$$P = P_{CO_2} + P_{CO} + P_{CH_4} + P_{O_2} + P_{H_2} + P_{H_2O}, \tag{14}$$

or, from equation (3) above,

$$P = \frac{f_{CO_2}}{\gamma_{CO_2}} + \frac{f_{CO}}{\gamma_{CO}} + \frac{f_{CH_4}}{\gamma_{CH_4}} + \frac{f_{O_2}}{\gamma_{O_2}} + \frac{f_{H_2}}{\gamma_{H_2}} + \frac{f_{H_2O}}{\gamma_{H_2O}}. \tag{15}$$

A final restriction may be imposed on this system by the addition of the assemblage quartz–fayalite–magnetite (QFM), which, through the equilibrium

$$3Fe_2SiO_4 + O_2 = 3SiO_2 + 2Fe_3O_4 \tag{f}$$

fixes the fugacity of oxygen. This is referred to as a 'buffer assemblage', and it imposes the following relation, as given by Eugster and Wones (1962):

$$\log_{10} f_{O_2} = -(26\,494/T) + 9.69. \tag{16}$$

The variance of the gas–graphite system, as noted above, is three. Adding the buffer system fixes f_{O_2} and reduces the variance to two. Setting the total pressure constant at say 2000 bar, reduces the variance to one. Now the fugacity of all species must be a function of temperature. French (1966) has combined the four equations of type (13) with equations (15) and (16), and calculated fugacity values for the species as a function of temperature (at 2000 bar), converting these to percentage concentration. The results, which are for the presence of the QFM buffer and for two other buffers, are shown in Figure 3.13. Under oxidizing conditions, determined by the hematite–magnetite (HM) buffer, the gas consists almost entirely of CO_2 and CO. Under the more reducing conditions which are characteristic of the medium- to high-grade metamorphism of marble, as represented possibly by the other two buffer systems (QFM; nickel–nickel oxide (NNO)), H_2O and CO_2 are the dominant molecular species present between 400 and 800 °C, but methane

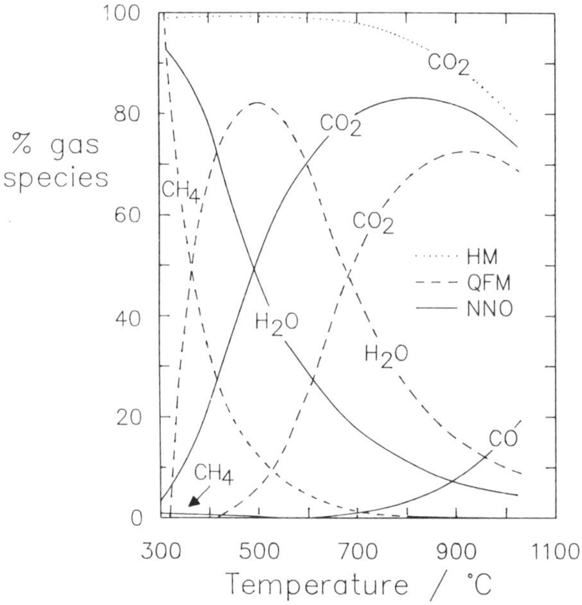

Figure 3.13 Proportions of gas species (CH_4, H_2O, CO_2, CO) in a C–O–H gas at 2.0 kbar, and in the presence of graphite and one of the indicated buffers; curve CO is the same for each buffer. From Eugster and Skippen (1967). Reproduced by permission of John Wiley & Sons, Inc.

may form a large fraction of the gas at lower temperatures. Figure 2.10 (Chapter 2) shows that at 2000 bar and 500 °C, a CO_2–H_2O mixture departs somewhat from ideal behavior, and the curves in Figure 3.13, which were located on the assumption of ideal mixing, may be displaced slightly from their correct position.

Some of the experimental results to be described below were obtained in the presence of graphite and a QFM or NNO buffer, and it seems likely, in view of Figure 3.13, that in the range of interest, ~400–800 °C, the gas consisted primarily of CO_2 and H_2O. The mineral equilibria can therefore be represented by a boundary curve expressed in terms of the mole fraction of CO_2 ($X_{CO_2} = CO_2/(CO_2 + H_2O)$) and temperature as variables, for a fixed total pressure.

The general form of the boundary curves in T–X diagrams was considered by Greenwood (1967a). With regard to the talc-forming reaction (equation (a) above), where H_2O is a reactant and CO_2 a product, the form of the curve, expressed in terms of its slope at different values of X_{CO_2}, is obtained as follows. At equilibrium,

$$\Delta G = 0 = G^{\mathrm{Tc}} + 3G^{\mathrm{Cal}} + 3\mu_{CO_2} - 3G^{\mathrm{Dol}} - 4G^{\mathrm{Qtz}} - \mu_{H_2O}. \tag{17}$$

This equation disregards the small amount of $CaMgC_2O_6$ dissolved in calcite. By substitution of

$$\mu_{CO_2} = G^{CO_2} + RT \ln X_{CO_2}, \tag{18}$$

and an analogous expression for H_2O, (which assume ideal mixing), followed by a rearrangement of terms,

$$3 \ln X_{CO_2} - \ln X_{H_2O} = -\Delta G^\circ/RT. \tag{19}$$

Differentiating with respect to temperature at constant pressure,

$$\frac{\mathrm{d}(3 \ln X_{CO_2} - \ln X_{H_2O})}{\mathrm{d}T} = -\frac{1}{R}\frac{\mathrm{d}(\Delta G^\circ/T)}{\mathrm{d}T} = \frac{\Delta H^\circ}{RT^2},$$

$$3\mathrm{d}(\ln X_{CO_2}) - \mathrm{d}(\ln X_{H_2O}) = (\Delta H^\circ/RT^2)\,\mathrm{d}T,$$

$$\frac{3\mathrm{d}X_{CO_2}}{X_{CO_2}} - \frac{1\mathrm{d}X_{H_2O}}{X_{H_2O}} = (\Delta H^\circ/RT^2)\,\mathrm{d}T. \tag{20}$$

But, because the fluid is a two-component mixture.

$$\mathrm{d}X_{H_2O} = -\mathrm{d}X_{CO_2}. \tag{21}$$

Thus,

$$\frac{3\mathrm{d}X_{CO_2}}{X_{CO_2}} + \frac{1\mathrm{d}X_{CO_2}}{X_{H_2O}} = (\Delta H^\circ/RT^2)\,\mathrm{d}T,$$

$$\frac{\mathrm{d}X_{CO_2}}{\mathrm{d}T}\left(\frac{3}{X_{CO_2}} + \frac{1}{X_{H_2O}}\right) = \Delta H^\circ/RT^2,$$

$$\frac{\mathrm{d}X_{CO_2}}{\mathrm{d}T} = (\Delta H^\circ/RT^2) \Bigg/ \left(\frac{3}{X_{CO_2}} + \frac{1}{X_{H_2O}}\right),$$

$$\frac{dT}{dX_{CO_2}} = \left(\frac{3}{X_{CO_2}} + \frac{1}{X_{H_2O}}\right)\frac{RT^2}{\Delta H^\circ}. \tag{22}$$

The slope of the boundary curve is determined largely by the term in brackets, which extends from $+\infty$ when $X_{CO_2} = 0$ to $+\infty$ when $X_{CO_2} = 1.0$, with an inflection point near $X_{CO_2} = 0.63$, $X_{H_2O} = 0.37$. Equation (22) is plotted in Figure 3.14, curve a.

A second example of a reaction involving CO_2 and H_2O is

$$1 \text{ tremolite} + 3 \text{ calcite} + 2 \text{ quartz} = 5 \text{ diopside} + 3CO_2 + 1H_2O. \tag{g}$$

This was investigated by Metz (1970) and Skippen (1971) and differs from the previous reaction in that both of H_2O and CO_2 are products. By the same procedure that was outlined above, one may obtain for this equilibrium,

$$\frac{dT}{dX_{CO_2}} = \left(\frac{3}{X_{CO_2}} - \frac{1}{X_{H_2O}}\right)\frac{RT^2}{\Delta H^\circ}. \tag{23}$$

The curve passes through a maximum, where

$$\left(\frac{3}{X_{CO_2}} - \frac{1}{X_{H_2O}}\right) = 0,$$

$$X_{CO_2} = 0.75.$$

This equation is plotted as curve b in Figure 3.14.

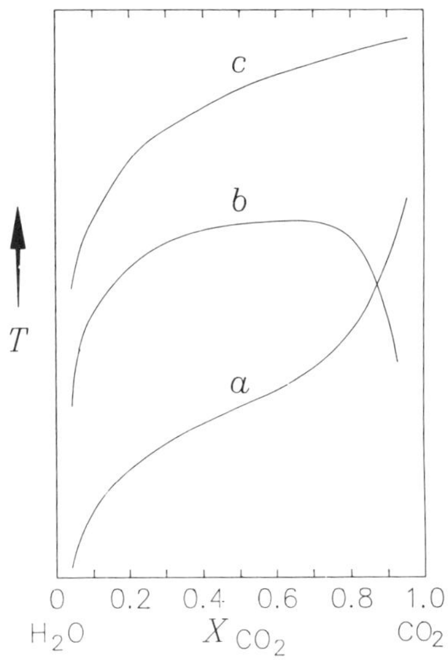

Figure 3.14 Three forms for boundary curves at constant pressure, and with temperature (T) and $CO_2/(H_2O + CO_2)$ as variables. From Greenwood (1967a). Reproduced by permission of John Wiley & Sons, Inc.

A third example is the wollastonite-forming reaction,

$$\text{calcite} + \text{quartz} = \text{wollastonite} + CO_2. \tag{h}$$

The boundary was located from thermochemical data by Danielsson (1950) and experimentally by Harker and Tuttle (1956) and Greenwood (1967b). Although H_2O is not involved directly, H_2O if present will mix with CO_2 and displace the equilibrium to lower temperatures. At equilibrium,

$$\ln X_{CO_2} = -\Delta G^\circ/RT,$$

$$\frac{d \ln X_{CO_2}}{dT} = \Delta H^\circ/RT^2,$$

$$\frac{(1/X_{CO_2})dX_{CO_2}}{dT} = \Delta H^\circ/RT^2,$$

$$\frac{dT}{dX_{CO_2}} = \frac{RT^2}{\Delta H^\circ}\left(\frac{1}{X_{CO_2}}\right). \tag{24}$$

Equilibria of this kind plot as curve c in Figure 3.14. Equations (22), (23), and (24) are the Greenwood equations.

With regard to components CaO, MgO, SiO_2, CO_2, H_2O, very many reaction equations may be written, including ones to produce talc, tremolite, diopside, and forsterite, which occur commonly in impure marble. Some of these were considered by Bowen (1940) and Turner (1967). Presented here are some results obtained by Skippen (1971), which have greatly clarified the phase relations that exist in this rather complex system. Similar results were obtained by Slaughter, Kerrick and Wall (1975).

Of the many equations that may be written with three or more of calcite, dolomite, quartz, talc, tremolite, diopside, enstatite, and forsterite, George Skippen chose five key equilibria for experimental study, and then from the equilibrium constants for these equilibria, he calculated the T–X_{CO_2} boundaries for all the others, finally eliminating those that are metastable. The resulting T–X_{CO_2} phase diagram at 2000 bar is here reproduced as Figure 3.15; the general form of the boundary curves is as predicted by the Greenwood equations (Figure 3.14). Details of the sophisticated experimental procedures and the calculations that were required to produce the phase diagram are described by Skippen (1971).

Some characteristics of the phase diagram may be noted:

(1) At low temperatures and in the presence of graphite, some methane is expected to be present (Figure 3.13), which would displace the boundaries slightly.

(2) For those equilibria that involve calcite and dolomite, a slight reduction in the activity of $CaCO_3$ in calcite, resulting from the presence of some $CaMgC_2O_6$ in solid solution, was disregarded, but was later taken into consideration (Skippen 1974).

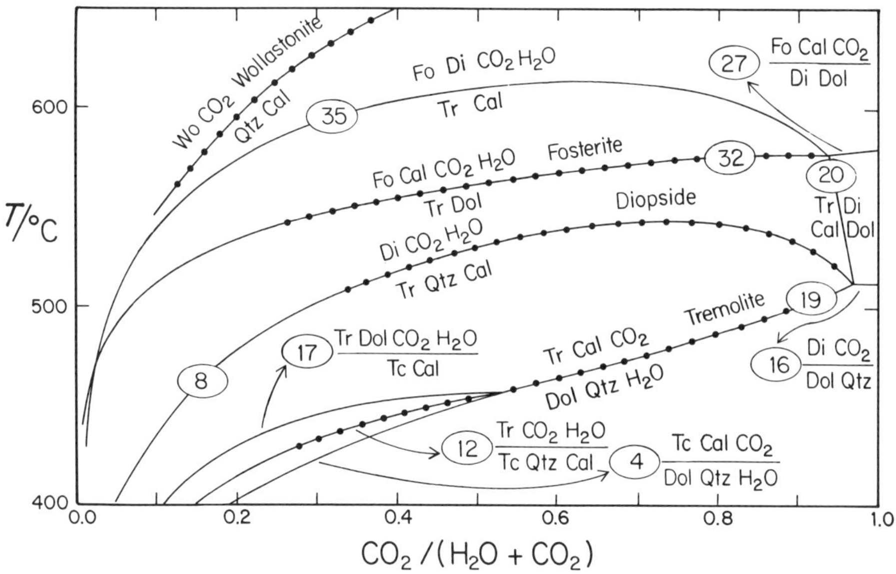

Figure 3.15 Phase relations involving calcite (Cal), dolomite (Dol), talc (Tc), tremolite (Tr), diopside (Di), forsterite (Fo), wollastonite (Wo), and a CO_2–H_2O gas mixture, at 2.0 kbar. From Skippen (1971, *Journal of Geology*). © University of Chicago Press, reproduced by permission

(3) Equilibria involving talc and tremolite are displaced if some F replaces OH in the crystals.

(4) The diagram applies to systems in which Ca > Mg, i.e., magnesite and enstatite are absent; systems with Ca < Mg, which are much less common, are considered elsewhere (Skippen 1974).

(5) Aside from methane, other components may be present in natural systems; for example, the influence of NaCl on the equilibria of Figure 3.15 was considered by Skippen and Trommsdorff (1986).

(6) With increasing pressure, the curves migrate upwards, and at 3 kbar, boundary 8 intersects 32 (Skippen 1974) providing an opportunity for forsterite to crystallize at a lower temperature than diopside.

In addition, when viewing the results in relation to equilibria within the Earth, H_2O and CO_2 may not be as accessible as they are in the experimental situation. If a fluid phase is not present, one can still refer to the fugacity of H_2O and CO_2 within a crystal-boundary phase, but the values could be somewhat reduced, and under these conditions CO_2 and H_2O molecules could presumably move, by diffusion, independently.

Despite numerous uncertainties, Figure 3.15 is extremely helpful in the interpretation of natural mineral assemblages in marble. Viewing this diagram at $X_{CO_2} \sim$

0.4–0.6, one might anticipate, with increasing grade, the sequence talc, tremolite, diopside, forsterite. This is precisely the sequence that occurs in the Alps, as was described by Trommsdorff (1966, 1972) and is illustrated in Figure 3.16. A similar sequence occurs in Russia (Kitsul 1964), the eastern Pyrenees (Guitard 1966), the Damara Belt of Namibia (Puhan and Hoffer 1973), the Alta aureole in Utah (Moore and Kerrick 1976), a contact aureole in Gifu-ken, Japan (Suzuki 1977) and in Beinn an Dubhaich aureole in Scotland (Hoersch 1981).

In the Alps and elsewhere, e.g., in the Grenville marbles (Hutcheon and Moore 1973, Kretz 1980), isobaric univariate assemblages are common, and invariate assemblages (e.g., calcite, dolomite, tremolite, diopside, forsterite; Figure 3.15) occur locally. Puhan and Hoffer (1973), Skippen (1974), and Greenwood (1975) have interpreted this to mean that the CO_2–H_2O phase was, to a certain extent, buffered by the assemblage. For example, beginning at $X_{CO_2} = 0.7$, as the tremolite curve is encountered, the reaction consumes H_2O and releases CO_2, which is expected to increase X_{CO_2}. With a further increase in temperature, the CO_2:H_2O ratio, by this interpretation, is controlled by the reaction, which results in persistence of the univariate assemblage over a range of temperature, leading possibly to the invariant point, calcite–dolomite–quartz–tremolite–diopside.

In general, because of curvature in the boundaries and because of their displacement with confining pressure, the reaction temperature for a particular reaction can vary considerably. But the two forsterite-forming reactions which involve dolomite as a reactant, do not vary much in temperature (for $X_{CO_2} > 0.5$ and a given confining pressure), assuming the reaction occurs on or slightly above the boundary, and these may provide a suitable reaction isograd. In general the appearance of forsterite in marble coincides approximately with the appearance of orthopyroxene in gneisses and amphibolites, and marks the lower boundary of the granulite facies.

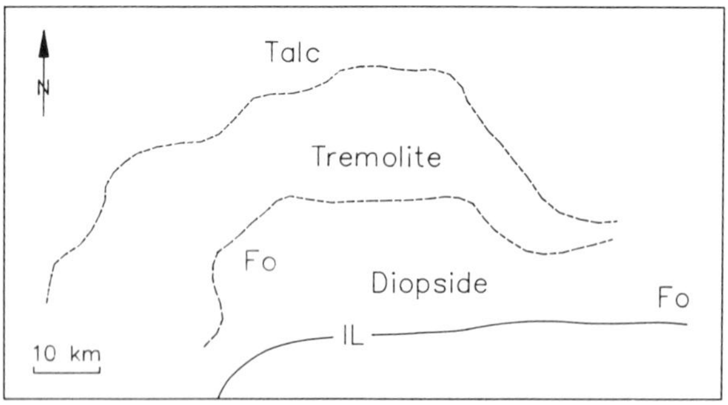

Figure 3.16 Distribution of talc, tremolite, diopside, and forsterite in the central Alps; tremolite is absent in the talc zone, and diopside in the tremolite zone; forsterite (Fo) occurs locally in the diopside zone. IL denotes the Insubric Line. From Trommsdorff (1972)

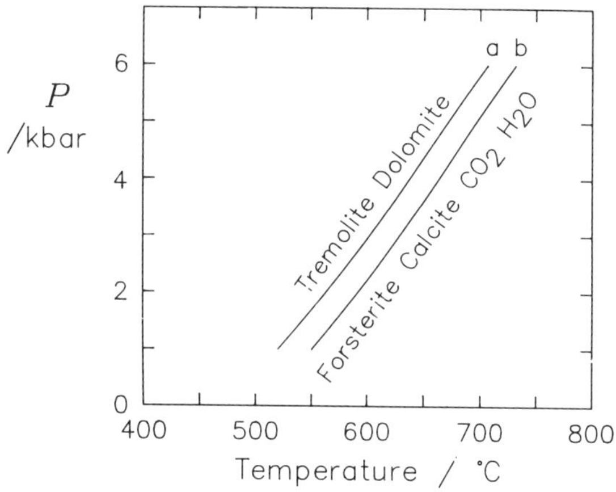

Figure 3.17 Boundary curve for a forsterite-forming reaction at $X_{CO_2} = 0.3$ (curve a) and $X_{CO_2} = 0.9$ (curve b). From Metz (1976). Reproduced by permission of Springer-Verlag

The location of the boundaries,

$$\text{tremolite} + \text{dolomite} = \text{forsterite} + \text{calcite} + CO_2 + H_2O,$$

$$\text{diopside} + \text{dolomite} = \text{forsterite} + \text{calcite} + CO_2,$$

as determined by Metz (1976) and Käse and Metz (1980) is therefore of interest; the former boundary is shown in Figure 3.17. These data indicate a temperature of some 700 °C for the amphibolite–granulite facies transition.

3.3 ORDER–DISORDER TRANSFORMATIONS

The term 'order–disorder' is here used to refer to a change in the arrangement of two kinds of atoms on two or more similar structural sites in a crystal. This 'substitutional' disorder is part of a large and rapidly growing field (Nagle 1975, Parsonage and Staveley 1978) which includes the study of positional disorder, magnetic transitions, and other forms of disorder in crystals. Changes of this kind are sometimes referred to as second-order transformations, to distinguish them from first-order transformations, which were examined in Sec. 3.1. We are at present concerned primarily with the concept of an equilibrium arrangement or distribution of atoms and the effect of temperature on this distribution rather than with reaction rates and the mechanisms by which the changes occur.

A review of the theory of substitutional disorder in metals was presented by Nix and Shockley (1938). Disorder in minerals has been reviewed by Burnham (1973), Ghose (1982), Carpenter (1985), Ganguly and Saxena (1987), and Putnis (1992).

The subject is here discussed under three headings: (1) the distribution of Cu and Zn atoms in the alloy CuZn, as a demonstration of order–disorder phenomena in general, followed by a discussion of disorder in dolomite, (2) the distribution of Al and Si on the tetrahedral sites in K-feldspar, and (3) the distribution of Mg and Fe between the M1 and M2 octahedral sites of orthopyroxene.

3.3.1 Order–disorder in CuZn and dolomite ($CaMgC_2O_6$)

Order–disorder transformations in metals were detected in 1925 by Johannson and Linde (by use of X-rays), but the phenomenon was anticipated in 1919 by Tammann (Bragg and Williams 1934). One of the metals that has been studied in detail is β-brass of composition CuZn (Chipman and Warren 1950), and an examination of this phase will serve as an introduction to the concept of order and disorder in crystals.

At temperatures below 200 °C, the Cu atoms in CuZn occupy the corners of a body-centered cubic lattice, while the Zn atoms occupy equivalent positions at the center of the cube. The structure is ordered, as shown in the two-dimensional analogue of Figure 3.18a. As temperature increases, some Cu atoms move into the Zn sites, and vice versa, and the structure becomes somewhat disordered, as in Figure 3.18b. At 465 °C, the Cu:Zn ratios on each of the original Cu and Zn sites are very nearly equal, and the structure is disordered or random, as in Figure 3.18c.

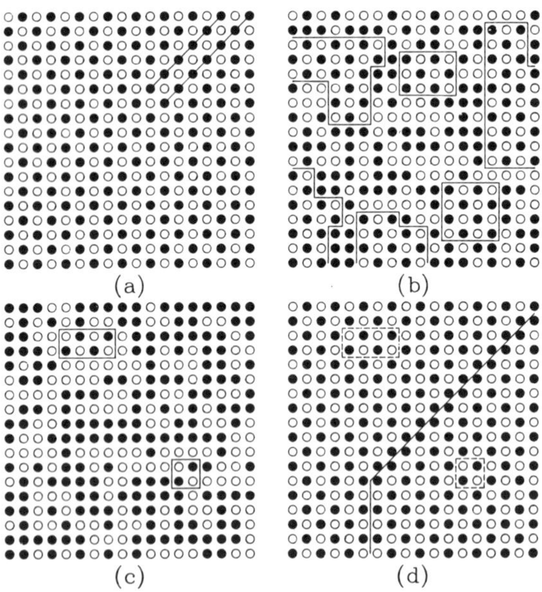

Figure 3.18 Two-dimensional atomic models of (a) order, (b) partial disorder (with domains of local order), (c) complete disorder (note two small domains of order), and (d) pattern produced by ordering in (c), i.e., by growth of the two small domains; note out-of-step boundary

This change can be detected by use of X-rays. In the ordered crystal, alternate planes of like and unlike atoms (Figure 3.18a) produce X-ray reflections which do not appear when the disordered crystal is examined. The ordered crystal is said to 'possess a superstructure'. With increasing temperature and disorder, the superstructure reflections progressively fade and disappear.

When the structure is disordered and atoms are arranged at random, small domains of order will necessarily exist (an integral part of a random array) and as temperature falls and ordering occurs, these will grow in size. In general the crystal cannot, however, become completely reordered because not all of the small domains will be 'in step'. Thus, ordered domains are created (Figure 3.18d), separated by out-of-step boundaries. This 'anti-phase structure' was anticipated by Bethe (1935) and has been observed in some minerals. Complete order can now be realized, if at all, by the migration of these boundaries, causing the growth of some domains and the disappearance of others.

A theoretical treatment of the order–disorder problem was presented by Bragg and Williams (1934), Bethe (1935), and others (Nix and Shockley 1938). Ideally, we would like to know how the thermodynamic properties of a crystal change as the transformation takes place, for it would then be possible to apply the concept of equilibrium as it was applied above in dealing with first-order phase transformations.

Most of the existing theory is 'quasi-chemical' in nature, i.e., it combines knowledge of crystal structure with thermodynamic concepts. Following is a brief examination of this theory, as presented by Slater (1939) and Swalin (1962).

First we need a measure of long-range order, i.e., a disorder parameter (L), then a relation between the Gibbs energy of a crystal and L, and finally a relation between L and temperature (T). The CuZn example will be retained, and although this phase differs considerably from natural minerals, the results will nevertheless form a guide in dealing with disorder in some minerals.

Let A and B refer to two kinds of atoms which in a crystal are present in equal proportions ($X_A = X_B = 0.5$) and are distributed between two equivalent sites, α and β. Let x_A^α denote the fraction of A atoms on α, and x_B^β the fraction of B atoms on β. Following Bragg and Williams (1934) we define a long-range order parameter as follows:

$$L = (x_A^\alpha - X_A)/(1 - X_A) = (x_A^\alpha - 0.5)/0.5,$$

$$L = (x_B^\beta - X_B)/(1 - X_B) = (x_B^\beta - 0.5)/0.5.$$

When the crystal is completely disordered ($x_A^\alpha = x_B^\beta = 0.5$), $L = 0$; when it is completely ordered ($x_A^\alpha = x_B^\beta = 1.0$), $L = 1.0$, and when it is partially ordered L lies between 0 and 1.

From the above equations, by rearrangement,

$$x_A^\alpha = x_B^\beta = (1 + L)/2,$$

$$x_B^\alpha = (1 - x_A^\alpha) = x_A^\beta = (1 - x_B^\beta) = (1 - L)/2.$$

An expression for the enthalpy (H) of the crystal as a function of L is obtained as follows. In the CuZn structure, each α site is surrounded by eight β sites. Thus the number of pairs of atoms is $8N(1/2)$ where N is the total number of atoms, and the factor (1/2) results from each pair being counted twice. The enthalpy of the crystal is determined approximately by the number of A–A, B–B, and A–B bonds, and the enthalpy associated with each bond type. The probability that a particular α site is occupied by A is x_A^α. The probability that an adjacent β site is occupied by A is x_A^β Thus the probability for an A–A pair is

$$p_{AA} = x_A^\alpha x_A^\beta.$$

By substitution,

$$\begin{aligned} p_{AA} &= \frac{(1+L)}{2} \cdot \frac{(1-L)}{2}, \\ &= (1-L^2)/4. \end{aligned}$$

The total number of A–A pairs is

$$\begin{aligned} P_{AA} &= p_{AA}(4N) \\ &= N(1-L^2). \end{aligned}$$

Similarly, the number of B–B pairs is

$$\begin{aligned} P_{BB} &= p_{BB}(4N) \\ &= N(1-L^2). \end{aligned}$$

The number of A–B pairs is

$$\begin{aligned} P_{AB} &= 4N - P_{AA} - P_{BB} \\ &= 2N(1+L^2). \end{aligned}$$

Now the enthalpy of a gram-atom of the crystal is

$$H = P_{AA}H_{AA} + P_{BB}H_{BB} + P_{AB}H_{AB},$$

where H_{AA}, H_{BB}, H_{AB} refer to the energy (enthalpy) per bond at the temperature of interest. By substitution,

$$H = N_0(1-L^2)(H_{AA} + H_{BB}) + 2N_0(1+L^2)H_{AB},$$

where N_0 is the number of atoms in one gram-atom of crystal.

We shall now obtain an expression for the entropy of the crystal in terms of L. The number of atoms of A and B on each site, α, β, is as follows:

$$N_A^\alpha = x_A^\alpha(N/2) = [(1+L)N]/4,$$

$$N_B^\alpha = x_B^\alpha(N/2) = [(1-L)N]/4,$$

$$N_B^\beta = x_B^\beta(N/2) = [(1+L)N]/4,$$

$$N_A^\beta = x_A^\beta(N/2) = [(1-L)N]/4.$$

The configurational entropy (Sec. 2.9) is

$$S_c = k \ln \Omega,$$

where

$$\Omega = \frac{(N/2)!}{N_A^\alpha! N_B^\alpha!} \cdot \frac{(N/2)!}{N_A^\beta! N_B^\beta!},$$

$$= \left(\frac{(N/2)!}{\left[(1+L)\frac{N}{4}\right]! \left[(1-L)\frac{N}{4}\right]!} \right)^2.$$

By use of Stirling's approximation,

$$S_c = Nk \left[\ln 2 - \frac{1}{2}[(1+L)\ln(1+L) + (1-L)\ln(1-L)] \right].$$

For complete disorder ($L = 0.0$) this equation gives

$$S_c = R \ln 2,$$

and for complete order ($L = 1.0$),

$$S_c = 0.$$

The total entropy at temperature T is the sum of the configurational and vibrational terms,

$$S = S_c + \int_0^T (C_P/T)\,dT.$$

The Gibbs energy per gram-atom of crystal is obtained by combining the above values of H and S,

$$G = H - TS,$$

$$G_0 = N_0(1-L^2)(H_{AA} + H_{BB}) + 2N_0(1+L^2)H_{AB}$$
$$- T\left[N_0 k \left\{ \ln 2 - \frac{1}{2}[(1+L)\ln(1+L) + (1-L)\ln(1-L)] \right\} \right.$$
$$\left. + \int_0^T (C_P/T)\,dT \right].$$

In general, ordering at low temperatures will only occur if there is a greater attraction between unlike than between like atoms, i.e., $H_{AB} - (H_{AA} + H_{BB})/2$ is negative. (Exceptions to this statement could arise when A and B differ in size to create an appreciable strain energy, which was not considered.) On this premise, Slater (1939) plotted the above equation for different values of T and the results are shown in Figure 3.19. At low temperature, $L = 1.0$ (negative values of L may be disregarded) and as temperature increases, L decreases. At T_c (the critical temperature), L becomes zero, and at $T > T_c$, L remains zero.

By minimizing G with respect to L, Swalin (1962) obtained the equation

$$\ln\left(\frac{1+L}{1-L}\right) = \frac{2LT_c}{T},$$

which is plotted in Figure 3.20. Thus the theory predicts that disordering in a crystal is a progressive change, occurring over a range of temperature, and once the process begins, it proceeds with increasing speed. There is moderately good agreement between theory and the experimental results on ordering in CuZn (Figure 3.20).

Short-range order is a tendency for one kind of atom (A) to be surrounded by the other kind (B) over and above the A–B association that occurs in a random distribution. The quasi-chemical approach can also deal with this problem (Swalin 1962). Theory predicts that some short-range order will persist at $T > T_c$, in agreement with experimental results for metals.

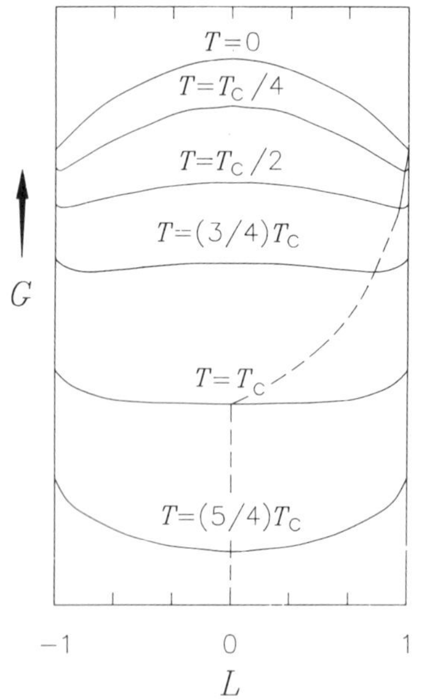

Figure 3.19 Gibbs energy of a crystal (G) as a function of the ordering parameter, L, for different temperatures (T_c is the temperature at which complete disorder is obtained, i.e., $L = 0$). The broken line is the path of minimum G. From Slater (1939)

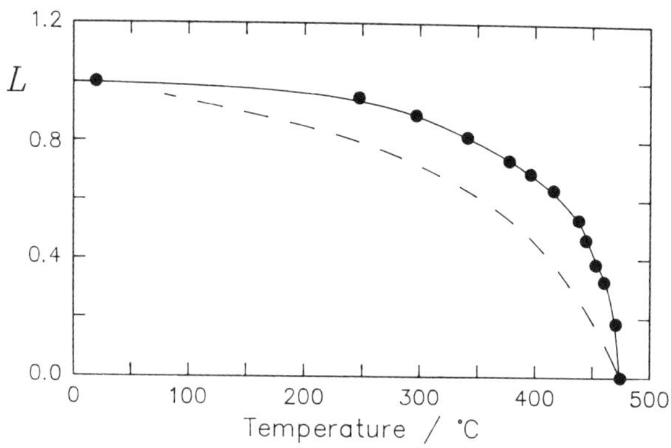

Figure 3.20 Order parameter (L) as a function of temperature; data on CuZn, from Chipman and Warren (1950); broken-line curve is for the theoretical expression, $\ln[(1+L)/(1-L)] = 2LT_c/T$

In oxide, carbonate, and silicate crystals, the atoms that experience order and disorder are not in direct contact, as they are in metals, and as ordering proceeds, changes in the framework structure may occur simultaneously. But despite these complications, a behavior somewhat similar to that prescribed by the above theory might be anticipated.

For example, the disordering process in dolomite, $CaMgC_2O_6$, may resemble the process that occurs in CuZn. In most natural dolomite, the Ca:Mg ratio is close to 1:1, and the structure is ordered, with Ca and Mg occupying alternate planes, parallel to the base. This is a superstructure ($R\bar{3}$) derived from calcite or magnesite ($R\bar{3}c$). Goldsmith and Heard (1961), in an X-ray study of annealed dolomite, observed a fading of the superstructure reflections, and concluded that disordering of Ca and Mg occurred progressively from about 1000 °C to 1200 °C. Reeder and Wenk (1983), by single-crystal X-ray refinement, determined a critical temperature near 1150 °C (some ordering on quenching could not be avoided), and confirmed that disordering occurs progressively, with $L < 0.85$ at 1050 °C and $L < 0.76$ at 1100 °C (Reeder 1983). These results are in agreement with disorder theory.

In dolomite, a disordering of Ca and Mg produces a volume change of +0.46% (Reeder and Wenk 1983). Thus an increase in pressure of a few kilobars could increase the critical temperature noticeably.

Disordered or poorly ordered dolomite is common in sediments (Reeder 1983) where it obviously represents non-equilibrium. During diagenesis and greenschist metamorphism, dolomite becomes ordered (Goldsmith and Graf 1958), either by an ordering process or by recrystallization. By the above experimental results, detectable disordering does not occur below ~1000 °C, and dolomite at medium- to high-grade temperatures is expected to remain ordered. Very minor disorder at upper granulite temperatures (~800 °C) cannot, however, be ruled out. The experiments have demonstrated that ordering at ~1000 °C is rapid, and if some disorder was created in very high-grade dolomitic marble or in carbonatite, it would be eliminated during cooling.

3.3.2 Order–disorder in K-feldspar ($KAlSi_3O_8$)

K-feldspar occurs in different forms that in the past have been referred to as microcline, orthoclase, and sanidine. Barth (1934) proposed that these forms represent differences in the degree of Al–Si disorder in the crystal structure.

The structure of K-feldspar, as described by Ribbe (1984) consists of a framework of linked (Si, Al)O_4 tetrahedra in which cavities exist and are occupied by potassium ions. One quarter of the tetrahedra are centered by Al, the remainder by Si. In sanidine, which is monoclinic, there are two structurally distinct tetrahedral sites, labelled T_1 and T_2; in high-temperature sanidine the Al and Si atoms are distributed at random on these sites. At about 500 °C, monoclinic sanidine changes to triclinic microcline, which contains four structurally distinct tetrahedral sites, labelled T_1O, T_1m, T_2O, and T_2m; Al is now concentrated in T_1O. The relative positions of the sites are shown in Figure 3.21 (inset).

Numerous X-ray investigations have been carried out to show that differences exist in the distribution of Al and Si atoms in sanidine and microcline. Although the distribution of these atoms represents an order–disorder phenomenon more complex than that found in CuZn, as discussed above, a comparison with metals is nevertheless possible. Thus Hafner and Laves (1957) have proposed the disordering scheme shown in Figure 3.21 (which may be compared with Figure 3.20),

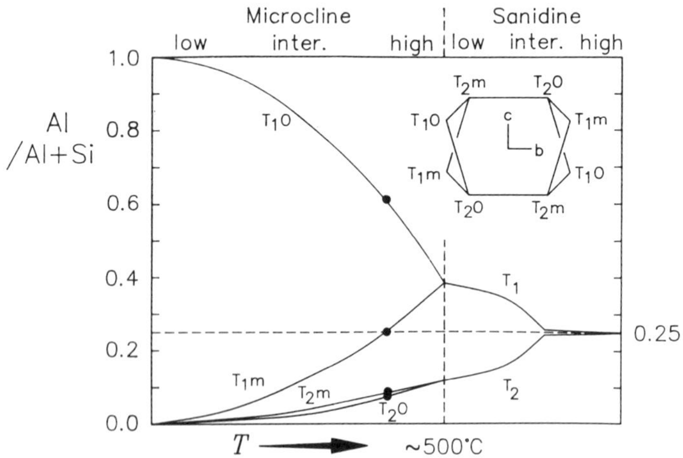

Figure 3.21 Order–disorder scheme proposed by Hafner and Laves (1957); dots are data for an intermediate microcline. Inset (from Ribbe 1984) shows relative positions of tetrahedral sites; in sanidine, T_1O and T_1m are equivalent, as are T_2O and T_2m. Inset by permission of Kluwer Academic Publishers

defining the disorder parameter as the atomic fraction Al/(Si+Al) on each class of site. Beginning with low (maximum) microcline which is completely ordered (T_1O filled with Al, the others filled with Si), an increase in temperature causes a redistribution of the atoms to a more random pattern (partial disorder). At about 500 °C, the structure becomes monoclinic, and beyond 500 °C disordering continues progressively, until at some high temperature disorder is complete.

Numerous heating experiments have been carried out on natural and synthetic $KAlSi_3O_8$ to provide information on the relation between crystal structure, including degree of order, and temperature; the results were reviewed by Brown and Parsons (1989). In this kind of work it is important to have a relatively rapid means of determining Al:Si concentration on the tetrahedral sites; the various methods that are available are described by Ribbe (1984) and Smith and Brown (1988). It is also necessary to define a disorder parameter and several have been proposed; one adopted by Brown and Parsons (1989) is

$$t_1\text{O} + t_1\text{m},$$

where t_1O and t_1m denote atomic fractions Al/(Si+Al) on sites T_1O and T_1m respectively. This parameter ranges from 0.5 (complete disorder) to 1.0 (complete order), with a value near 0.75 at the sanidine–microcline transition. The Al:Si ratios in T_2O and T_2m are nearly equal (Figure 3.21).

The approximate position of the equilibrium disorder curve, based on numerous experimental data, is shown in Figure 3.22. The experimental results, combined with information on natural occurrences and recently obtained microstructural (TEM) images, can now be integrated to provide the following partial understanding of the behavior of $KAlSi_3O_8$ within the Earth. A major outstanding problem is the nature and status of common orthoclase.

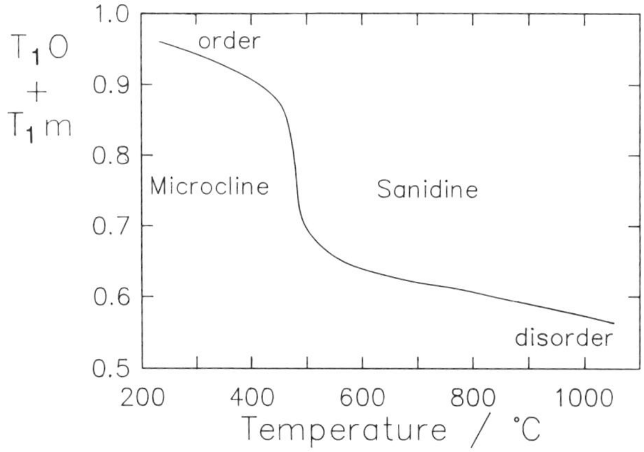

Figure 3.22 Relation between order parameter ($T_1O + T_1m$) in K-feldspar and temperature, based on numerous experimental results. From Brown and Parsons (1989). Reproduced by permission of the Mineralogical Society, London

Moving from high to low temperature, there is no difficulty with high sanidine, which occurs in some volcanic rocks (normally 10–20% of the K is replaced by Na), and is stable in the vicinity of 1000 °C. Sanidine may persist as a disordered phase during cooling, which implies that the rate of ordering is considerably slower than in metals and dolomite.

The dominant form of $KAlSi_3O_8$ in high-grade gneisses and granulites is 'orthoclase'. From the studies of Laves (1952), McConnell (1965), Eggleton and Buseck (1980), and others, it is now clear that 'orthoclase' consists of a very fine microstructure of alternating ordered and anti-ordered domains of triclinic symmetry. In high-grade metamorphic terranes and granite plutons, K-feldspar evidently crystallized as low sanidine. During cooling, the domain microstructure forms as the symmetry changes from monoclinic to triclinic, and the domain microstructure evidently provides an impediment to further ordering (Brown and Parsons 1989). Shear stress and possibly the presence of a vapor phase can promote ordering and cause unstable 'orthoclase' to change to low microcline. Conditions which control the rate of ordering are not fully understood, however, because widely different states of disorder are found within small volumes of rock (Smithson 1962, Dal Negro, De Pieri and Quareni 1978).

Presumably there is only one *equilibrium* ordering path but different paths can be taken during non-equilibrium ordering, i.e., when the reaction lags behind temperature. This was discussed by Hafner and Laves (1957) who considered single-step ordering (as compared with the two-step ordering of Figure 3.21), in which all Al atoms move directly into T_1O. Cherry and Trembath (1979) presented evidence for two-step ordering in natural K-feldspar crystals from granite plutons and metamorphic rocks, and found evidence for two-step disordering in the heating of crystals, and a two-step process of both ordering and disordering is evidently demanded by equilibrium.

Several studies have been carried out on natural 'orthoclase' and microcline in metamorphic terranes. Results from various localities, referred to by Collerson (1976), show that low microcline is the dominant form in the amphibolite facies, and 'orthoclase' in the granulite facies, but much variation in the degree of disorder is found locally, especially in the upper amphibolite and granulite facies. These trends and variations are interpreted in terms of changes that occurred during cooling, i.e., conditions that existed at the peak of metamorphism were not preserved during cooling.

In the Front Range of Colorado, the Eldora quartz–monzonite stock of Tertiary age was emplaced into Precambrian schist and amphibolite which enclose lenses of K-feldspar-bearing pegmatite. Steiger and Hart (1967) and Wright (1967) examined the structural state of K-feldspar from the pegmatite veins along a traverse that approached the stock from a point 7 km away, i.e., in a direction which obviously represented, during stock emplacement, an increase in temperature. Far from the stock, the K-feldspar was found to be low microcline. Between 840 m and 340 m from the stock, a mixture of low microcline and 'orthoclase' was found, the proportion of 'orthoclase' increasing toward the stock. From 340 m to the contact, 'orthoclase' is the only form present. Here, as elsewhere, the presence of some 'orthoclase' is taken as evidence that the microcline–sanidine inversion temperature was exceeded. Heat-flow calculations for the Eldora stock and aureole place the transition temperature about 100 °C below the present consensus of ~500 °C (Heier 1957, Brown and Parsons 1989) for the inversion.

In the schist–gneiss terrane of the Hohe Tauern area (Eastern Alps), Raase and Morteani (1976) and Bernotat and Morteani (1982) located an isograd separating highly ordered microcline in rocks of the greenschist facies from dominantly intermediate microcline and 'orthoclase' in a large central amphibolite-facies zone. Oxygen-isotope temperatures on the isograd are approximately 500 °C. The authors have interpreted the results in terms of a temperature gradient, and suggest that the central zone experienced peak metamorphic temperatures of 500 to 650 °C, which transformed the low microcline initially present to low sanidine. Assuming that the temperature gradient across the isograd was smooth, K-feldspar in the marginal (greenschist-facies) zones must have reached temperatures just below 500 °C, but because it was not transformed to monoclinic sanidine, the crystals ordered completely during subsequent cooling. The importance of deformation as an additional control on ordering of K-feldspar in Alpine terranes was emphasized by Merz, Thelin and Persoz (1989).

3.3.3 Order–disorder in orthopyroxene, $((Mg, Fe)^{M2}(Mg,Fe)^{M1}Si_2O_6)$

The crystal structure of orthopyroxene was reviewed by Ghose (1982). Parallel chains of linked SiO_4 tetrahedra are held together by Mg and Fe atoms that occupy two kinds of octahedral sites. The M1 octahedron, which corresponds to the Mg octahedron in diopside, $CaMgSi_2O_6$, is nearly regular, with average cation–oxygen distances of 2.075 Å (in enstatite) and 2.135 Å (in ferrosilite). The M2 octahedron, which corresponds to the Ca polyhedron in diopside, is distorted, with average distances to four oxygen ions at 2.042 Å (enstatite) and 2.076 Å (ferrosilite), while distances to the remaining two oxygen ions are much greater, at 2.290 Å and 2.447 Å (enstatite), and 2.460 Å and 2.600 Å (ferrosilite). In enstatite, $Mg^{M2}Mg^{M1}Si_2O_6$, both sites are occupied by Mg; in common orthopyroxene which contains Mg and Fe, an interesting question arises concerning the distribution of Mg and Fe atoms between the two sites, and the effect of temperature on this distribution.

Ghose (1965) carried out a single-crystal X-ray study of an orthopyroxene sample of intermediate Fe:Mg ratio from a granulite-facies terrane in Greenland, and he found that Fe was concentrated in M2 (x^{M2} = Fe/(Mg+Fe) = 0.90) and Mg in M1 ($x^{M2} = 0.15$). Subsequently, many metamorphic and igneous orthopyroxene crystals of different Mg:Fe composition have been analyzed, principally by Mössbauer resonance spectroscopy (Ghose 1982), and in all cases, Fe atoms show a preference for M2, Mg for M1. Cases of complete disorder, $x^{M2} = x^{M1}$, were not encountered.

The following information shows that disorder in orthopyroxene differs considerably from disorder in metals (CuZn) and to some extent from disorder in K-feldspar, as discussed above. (1) The two sites (M2, M1), although similar, remain distinctly different with regard to bond length and bond angle, regardless of their occupancy, (2) the relative numbers of the two atoms (Mg, Fe) can vary greatly and continuously, depending on choice of sample, and (3) complete order and complete disorder evidently do not occur. It is therefore convenient to redefine order–disorder when dealing with orthopyroxene and other Fe–Mg silicates.

Mueller (1967a) introduced the disorder parameter

$$\frac{x^{M2}}{1-x^{M2}} \cdot \frac{1-x^{M1}}{x^{M1}},$$

where x is the atomic ratio Fe/(Mg+Fe) and the quotient is denoted k_D. This is based on the proposal that the order–disorder process in orthopyroxene may be viewed as a within-crystal exchange reaction,

$$Fe^{M2} + Mg^{M1} = Mg^{M2} + Fe^{M1}.$$

For example, an orthopyroxene crystal with $x^{M2} = 0.76$, $x^{M1} = 0.24$ ($k_D = 10.0$) may be viewed as the product of an ordering process (1) or a disordering process (2) as shown in Figure 3.23.

In view of the difficulty of treating this problem in terms of thermodynamic components and chemical potentials, Mueller (1967a) adopted a kinetic approach,

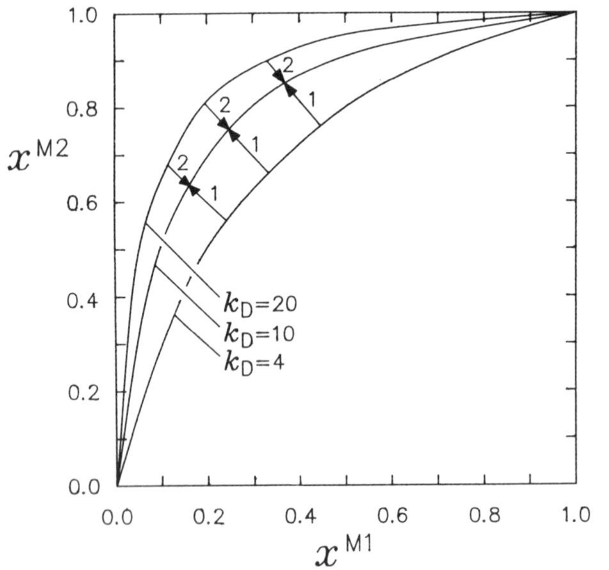

Figure 3.23 Illustrative plot of the equation $k_D = x^{M2}(1 - x^{M1})/(1 - x^{M2})x^{M1}$ for k_D = 20, 10, 4, representing differences in the distribution of Mg and Fe (x = Fe/(Mg + Fe)) on sites M1 and M2 in orthopyroxene. Arrows represent ordering (1) and disordering (2)

assuming that in a disordering process, the rate of change of concentration of Fe on M1 is expressible by the equation

$$-\frac{dx^{M1}}{dt} = \frac{1}{2}C_0(K_{M1M2}\phi_{M1M2}x^{M1}(1-x^{M2}) - K_{M2M1}\phi_{M2M1}x^{M2}(1-x^{M1})),$$

where t is time, C_0 is concentration of total sites (M1 and M2), K denotes a rate constant, and ϕ represents a coefficient comparable to a macroscopic activity coefficient. At equilibrium, $dx^{M1}/dt = 0$, and

$$\frac{x^{M2}}{1-x^{M2}} \cdot \frac{1-x^{M1}}{x^{M1}} = \frac{K_{M1M2}\phi_{M1M2}}{K_{M2M1}\phi_{M2M1}}.$$

The quotient on the left is the disorder parameter, k_D.

Initially we shall assume that the two ϕ terms equal 1.0; in this case k_D at a given temperature and pressure is a constant, i.e., independent of Fe/(Mg+Fe) in the crystal as a whole. Illustrative plots of the equation

$$\frac{x^{M2}}{1-x^{M2}} \cdot \frac{1-x^{M1}}{x^{M1}} = k_D,$$

which may be written

$$x^{M2} = k_D/(k_D - 1 + (1/x^{M1})),$$

for k_D = 20, 10, and 4 are shown in Figure 3.23.

From a general thermodynamic viewpoint, k_D may be regarded as an equilibrium constant:

$$\ln k_D = -\Delta G^\circ/RT,$$

where ΔG° is the exchange energy. Thus,

$$\frac{\partial \ln k_D}{\partial(1/T)} = -\Delta H^\circ/R,$$

where ΔH° is the near-constant exchange enthalpy. Consequently a linear $\ln k_D$ versus $(1/T)$ relation is expected.

Saxena and Ghose (1971) carried out a detailed experimental study of Mg–Fe disorder in orthopyroxene crystals of different total Mg:Fe composition, and the results are shown in Figure 3.24. Natural crystals were annealed at set temperatures for one to three weeks, cooled rapidly, and then examined by Mössbauer spectroscopy to determine the concentration of Fe on M2 and M1. Disregarding the most Fe-rich crystals, the data can be represented exactly by use of the parameter k_D as defined above (Figure 3.24), i.e., with k_D independent of Mg:Fe in the crystal

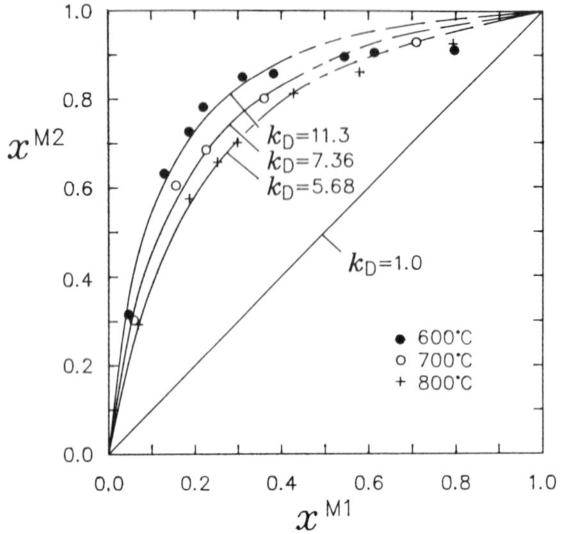

Figure 3.24 Distribution of Fe and Mg between M1 and M2 sites of orthopyroxene at 600, 700, and 800 °C. Curves of $k_D = 11.3$, 7.36, and 5.68 are drawn to fit the data, barring the crystals rich in Fe. Experimental data are from Saxena and Ghose (1971)

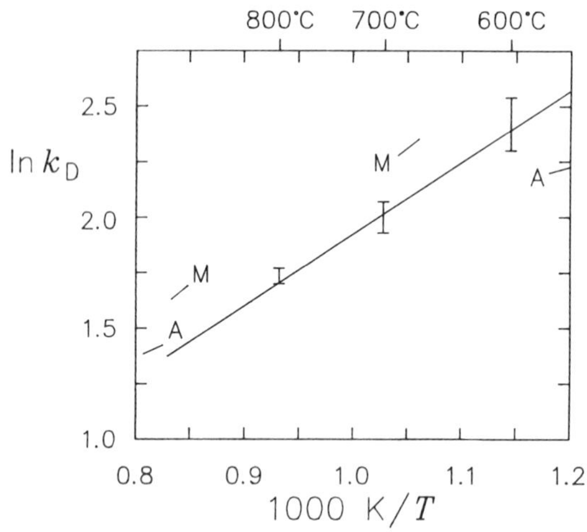

Figure 3.25 Linear ln k_D versus $(1/T)$ plot for the data from Figure 3.24; error bars are 0.95 confidence limits; the relation is, $\ln k_D = (3250/T) - 1.31$. Lines A–A and M–M are results of Anovitz, Essene and Dunham (1988) and Molin, Saxena and Brizi (1991)

as a whole. The change of k_D with temperature is shown in Figure 3.25; a linear relation of $\ln k_D$ with $1/T$ is observed, as expected.

In the above treatment we have taken the ϕ terms to equal 1.0. It is possible to estimate values for these terms, and thereby obtain a better fit to the experimental data in the Fe-rich portion (upper right) of Figure 3.24. This was done by Saxena and Ghose (1971) and Ganguly (1982).

Some more recent experimental data are available. Those obtained by Anovitz, Essene and Dunham (1988) are represented by line A–A in Figure 3.25 and those obtained by Molin, Saxena and Brizi (1991), by line M–M These differ slightly from the earlier results of Saxena and Ghose (1971).

Several orthopyroxene crystals from granulite-facies rocks were analyzed by Virgo and Hafner (1970) and were found to fall below the 600 °C isotherm shown in Figure 3.24. Hence, some exchange evidently occurred within these crystals as they slowly cooled from the peak of metamorphism, which was 700–800 °C. Crystals from volcanic rocks were found to fall near the 600 °C isotherm, indicating that the more rapid cooling of these crystals did not permit the exchange reaction to proceed as far as in the metamorphic crystals. Similar results were obtained by Johnston and Knedler (1979). These results reflect the decreasing mobility of atoms with decreasing temperature, preventing the reaction from proceeding at temperatures lower than about 500 °C.

Within-crystal distribution data are available for other Fe–Mg minerals, including Ca pyroxene, the amphibole minerals, olivine, biotite, and spinel. Sources to this literature are given by Navrotsky and Kleppa (1967), Perchuk and Surikov (1971), Ghose (1982), and Ganguly (1982).

One may also deal with disorder in Fe–Mg minerals in terms of crystal chemistry. Burns (1970) has used optical absorption spectra to calculate the crystal-field stabilization energy of Fe in different sites, thereby providing, for example, an explanation for the concentration of Fe in M2 of orthopyroxene. Ohashi and Burnham (1972), from electrostatic and repulsive energy considerations, determined that the bond energy in orthopyroxene is minimized when Fe resides in M2, Mg in M1. O'Nions and Smith (1973) applied molecular orbital theory, which takes into consideration the partial covalent nature of Fe–O bonds, to deduce that Fe–O bonds in M2 are on average more covalent than in M1, as proposed by Ghose (1965), and therefore Fe in M2 is favored. Qualitatively, these theories are in agreement with observation but they are not yet capable of predicting k_D values and k_D–T relations.

In this section we have touched on some of the interesting problems that exist regarding order–disorder phenomena in minerals. For further and more advanced discussions of the subject, the reader is referred to Thompson (1969), Burns (1970), Navrotsky (1971), Ganguly and Saxena (1987), and Truskinovskiy et al. (1988).

3.4 SHARING OF TWO COMPONENTS BY TWO PHASES (SOLVUS EQUILIBRIA)

Several examples exist of two minerals of similar or identical structure which at metamorphic temperatures and pressures are mutually miscible to a limited degree. With increasing temperature, miscibility increases and may become complete or may be interrupted by melting or other transformations. Familiar examples include albite–K-feldspar, calcite–dolomite, muscovite–paragonite, enstatite–diopside, and the discontinuities that exist in the plagioclase solid-solution series. The equilibrium composition of the two associated phases, as expressed in a temperature–composition phase diagram, is here referred to as a 'solvus', regardless of whether or not the structure of the two phases is identical, and the equilibrium relations are referred to as 'solvus equilibria'. Miscibility in the alkali feldspar and the calcite–dolomite systems are here chosen to illustrate mineral reactions and equilibria of this kind. Finally, we shall briefly view immiscibility in terms of atomic interactions.

3.4.1 The alkali feldspar, $(Na,K)AlSi_3O_8$, solvus

It is well known from the study of felsic volcanic rocks that despite a structural discontinuity, a continuous solid-solution series exists at high temperatures between high albite ($NaAlSi_3O_8$, triclinic) and sanidine ($KAlSi_3O_8$, monoclinic). This was confirmed by Bowen and Tuttle (1950) who observed a continuous change in $(\bar{2}01)$ X-ray reflections for crystals which were prepared synthetically at 900 °C. In rocks formed at lower temperatures (e.g., granite, pegmatite, gneiss) a Na-rich and a K-rich phase are commonly found side by side, and a region of immiscibility must therefore exist. Thus Barth (1956, 1969), from a study of natural occurrences and temperature estimates, proposed the limits of miscibility, i.e., the solvus, shown in Figure 3.26 (line B).

Where Na-rich crystals (Ab) and K-rich crystals (Kf) are associated, the reaction of interest is

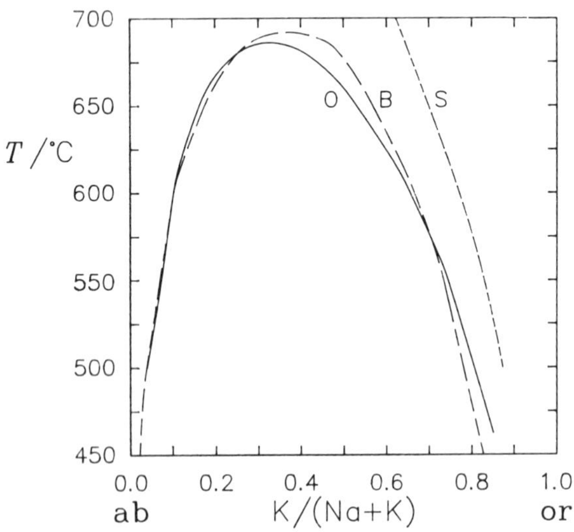

Figure 3.26 Alkali feldspar solvus, proposed by Barth (1969, curve B), and located experimentally at 2.0 kbar by Orville (1963, curve O). The K-feldspar limb at 10.0 kbar (curve S) is from Seck (1972)

$$NaAlSi_3O_8 \text{ (in Ab)} = NaAlSi_3O_8 \text{ (in Kf)},$$

$$KAlSi_3O_8 \text{ (in Ab)} = KAlSi_3O_8 \text{ (in Kf)},$$

and the energy-balance equations are

$$\mu_{ab}^{Ab} = \mu_{ab}^{Kf},$$

$$\mu_{or}^{Ab} = \mu_{or}^{Kf},$$

where ab and or stand for the solution components $NaAlSi_3O_8$ and $KAlSi_3O_8$, and μ is chemical potential.

Orville (1963) located the solvus by carrying out a number of (Na,K) ion-exchange experiments in which synthetic alkali feldspar crystals (annealed at 1050 °C) were brought into equilibrium with an aqueous vapor containing 2 molar (total) NaCl and KCl in solution. Measured amounts of crystal and solution of known composition were sealed in Pt capsules and subjected to a pressure of 2.0 kbar in a cold-seal pressure vessel. Experiments were carried out in the range of 350 to 700 °C and for durations of 1.5 to 720 h. Upon completion of an experiment, the composition of the quenched crystals was determined by X-rays and that of the fluid by chemical analysis. The results obtained at 600 °C for example, revealed a discontinuity in the composition, $X_{or} = K/(Na + K)$, of the crystals extending from $X_{or} = 0.12$ to 0.65, thus providing a point on each limb of the solvus. By this procedure, the miscibility gap at different temperatures was determined, and the resulting 2.0 kbar solvus is shown in Figure 3.26 (line O). The peak on the solvus lies at a temperature (T_c) of 686 °C and a composition of X_{or} = 0.33. The Orville solvus differs slightly from one obtained earlier by Bowen and Tuttle (1950) who measured the composition of unmixed alkali feldspar in hydrothermal experiments, and is almost identical to the solvus proposed by Barth (1969; Figure 3.26, line B).

In Chapter 2, Sec. 2.10, we noted that in the Margules symmetrical solution model (II) values of the Margules parameter (W) greater than $2RT_c$ result in immiscibility (Figure 2.12), in the form of a symmetrical solvus. Similarly, large positive values of W_A and W_B in the asymmetrical model (III) give rise to an asymmetrical solvus (Figure 2.13). Although the alkali feldspar solvus is obviously not symmetrical about X_{or} = 0.5, the *approximate* magnitude of the two Margules parameters, W_{ab} and W_{or} may be obtained by use of the equation

$$W = 2RT_c$$

which would apply exactly if the solvus were symmetrical. This equation is derived by Guggenheim (1977, p. 194). Thus for the Orville solvus (T_c = 960 K), we anticipate W values of approximately 16 000 J/mol.

Thompson and Waldbaum (1968) have used the Orville (1963) experimental data in combination with Margules asymmetrical equations and the energy-balance restrictions, as stated above, to calculate W_{ab} and W_{or} parameters. (A slightly different procedure is described by Luth and Fenn (1973).) For example, at 650 °C, the parameters are $W^G_{ab} = 13\,140$ J/mol and $W^G_{or} = 17\,620$ J/mol. Thus the molar Gibbs energy of mixing ($\Delta_m G$) at 650 °C is given by the equation (Sec. 2.10, Model II),

$$\Delta_m G = [X_{ab}RT \ln X_{ab} + X_{or}RT \ln X_{or})] + [(X_{ab}X_{or})(X_{ab}W^G_{or} + X_{or}W^G_{ab})],$$

with $R = 8.314$ J/K·mol, $T = 923$ K, and W^G_{ab} and W^G_{or} as noted above. The equation is plotted in Figure 3.27, where the common tangent (a–b) defines the 650 °C, 2.0 kbar miscibility limits as shown in Figure 3.26.

Luth and Tuttle (1966) investigated the effect of bulk composition on the position of the solvus. Three solvus curves were located (see Luth 1974) corresponding to a stoichiometric (ab, or) starting composition, one with excess Al (peraluminous) and one with excess Na and K (peralkaline); the latter is nearly identical to that obtained by Orville (1963). It is possible that in the three sets of Luth–Tuttle experiments, structural or compositional differences existed in the feldspar crystals which, although small, were sufficiently large to displace the solvus noticeably.

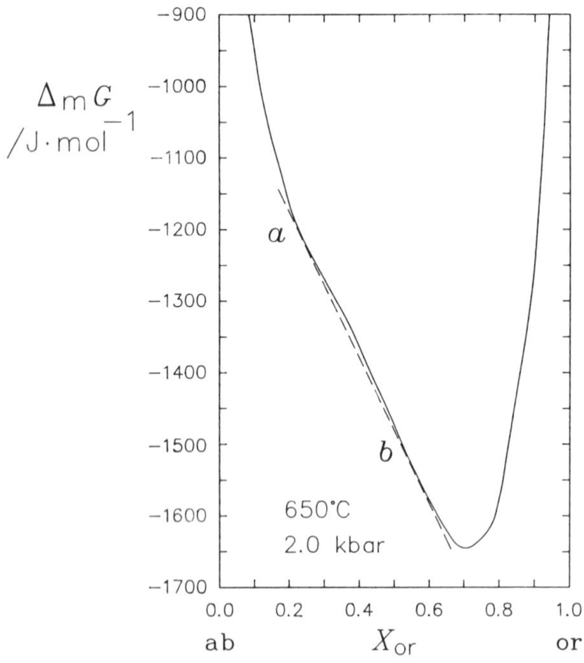

Figure 3.27 Gibbs energy of mixing ($\Delta_m G$) in alkali feldspar at 2.0 kbar and 650 °C. Margules parameters ($W^G_{ab} = 13\,140$ J/mol; $W^G_{or} = 17\,620$ J/mol) are from Thompson and Waldbaum (1968); the common tangent, a–b, defines the miscibility gap (Figure 3.26) at 650 °C

The effect of pressure on the solvus was investigated by Luth and Tuttle (1966) and Seck (1972) who showed that increasing pressure causes the crest of the solvus to rise. For example, the K-feldspar limb of the solvus at 10.0 kbar pressure as determined by Seck (1972) is shown in Figure 3.26 (line S). Numerous additional data have been obtained and are compiled by Merkel and Blencoe (1982) and by Haselton et al. (1983).

In both albite and K-feldspar a number of structural and substitutional changes occur (or are demanded by equilibrium) as temperature increases from ~300 °C to ~800 °C (see Sec. 3.3.2 for information on K-feldspar) and the thermodynamic properties of the crystals, as reviewed by Helgeson et al. (1978) must vary accordingly. Some of these changes occur slowly, and hence complete equilibrium with regard to the solvus and other reactions was possibly not achieved in all of the experimental investigations. This was recognized by Orville (1963) who noted that by using high-temperature crystals as starting materials, complete equilibrium was possibly not achieved at lower temperatures. Thus the problem of solvus equilibrium in the alkali feldspar system is complex and not yet fully resolved.

Some data on associated albite and K-feldspar in medium-grade metamorphic rocks (staurolite–kyanite zone) from the Appalachians may be examined in relation to the experimental solvus. Scotford (1973) reported analyses of albite (X_{or} = 0.013 to 0.040, mean 0.028; one extreme omitted) and K-feldspar (X_{or} = 0.82 to 0.92, mean 0.85), considering only Ca-poor (<1.3 wt % CaO in plagioclase) mineral pairs. These results, in relation to the Orville solvus, represent equilibrium at approximately 470 °C, in agreement with other estimates of temperature in the mid-amphibolite facies.

Exsolution, to form an intergrowth of two feldspar phases (perthite) is a process that can be understood to a large extent by referring to phase relations in the alkali feldspar system. This reaction, which requires a modified solvus, will be discussed in Chapter 5.

3.4.2 The calcite ($CaCO_3$)–dolomite ($Ca_{0.5}Mg_{0.5}CO_3$) solvus

Calcite (rhombohedral $R\bar{3}c$) and dolomite (rhombohedral $R\bar{3}$) commonly occur together in metamorphosed limestone. Goldsmith and Heard (1961) and Goldsmith and Newton (1969) have determined experimentally the Mg:Ca ratio of associated calcite and dolomite between 400 and 1200 °C, at pressures between 10 bar and 21 kbar. The results show that the equilibrium Mg/(Ca+Mg) content of calcite increases conspicuously with increasing temperature, while the same ratio in associated dolomite decreases slightly. The relations at 5 kbar pressure and at temperatures from 400 to 800 °C are presented in Figure 3.28.

The temperature-dependent mutual solubility of solution components $CaCO_3$ (cal) and $Ca_{0.5}Mg_{0.5}CO_3$ (dol) may be expressed by the two reactions,

$$Ca_{0.5}Mg_{0.5}CO_3 \text{ (in dolomite)} = Ca_{0.5}Mg_{0.5}CO_3 \text{ (in calcite)},$$

$$CaCO_3 \text{ (in dolomite)} = CaCO_3 \text{ (in calcite)}.$$

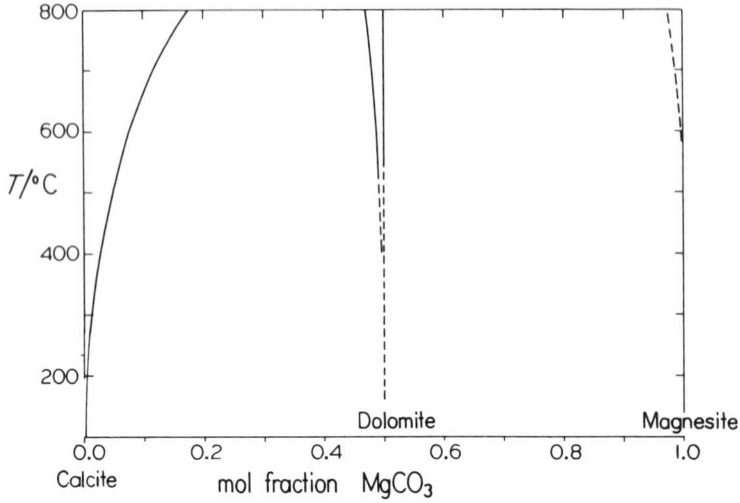

Figure 3.28 Phase relations in the system $CaCO_3$–$MgCO_3$. The calcite limb of the solvus is a plot of 5.0 kbar data of Goldsmith and Newton (1969); the remaining results are from Goldsmith and Heard (1961). Reproduced by permission of *American Journal of Science*

The first of these reactions is displaced to the right with increasing temperature, and the second is displaced slightly to the left, until at equilibrium,

$$\mu_{\text{dol}}^{\text{Dol}} = \mu_{\text{dol}}^{\text{Cal}}, \tag{25}$$

$$\mu_{\text{cal}}^{\text{Dol}} = \mu_{\text{cal}}^{\text{Cal}}, \tag{26}$$

where Dol and Cal stand for the phases dolomite and calcite and μ is chemical potential.

In the range of 400 to 800 °C (corresponding to amphibolite and granulite facies metamorphism), the Mg/(Ca+Mg) ratio of dolomite in equilibrium with calcite is only marginally less than 0.500. Thus Goldsmith and Heard (1961) determined a ratio of 0.488 in their experimental results at 700 °C (Figure 3.28) and Kretz (1980) reported a mean ratio of 0.495 (standard deviation 0.003) for dolomite from marble that crystallized at 650 to 700 °C. Therefore, as a simplification, the second in the above pairs of equations may be disregarded.

Recall the relation between chemical potential (μ) and activity (a),

$$\mu_{\text{dol}}^{\text{Cal}} = G_{\text{dol}}^{\text{Cal}} + RT \ln a_{\text{dol}}^{\text{Cal}}. \tag{27}$$

Now, by adopting the above simplification, i.e.,

$$\mu_{\text{dol}}^{\text{Dol}} = G_{\text{dol}}^{\text{Dol}}, \tag{28}$$

we have by substitution into equation (25),

$$\ln a_{\text{dol}}^{\text{Cal}} = -(G_{\text{dol}}^{\text{Cal}} - G_{\text{dol}}^{\text{Dol}})/RT = -\Delta G_{\text{dol}}^{\circ}/RT. \tag{29}$$

Here $G_{\text{dol}}^{\text{Dol}}$ is the Gibbs energy of phase dolomite at the temperature and pressure of interest and $G_{\text{dol}}^{\text{Cal}}$ is the Gibbs energy of a fictitious standard state ($Ca_{0.5}Mg_{0.5}CO_3$ with a calcite structure) at the same temperature and pressure.

The activity of $Ca_{0.5}Mg_{0.5}CO_3$ in calcite ($a_{\text{dol}}^{\text{Cal}}$) is related to composition by

$$a_{\text{dol}}^{\text{Cal}} = \gamma_{\text{dol}}^{\text{Cal}} X_{\text{dol}}^{\text{Cal}}, \tag{30}$$

where X is the mole fraction $Ca_{0.5}Mg_{0.5}CO_3/(CaCO_3 + Ca_{0.5}Mg_{0.5}CO_3)$ which extends from zero in pure calcite to 1.0 in pure dolomite, and γ is the activity coefficient. Thus,

$$\ln \gamma_{\text{dol}}^{\text{Cal}} + \ln X_{\text{dol}}^{\text{Cal}} = -\Delta G_{\text{dol}}^{\circ}/RT. \tag{31}$$

Differentiating with respect to $(1/T)$ at constant pressure,

$$\frac{\partial(\ln \gamma_{\text{dol}}^{\text{Cal}})}{\partial(1/T)} + \frac{\partial(\ln X_{\text{dol}}^{\text{Cal}})}{\partial(1/T)} = -\frac{1}{R}\left(\frac{\partial(\Delta G_{\text{dol}}^{\circ}/T)}{\partial(1/T)}\right) = -\frac{\Delta H_{\text{dol}}^{\circ}}{R},$$

or

$$\frac{\partial(\ln X_{\text{dol}}^{\text{Cal}})}{\partial(1/T)} = -\frac{\Delta H_{\text{dol}}^{\circ}}{R} - \frac{\partial(\ln \gamma_{\text{dol}}^{\text{Cal}})}{\partial(1/T)}. \tag{32}$$

Here $\Delta H_{\text{dol}}^{\circ}$ is the enthalpy difference, corresponding to $\Delta G_{\text{dol}}^{\circ}$.

Gordon and Greenwood (1970) calculated the activity coefficient of $MgCO_3$ in calcite at 1 bar pressure and at 600 to 850 °C, and could not detect a change in this coefficient with temperature. We shall assume therefore that the term $\partial \ln \gamma/\partial(1/T)$ in equation (32) is zero, and because ΔH is normally independent of temperature, we anticipate a linear relation between $\ln X_{\text{dol}}^{\text{Cal}}$ and $1/T$, or, in terms of a more convenient concentration term, a linear relation between $\ln X_{MgCO_3}^{\text{Cal}}$ and $1/T$.

A replotting of the Goldsmith and Newton (1969) 5 kbar calcite limb of the solvus (Figure 3.28) produces the two linear segments shown in Figure 3.29. The equations for these segments are

$$X_{MgCO_3}^{\text{Cal}} = 0.0283 \text{ to } 0.0683:$$

$$T = 2860/(0.689 - \ln X_{MgCO_3}^{\text{Cal}}), \tag{33}$$

$$X_{MgCO_3}^{\text{Cal}} = 0.0683 \text{ to } 0.161:$$

$$T = 3740/(1.720 - \ln X_{MgCO_3}^{\text{Cal}}). \tag{34}$$

Note that T denotes absolute temperature and $X_{MgCO_3}^{\text{Cal}}$ is the $Mg/(Ca+Mg)$ atomic ratio in calcite. The first of these equations was used to extrapolate the solvus from 400 °C to 200 °C in Figure 3.28. The cause of the change in slope in Figure 3.29 is not known but could result from the term $\partial \ln \gamma_{\text{dol}}^{\text{Cal}}/\partial(1/T)$ being non-zero at temperatures below 575 °C.

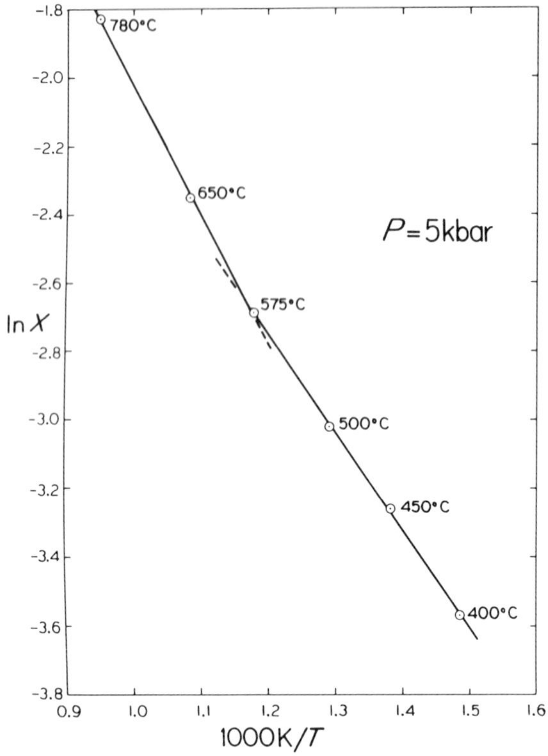

Figure 3.29 The calcite limb of the calcite–dolomite solvus (Figure 3.28) replotted with $\ln X$ and $1/T$ as variables; X = Mg/(Mg+Ca) in calcite; experimental data from Goldsmith and Newton (1969). From Kretz (1980). Reproduced by permission of Oxford University Press

Goldsmith and Newton (1969) determined that the effect of pressure on the solvus is small. For example, a 1.0 kbar increase of pressure at 450–780 °C produces an increase in the $MgCO_3$ content of calcite ($X_{MgCO_3}^{Cal}$) of only 0.09 to 0.15 mole %, corresponding to a temperature difference < 6 °C.

In the presence of more than trace amounts of Fe, which enters both calcite and dolomite, the above equations no longer hold. This problem was investigated experimentally by Goldsmith et al. (1962), Rosenberg (1967), and Powell, Condliffe and Condliffe (1984), and was discussed by Anovitz and Essene (1987).

The relation between the Mg content of calcite (in the presence of dolomite) and temperature, as defined by Figure 3.28 or Figure 3.29 and equations (33) and (34) can be used to estimate the temperature of metamorphism. The principal difficulty in doing so is the reverse (exsolution) reaction which may occur during cooling, resulting in the common presence of dolomite plates and rods. If these are present it is necessary to attempt a reconstruction of the composition of calcite as it existed prior to exsolution.

Methods that may be used to estimate the original Mg content of calcite are reviewed by Essene (1983). Hutcheon and Moore (1973) obtained numerous microprobe analyses on each of several crystals in a single specimen and selected that crystal with the highest Mg content as the best estimate of the original composition of calcite in that specimen. Puhan (1976) added the amount of exsolved dolomite (as measured by point-count analysis) to the Mg content of calcite, obtained by microprobe analysis, to recover the original Mg content. Kretz (1980) separated crystals of calcite together with their dolomite inclusions, and the separates were then analyzed by atomic absorption analysis. The separates gave different Mg concentrations for different specimens from the same metamorphic terrane and grade, which indicated that in some specimens, dolomite migrated beyond crystal boundaries and was therefore not recovered; reasonable temperatures were obtained, however, by selecting that specimen from a group of specimens collected from a restricted area and uniform grade which gave the highest Mg concentration.

The study of a contact aureole in Japan by Suzuki (1977) provides an example of the application of the calcite–dolomite geothermometer. Minerals that appear in siliceous dolomite on approaching the granite contact at this locality are talc, tremolite, diopside, and forsterite; the forsterite isograd is shown in Figure 3.30.

Exsolution was detected locally, and to obtain an estimate of the (peak) metamorphic Mg content of calcite for a particular sample, many calcite crystals were analyzed, and the analysis that gave the highest Mg concentration was then selected.

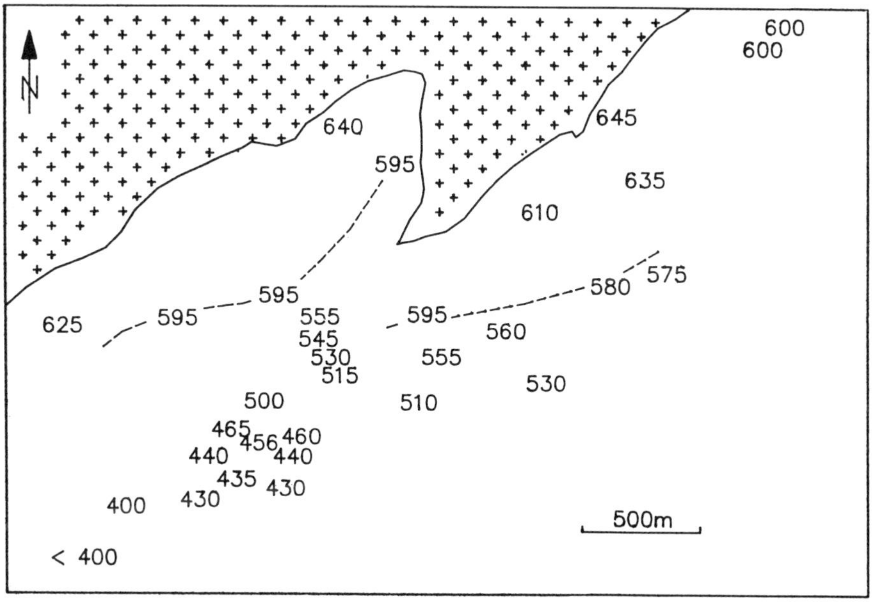

Figure 3.30 Calcite–dolomite temperature estimates (°C) in the contact aureole of the Kaizuki-Yama granite pluton in Japan; broken line is the forsterite isograd. From Suzuki (1977). Reproduced by permission of Springer-Verlag

This 'highest value' should possibly be reduced slightly because analytical precision (counting error) as defined by Potts, Tindle and Isaacs (1983) was not taken into consideration.

The results obtained by Suzuki (1977) are shown in Figure 3.30, where calcite–dolomite temperatures are given to the nearest 5 °C. Four temperatures on the forsterite isograd lie near 595 °C, which agrees well with the experimental results of Figure 3.17. The range extends from <400 °C below the talc isograd to 650 °C at the granite contact.

3.4.3 The quasi-chemical view of solvus equilibria

An examination of metallic crystalline solutions from the viewpoint of atomic bonds and the distribution of atoms has led to an improved understanding of limited miscibility and solvus equilibria in these systems. The semi-quantitative theory, as presented by Swalin (1962) and summarized below, may also apply, in part, to minerals, where, although pairs of exchangeable atoms do not form bonds with each other (as in metals), interactions between them nevertheless do occur.

Consider one gram formula weight of a crystalline solution, (A, B), containing N_A atoms of A and N_B atoms of B. Three kinds of bonds exist, namely A–A, B–B, and A–B bonds. Let the numbers of these bonds be N_{AA}, N_{BB} and N_{AB} respectively, and let Z denote the number of nearest neighbors.

It is possible to express the number of A atoms (N_A) in terms of the number of A–A and A–B bonds, as follows:

$$N_A = N_{AA}(2/Z) + N_{AB}(1/Z) \tag{35}$$

This is readily confirmed by imagining a one-dimensional crystal for which $Z = 2$. Rearranging terms,

$$N_{AA} = (Z/2)[N_A - (N_{AB}/Z)]. \tag{36}$$

Similarly,

$$N_{BB} = (Z/2)[N_B - (N_{AB}/Z)]. \tag{37}$$

A quantity of enthalpy may be assigned to each bond; the total enthalpy for the crystal is therefore

$$H^{(A,B)} = N_{AA}H_{AA} + N_{BB}H_{BB} + N_{AB}H_{AB}, \tag{38}$$

and by substitution of (36) and (37) in (38),

$$H^{(A,B)} = (Z/2)N_A H_{AA} + (Z/2)N_B H_{BB} + N_{AB}[H_{AB} - 0.5(H_{AA} + H_{BB})]. \tag{39}$$

We now assume that the enthalpy of A–A bonds in pure A is equal to that in the (A,B) mixture, and similarly for B–B bonds. Thus, the enthalpy change upon mixing of pure A and pure B is

$$\Delta_m H^{(A,B)} = H^{(A,B)} - N_{AA}^{A} H_{AA} - N_{BB}^{B} H_{BB}, \tag{40}$$

where N_{AA}^{A} and N_{BB}^{B} stand for the numbers of A–A and B–B bonds in pure A and B. These are equal to $(Z/2)N_A$ and $(Z/2)N_B$ respectively. Thus,

$$\Delta_m H^{(A,B)} = H^{(A,B)} - (Z/2)N_A H_{AA} - (Z/2)N_B H_{BB}. \tag{41}$$

By substitution of (39) in (41),

$$\Delta_m H^{(A,B)} = N_{AB}[H_{AB} - 0.5(H_{AA} + H_{BB})]. \tag{42}$$

For an ideal solution, $\Delta_m H = 0$, and consequently,

$$H_{AB} = 0.5(H_{AA} + H_{BB}), \tag{43}$$

i.e., the energy associated with the A–B bonds is the mean of the energies associated with the A–A and B–B bonds within the crystal.

For certain mixtures, $\Delta_m H$ is not zero but the excess entropy of mixing (S^E) remains at zero, which implies that atoms are randomly distributed in the crystal. Suppose this applies to the (A, B) crystal under discussion. N_{AB} can then be evaluated from statistical considerations, as follows.

Let N_0 be the total number of A and B atoms present. Then the atomic fractions are,

$$X_A = N_A/N_0,$$

$$X_B = N_B/N_0.$$

The probability that a particular site will be occupied by A is X_A, and the probability that a neighboring site will be occupied by B is X_B. Hence the probability for an AB pair is $X_A X_B$, for a BA pair is $X_B X_A$, and the probability that two adjacent sites will be occupied by unlike atoms is $2X_A X_B$. The number of A–B bonds is therefore,

$$N_{AB} = (2X_A X_B)(Z/2)(N_0),$$

or

$$N_{AB} = X_A X_B Z N_0. \tag{44}$$

Substitution of (44) in (42) gives,

$$\Delta_m H = X_A X_B Z N_0[H_{AB} - 0.5(H_{AA} + H_{BB})]. \tag{45}$$

Now, let

$$\Omega = Z N_0[H_{AB} - 0.5(H_{AA} + H_{BB})]. \tag{46}$$

Thus,

$$\Delta_m H = X_A X_B \Omega. \tag{47}$$

By recalling the earlier equations for a symmetrical solution (model II, Sec. 2.10),

$$\Delta_m H = H^E, \tag{48}$$

and

$$H^{\mathrm{E}} = X_{\mathrm{A}}X_{\mathrm{B}}W - T\left(\frac{\partial W}{\partial T}\right), \tag{49}$$

which at constant temperature is

$$H^{\mathrm{E}} = X_{\mathrm{A}}X_{\mathrm{B}}W, \tag{50}$$

we see that Ω may be identified with the Margules parameter, W.

We are now in a position to view departures from ideal mixing in terms of attractions between atoms in a crystal. Departure may occur in two directions, (1) deviation resulting from a greater attraction between 'unlike' atoms, and (2) deviation resulting from a greater attraction between 'like' atoms. In (1), H_{AB} bond energy is large and negative (equation (46)), W is negative, and $\Delta_{\mathrm{m}}H$ is negative (exothermic). The mixture is *more* stable than an ideal solution. In (2), H_{AA} and H_{BB} bond energies are large and negative (equation (46)), W is positive, and $\Delta_{\mathrm{m}}H$ is positive (endothermic). The mixture is *less* stable than an ideal mixture, and if Ω or W is sufficiently large, a miscibility gap will appear in the solid-solution series (Figure 2.12).

Swalin (1962) has also examined the interesting interplay of enthalpy and entropy that occurs in solid solutions. In these crystals, the Gibbs energy of mixing is reduced when the enthalpy of mixing *decreases* and when the entropy of mixing *increases*, i.e.,

$$\Delta_{\mathrm{m}}G = \Delta_{\mathrm{m}}H - T\Delta_{\mathrm{m}}S. \tag{51}$$

When the A–A and B–B bond enthalpies are small relative to A–B (i.e., large negative numbers) domains of AAA and BBB will tend to form, thereby increasing the number of A–A and B–B bonds and decreasing the enthalpy of mixing ($\Delta_{\mathrm{m}}H$). This tendency is opposed, however, by the entropy of mixing which decreases as like atoms form clusters. Thus a balance is achieved between two tendencies, increasing clustering, which reduces heat energy, and increasing randomness which increases entropy, both of which reduce the Gibbs energy of the crystal. Because the entropy term is multiplied by temperature, entropy dominates at higher temperature, and a miscibility gap, if present, decreases in width with increasing temperature, as it does in the feldspar and carbonate systems examined above.

3.5 SOLVUS EQUILIBRIA IN THREE-COMPONENT SYSTEMS

In Sec. 3.4 we studied examples of restricted miscibility in two-component systems. We now extend this study to restricted miscibility in three-component systems, choosing as examples the $MgSiO_3$–$FeSiO_3$–$CaSiO_3$ pyroxene and the $KAlSi_3O_8$–$NaAlSi_3O_8$–$CaAl_2Si_2O_8$ feldspar systems. The composition–temperature space is now a volume, and the solvus is a curved surface.

3.5.1 Solvus equilibria in the pyroxene quadrilateral

Orthopyroxene (Opx; $(Mg,Fe)_2Si_2O_6$; Pbca) and monoclinic Ca-rich pyroxene (Cpx; $Ca(Mg,Fe)Si_2O_6$; C2/c) commonly occur together in felsic to mafic rocks of granulite-facies terranes. In view of the simplified formulas, as written above, these minerals can be plotted in a $MgSiO_3$–$FeSiO_3$–$CaSiO_3$ triangle. For example, mineral pairs from Quairading, Australia, which were analyzed by Davidson (1968), are plotted in Figure 3.31a; all points fall in a portion of the triangle, that is referred to as the 'pyroxene quadrilateral'. When plotting these data, Na, Al, Fe^{3+}, Ti, and other elements present in small amounts, are disregarded.

The Quairading minerals show clearly that minor Ca is present in orthopyroxene, and that Ca pyroxene contains more Mg and Fe than is indicated by the simplified diopside–hedenbergite formula. Thus the two minerals have evidently experienced mutual dissolution to a limited extent, and the tie lines in Figure 3.31a define a region of immiscibility that refers to the particular temperature and pressure of crystallization.

In gabbro and basalt, monoclinic pigeonite ($(Ca,Mg,Fe)_2Si_2O_6$; $P2_1/c$) may appear in place of or in addition to orthopyroxene. Thus in the Skaergaard complex (Brown 1957), Ca pyroxene and pigeonite occur commonly (the latter has largely inverted to orthopyroxene with Ca pyroxene lamellae), and three of these mineral pairs together with one Ca-pyroxene–orthopyroxene pair are also plotted in Figure 3.31a. These data, when compared with those from Quairading clearly demonstrate that the solubility of $(Mg,Fe)SiO_3$ in Ca pyroxene increases with increasing temperature.

In Fe-rich rocks of metamorphic or igneous origin, Fe olivine and quartz appear in place of $FeSiO_3$, and the solvus surface does not extend to the $CaFeSi_2O_6$–$FeSiO_3$ side of the quadrilateral; equilibrium relations in this region will not be considered at present.

The solvus in the Fe-free enstatite (en)–diopside (di) system was located experimentally by Atlas (1952), Boyd and Schairer (1964), Davis and Boyd (1966), Warner and Luth (1974), Nehru and Wyllie (1974), Howells and O'Hara (1975) and Mori and Green (1975). The results presented here (Figure 3.32) were obtained by Lindsley and Dixon (1976) who carried out experiments in the range of 800 to 1400 °C and 5 to 35 kbar. The composition of the synthetic starting crystals was chosen so that mixing occurred in some experiments, unmixing in others, and the reactions were therefore reversed (Figure 3.32). Run products were examined by X-rays, and the Mg:Ca ratios were determined by microprobe analysis.

Lindsley and Dixon (1976) concluded from their results and those obtained previously that the effect of pressure on the solvus is small. Increasing pressure causes the miscibility gap to widen slightly, for example at 1200 °C, a 1 kbar increase in pressure causes the $CaSiO_3$ content of the diopside phase to increase by 0.10 to 0.12 mole %. According to Mori and Green (1975), at temperatures below 900 °C there is 'no perceptible' pressure effect on the Ca pyroxene limb of the solvus. Some of the Lindsley–Dixon run products contained quartz in addition to

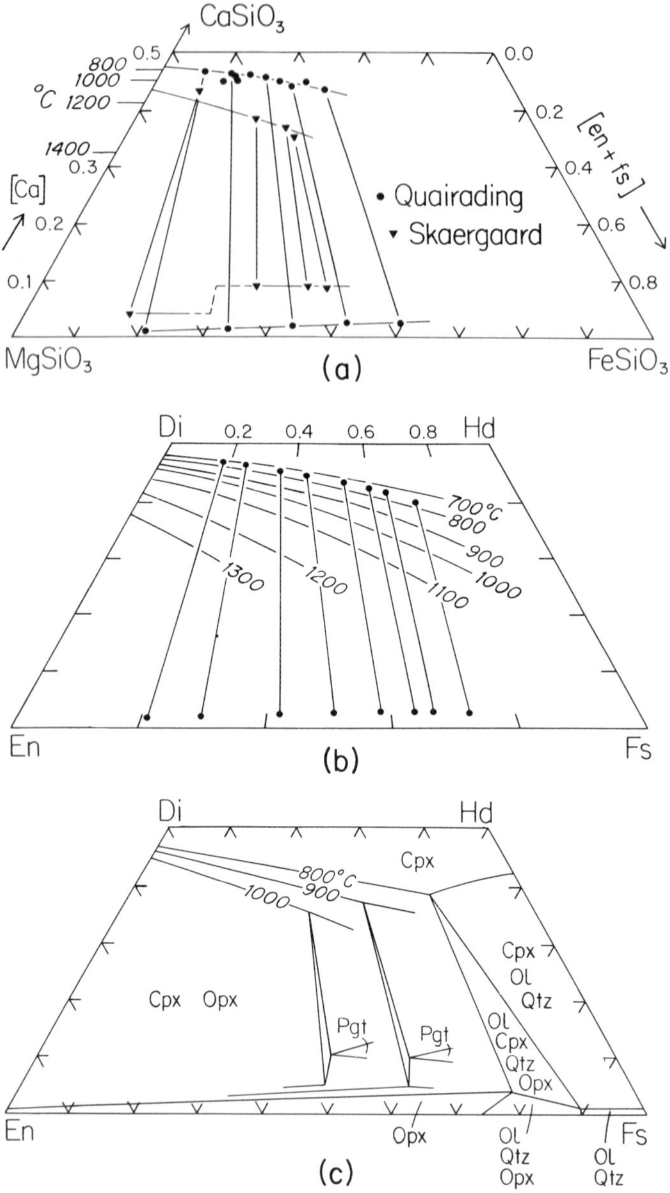

Figure 3.31 The pyroxene quadrilateral, with corners $MgSiO_3$ (En), $FeSiO_3$ (Fs), $Ca_{0.5}Mg_{0.5}SiO_3$ (Di), $Ca_{0.5}Fe_{0.5}SiO_3$ (Hd). (a) Associated Ca pyroxene and Ca-poor pyroxene from the Quairading granulite-facies terrane (Davidson 1968) and the Skaergaard igneous complex (Brown 1957); temperatures on the Mg-Ca side are solvus temperatures, from Lindsley and Dixon (1976). (b) Some of the experimental results (tie lines) of Fonarev and Graphchikov (1982) at 800 °C and 2.94 kbar; the isotherms are from an earlier model (equations (52) in text), based on the data of (a). (c) phase relations at 3.0 kbar and 800, 900, 1000 °C (excluding some data in the Fe-rich region), from Lindsley and Frost (1992).

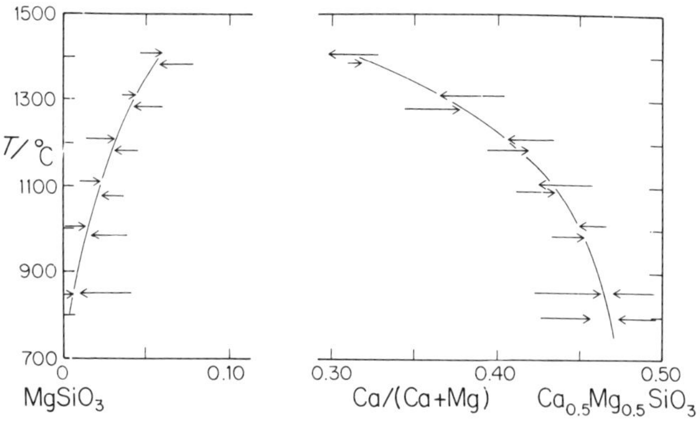

Figure 3.32 The 15 kbar enstatite–diopside solvus as located by Lindsley and Dixon (1976); arrows indicate direction of approach to equilibrium. Reprinted by permission of *American Journal of Science*

the pyroxene phases, and others contained olivine, but the solvus was not displaced noticeably by the presence of these additional phases.

Most of the experimental data now available on the composition of associated pyroxene phases *within* the quadrilateral were obtained by Fonarev and Graphchikov (1982), Lindsley (1983), and Lindsley and Andersen (1983). Some of the results of Fonarev and Graphchikov (1982) at 800 °C and 2.94 kbar are plotted in Figure 3.31b. The numerous data of Lindsley (1983) and Lindsley and Andersen (1983) have produced quadrilateral phase diagrams for pressures of 1 bar and 5, 10, and 15 kbar; and some of the more recent results, obtained by Lindsley and Frost (1992) are presented in Figure 3.31c.

Before adequate experimental data were available on relations within the quadrilateral, the writer (Kretz 1982a) attempted to define the Ca pyroxene slope of the solvus surface and the relative Mg:Fe ratio of associated pyroxene phases by combining data on the en–di solvus of Lindsley and Dixon (1976) with the information from Quairading and Skaergaard as shown in Figure 3.31a. By drawing an analogy with the Ca–Mg carbonate system (Sec. 3.4, Figure 3.29) the Lindsley and Dixon (1976) diopside limb of the en–di solvus was re-plotted as two linear segments, with equations,
1400 to 1080 °C:

$$1000/T = -0.123 \ln(1 - 2[\mathrm{Ca}]^{\mathrm{Cpx}}) + 0.468,$$

1080 to 800 °C:

$$1000/T = -0.304 \ln(1 - 2[\mathrm{Ca}]^{\mathrm{Cpx}}) + 0.054,$$

where $[\mathrm{Ca}]^{\mathrm{Cpx}}$ is Ca/(Ca + Mg) in Ca pyroxene and T is absolute temperature. Maintaining a logarithmic scale for the Ca content of Ca pyroxene, the Quairading data of Figure 3.31a produced a straight line with an intercept on the en–di solvus of 730 °C; a parallel line through the Skaergaard data produced an intercept of 1100 °C. These two lines were then taken as 730 and 1100 °C isotherms on the solvus surface. Thus the following two equations were obtained for a portion of this surface:

1400 to 1080 °C:

$$T = 1000/[0.468 + 0.246X^{\mathrm{Cpx}} - 0.123\ln(1 - 2[\mathrm{Ca}]^{\mathrm{Cpx}})], \tag{52}$$

1080 to 800 °C:

$$T = 1000/[0.054 + 0.608X^{\mathrm{Cpx}} - 0.304\ln(1 - 2[\mathrm{Ca}]^{\mathrm{Cpx}})],$$

where T is absolute temperature, X^{Cpx} is $Fe^{2+}/(Mg + Fe^{2+})$ in Ca pyroxene and $[Ca]^{\mathrm{Cpx}}$ is $Ca/(Ca + Mg + Fe^{2+})$ in Ca pyroxene. The resulting isotherms are plotted in Figure 3.31b, where the 800 °C isotherm may be compared with the 800 °C experimental results of Fonarev and Graphchikov (1982).

Early mineral analyses by Howie (1955) showed that the ratio $Fe^{2+}/(Mg + Fe^{2+})$ in orthopyroxene of high-grade metamorphic rocks is slightly greater than that of associated Ca pyroxene. This ratio (denoted X) in one mineral relative to that in the other is best expressed by use of the distribution coefficient, defined as

$$K_{\mathrm{D}} = \frac{X^{\mathrm{Opx}}}{1 - X^{\mathrm{Opx}}} \cdot \frac{1 - X^{\mathrm{Cpx}}}{X^{\mathrm{Cpx}}}.$$

Bartholomé (1961) and Kretz (1961) detected a smaller Fe:Mg contrast in gabbroic rocks as compared with metamorphic rocks, i.e., K_{D} decreases from 1.86 to 1.38, and they attributed this difference to the temperature of crystallization. The experimental data of Lindsley (1983; plotted in Figure 3.33) confirm that K_{D} does change with temperature, but whether the change must be attributed entirely to temperature or to a change in the Ca content of Ca pyroxene (Huebner 1980, Davidson and Lindsley 1985) or to both remains uncertain. The important point is that this change, although small in magnitude, must be taken into consideration when presenting a description of the pyroxene quadrilateral.

The experimental data on Fe–Mg distribution do not agree well with the natural data. For example, at Quairading (~730 °C), $K_{\mathrm{D}} = 1.86$ ($X^{\mathrm{Opx}} < 0.64$), whereas the value obtained by an extrapolation of the experimental data gives $K_{\mathrm{D}}(730\,°\mathrm{C}, X^{\mathrm{Opx}} \leq 0.64) = 1.6$. A proposal for a relation between K_{D} and T in *natural* pyroxenes, based largely on the assumption that exchange equilibrium in the Quairading terrane was achieved and preserved at 730 °C and in the Skaergaard complex at 1100 °C, is

$$T = 1130/(\ln K_{\mathrm{D}} + 0.505).$$

Here T is absolute temperature and K_{D} is calculated from mineral pairs with $X \leq 0.64$, the range in which K_{D} is independent of (Fe, Mg) composition.

We are now in a position to define the chemical reactions that occur between orthopyroxene and Ca pyroxene in a system that experiences a change in temperature (for example from 810 °C to 990 °C) at constant pressure. Reactions to produce

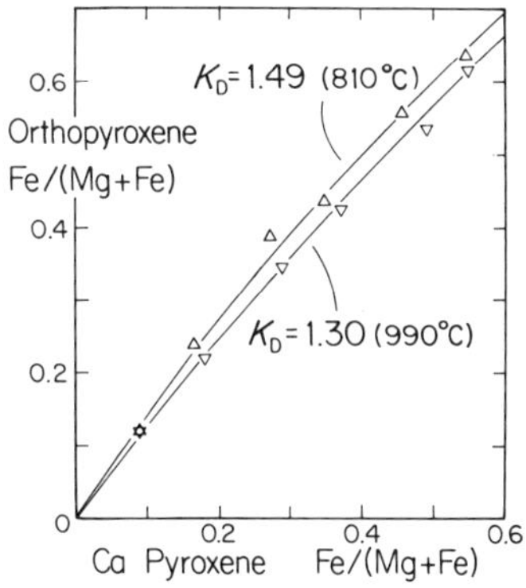

Figure 3.33 Distribution of Mg and Fe between Ca pyroxene and orthopyroxene at 810 °C and 990 °C (15 kbar); from tabulated data of Davidson and Lindsley (1985)

pigeonite will not be considered. It is convenient to begin with the Fe-free system, where the reaction can be described as the mutual mixing of two components, $MgSiO_3$ (en) and $Ca_{0.5}Mg_{0.5}SiO_3$ (di):

$$MgSiO_3 \text{ (in En)} = MgSiO_3 \text{ (in Di)},$$

$$Ca_{0.5}Mg_{0.5}SiO_3 \text{ (in En)} = Ca_{0.5}Mg_{0.5}SiO_3 \text{ (in Di)}.$$

With increasing temperature miscibility increases, i.e., the first reaction is displaced to the right and the second to the left, until at equilibrium,

$$\mu_{en}^{En} = \mu_{en}^{Di},$$

$$\mu_{di}^{En} = \mu_{di}^{Di}.$$

Note that en and di refer to solution components and En and Di to solid-solution phases.

Moving into the quadrilateral, the natural and experimental data show that temperature-dependent changes occur in the Ca content of the crystals as well as the Fe:Mg ratios. If, for a particular 810 °C tie line, the Fe:Mg ratio was found to be the same in both minerals and remained the same to 990 °C, the equilibrium could be described completely in terms of transfer reactions, as in the en–di system. Or, if the Ca content of both minerals remained unchanged with increasing Fe/(Mg + Fe) and with increasing temperature, the equilibrium could be described completely in terms of an Fe–Mg exchange reaction. But neither of these conditions is encountered alone, and the equilibrium must therefore be described in terms of a transfer and an exchange reaction occurring simultaneously.

Two of the Lindsley (1983) 15 kbar experiments, one at 810 °C and the other at 990 °C, both of which employed the same starting material (labelled K) will be used to demonstrate compositional changes that occur within the quadrilateral in response to an increase in temperature. Figure 3.34 shows, by use of arrows, the

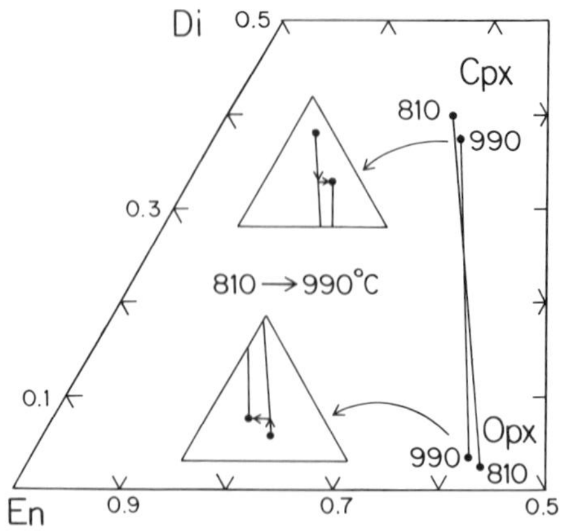

Figure 3.34 Ca-pyroxene–orthopyroxene tie lines at 810 °C and 990 °C (15 kbar), as located experimentally by Lindsley (1983) from the same starting material. Inset triangles show that changes in composition, resulting from an increase in temperature may be viewed as mutual mixing (on tie-line arrows) and Fe–Mg exchange (horizontal arrows). During slow heating, near-equilibrium changes would be represented by arrows that are the resultants of those shown

required changes in Ca content of both minerals (on tie-line arrows), and in the Fe:Mg ratio (horizontal arrows) as temperature increases to 990 °C. The required Fe–Mg exchange is also shown in Figure 3.33.

The transfer part of the reaction may be expressed as follows:

$$(Mg,Fe)SiO_3 \text{ (in Opx)} = (Mg,Fe)SiO_3 \text{ (in Cpx)},$$

$$Ca_{0.5}(Mg,Fe)_{0.5}SiO_3 \text{ (in Opx)} = Ca_{0.5}(Mg,Fe)_{0.5}SiO_3 \text{ (in Cpx)},$$

$$\mu^{\mathrm{Opx}}_{(\mathrm{en,fs})} = \mu^{\mathrm{Cpx}}_{(\mathrm{en,fs})},$$

$$\mu^{\mathrm{Opx}}_{(\mathrm{di,hd})} = \mu^{\mathrm{Cpx}}_{(\mathrm{di,hd})}.$$

The exchange part of the reaction is expressed as

$$MgSiO_3 \text{ (in Opx)} + CaFeSi_2O_6 \text{ (in Cpx)}$$

$$= FeSiO_3 \text{ (in Opx)} + CaMgSi_2O_6 \text{ (in Cpx)},$$

$$\mu^{\mathrm{Opx}}_{\mathrm{en}} + \mu^{\mathrm{Cpx}}_{\mathrm{hd}} = \mu^{\mathrm{Opx}}_{\mathrm{fs}} + \mu^{\mathrm{Cpx}}_{\mathrm{di}}.$$

The Gibbs energy surface within the volume in which the quadrilateral is the base and G is the vertical axis, is a folded surface with two troughs and an intervening crest. At a given bulk composition, the equilibrium composition is given by a common tangent to the two troughs, passing through the bulk composition. But there are many possible tangent lines; the one that is chosen at equilibrium is determined by another surface, concave upward, which intersects the first surface and governs the exact position of the tangent and the relative Fe:Mg ratios of the crystals.

It is interesting to consider also the changes that occur in each of the two sites, M1 and M2, that exist in equal proportions in both orthopyroxene and Ca pyroxene. The writer (Kretz 1982a) considered this problem by drawing on data obtained independently on the relative Mg:Fe ratios on the two sites as a function of temperature, assuming that Ca is confined to M2. Such information for orthopyroxene was presented in Sec. 3.3.3 (above); information on Ca pyroxene was obtained by McCallister, Finger and Ohashi (1976). The calculations showed that for a particular metamorphic mineral pair the Mg:Fe ratios in M1 of the orthopyroxene and Ca pyroxene crystals are equal, as was proposed by Mueller (1962). Moreover an increase in temperature to the region of 1000 °C evidently leaves these ratios unchanged, i.e., nearly all of the change in the Mg and Fe content of the crystals, as well as the Ca content, take place in M2.

Several thermodynamic models have been constructed, based on the Margules asymmetrical equations and on order-disorder relations in the pyroxene crystals. The reader is referred to papers by Holland, Navrotsky and Newton (1979), Lindsley, Grover and Davidson (1981), Davidson and Lindsley (1985), Saxena, Sykes and Eriksson (1986), and Lindsley and Frost (1992) for information on these models.

The composition of associated orthopyroxene and Ca pyroxene may be used to obtain estimates of the temperature of crystallization within the granulite facies or at higher temperatures. The various 'thermometers' that have been proposed are reviewed by Lindsley (1983) and Fonarev and Graphchikov (1991).

Examples of temperature estimates obtained by use of the Ca pyroxene slope of the solvus surface, as defined by equations (52) above, and Figure 3.31b are listed in Table 3.2. Applications of the Lindsley and Frost (1992) program to

Table 3.2 Temperature estimates (°C) for associated Ca pyroxene and orthopyroxene, obtained by use of the solvus equations (52) in text

	Specimens	$T/°C$
Lower granulite facies, Broken Hill (Binns 1962)	3	690
Higher granulite facies, Broken Hill (Binns 1962)	4	770
Granular ultramafic nodules in kimberlite, Lesotho (Nixon and Boyd 1973)	6	890
Chondrite meteorites, H6	7	880
L6	8	920
LL6	6	950
(Bunch and Olsen 1974)		
Bushveld mafic intrusion (Atkins 1969)	7	1020
Calc-alkaline volcanics, New South Wales (Wilkinson 1971)	4	1090
Peridotite intrusion, Cornwall (Green 1964)	2	1060
Lunar metaclastic rocks (Albee, Gancarz and Chodos 1973)	3	1160
Sheared ultramafic nodules in kimberlite, Lesotho (Nixon and Boyd 1973)	7	1360

For sources of data, see Kretz (1982a)

high-temperature pyroxene assemblages (basalt, tuff, gabbro) are reported by Frost and Lindsley (1992).

3.5.2 Solvus equilibria in the feldspar system

Plagioclase (Pl) and K-feldspar (Kf) commonly occur together in crystalline terranes of medium-to-high metamorphic grade and are particularly abundant in various granitic rocks, gneisses, and granulites. Apart from minor Ba, Sr, Rb, and other elements (Table 3.3) the chemical composition of common plagioclase and K-feldspar can be expressed in terms of three solution components, albite ($NaAlSi_3O_8$), anorthite ($CaAl_2Si_2O_8$), and orthoclase ($KAlSi_3O_8$) which are here abbreviated as ab, an, and or. Thus, simplified mineral compositions can be plotted in the ab–an–or triangle shown in Figure 3.35a, where three mineral pairs (joined by tie lines) from a high-grade metamorphic terrane are compared with two mineral pairs from volcanic trachyte. It is apparent from these natural data that the solvus which exists on the albite–orthoclase join, as described in Sec. 3.4.1 above, extends into the composition–temperature volume, where it forms a curved surface, the crest of which rises as more and more anorthite component enters the feldspar phases. The 650 and 750 °C (1 kbar) isotherms on this surface, as located experimentally by Seck (1971a, b), are also shown in Figure 3.35a, as is the 650 °C, 10 kbar isotherm.

It is apparent from this figure that in both rock types, the anorthite content of K-feldspar remains very low (see also Table 3.3), while the orthoclase content of plagioclase is greater in the volcanic than in the metamorphic rocks. But the most striking difference between the metamorphic and volcanic mineral pairs is a difference in the orientation of the tie lines, which implies a difference in the albite content

Table 3.3 Analyses of associated plagioclase and K-feldspar from biotite–garnet–sillimanite gneiss, Otter Lake, Quebec

	Plagioclase	K-feldspar
SiO_2 wt%	61.0	63.5
Al_2O_3	23.4	18.3
TiO_2	0.06	0.03
FeO	0.25	0.08
MgO	0.09	0.04
CaO	4.75	0.11
BaO	0.035	0.30
SrO	0.026	0.027
Na_2O	7.90	1.43
K_2O	1.28	14.3
Rb_2O	0.002	0.037
ab mole fraction	0.695	0.131
an	0.231	0.0056
or	0.074	0.863
X_{ab}^{Pl} = ab/(ab + an)	0.751	
X_{ab}^{Kf} = ab/(ab + or)		0.132

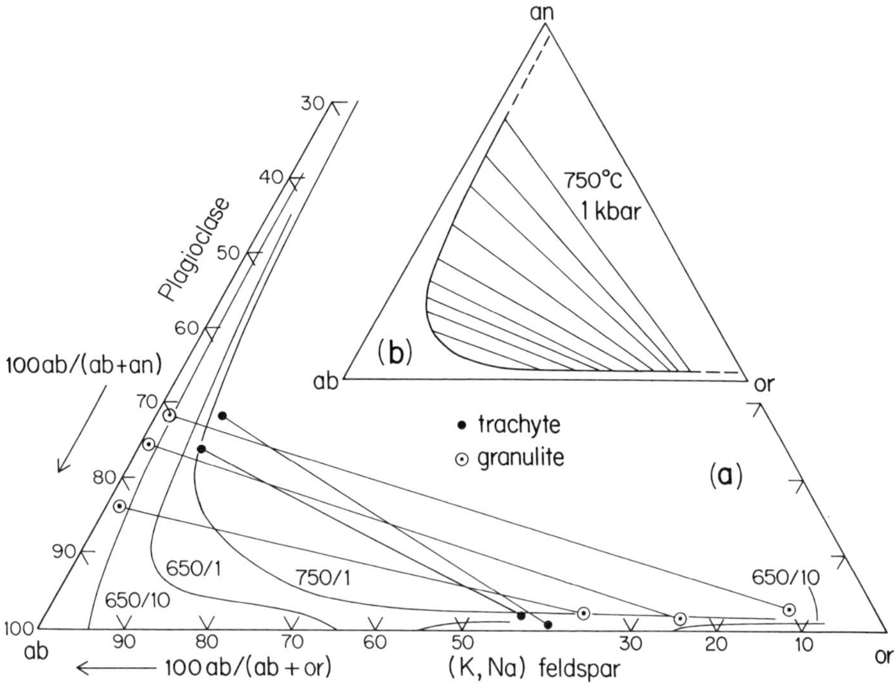

Figure 3.35 The $NaAlSi_3O_8$ (ab)–$CaAl_2SiO_8$ (an)–$KAlSi_3O_8$ (or) feldspar triangle. (a) Three mineral pairs from a high-grade metamorphic terrane (Bohlen and Essene 1977, open symbols) compared with two from volcanic trachyte (I.S.E. Carmichael, tabulated by Ghiorso (1984), closed symbols); isotherms on the solvus (e.g., 650 °C, 1 kbar) are from Seck (1971a, b). (b) Inset, some of the experimental results of Seck (1971a)

of plagioclase relative to that of K-feldspar. Differences of this kind were observed by Barth (1951, 1962) who proposed that equilibrium in the feldspar system is most readily examined in terms of the distribution of the albite component between the two feldspar phases, and the effect of temperature on the distribution ratio.

Experimental information on plagioclase–K-feldspar equilibria was obtained by Iiyama, Wyart and Sabatier (1963), Iiyama (1966), Seck (1971a, b), Johannes (1979), and Elkins and Grove (1990), and early results were examined from a theoretical viewpoint by Perchuk and Ryabchikov (1968). The most numerous data were obtained by Seck (1971a, b) who carried out experiments at 650 to 900 °C and 0.5 to 10 kbar. Gels of known composition were crystallized to plagioclase and K-feldspar, but crystalline starting materials were also used, with similar results. First, isotherms on the solvus were located and these were then used to assist in the determination of feldspar composition in the run products. Some of the results obtained at 750 °C and 1 kbar are shown, for example, in Figure 3.35b.

An example of the change in tie-line orientation (at 1.0 kbar) encountered by Seck (1971a) is shown in Figure 3.36. In this case, an increase in temperature

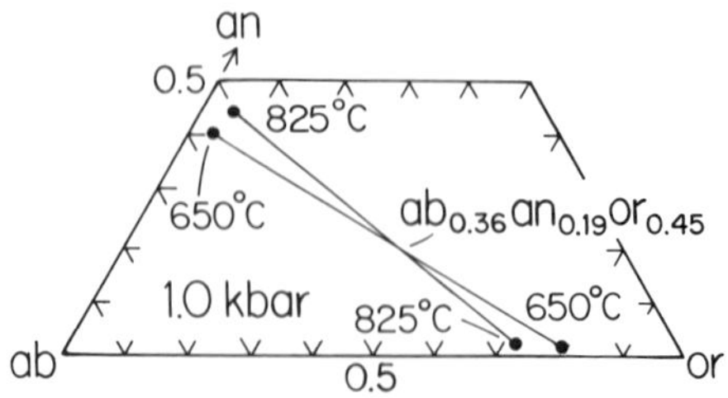

Figure 3.36 Comparison of 650 °C and 825 °C tie lines in the feldspar triangle (P = 1 kbar); data from Seck (1971a)

from 650 to 825 °C produces an increase in the albite content of K-feldspar and a decrease in the albite content of plagioclase, while the K-feldspar:plagioclase molar mass ratio changes from 54:46 to 61:39. The anorthite content of K-feldspar and the orthoclase content of plagioclase also increase, but only slightly. A mass-balance calculation leads to the following mass-transfer expressions; the numbers refer to the numbers of moles of component or phase relative to 1.0 mole of initial feldspar ($ab_{0.36}an_{0.19}or_{0.45}$) defined by the tie-line intersection:
650 → 825 °C:

$$0.54[ab_{0.189}an_{0.015}or_{0.796}]^{Kf} + 0.46[ab_{0.560}an_{0.402}or_{0.039}]^{Pl}$$
$$\rightarrow 0.61[ab_{0.266}an_{0.030}or_{0.704}]^{Kf} + 0.39[ab_{0.505}an_{0.442}or_{0.053}]^{Pl} \quad (53)$$

$$[0.060\ ab]^{Pl} \rightarrow [0.060\ ab]^{Kf}$$
$$[0.010\ an]^{Pl} \rightarrow [0.010\ an]^{Kf} \quad (54)$$
$$[0.002\ or]^{Pl} \leftarrow [0.002\ or]^{Kf}$$

The temperature-dependent compositional changes are therefore transfer (not exchange) reactions, in which components are transferred from one phase to the other until at equilibrium

$$\mu_{ab}^{Pl} = \mu_{ab}^{Kf}, \quad (55)$$

$$\mu_{an}^{Pl} = \mu_{an}^{Kf}, \quad (56)$$

$$\mu_{or}^{Pl} = \mu_{or}^{Kf}, \quad (57)$$

where μ denotes chemical potential. The small but noticeable increase in the orthoclase content of plagioclase (Figure 3.36) results principally from the removal of albite from these crystals; the calculated transfer of orthoclase (above) is so small that it cannot be stated with confidence.

At equilibrium, all three energy-balance equations, (55) to (57), must be satisfied, but from the above calculation (equation (53)), it is clear that for a given change in temperature, the largest change required of K-feldspar is a change in the ab:or ratio, and of plagioclase is a change in the ab:an ratio, i.e., in the ab content of the two phases. Therefore the first of the energy-balance equations, (55), which focuses on the chemical potential of ab in the crystals, is now chosen to define relationships between feldspar composition and temperature and pressure. Thus, by substitution of the following two equations into (55),

$$\mu_{ab}^{Pl} = G_{ab}^{Pl} + RT \ln \gamma_{ab}^{Pl} X_{ab(ab-an)}^{Pl}, \tag{58}$$

$$\mu_{ab}^{Kf} = G_{ab}^{Kf} + RT \ln \gamma_{ab}^{Kf} X_{ab(ab-or)}^{Kf}, \tag{59}$$

we obtain

$$X_{ab(ab-an)}^{Pl} / X_{ab(ab-or)}^{Kf} = (\gamma_{ab}^{Kf} / \gamma_{ab}^{Pl}) \exp(-(G_{ab}^{Pl} - G_{ab}^{Kf})/RT). \tag{60}$$

Here G_{ab}^{Pl} and G_{ab}^{Kf} are the molar Gibbs energy of pure albite in the form of plagioclase and of K-feldspar respectively, and $X_{ab(ab-an)}^{Pl}$ and $X_{ab(ab-or)}^{Kf}$ are defined as the mole fractions ab/(ab + an) in plagioclase and ab/(ab + or) in K-feldspar respectively. Thus in a large part of the composition triangle, where components 'or' in plagioclase and 'an' in K-feldspar are low, these minor components are here regarded as dilutents (together with $RbAlSi_3O_8$, $BaAl_2Si_2O_8$, $SrAl_2Si_2O_8$) which will have only a slight effect on relations between chemical potential and concentration for the major components. Note that the expression on the right of equation (60) is a function of temperature and pressure, and therefore the quotient on the left (the distribution coefficient, denoted K_D) is expected to vary with both temperature and pressure.

Figure 3.37 presents the 1.0 kbar results of Seck (1971a) in terms of $X_{ab(ab-an)}^{Pl}$ relative to $X_{ab(ab-or)}^{Kf}$. In the range of $X_{ab(ab-an)}^{Pl}$ from 0.35 to 0.70, which covers a large part of the composition range of interest, K_D is in fact constant at constant temperature:

kbar	°C	K_D
1.0	650	3.02 3.03 2.97 2.91 3.00; mean: *2.99*
1.0	750	2.35 2.46 2.43 2.40 2.36 2.48 2.41; mean: *2.41*
1.0	825	1.98 1.97 2.06 2.15 2.08 1.95 2.15 1.96; mean: *2.04*

The relation between $\ln K_D$ ($X_{ab(ab-an)}^{Pl} = 0.35$ to 0.70) and $1/T$ is shown in Figure 3.38a, and is described by the equation

$$\ln K_D = 2.08(1000/T) - 1.16 \tag{61}$$

The distribution coefficient also remains independent of $X_{ab(ab-an)}^{Pl}$ at the higher pressures of 5 and 10 kbar (650 °C), again within the range of $X_{ab(ab-an)}^{Pl} = 0.35$ to 0.70, and possibly to 0.80. The K_D values at these pressures, as calculated from the Seck (1971b) data are as follows,

kbar	°C	K_D
1.0	650	2.99
5.0	650	3.45
10.0	650	3.67

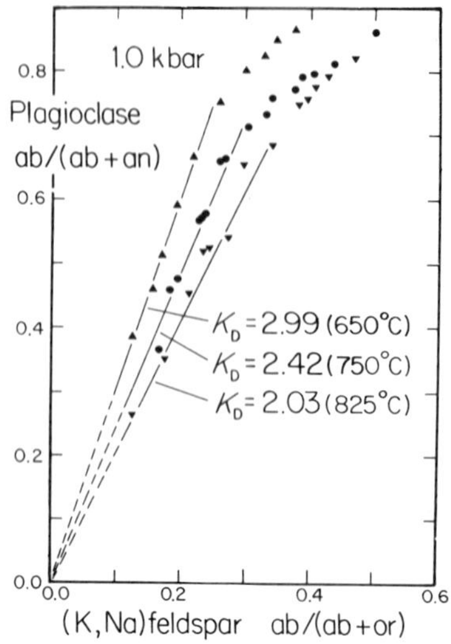

Figure 3.37 Distribution of $NaAlSi_3O_8$ (ab) between plagioclase (Pl) and (K, Na) feldspar (Kf) at 1.0 kbar and 650, 750 and 825 °C; note that $K_D = X^{Pl}_{ab(ab-an)}/X^{Kf}_{ab(ab-or)}$ is constant at a given temperature provided that $X^{Pl}_{ab(ab-an)}$ lies between 0.3 and 0.7. Data from Seck (1971a)

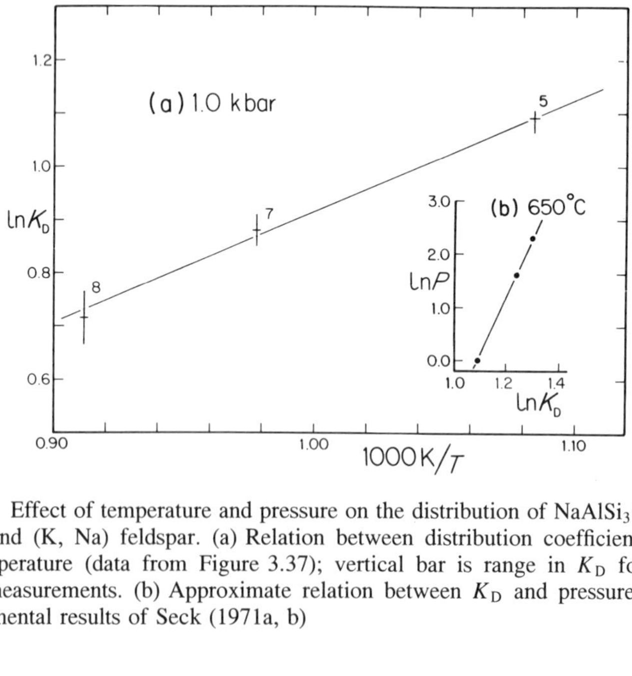

Figure 3.38 Effect of temperature and pressure on the distribution of $NaAlSi_3O_8$ between plagioclase and (K, Na) feldspar. (a) Relation between distribution coefficient (K_D) and absolute temperature (data from Figure 3.37); vertical bar is range in K_D for indicated number of measurements. (b) Approximate relation between K_D and pressure at 650 °C. From experimental results of Seck (1971a, b)

(Figure 3.38b), and the K_D-pressure relation can be expressed approximately as

$$\ln P = 11.2 \ln K_D - 12.3. \tag{62}$$

Assuming equations (61) and (62) hold in the range of interest, i.e., 600 to 900 °C, 3 to 10 kbar, $X^{Pl}_{ab(ab-an)}$ ~0.35 to 0.75 (appropriate to mid-to-deep crustal metamorphic and plutonic crystallization), we obtain by combining these two equations,

$$T = 2080/(\ln K_D - 0.091 \ln P + 1.16), \tag{63}$$

where T is absolute temperature and P is in kbar units.

Rearranging (63), we also have

$$P = \exp\{[\ln K_D - (2080/T) + 1.16]/0.091\}. \tag{64}$$

Equation (63), which is here referred to as the 'Seck relation', may be used to obtain estimates of equilibration temperatures for various crystalline rocks. An uncertainty in pressure does not create a large uncertainty in the temperature estimate; for example, with $X^{Pl}_{ab(ab-an)} = 0.70$, $X^{Kf}_{ab(ab-or)} = 0.30$ (i.e., $K_D = 2.33$), temperature estimates at 5, 6, 7, and 8 kbar are 840, 850, 860, and 870 °C. Equation (64) cannot, however, be expected to produce pressure estimates unless very accurate analytical data are available and temperature is well known. For example, at $K_D = 3.5$, temperatures of 640, 650, and 660 °C yield pressures of 4.3, 5.8, and 7.3 kbar.

Thermodynamic relations in the feldspar system are of interest and some progress has been made in the construction of thermodynamic models to define relations between phase composition, temperature, and pressure.

Stormer (1975) developed a model by assuming that $(G^{Pl}_{ab} - G^{Kf}_{ab}) = 0$ and that plagioclase forms an ideal solution ($\gamma^{Pl}_{ab} = 1.0$). Thus equation (60) is simplified to

$$RT \ln K_D = + \ln \gamma^{Kf}_{ab}. \tag{65}$$

A Margules asymmetrical equation (Sec. 2.10) was then adopted for the albite–orthoclase join, i.e.,

$$RT \ln \gamma_{ab} = W_{ab} X^2_{or} + 2X^2_{or} X_{ab}(W_{or} - W_{ab}), \tag{66}$$

with the following Margules parameters, taken from Thompson and Waldbaum (1969),

$$W_{ab} = 6327 + 0.0925P - 4.632T, \tag{67}$$

$$W_{or} = 7672 + 0.112P - 3.857T.$$

On combining these equations and rearranging terms we obtain,

$$T = \frac{0.925 + 0.0392(X^{Kf}_{or})^2 P + (6327 + 2690X^{Kf}_{ab})(X^{Kf}_{or})^2}{1.987 \ln K_D + (4.632 - 1.550X^{Kf}_{ab})(X^{Kf}_{or})^2}. \tag{68}$$

This is equivalent to the Stormer (1975) equation which has been used extensively to obtain temperature estimates. Note that X_{ab}, X_{an}, and X_{or} are here defined as ternary mole fractions, e.g., $X_{ab} = ab/(ab + an + or)$, and that K_D in equation (68) is

$$K_D = \frac{[ab/(ab + an + or)]^{Pl}}{[ab/(ab + an + or)]^{Kf}}$$

i.e., $K_D > 1.0$. Although we know that at 600–900 °C plagioclase is not an ideal solution (Sec. 2.11) and although revised Margules parameters for high-temperature alkali feldspar could be adopted (Thompson and Hovis 1979), the Stormer equation (68) does agree well with the experimental results of Seck (1971a, b).

Haselton et al. (1983) have constructed a model in which mixing on both the ab–or and ab–an binaries is expressible in terms of the Margules asymmetrical model, as follows,

$$a_{ab}^{Kf} = X_{ab}^{Kf} \exp\left[\frac{(X_{or}^{Kf})^2(18\,810 + 17\,030X_{ab}^{Kf} - 10.3T + 0.364P)}{8.314\ T}\right], \tag{69}$$

$$a_{ab}^{Pl} = (X_{ab}^{Pl})^2(2 - X_{ab}^{Pl}) \exp\left[\frac{(X_{an}^{Pl})^2(28\,230 - 39\,520X_{ab}^{Pl})}{8.314\ T}\right], \tag{70}$$

where the exponential terms are activity coefficients and the X terms are ternary mole fractions. By substitution of these equations in

$$a_{ab}^{Pl} = a_{ab}^{Kf},$$

and rearranging terms we obtain

$$T = \frac{((X_{or}^{Kf})^2(18\,810 + 17\,030X_{ab}^{Kf} + 0.346P) - (X_{an}^{Pl})^2(28\,230 - 39\,520X_{ab}^{Pl}))}{10.3(X_{or}^{Kf})^2 + 8.314\ln((X_{ab}^{Pl})^2(2 - X_{ab}^{Pl})/X_{ab}^{Kf})}, \tag{71}$$

with T as absolute temperature and P in bars. The authors state that this equation does not hold at temperatures above 850–900 °C and Ghiorso (1984) has confirmed that when applied to volcanic rocks it gives unreasonably high temperatures. For one mineral assemblage from the Adirondacks (Bohlen and Essene 1977; Figure 3.35a; 8 kbar) equation (71) gives 680 °C, compared with 640 °C obtained by both the Stormer equation (68) and the Seck relation (63).

Thermodynamic models which employ three-component Margules equations were constructed by Ghiorso (1984), Green and Usdansky (1986), and Fuhrman and Lindsley (1988). Recall (Sec. 2.10) that the activity coefficient (γ) of a component (e.g., ab) in an asymmetrical binary mixture (e.g., ab–an) is expressed by the equation

$$RT \ln \gamma_{ab} = W_{ab-an}X_{an}^2 + 2X_{an}^2X_{ab}(W_{an-ab} - W_{ab-an}). \tag{72}$$

Wohl (1946, 1953) derived the following equation for the activity coefficient of the same component in a three-component mixture (e.g., ab–an–or):

$$\begin{aligned} RT \ln \gamma_{ab} = {} & X_{or}^2[W_{ab-or} + 2X_{ab}(W_{or-ab} - W_{ab-or})] \\ & + X_{an}^2[W_{ab-an} + 2X_{ab}(W_{an-ab} - W_{ab-an})] \\ & + X_{an}X_{or}[0.5(W_{an-ab} + W_{ab-an} + W_{or-ab} + W_{ab-or} - W_{an-or} - W_{or-an}) \\ & + X_{ab}(W_{an-ab} - W_{ab-an} + W_{or-ab} - W_{ab-or}) \\ & + (X_{an} - X_{or})(W_{an-or} - W_{or-an}) - (1 - 2X_{ab})C^*]. \end{aligned} \tag{73}$$

This equation contains six binary Margules parameters in W^G, three ternary mole fractions, and a constant (C^*). Corresponding equations for γ_{an} and γ_{or} can be obtained by use of the rotation principle. Each of the W terms in equation (73) consists of three parts:

$$W^G = W^H - TW^S + (P - 1)W^V. \tag{74}$$

The W values chosen by Fuhrman and Lindsley (1988) are listed in Table 3.4; they differ slightly from the values used by Green and Usdansky (1986) and Ghiorso (1984).

Recall (Sec. 2.11) that with regard to plagioclase, Newton, Charlu and Kleppa (1980) replaced X in the equation

$$a = X\gamma \tag{75}$$

by a' which is referred to as an 'ideal' activity, and they used the Al-avoidance principle to define a'_{an} as

$$a'_{an} = (1/4)(X_{an})(1 + X_{an})^2. \tag{76}$$

Also, for the other components (Ghiorso 1984),

$$\begin{aligned} a'_{ab} &= (X_{ab})(2 - X_{ab} - X_{or})(X_{ab} + X_{or}) \\ &= X_{ab}(1 - X_{an}^2), \end{aligned} \tag{77}$$

Table 3.4 Margules parameters used by Fuhrman and Lindsley (1988)

Interaction	W^H J/mol	W^S J/mol · K	W^V J/mol · bar
an–ab	8 471	0	0
ab–an	28 226	0	0
or–ab	27 320	10.3	0.394
ab–or	18 810	10.3	0.394
an–or	52 468	0	−0.120
or–an	47 396	0	0
C^*	8 700	0	−1.094

$$a'_{or} = (X_{or})(2 - X_{ab} - X_{or})(X_{ab} + X_{or})$$
$$= X_{or}(1 - X_{an}^2). \tag{78}$$

Thus, the activity of albite in a ternary feldspar mixture (Pl or Kf) is obtained by substitution of equations (73) and (77) in

$$a_{ab} = a'_{ab}\gamma_{ab}, \tag{79}$$

written as

$$RT \ln a_{ab} = RT \ln a'_{ab} + RT \ln \gamma_{ab}. \tag{80}$$

Analogous expressions are obtained for a_{an} and a_{or}.

Now by substitution of these expressions for activity in the equations,

$$a_{ab}^{Pl} = a_{ab}^{Kf},$$
$$a_{an}^{Pl} = a_{an}^{Kf}, \tag{81}$$
$$a_{or}^{Pl} = a_{or}^{Kf},$$

as described by Ghiorso (1984), equilibrium relations can be determined for a given temperature and pressure. Alternatively, given the composition of associated plagioclase and K-feldspar, and a pressure estimate, three temperatures can be calculated (T_{ab}, T_{an}, T_{or}). The computer program prepared by Fuhrman and Lindsley (1988), which employs the Margules parameters listed in Table 3.4, when applied to three samples of granulite, produces the temperature estimates listed in Table 3.5. This table also shows that calculated values of T_{ab} agree well with temperatures obtained by use of the Seck relation (equation (63)).

As an example of feldspar thermometry in a metamorphic terrane we shall examine the data collected by Bohlen and Essene (1977) from granulites in the

Table 3.5 Composition of granulite-facies feldspar minerals, and temperature estimate

Composition (mole fraction)

	Plagioclase			K-feldspar			Plagioclase	K-feldspar
Sample	ab	or	an	ab	or	an	$X_{ab(ab-an)}$	$X_{ab(ab-or)}$
1. HA260	0.697	0.008	0.295	0.192	0.798	0.010	0.700	0.194
2. HA309	0.728	0.013	0.259	0.235	0.751	0.014	0.738	0.238
3. HA349	0.735	0.009	0.256	0.223	0.760	0.017	0.741	0.227

Temperature estimates (°C)

	Fuhrman and Lindsley (1988)			Seck relation (Eq. (63))	
	T_{ab}	T_{or}	T_{an}	K_D	T
1	650	590	660	3.61	650
2	700	610	700	3.10	710
3	700	590	680	3.26	690

Note: Analyses, obtained by L.E. Ehrhard, are reported by Fuhrman and Lindsley (1988); P = 7.2 kbar.

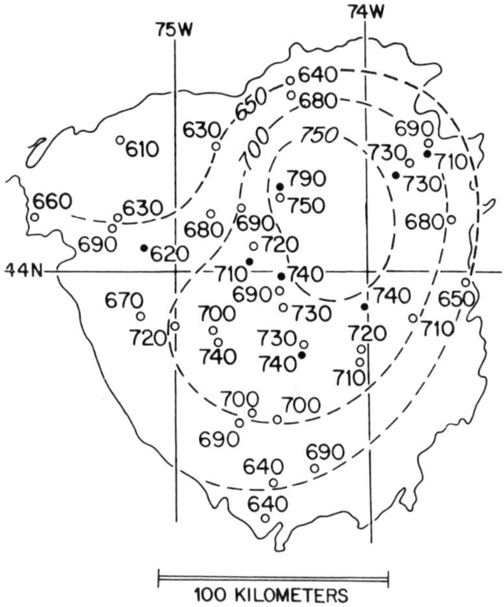

Figure 3.39 Temperature estimates, Adirondack Highlands, obtained from feldspar analyses (open symbols) and oxide analyses (closed symbols); error estimate is ±50 °C. From Bohlen and Essene (1977). Reproduced by permission of Springer-Verlag

Adirondack Highlands. In addition to plagioclase and K-feldspar, the rocks contain various combinations of quartz, orthopyroxene, Ca pyroxene, garnet, hornblende, magnetite, and ilmenite. Exsolution has occurred during cooling, but by obtaining a large number of point microprobe analyses (100 to 200 on each of several feldspar grains in a thin-section) it was possible to calculate with confidence the original (peak metamorphic) feldspar composition. Three of these mineral pairs are plotted in Figure 3.35a. The Stormer (1975) equation (68) was then used to derive temperature estimates, assuming a pressure of 8 kbar. The results of this study, together with temperature estimates obtained from the composition of magnetite and ilmenite, are plotted in Figure 3.39. The 700 and 750 °C isotherms are concentric about the central anorthosite massif. The feldspar and oxide temperatures are nearly identical.

3.6 COMPLEX TRANSFORMATIONS INVOLVING MINERALS OF VARIABLE COMPOSITION

In Sections 3.1 and 3.2, above, we studied transformations involving minerals of fixed or nearly fixed composition. The most common mineral-forming reactions within the solid Earth are more complex in the sense that ferromagnesian and other minerals that vary widely in chemical composition participate as reactants and products. These have been referred to as 'continuous reactions' because they proceed over a range of temperature and pressure. In many cases, reactant and product crystals must change in composition as the reaction progresses, provided a condition of near-equilibrium is maintained. Three examples of mineral equilibria of this kind are now examined.

3.6.1 The equilibrium, biotite + sillimanite + quartz = garnet + K-feldspar + H_2O

$$K_{1/3}(Mg,Fe)Al_{1/3}SiO_{10/3}(OH)_{2/3} + 1/3\ Al_2SiO_5 + 2/3SiO_2$$
$$= (Mg,Fe)Al_{2/3}SiO_4 + 1/3\ KAlSi_3O_8 + 1/3\ H_2O$$

Biotite garnet sillimanite gneisses occur in many high-grade metamorphic terranes, and when quartz and K-feldspar are present the five minerals are interrelated by the above provisional reaction equation which was first proposed by Ramberg (1948). The mineral assemblage occurs, for example, in the Alps (northern Italy) where Schmid and Wood (1976) observed a progressive up-grade increase in the abundance of garnet and K-feldspar at the expense of biotite and sillimanite; thus in an initial analysis the above reaction may be assumed to move to the right with increasing temperature.

In the common five-phase assemblages, garnet contains minor Mn and Ca, K-feldspar contains some Na, and the actual biotite formula differs considerably from that written above (e.g., Chapter 1, Table 1.2). Also, one must assume that H_2O was present during metamorphism, as a grain-boundary phase or a component of such a phase. We shall therefore begin by constructing an elementary model in which minerals possess simplified formulas and pure H_2O is present, and then consider possible perturbations resulting from the presence of additional elements and reduced H_2O activity.

It is well known that the Fe/(Mg + Fe) ratio of garnet is greater than that of associated biotite, and the above equation is not balanced for Mg and Fe individually. Thus two mass-balance equations are needed to describe the equilibrium:

$$\underset{\text{phl in biotite}}{K_{1/3}MgAl_{1/3}SiO_{10/3}(OH)_{2/3}} + 1/3\ Al_2SiO_5 + 2/3\ SiO_2$$

$$= \underset{\text{prp in garnet}}{MgAl_{2/3}SiO_4} + 1/3\ KAlSi_3O_8 + 1/3\ H_2O, \qquad \text{(a)}$$

$$\underset{\text{ann in biotite}}{K_{1/3}FeAl_{1/3}SiO_{10/3}(OH)_{2/3}} + 1/3\ Al_2SiO_5 + 2/3\ SiO_2$$

$$= \underset{\text{alm in garnet}}{FeAl_{2/3}SiO_4} + 1/3\ KAlSi_3O_8 + 1/3\ H_2O. \qquad \text{(b)}$$

These equations define the components phlogopite (phl) and annite (ann) of biotite (Bt), and pyrope (prp) and almandine (alm) of garnet (Grt). The two corresponding energy-balance equations are

$$\mu_{\mathrm{phl}} + (1/3)G^{\mathrm{Sil}} + (2/3)G^{\mathrm{Qtz}} = \mu_{\mathrm{prp}} + (1/3)G^{\mathrm{Kf}} + (1/3)G^{\mathrm{H_2O}}, \tag{82}$$

$$\mu_{\mathrm{ann}} + (1/3)G^{\mathrm{Sil}} + (2/3)G^{\mathrm{Qtz}} = \mu_{\mathrm{alm}} + (1/3)G^{\mathrm{Kf}} + (1/3)G^{\mathrm{H_2O}}. \tag{83}$$

We now assume ideal mixing in biotite and garnet, and substitute into (82) and (83), four equations of the kind

$$\mu_{\mathrm{phl}} = G^{\mathrm{Phl}} + RT \ln X_{\mathrm{phl}}.$$

Proceeding as in Sec. 2.5 of Chapter 2, a differentiation with respect to temperature at constant pressure, followed by an integration from the end-member equilibrium temperatures (T_{a}, T_{b}) to a temperature of interest (T), yields

$$X_{\mathrm{prp}}/X_{\mathrm{phl}} = \exp -[(\Delta H_{\mathrm{a}}/R)(T^{-1} - T_{\mathrm{a}}^{-1})], \tag{84}$$

$$(1 - X_{\mathrm{prp}})/(1 - X_{\mathrm{phl}}) = \exp -[(\Delta H_{\mathrm{b}}/R)(T^{-1} - T_{\mathrm{b}}^{-1})], \tag{85}$$

where X_{phl} = phl/(phl + ann) = Mg/(Mg + Fe) in biotite, X_{prp} = prp/(prp + alm) = Mg/(Mg + Fe) in garnet, and ΔH_{a}, ΔH_{b} are enthalpy changes for (a) and (b) at the temperature and pressure of interest. Solving explicitly for X_{prp} and X_{phl} we obtain the van Laar (1908) equations,

$$X_{\mathrm{prp}} = \left(\frac{\lambda_{\mathrm{a}}(\lambda_{\mathrm{b}} - 1)}{\lambda_{\mathrm{b}} - \lambda_{\mathrm{a}}}\right), \tag{86}$$

$$X_{\mathrm{phl}} = \left(\frac{\lambda_{\mathrm{b}} - 1}{\lambda_{\mathrm{b}} - \lambda_{\mathrm{a}}}\right), \tag{87}$$

where λ_{a} and λ_{b} represent the exponential terms in (84) and (85) respectively.

The enthalpy change for dehydration reactions does not vary greatly with changes in temperature and pressure (Sec. 3.2.1) and a numerical estimate for ΔH_{a} is obtained from the data listed in Table 2.1, to give $\Delta H_{\mathrm{a}(T,P)} \approx \Delta H_{\mathrm{a(STP)}} = +38.8$ kJ/mol. Assuming an equal value for ΔH_{b} implies that the heat of formation of annite (which is not well known) is $\Delta_{\mathrm{f}}H = -5190$ kJ/mol, which compares with -5160 and -5150 kJ/mol of $KFe_3AlSi_3O_{10}(OH)_2$, as listed by Helgeson et al. (1978) and Holland and Powell (1990) respectively. Thus, we shall assume $\Delta H_{\mathrm{a}} = \Delta H_{\mathrm{b}} = 39.0$ kJ/mol.

Analyses of biotite and garnet from an upper amphibolite to (locally) lower granulite-facies terrane in the Grenville Province (Canadian Shield) are now used to complete the construction of our model. The estimated pressure (Sec. 3.6.3, below) is 7 kbar. Figure 3.40 shows that when the end-member equilibrium temperatures are taken as $T_{\mathrm{a}} = 1160$ K and $T_{\mathrm{b}} = 860$ K, the van Laar equations (86) and (87) provide a satisfactory fit to the natural data. One anomalous mineral pair can be traced to an exceptionally low Al and high Cl content of the biotite. In terms of the above model, the chosen samples achieved equilibrium at different temperatures in the range of 650 to 700 °C.

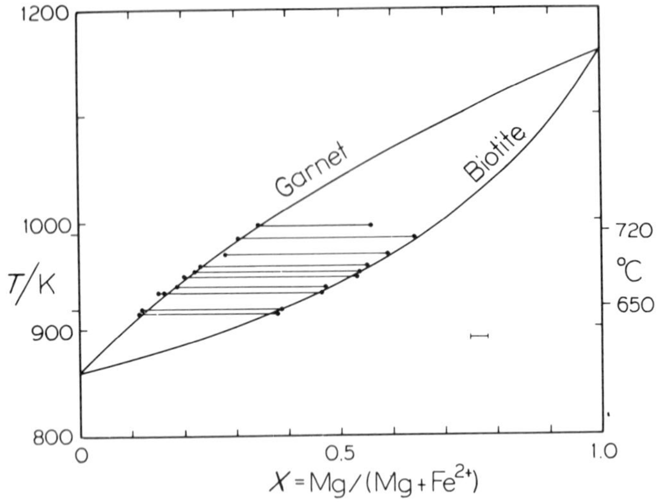

Figure 3.40 Model for the equilibrium, biotite + sillimanite + quartz = garnet + K-feldspar + H_2O at ~7 kbar; data from upper amphibolite facies gneisses are plotted (one sample is anomalous); bar represents analytical error ($\pm 2s$). From Kretz (1990). Reprinted by permission of Blackwell Scientific Publications, Inc.

We have constructed above an idealized model, and then examined some natural data in terms of this model. In doing so we have introduced an approximation which results from the view that the minor components in biotite, garnet, K-feldspar, and sillimanite cause, by their presence, only minor disturbances in the standard-state terms G^{Phl}, G^{Ann}, G^{Prp}, G^{Alm} and in the quantities G^{Kf} and G^{Sil} as they appear in the energy-balance equations for the idealized model. The justification for this approximation lies mainly in the small-magnitude decrease in the chemical potential of a major component due to the presence of a minor component, as shown in Figure 2.7. This approximation, which in a sense ignores minor and trace components, is commonly if not invariably employed in the thermodynamic treatment of natural minerals. In the present case, concentrations of Ti, Fe^{3+}, Al and vacancies in the biotite crystals, and Mn and Ca in garnet, and Na in K-feldspar are sufficiently low or sufficiently uniform that perturbations of the van Laar equilibrium (Figure 3.40) cannot be detected.

We see from the above analysis that when the reaction biotite + sillimanite + quartz = garnet + K-feldspar + H_2O is displaced to the right in response to a slow increase in temperature, garnet is produced at the expense of biotite, and in addition, barring large departures from equilibrium, all of the biotite and garnet crystals present at the reaction site must become enriched in Mg relative to Fe. Thus the transfer of Mg and Fe atoms from biotite to garnet is accompanied by the diffusion of Mg to interiors of biotite and garnet crystals in exchange for Fe. The overall change is therefore viewed as a transfer or van Laar reaction, and not an Fe–Mg exchange *between* biotite and garnet as is commonly assumed.

In the idealized model, pure H_2O is present. It seems more likely that because of escape from reaction sites or mixing of H_2O molecules with other components, the activity of H_2O will be somewhat reduced. Froese (1973) in particular has suggested that the equilibrium can act as a sensor of H_2O activity.

A reduced activity of H_2O does not change ΔH_a and ΔH_b in equations (84) and (85), but will reduce T_a and T_b, i.e., the univariant end-member boundary curves will shift to lower temperatures. This problem is here dealt with by use of the modified Clapeyron equation (Sec. 3.2.1) which in its inverted form is

$$\frac{d(1/T)}{d \ln f_{T'P'}} = -\frac{R}{\Delta H_{T'1} + \Delta V^{S}_{STP}(P')}, \tag{88}$$

where (T', P') represents a point on the boundary curve in $\ln f$, $1/T$ space, and f is the fugacity of H_2O, ΔH is the enthalpy change, ΔV^S is the solid-volume change assumed equal to that at 298 K, 1 bar (STP), and R is the gas constant. Taking the Mg end-member reaction for example (reaction (a) above), we obtain by integration,

$$\int_{1/T_{a(1)}}^{1/T_{a(2)}} d(1/T) = -\left(\frac{R}{\Delta H_{a(T'1)} + \Delta V^{S}_{a(STP)}(P')}\right) \int_{f_1}^{f_2} d \ln f_{T'P'}, \tag{89}$$

$$\left(\frac{1}{T_{a(2)}} - \frac{1}{T_{a(1)}}\right) = -\left(\frac{R}{\Delta H_{a(T'1)} + \Delta V^{S}_{a(STP)}(P')}\right) \ln(f_2/f_1). \tag{90}$$

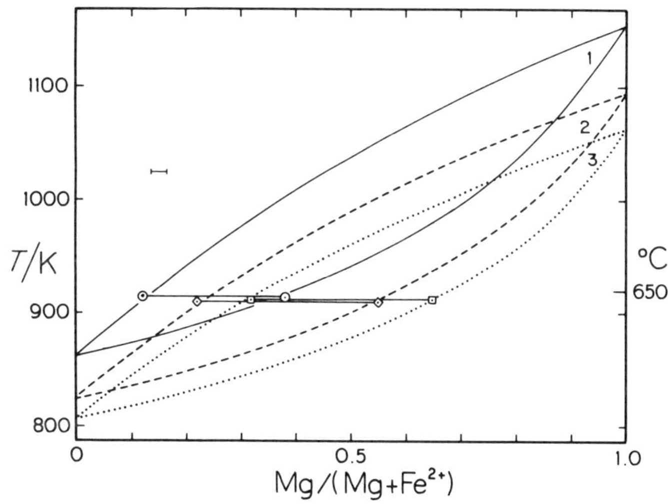

Figure 3.41 Displacement of the biotite–garnet loop in Figure 3.40 (loop 1) by a reduction in the fugacity of H_2O by a factor of 0.82 (loop 2) and 0.73 (loop 3); three of the tie lines in Figure 3.40 are plotted, to show that all could have obtained equilibrium at the same temperature. From Kretz (1990). Reprinted by permission of Blackwell Scientific Publications, Inc.

By adopting the values $T_{a(1)} = 1160$ K, $\Delta H_{a(T',1)} = \Delta H_{a(T'P')} = 39$ kJ/mol, $\Delta V_{a(STP)} = -0.777$ J/bar (from data of Table 2.1), $P' = 6000$ bar, $f_2/f_1 = 0.82$, we calculate $T_{a(2)} = 1100$ K. Similarly, $T_{b(2)} = 825$ K. Thus a reduction in H_2O fugacity by a factor of 0.82 produces loop 2 shown in Figure 3.41; a further reduction to a factor of 0.73 produces loop 3. We see then that the observed variation in biotite and garnet Fe:Mg composition could be the consequence of a variation from place to place within the metamorphic terrane of the fugacity of H_2O, rather than a variation in temperature. This interpretation is preferred.

Other equilibria of this kind were examined by Thompson (1976).

3.6.2 The equilibrium, cordierite = garnet + sillimanite + quartz

$$(\text{Mg,Fe})\text{Al}_2\text{Si}_{5/2}\text{O}_9 = (\text{Mg,Fe})\text{Al}_{2/3}\text{SiO}_4 + (2/3)\text{Al}_2\text{SiO}_5 + (5/6)\text{SiO}_2$$

Cordierite, garnet, sillimanite, and quartz occur together in upper amphibolite and granulite facies terranes of relatively low pressure, and an equilibrium involving these minerals, as expressed by the above simplified equation, was recognized by Korzhinskii (1936a). The volume change for the end-member reaction,

$$\begin{aligned}\text{FeAl}_2\text{Si}_{5/2}\text{O}_9 \cdot (n/2)\text{H}_2\text{O} = {} & \text{FeAl}_{2/3}\text{SiO}_4 + (2/3)\text{Al}_2\text{SiO}_5 \\ & + (5/6)\text{H}_2\text{O} + (n/2)\text{H}_2\text{O},\end{aligned}$$

at moderate temperature and pressure (e.g., 740 °C, 4.0 kbar) is negative regardless of whether the cordierite is anhydrous or contains channel H_2O (Holdaway and Lee 1977), and increasing pressure must displace the equilibrium to the right. Also, because the Mg/(Mg + Fe) ratio of cordierite is greater than that of associated garnet, the reaction is continuous and similar to the biotite-to-garnet reaction considered in Sec. 3.6.1, above.

Data on the channel occupants of natural cordierite are reviewed by Vry, Brown and Valley (1990) who obtained many new analyses from various amphibolite and granulite terranes. The principal constituents found are H_2O and CO_2 but others, including He, Ar, Ne, and H_2S and various cations have been detected in the past. In relation to the cordierite formula expressed as

$$(\text{Mg,Fe})_2\text{Al}_4\text{Si}_5\text{O}_{18} \cdot n(\text{H}_2\text{O}, \text{CO}_2),$$

Johannes and Schreyer (1981) have shown that n cannot exceed 1.0, corresponding to 3.0 wt % H_2O or 7.0 wt % CO_2. An illustration of channel occupancy is provided by nine samples from the Pikwitonei granulite terrane in Manitoba (Canadian Shield) which contain both H_2O and CO_2 as follows: $n(H_2O) = 0.11$ to 0.29; $n(CO_2) = 0.09$ to 0.27; $n(H_2O, CO_2) = 0.32$ to 0.47 (Vry, Brown and Valley 1990).

Newton and Wood (1979) and Vry, Brown and Valley (1990) have summarized existing experimental data on the H_2O content of cordierite. In general, the concentration term, n, increases with H_2O pressure and decreases with temperature. For example, at 600 °C, $n(H_2O)$ increases from 0.4 to 0.9 as $P(H_2O)$ increases from

1 to 10 kbar. Although one cannot state with confidence that the H_2O and CO_2 now found in cordierite were present in the same concentration at the time of metamorphism, it seems likely that some of these molecules were present and their influence on equilibria involving cordierite must therefore be considered.

The effect of temperature on the cordierite–garnet–sillimanite–quartz equilibrium can be described by use of the van Laar equations which were developed and applied in Sec. 3.6.1, above. Our concern at present is to describe the effect of pressure. Consider a system in which the only channel occupants of cordierite are H_2O. This crystal may be viewed as a mixture of anhydrous Mg cordierite (Mg-crd), anhydrous Fe cordierite (Fe-crd), and H_2O, to produce the following three mass-balance equations for the cordierite–garnet–sillimanite–quartz equilibrium:

$MgAl_2Si_{5/2}O_9$	=	$MgAl_{2/3}SiO_4$	+	$(2/3)Al_2SiO_5$	+	$(5/6)SiO_2$,	(a)
Mg-crd component in cordierite		prp component in garnet					
$FeAl_2Si_{5/2}O_9$	=	$FeAl_{2/3}SiO_4$	+	$(2/3)Al_2SiO_5$	+	$(5/6)\ SiO_2$,	(b)
Fe-crd component in cordierite		alm component in garnet					
H_2O	=	H_2O					(c)
in cordierite		in vapor or crystal-boundary phase					

The corresponding energy-balance equations are

$$\mu_{\text{Mg-crd}} = \mu_{\text{prp}} + (2/3)G^{\text{Sil}} + (5/6)G^{\text{Qtz}}, \tag{91}$$

$$\mu_{\text{Fe-crd}} = \mu_{\text{alm}} + (2/3)G^{\text{Sil}} + (5/6)G^{\text{Qtz}}, \tag{92}$$

$$\mu_{H_2O}^{\text{Crd}} = \mu_{H_2O}^{\text{v/cb}}, \tag{93}$$

where Crd refers to the cordierite phase, and v/cb to the vapor phase in the experimental setting or the crystal-boundary phase in a natural setting.

At equilibrium, a small increase in pressure will cause a small displacement of equations (a) and (b) to the right, removing both Mg-crd and Fe-crd components from the cordierite crystals; meanwhile H_2O is released, but if the cordierite is not saturated some could migrate into the remaining crystals, in response to the increase in $P(H_2O)$, until equation (93) is again satisfied. Thus H_2O is not necessarily released, and in this respect the equilibrium differs from one involving a hydroxyl-bearing phase (e.g., biotite) as discussed in Sec. 3.6.1, above.

H_2O in cordierite is now viewed as a dilutent which does not affect appreciably the Fe-crd–Mg-crd mixing process, i.e., we write

$$\mu_{\text{Mg-crd}} = G^{\text{Mg-Crd}} + RT \ln a_{\text{Mg-crd}}, \tag{94}$$

$$\mu_{\text{Fe-crd}} = G^{\text{Fe-Crd}} + RT \ln a_{\text{Fe-crd}}, \tag{95}$$

where the G values, which refer to anhydrous cordierite, are decreased slightly by the presence of H_2O. Assuming further that Fe–Mg mixing in cordierite and in garnet is nearly ideal, we obtain from (91) and (92),

$$\ln(X_{\text{prp}}/X_{\text{Mg-crd}}) = -(\Delta G_{\text{a}}/RT), \tag{96}$$

$$\ln(X_{\text{alm}}/X_{\text{Fe-crd}}) = -(\Delta G_{\text{b}}/RT), \tag{97}$$

where a and b identify the Mg and Fe end-member reactions, respectively. Differentiating (96) with respect to pressure at constant temperature,

$$\frac{\partial \ln(X_{\text{prp}}/X_{\text{Mg-crd}})}{\partial P} = -\Delta V_{\text{a}}/RT, \tag{98}$$

followed by an integration from the equilibrium pressure (P_{a}) to a pressure of interest (P),

$$\int_{P_{\text{a}}}^{P} \mathrm{d}\ln(X_{\text{prp}}/X_{\text{Mg-crd}}) = -\int_{P_{\text{a}}}^{P} (\Delta V_{\text{a}}/RT)\,\mathrm{d}P, \tag{99}$$

we obtain, assuming ΔV_{a} is constant,

$$(X_{\text{prp}}/X_{\text{Mg-crd}}) = \exp -[(\Delta V_{\text{a}}/RT)(P - P_{\text{a}})], \tag{100}$$

and similarly,

$$(X_{\text{alm}}/X_{\text{Fe-crd}}) = \exp -[(\Delta V_{\text{b}}/RT)(P - P_{\text{b}})]. \tag{101}$$

Equations (100) and (101), which in their general form were applied to mineral systems by Mueller (1963), may now be combined to solve explicitly for X_{prp} and $X_{\text{Mg-crd}}$:

$$X_{\text{prp}} = \frac{\pi_{\text{a}}(\pi_{\text{b}} - 1)}{\pi_{\text{b}} - \pi_{\text{a}}} \tag{102}$$

$$X_{\text{Mg-crd}} = \frac{\pi_{\text{b}} - 1}{\pi_{\text{b}} - \pi_{\text{a}}} \tag{103}$$

where π_{a} and π_{b} stand for the terms on the right-hand sides of (100) and (101) respectively. Equations (102) and (103) are forms of the van Laar equations of Sec. 3.6.1.

Solutions to equations (102) and (103) were obtained by use of the following data: $T = 973$ K (700 °C), $\Delta V_{\text{a}} = -2.587$ J/bar, $\Delta V_{\text{b}} = -2.783$ J/bar (1 cation mole, allowance made for compressibility and expansion to 700 °C, 7 kbar), $P_{\text{b}} =$ 3.0 kbar (the lower limit of uncertainty in the determination of Holdaway and Lee (1977); P_{a} (metastable) = 9.5 kbar (assumed). Then equations (102) and (103) plot as the garnet and cordierite curves in Figure 3.42.

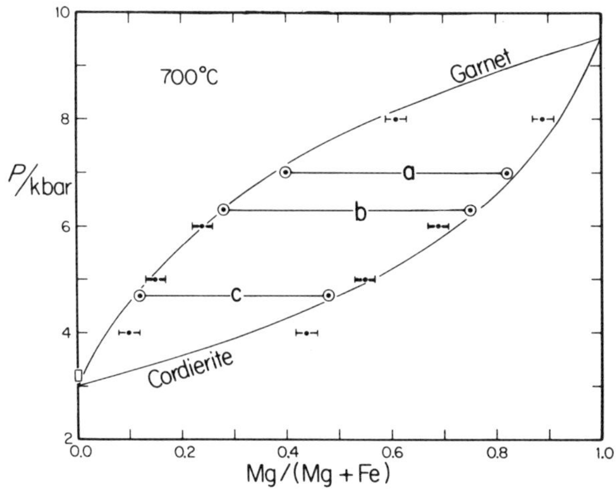

Figure 3.42 The equilibrium, cordierite = garnet + sillimanite + quartz at 700 °C; horizontal bars are experimental results of Aranovich and Podlesskii (1983), heavy symbols-reversals achieved; vertical rectangle at Fe end is from Holdaway and Lee (1977). Curves are plots of equations (102) and (103) in text, which are based on volume changes, ideal mixing, and end-member pressures of 3.0 and 9.5 kbar. Tie lines are for natural minerals from Antarctica (a and b) and California (c), from compilation of Hensen and Green (1973)

Several experimental data are available on the cordierite = garnet + sillimanite + quartz equilibrium. Those presented here are by Aranovich and Podlesskii (1983), who used mainly natural garnet (containing minor Ca and Mn) and other minerals. H_2O was added (some presumably entered the cordierite channels) and oxidation was controlled with the QFM buffer. The composition of the product crystals, as determined by microprobe analysis, was found to be somewhat variable, even after run times of 120 to 670 h. The results at 700 °C are plotted as data bars in Figure 3.42. After the data were plotted, the end points (P_a and P_b) were adjusted in five iterations, to give an improved fit of the van Laar loop to the experimental data, especially those (heavy symbols) that were obtained by reversal. Thus the phase diagram is based in part on thermodynamic data (ΔV) and assumptions (ideal mixing) and in part on experimental results.

Also plotted in this figure are three natural metamorphic mineral pairs (associated with quartz and sillimanite), from a compilation of Hensen and Green (1973); a and b are from Antarctica and c from the Sierra Nevada in California.

The Aranovich and Podlesskii (1983) results for a slightly higher temperature of 750 °C indicate that the $P-X$ loop of Figure 3.42 must rise slightly with increasing temperature (implying that ΔS_b is negative), but this does not agree with the negative $\mathrm{d}P/\mathrm{d}T$ slope for the Fe end-member boundary as determined by Holdaway and Lee (1977). Additional uncertainties and conflicts in existing experimental results on the cordierite-garnet-sillimanite-quartz equilibrium are reviewed by Lonker (1981) and Newton (1985). As these become resolved, the equilibrium should provide a useful geobarometer.

3.6.3 The equilibrium, anorthite (in plagioclase) = grossular (in garnet) + Al silicate + quartz

Plagioclase, garnet, kyanite (or sillimanite), and quartz are commonly found together in schist and gneiss of medium- and high-grade metamorphic terranes. In a study of gneisses from Otter Lake (Grenville Province, Canadian Shield) Kretz (1964) pointed out that the four minerals are interrelated by a chemical equation,

$CaAl_2Si_2O_8$	=	$CaAl_{2/3}SiO_4$	+	$(2/3)Al_2SiO_5$	+	$(1/3)SiO_2$,	(a)
an component in plagioclase		grs component in garnet		kyanite or sillimanite		quartz	

which, when displaced to the right, results in a large decrease in volume ($-18\ cm^3$/mol of $CaAl_2Si_2O_8$), and he suggested that the equilibrium could therefore form a suitable pressure indicator. Ghent (1976) derived an equation for equilibrium (a) based on experimental data on the end-member boundary with kyanite, obtained by Hays (1967) and Hariya and Kennedy (1968), and he demonstrated, by referring to mineral analyses from several metamorphic terranes, that the equilibrium is indeed capable of providing reasonable estimates of pressure, provided temperature is known approximately. Newton and Haselton (1981) adopted the revised end-member boundary equation of Goldsmith (1980) and existing activity and volume relations for plagioclase and garnet to produce an equation which has been used widely to obtain estimates of metamorphic pressure. The present discussion of equilibrium (a) is based largely on new experimental data on the end-member boundary and on the activity of grossular in Ca–Fe–Mg garnet, obtained by Koziol and Newton (1988, 1989).

Consider equation (a) with plagioclase (an, ab), garnet (grs, alm, prp), pure sillimanite, and pure quartz. Then at equilibrium,

$$\Delta G_{T,P} = (\mu_{\text{grs}})_{T,P} + (2/3)G_{T,P}^{\text{Sil}} + (1/3)G_{T,P}^{\text{Qtz}} - (\mu_{\text{an}})_{T,P} = 0. \tag{104}$$

Recall (Sec. 2.1) that for a phase of fixed composition at constant temperature,

$$\mathrm{d}G = V\,\mathrm{d}P.$$

Integrating from $P = 1$ bar to $P \gg 1$ bar,

$$\int_{P=1}^{P=P} \mathrm{d}G = \int_{P=1}^{P=P} V\,\mathrm{d}P,$$

we obtain

$$G_{T,P} = G_{T,1} + V(P), \tag{105}$$

provided that V is independent of pressure. Similarly, for a mixture of components A and B,

$$\mathrm{d}G_A = V_A\,\mathrm{d}P,$$

where G_A is the partial Gibbs energy of A in the mixture (i.e., the chemical potential of A, μ_A) and V_A is the partial molar volume of A. Thus, at constant temperature,

$$\int_{P=1}^{P=P} \mathrm{d}\mu_A = \int_{P=1}^{P=P} V_A \,\mathrm{d}P,$$

and

$$(\mu_A)_{T,P} = (\mu_A)_{T,1} + V_A(P), \tag{106}$$

provided that V_A is independent of pressure. By substitution of equations of kind (105) and (106) in (104), we obtain

$$\Delta G_{T,P} = (\mu_{\text{grs}})_{T,1} + V_{\text{grs}}P + (2/3)G_{T,1}^{\text{Sil}} + (2/3)V^{\text{Sil}}P$$
$$+ (1/3)G_{T,1}^{\text{Qtz}} + (1/3)V^{\text{Qtz}}P - (\mu_{\text{an}})_{T,1} - V_{\text{an}}P = 0. \tag{107}$$

Recall (Sec. 2.4) that for a mixture of components A and B,

$$(\mu_A)_{T,P} = G_{T,P}^{A} + RT\ln(a_A)_{T,P} \tag{108}$$

A further substitution of two equations of kind (108) in (107) yields

$$\Delta G_{T,P} = G_{T,1}^{\text{Grs}} + RT\ln(a_{\text{grs}})_{T,1} + V_{\text{grs}}P + (2/3)G_{T,1}^{\text{Sil}} + (2/3)V^{\text{Sil}}P$$
$$+ (1/3)G_{T,1}^{\text{Qtz}} + (1/3)V^{\text{Qtz}}P - G_{T,1}^{\text{An}} - RT\ln(a_{\text{an}})_{T,1} - V_{\text{an}}P = 0,$$

or

$$RT\ln(a_{\text{grs}}/a_{\text{an}})_{T,1} + \Delta G_{T,1}^{\circ} + (\Delta V)P = 0. \tag{109}$$

Now by analogy with (105), we have for the end-member reaction,

$$\Delta G_{T,P^\circ}^{\circ} = \Delta G_{T,1}^{\circ} + (\Delta V^\circ)P^\circ.$$

But $\Delta G_{T,P^\circ}^{\circ} = 0$, representing a point on the boundary curve, and

$$\Delta G_{T,1}^{\circ} = -(\Delta V^\circ)P^\circ. \tag{110}$$

Placing (110) in (109),

$$RT\ln(a_{\text{grs}}/a_{\text{an}})_{T,1} - (\Delta V^\circ)P^\circ + (\Delta V)P = 0. \tag{111}$$

In the above integrations, all volume terms were taken as independent of pressure and, because the integrations were carried out at a constant but arbitrary temperature, the assumption of constant volume extends, implicitly, to temperature as well. These simplifications are now removed by adding that ΔV in (111) refers to the temperature (T') and pressure (P') of interest (i.e., with regard to equilibrium (a)), and ΔV° refers to T' and P°, where P° is the pressure on the end-member boundary curve at T'. Thus,

$$RT'\ln(a_{\text{grs}}/a_{\text{an}})_{T',1} - (\Delta V^\circ)_{T'P^\circ}P^\circ + (\Delta V)_{T',P'}P' = 0. \tag{112}$$

This is the Newton–Haselton (1981) equation. Note that $\Delta V°$ is the volume change for the end-member reaction (in this case, with sillimanite), while ΔV is the volume change for (a), involving partial molar volumes of the components grs in garnet and an in plagioclase. Before this equation can be used to estimate metamorphic pressure (P'), information is needed on $P°$, T', $\Delta V°$, ΔV, a_{grs}, and a_{an}. These terms are now examined individually.

$P°$ stands for the equilibrium pressure of the end-member reaction at the temperature of interest (T'). The three reactions of importance are

$$\underset{}{CaAl_2Si_2O_8} = CaAl_{2/3}SiO_4 + (2/3)\underset{\text{kyanite}}{Al_2SiO_5} + (1/3)SiO_2, \tag{b}$$

$$\underset{\text{kyanite}}{Al_2SiO_5} = \underset{\text{sillimanite}}{Al_2SiO_5}, \tag{c}$$

$$CaAl_2Si_2O_8 = CaAl_{2/3}SiO_4 + (2/3)\underset{\text{sillimanite}}{Al_2SiO_5} + (1/3)SiO_2. \tag{d}$$

The reaction with kyanite (b) has been located by several investigators, most recently by Koziol and Newton (1988), who obtained eight reversed brackets in the range of 900 to 1250 °C and 19–28 kbar. A piston-cylinder apparatus was used in which NaCl provided the pressure medium. Starting materials consisted of natural quartz and kyanite and synthetic grossular and anorthite. An additional point at 650 °C and with pressure between 12.8 and 14.1 kbar was located by calculation. The equation for the boundary curve is

$$P^{\mathrm{b}} = 22.80\ T - 7317 \text{ (with kyanite)}, \tag{113}$$

where T is absolute temperature.

The equation for the metastable equilibrium with sillimanite (d) is obtained by combining reaction equations (b) and (c). The equation for (c) is obtained from the experimental data of Holdaway (1971) and the Clapeyron slope calculated by Robie and Hemingway (1984) as described in Sec. 3.1.2, and is

$$P^{\mathrm{c}} = 20.1\ T - 12\,060, \tag{114}$$

where T is absolute temperature. Additional information needed is the volume changes for (b), (c), and (d), which, following Koziol and Newton (1988), may be calculated at 1073 K and 10 000 bar; the present calculation is based on the data in Table 2.1, to give

$$\Delta V^{\mathrm{b}} = -2.098 \text{ J/bar/mol } CaAl_2Si_2O_8,$$

$$\Delta V^{\mathrm{c}} = +0.552 \text{ J/bar/mol } Al_2SiO_5,$$

$$\Delta V^{\mathrm{d}} = -1.729 \text{ J/bar/mol } CaAl_2Si_2O_8.$$

Now by use of equation (110),

$$\Delta G^{\mathrm{b}} = -\Delta V^{\mathrm{b}} P^{\mathrm{b}},$$

$$\Delta G^{c} = -\Delta V^{c} P^{c},$$

and from

$$\Delta G^{d} = \Delta G^{b} + (2/3)\Delta G^{c},$$

we obtain

$$P^{d} = 23.4\ T - 6310 \text{ (with sillimanite)}, \tag{115}$$

where T is absolute temperature.

An estimate of the metamorphic temperature (T') in equation (112) must be obtained independently, for example, by use of the garnet–biotite Fe^{2+}–Mg exchange equilibrium described in Sec. 3.7.1, below.

The volume change, ΔV°, refers to the end-member equilibrium at T' and P°. The volume change, ΔV, refers to the actual volume change of (a) at T', P'. The excess volume of mixing in plagioclase (Sec. 2.11) and in Ca–Mg and Ca–Fe garnet (Sec. 2.11) is very small or zero, and the partial molar volumes of anorthite and grossular are taken as equal to those of the pure phases. Coefficients of expansion and compressibility (Table 2.1) enable one to calculate ΔV at the estimated temperature and anticipated pressure.

The activity of grossular in Ca–(Mg,Fe) garnet (a_{grs}) as a function of concentration, X_{grs}(= Ca/(Ca + Mg + Fe)) and Mg:Fe ratio was measured by Koziol and Newton (1989), using the displaced-equilibrium technique described in Sec. 2.11. Synthetic garnet crystals with predetermined Mg/(Mg + Fe) ratios of 0.00, 0.14, 0.21, 0.29, 0.43, and 0.57 were used in the experiments. For example, the results obtained for the mixing of grossular in crystals with Mg/(Mg + Fe) = 0.21 are shown in Figure 3.43a, where the data were fitted to the Margules asymmetrical equation (Sec. 2.10),

$$RT \ln \gamma_{grs} = (1 - X_{grs})^2[W_{Ca} + 2X_{grs}(W_{FM} - W_{Ca})], \tag{116}$$

to give $W_{Ca} = 4030$ and $W_{FM} = -2870$ J/1 Ca mole. The Margules parameter W_{Ca} was then related to Mg/(Mg + Fe) in garnet by an empirical equation (Figure 3.43b),

$$W_{Ca} = -2060 + 3.57 \times 10^4 M - 4.95 \times 10^4 M^2, \tag{117}$$

where M = Mg/(Mg + Fe). The scatter in Figure 3.43b shows that W_{Ca} is not known accurately; solving equation (117) for W_{Ca} with $M = 0.21$ produces the broken line in Figure 3.43a. Similarly, the dependence of the accompanying Margules parameter, W_{FM}, on M is

$$W_{FM} = +3390 - 3.71 \times 10^4 M + 6.49 \times 10^4 M^2. \tag{118}$$

These relations refer to 900 to 1000 °C; whether or not they differ at lower temperatures is not yet known. Thus the activity of grossular in garnet,

$$a_{grs} = \gamma_{grs} X_{grs}, \tag{119}$$

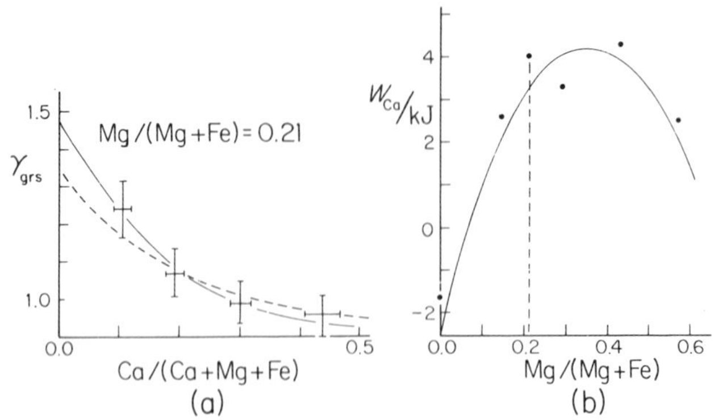

Figure 3.43 (a) Experimental results at 1000 °C for the activity coefficient of $CaAl_{2/3}SiO_4$ (γ_{grs}) in garnet with Mg/(Mg + Fe) = 0.21. Broken curve is relation with Margules parameter, W_{Ca}, taken from the smoothed curve in (b), and the same for W_{FM}. (b) Empirical relation of Margules parameter, W_{Ca} (in kJ/mole $CaAl_{2/3}SiO_4$) to Mg/(Mg + Fe) in garnet. Data from Koziol and Newton (1989)

as it appears in equation (112) is obtained by solving (116) for γ_{grs}, using values of W_{Ca} and W_{FM} from (117) and (118), with M = Mg/(Mg + Fe) determined by chemical analysis, and then by solving (119) with X_{grs} determined by chemical analysis. Note that the formula for garnet is here written $(Ca,Mg,Fe)Al_{2/3}SiO_4$, and activity is defined as above, rather than as $a_{grs} = (\gamma_{grs}X_{grs})^3$. In many natural assemblages, X_{grs} is small (<0.1) and γ_{grs}, based on the Koziol–Newton experiments, is expected to fall in the range of 1.2 to 1.6. These are extrapolations, and it will be helpful to have, in the future, some experimental results from within the indicated concentration range. One may note in passing that the Margules model (equation (116), Figure 3.43a) appears not to permit γ_{grs} to assume a constant value (for given Mg/(Mg + Fe) as $X_{grs} \to 0$ (Henry's rule), and this may be a defect in the model.

Natural garnet crystals normally contain some Mn but in assemblages with Al_2SiO_5, these concentrations are low. Koziol (1990) has shown that mixing of Ca and Mn garnet at 1000 °C is ideal (Figure 2.20, above) and she has pointed out that with regard to equilibrium (a), the spessartine component is expected to act as a minor dilutent.

An estimate of the activity of anorthite in plagioclase (a_{an}) is obtained by use of the Newton–Charlu–Kleppa (1980) equation (Sec. 2.11),

$$a_{an} = (1/4)(X_{an})(1 + X_{an})^2 \exp((1/RT)(1 - X_{an})^2(2025 + 9440X_{an})) \qquad (120)$$

where X_{an} is the mole fraction of anorthite in plagioclase (= Ca/(Ca + Na)), R = 1.987 cal/K·mol, and T is absolute temperature.

Data from Otter Lake gneisses provide a convenient illustration of the application of the plagioclase–garnet–sillimanite–quartz equilibrium. The garnet and plagioclase analyses presented by Perkins, Essene and Marcotty (1982) and Kretz

(1990) have been reinterpreted, using the new Koziol and Newton experimental results described above, and some of the revised pressure estimates for this upper amphibolite to (locally) lower granulite facies terrane are listed in Table 3.6. The new mean is 7.2 kbar ($n = 17$), and the standard deviation (s) is 1.0 kbar.

Numerous systematic and random sources of error exist in the calculation of pressure by use of equation (112). Examples of systematic error include the uncertainty in pressure resulting from an error in the estimate of temperature (a 30 K error in temperature results in a 0.4 kbar error in pressure), and the uncertainty that remains in the activity coefficients for grossular at low concentrations. A variation in the composition of garnet and plagioclase within a thin-section represents a random error, as does variation from place to place within a metamorphic terrane (Table 3.6). The latter variation, which at Otter Lake cannot be related to regional gradients or to faulting, is greater than that found in a single specimen, and because usually our main concern is in the pressure of a metamorphic zone or terrane rather than a single sample, precision is improved by analyzing more samples rather than more crystals in a single sample. Thus, whereas the pressure estimate for a single sample from Otter Lake possesses an uncertainty (precision) of $\pm 1.96\ s$ or ± 2.0 kbar (0.95 confidence), the uncertainty in the mean, based on $n = 17$ samples, is

$$\bar{x} \pm 1.96(s/\sqrt{n}),$$

or

$$7.2 \pm 0.5 \text{ kbar}.$$

The cause of this between-sample variation is not understood, but it may result from variation in the approach to equilibrium from place to place, or variation in the preservation of equilibrium in the post-metamorphic history of the terrane. The presence of one obviously anomalous sample (Table 3.6) which can be traced to

Table 3.6 Some pressure estimates for a metamorphic terrane in the Grenville Province, Canadian Shield

	Garnet			Plagioclase			
	X_{grs}	Mg/(Mg + Fe)	a_{grs}	X_{an}	a_{an}	a_{grs}/a_{an}	P/kbar
1	0.0325	0.314	0.0516	0.270	0.400	0.129	6.6
2	0.0262	0.226	0.0386	0.066			
3	0.0168	0.207	0.0245	0.148	0.184	0.133	6.6
4	0.0167	0.240	0.0256	0.101	0.111	0.231	9.1
5	0.0352	0.189	0.0480	0.258	0.378	0.127	6.4
6	0.0306	0.227	0.0456	0.270	0.400	0.114	5.9
7	0.0337	0.129	0.0407	0.220	0.310	0.131	6.6
8	0.0342	0.122	0.0406	0.250	0.364	0.112	5.9
9	0.0590	0.154	0.0732	0.270	0.400	0.183	8.1
							6.9

From Kretz (1990); specimen 2 is anomalous in the sense that the anorthite content of plagioclase is very low.

a very low anorthite concentration in plagioclase (garnet is normal) implies that some rather large post-metamorphic changes have occurred locally.

3.7 EXCHANGE EQUILIBRIUM: THE DISTRIBUTION OF Mg AND Fe^{2+} BETWEEN FERROMAGNESIAN MINERALS

In this and the following section (3.8) we shall examine several examples of exchange equilibrium involving common metamorphic minerals. The term 'exchange equilibrium' is here defined as an equilibrium in the distribution of two or more kinds of atoms or molecules between two associated phases. Unlike the systems considered in Sec. 3.6 (above), the variance must be sufficiently large to permit concentrations of the exchangeable species to vary at a fixed temperature and pressure.

Exchange reactions involving clay and zeolite minerals have been known for more than a century. As experimental data became available, the concept of exchange equilibrium was developed, particularly by Roozeboom (1891), Ratner (1933), and Kielland (1935). For example, Yutzy and Kolthoff (1937) expressed the equilibrium distribution of Cl^- and Br^- between crystalline Ag(Cl,Br) (phase α) and the saturated aqueous solution (phase β) by use of the equation

$$\frac{X_{\mathrm{Br}}^{\alpha}}{1 - X_{\mathrm{Br}}^{\alpha}} \cdot \frac{1 - X_{\mathrm{Br}}^{\beta}}{X_{\mathrm{Br}}^{\beta}} = \frac{\gamma_{\mathrm{Cl}}^{\alpha}\gamma_{\mathrm{Br}}^{\beta}}{\gamma_{\mathrm{Br}}^{\alpha}\gamma_{\mathrm{Cl}}^{\beta}} \cdot K\,,$$

where X and γ denote mole fraction and activity coefficient, and K is the equilibrium constant, a function of temperature. The quotient on the left is here referred to as the 'distribution coefficient'. Numerous experimental results are now available on exchange equilibrium involving minerals and an aqueous phase, including those of Barrer and Sammon (1955), Eugster (1955), and Orville (1963).

Distribution relations involving associated crystalline phases do not differ from those referred to above. Thus Tubandt and Reinhold (1929) considered the equilibrium distribution of Cu and Ag between $(Cu,Ag)_2S$ and (Cu,Ag)I, and the displacement of the equilibrium by temperature.

The numerous opportunities that exist for studies on the distribution of Mg and Fe^{2+} between associated ferromagnesian minerals was recognized by Eskola (1922). The first detailed investigation of this kind was carried out by Ramberg and De Vore (1951) who examined the distribution of Mg and Fe^{2+} between olivine and orthopyroxene and presented a general thermodynamic analysis of the problem. They also foresaw the use of Mg–Fe distribution coefficients as temperature indicators.

The following discussions represent only a small sample of the large amount of information that now exists on Mg–Fe exchange involving all of the common ferromagnesian minerals. Reviews of some of this literature were prepared by Ganguly and Saxena (1987) and by Perchuk (1989).

3.7.1 The distribution of Mg and Fe^{2+} between biotite and garnet

Biotite and garnet are widely distributed in metamorphic terranes, persisting, as they do, throughout the entire amphibolite facies and into the lower granulite facies. The problem of the equilibrium distribution of Mg and Fe^{2+} between the crystals of these minerals is therefore one of special interest.

It is well known that garnet prefers Fe^{2+} and associated biotite prefers Mg, and therefore if crystals of Mg-rich garnet are brought together with those of Fe-rich biotite, an exchange reaction should occur, with Mg atoms moving from garnet to biotite and an equal number of Fe atoms moving from biotite to garnet, until an equilibrium distribution is achieved. Thus an exchange reaction and mass-balance equation may be expressed as follows,

$$\underset{\text{ann in biotite}}{K_{1/3}FeAl_{1/3}SiO_{10/3}(OH)_{2/3}} + \underset{\text{prp in garnet}}{MgAl_{2/3}SiO_4}$$

$$= \underset{\text{phl in biotite}}{K_{1/3}MgAl_{1/3}SiO_{10/3}(OH)_{2/3}} + \underset{\text{alm in garnet}}{FeAl_{2/3}SiO_4}, \qquad \text{(a)}$$

where simplified formulas are chosen for biotite (phl, ann) and garnet (prp, alm). At equilibrium,

$$\mu_{\text{ann}} + \mu_{\text{prp}} = \mu_{\text{phl}} + \mu_{\text{alm}}.$$

By substitution of four equations of the kind

$$\mu_{\text{alm}} = G^{\text{Alm}} + RT \ln \gamma_{\text{alm}} X_{\text{alm}},$$

we obtain

$$\left(\frac{X_{\text{phl}}}{1 - X_{\text{phl}}} \cdot \frac{1 - X_{\text{prp}}}{X_{\text{prp}}}\right)\left(\frac{\gamma_{\text{phl}}\gamma_{\text{alm}}}{\gamma_{\text{ann}}\gamma_{\text{prp}}}\right) = \exp\left(\frac{-\Delta G_{\text{a}}}{RT}\right), \qquad (121)$$

where ΔG_{a} is the Gibbs energy change for equation (a) with components pure (at the temperature and pressure of interest) and the first quotient on the left is the distribution coefficient, denoted K_{D}. Note that

$$X_{\text{phl}} = \text{phl}/(\text{phl} + \text{ann}) = \text{Mg}/(\text{Mg} + \text{Fe}^{2+}) = X_{\text{Mg}}^{\text{Bt}},$$

and

$$X_{\text{phl}} + X_{\text{ann}} = 1.0.$$

The nature of the distribution equation (121) is revealed by assuming that both phases, α and β, are ideal solutions of Mg and Fe components (all γ terms = 1.0) and that temperature and pressure remain constant. Then, the mole fraction of Mg (X = Mg/(Mg + Fe) in one mineral is related to that in the other mineral by

$$\frac{X^{\alpha}}{1 - X^{\alpha}} \cdot \frac{1 - X^{\beta}}{X^{\beta}} = K_{\text{D}} = \text{constant}. \qquad (122)$$

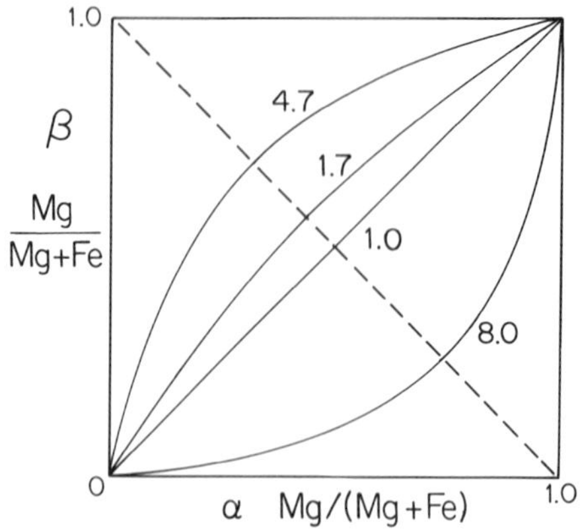

Figure 3.44 Illustration of the ideal distribution equation,

$$K_D = \frac{X_A^\alpha}{1 - X_A^\alpha} \cdot \frac{1 - X_B^\beta}{X_B^\beta}$$

where X_A is $Mg/(Mg + Fe^{2+})$ in mineral phases α and β, and the distribution coefficient (K_D) is a constant. The examples are $K_D = 4.7$ (α = staurolite, β = biotite; Hounslow and Moore (1967)); $K_D = 1.7$ (α = hornblende, β = Ca pyroxene; Saxena (1968)); $K_D = 1.0$ (α = actinolite, β = Ca pyroxene; Mueller (1960)); $K_D = 8.0$ (α = olivine, β = Fe^{3+} spinel; Jamieson and Roeder (1984))

This equation, which is referred to as the 'ideal distribution equation', is plotted, for four different values of K_D in Figure 3.44. The following points may be noted:

(1) The quotient in (122) could just as well be written

$$\frac{X^\beta}{1 - X^\beta} \cdot \frac{1 - X^\alpha}{X^\alpha},$$

to give the reciprocal of K_D. In the following discussions, the quotient is usually written so that $X^\alpha > X^\beta$ and $K_D > 1.0$.

(2) For ease of plotting, equation (122) can be rearranged to

$$X^\alpha = \frac{K_D}{(1/X^\beta) + (K_D - 1)}.$$

(3) The plotted curve (Figure 3.44) is symmetrical about the upper-left-to-lower-right diagonal, and where the curve intersects the diagonal, $X^\alpha = (K_D)^{1/2} X^\beta$.

(4) At low values of X^α and X^β, the distribution curve is nearly linear, and

$$\frac{X^\alpha}{X^\beta} \cong K_D.$$

(5) Where departure from ideal mixing occurs in one or both of α and β, the plotted curve is asymmetrical about the diagonal, unless, through cancellation,

the quotient in equation (121) that contains the four activity coefficients, remains constant. An example is found in α = actinolite, β = cummingtonite (Mueller 1961a). With X^{α} and X^{β} small, we still have, in consequence of the Henry and Raoult rules, X^{α}/X^{β} = constant.

The displacement of an exchange equilibrium by a change in temperature is expressed by first rewriting (121) as

$$\ln(K_{\mathrm{D}}Q) = -\Delta G_{\mathrm{a}}/RT,$$

where Q stands for the quotient that contains the four activity coefficients, and then differentiating with respect to $(1/T)$ at constant pressure,

$$\frac{\partial \ln(K_{\mathrm{D}}Q)}{\partial(1/T)} = -\frac{1}{R}\frac{\partial(\Delta G_{\mathrm{a}}/T)}{\partial(1/T)},$$

$$\frac{\partial \ln(K_{\mathrm{D}}Q)}{\partial(1/T)} = \frac{-\Delta H_{\mathrm{a}}}{R}, \tag{123}$$

where ΔH_{a} is the enthalpy change for (a). Bearing in mind that activity coefficients are in general functions of temperature and concentration, a complex relation between K_{D} and temperature could exist. But if Q remains constant over a temperature range of interest, a linear relation between $\ln K_{\mathrm{D}}$ and $(1/T)$ is expected.

Similarly, the effect of pressure is

$$\frac{\partial \ln(K_{\mathrm{D}}Q)}{\partial P} = \frac{-\Delta V_{\mathrm{a}}}{RT}. \tag{124}$$

Integrating from P_1 to P_2 while ΔV_{a} (the volume change for (a)) remains constant, we obtain

$$\ln\left(\frac{(K_{\mathrm{D}}Q)_2}{(K_{\mathrm{D}}Q)_1}\right) = -(\Delta V_{\mathrm{a}}/RT)(P_2 - P_1). \tag{125}$$

Finally, a Mg–Fe exchange equilibrium can be displaced by variation in the concentration of additional elements. Thus as Ca and Mn enter garnet and Fe^{3+}, Ti, Al^{VI}, and F enter biotite at constant temperature and pressure, a change in the Mg–Fe^{2+} distribution coefficient could conceivably result from these compositional changes. This problem with regard to Ca and Mn in garnet was discussed by Ganguly and Kennedy (1974) who chose a symmetrical mixing model for Mg–Fe–Ca–Mn garnet, which expresses the activity coefficients of the almandine and pyrope components in terms of concentration and Margules parameters, as follows:

$$\begin{aligned} RT\ln\gamma_{\mathrm{alm}} = {} & W_{\text{alm-prp}}(X_{\mathrm{prp}})^2 + W_{\text{alm-grs}}(X_{\mathrm{grs}})^2 + W_{\text{alm-sps}}(X_{\mathrm{sps}})^2 \\ & + (W_{\text{alm-prp}} - W_{\text{prp-grs}} + W_{\text{alm-grs}})X_{\mathrm{prp}}X_{\mathrm{grs}} \\ & + (W_{\text{alm-grs}} - W_{\text{grs-sps}} + W_{\text{alm-sps}})X_{\mathrm{grs}}X_{\mathrm{sps}} \\ & + (W_{\text{alm-prp}} - W_{\text{prp-sps}} + W_{\text{alm-sps}})X_{\mathrm{prp}}X_{\mathrm{sps}}, \end{aligned}$$

$$\begin{aligned} RT \ln \gamma_{\text{prp}} = {} & W_{\text{alm-prp}}(X_{\text{alm}})^2 + W_{\text{prp-grs}}(X_{\text{grs}})^2 + W_{\text{prp-sps}}(X_{\text{sps}})^2 \\ & + (W_{\text{alm-prp}} - W_{\text{alm-grs}} + W_{\text{prp-grs}})X_{\text{alm}}X_{\text{grs}} \\ & + (W_{\text{prp-grs}} - W_{\text{grs-sps}} + W_{\text{prp-sps}})X_{\text{grs}}X_{\text{sps}} \\ & + (W_{\text{alm-prp}} - W_{\text{alm-sps}} + W_{\text{prp-sps}})X_{\text{alm}}X_{\text{sps}}. \end{aligned}$$

By subtraction, we have

$$\begin{aligned} RT \ln(\gamma_{\text{alm}}/\gamma_{\text{prp}}) = {} & W_{\text{alm-prp}}(X_{\text{prp}} - X_{\text{alm}}) \\ & + (W_{\text{alm-grs}} - W_{\text{prp-grs}})X_{\text{grs}} + (W_{\text{alm-sps}} - W_{\text{prp-sps}})X_{\text{sps}}. \end{aligned} \tag{126}$$

Equation (121) is now rewritten as

$$\ln K_{\text{D}} + \ln(\gamma_{\text{phl}}/\gamma_{\text{ann}}) + \ln(\gamma_{\text{alm}}/\gamma_{\text{prp}}) = \ln K_{\text{a}}, \tag{127}$$

noting that $\exp(-\Delta G_{\text{a}}/RT)$ has been replaced by K_{a}, the equilibrium constant. Now by substitution of (126) in (127), we have

$$\begin{aligned} \ln K_{\text{D}} = {} & \ln K_{\text{a}} - \ln(\gamma_{\text{phl}}/\gamma_{\text{ann}}) \\ & - W_{\text{alm-prp}}(X_{\text{prp}} - X_{\text{alm}})/RT \\ & - (W_{\text{alm-grs}} - W_{\text{prp-grs}})X_{\text{grs}}/RT \\ & - (W_{\text{alm-sps}} - W_{\text{prp-sps}})X_{\text{sps}}/RT. \end{aligned} \tag{128}$$

The last two terms in this equation determine a progressive displacement of K_{D} with increasing concentration of Ca and Mn in garnet.

The distribution of Mg and Fe between biotite and garnet, and the effect of temperature on this distribution (equation (123)) were investigated experimentally by Ferry and Spear (1978) and by Perchuk and Lavrent'eva (1983). Ferry and Spear used synthetic crystals of simple formula (similar to the formulas written above) which were equilibrated at 2.07 kbar and at temperatures from 550 to 800 °C. The attainment of exchange equilibrium was facilitated by using a large volume of garnet relative to that of biotite (hence only minor Fe–Mg exchange diffusion was required to occur in garnet), and equilibrium was demonstrated by reversal of the exchange reaction. Changes in mineral composition (biotite only) were determined by microprobe analysis. The results, as plotted in Figure 3.45 show, for example, that at 700 °C, K_{D} (as defined by equation (122)) = 3.85 ($X_{\text{prp}} = 0.10$). The change in the distribution coefficient with absolute temperature (cf. equation (123)) was determined as

$$\ln K_{\text{D}}^{\text{FS}} = (2109/T) - 0.782. \qquad (P = 2.07 \text{ kbar}) \tag{129}$$

Analytical uncertainties in the biotite analyses, when propagated to K_{D}, produce the confidence limits shown in Figure 3.45 (arrows).

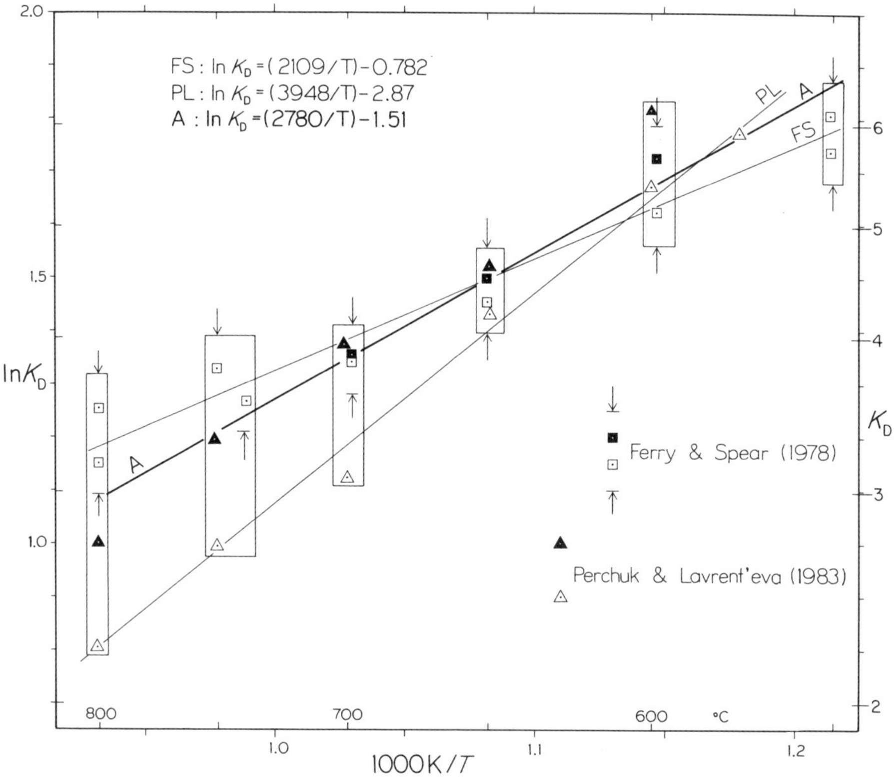

Figure 3.45 Experimental results on the effect of absolute temperature (T) on the distribution of Mg and Fe between garnet and biotite, as expressed by the distribution coefficient (K_D). Squares are from Ferry and Spear (1978) with arrows defining limits of uncertainty; open triangles are from Perchuk and Lavrent'eva (1983), closed triangles are recalculated K_D values, assuming 86% of iron in biotite is Fe^{2+}. Line A is a proposed compromise. From Kretz (1990). Reprinted by permission of Blackwell Scientific Publications, Inc.

Most of the Perchuk and Lavrent'eva experiments were carried out with natural crystals, which were equilibrated at 6.0 kbar and at temperatures from 575 to 950 °C. Reversals in the exchange reaction were achieved between 700 and 900 °C. Reaction products, which in general were not homogeneous, were characterized by microprobe analysis. The results of the many experimental runs are summarized in Figure 3.45. For example, at 700 °C, K_D (mean of 19 measurements) = 3.10 (X_{prp} = 0.15 to 0.56), compared with 3.85 as determined by Ferry and Spear (above). The relation of the distribution coefficient to absolute temperature, as determined by Perchuk and Lavrent'eva (Figure 3.45) is

$$\ln K_D^{PL} = (3948/T) - 2.87. \qquad (P = 6.0 \text{ kbar}) \qquad (130)$$

Where natural biotite was used in the experiments, it seems likely that some Fe^{3+} was present in the product crystals, and if K_D^{PL} is recalculated by assuming that $Fe^{2+}/(Fe^{2+}+Fe^{3+})$ in biotite is 0.86, the data points in Figure 3.45 move upwards, as shown by triangles.

The straight line A (Figure 3.45), which is an attempt to bring the two data sets into agreement, gives the relation between K_D and temperature as

$$\ln K_D^A = (2780/T) - 1.51 \qquad (P = 2 \text{ to } 6 \text{ kbar}) \tag{131}$$

The effect of pressure on the exchange equilibrium has not been investigated experimentally but can be estimated from ΔV_a (equation (125)). By use of data from Table 2.1 and with V (annite) $= 154.32\ cm^3/gfw$ from Holland and Powell (1990), $\Delta V_a(STP) = +0.087$ J/bar/mol $(Mg,Fe)_1$, i.e. increasing pressure is expected to produce a small decrease in K_D. For example, by use of equation (125) we find that at 973 K,

$$K_D(6000 \text{ bar}) = (0.96)K_D(2000 \text{ bar}),$$

assuming a negligible change in ΔV with increasing temperature and pressure. Thus an increase in pressure from 2.0 to 6.0 kbar at 700 °C decreases K_D as measured by Ferry and Spear from 3.85 to 3.70, and the smaller value of K_D (i.e., 3.10) measured by Perchuk and Lavrent'eva at 700 °C cannot be fully understood in terms of a pressure difference. In conclusion, the effect of pressure is small relative to that of temperature; a 4000 bar change in pressure produces the same displacement in K_D as does a 20 °C change in temperature.

The influence of extraneous elements on the distribution of Mg and Fe^{2+} between biotite and garnet is imperfectly understood, but Ellis and Green (1979) and others have found experimentally that an increase in the Ca content of garnet from 10 to 30 mole % grossular (at 1200 °C, 30 kbar) causes K_D (garnet–Ca-pyroxene) to increase from 1.6 to 3.3, and a similar effect is anticipated for K_D (garnet–biotite). The influence upon K_D of variable concentrations of Al, Ti, and F in biotite was considered by Indares and Martignole (1985) and by Aranovich, Lavrent'eva and Kosyakova (1988), but this problem now awaits a detailed experimental investigation.

The Otter Lake area (upper amphibolite to locally lower granulite facies) referred to in Secs. 3.6.1 and 3.6.3 provides a convenient source of natural data which may be examined in relation to the theoretical and experimental results presented above. All of the specimens chosen for this purpose are garnet–biotite and garnet–biotite–hornblende gneisses in which restrictions of the kind discussed in Sec. 3.6.1 are absent, and the Mg:Fe^{2+} ratio of the garnet and biotite crystals is therefore free to vary with changes in rock composition. The Ca content, $100\ Ca/(Mg + Fe^{2+} + Ca + Mn)$ of the garnet crystals (denoted [Ca]) ranges from 3 to 24, and [Mn] ranges from 1 to 12. Chemical analyses and formula considerations show that $Fe^{2+}/(Fe^{2+} + Fe^{3+})$ in garnet is very close to 1.0. In biotite, 66 to 75% of the octahedral sites are occupied by $Mg+Fe^{2+}$ (determined by analysis), the remainder by Fe^{3+}, Ti, Al^{VI}, and vacancies. Thus 34 to 25 mole % of the crystals consist of components other than phlogopite and annite; these are collectively referred to as 'β biotite' (Kretz 1990).

In one group of five mineral pairs, the garnet crystals contain only little Ca and Mn ([Ca] = 3.4 to 7.2; [Mn] = 1.5 to 2.9) and the distribution of Mg and Fe^{2+} between this garnet and the associated biotite can be described by use of equation (122), with K_D = 4.1 (Figure 3.46) and with Q constant. The preferred interpretation regarding Q is that Mg–Fe^{2+} mixing in the two minerals under upper-amphibolite facies conditions is very close to ideal, with all activity coefficients very close to 1.0 and Q close to 1.0.

In a second group of five mineral pairs, the garnet crystals contain more Ca ([Ca] = 11.4 to 16.0) and a displacement of K_D from 4.1 to 5.4 is apparent from Figure 3.46. The relation between K_D and [Ca], revealed by these and additional specimens from the study area, is shown in Figure 3.47 (sillimanite–K-feldspar zone), and if a linear $\ln K_D$ versus [Ca] relation is assumed, the resulting least-squares equation is

$$\ln K_D = 1.30 + 0.025[\text{Ca}]. \tag{132}$$

Information regarding a possible dependence of the Mg–Fe^{2+} distribution coefficient upon the Mn content of garnet is obtained by comparing the following two groups of specimens:

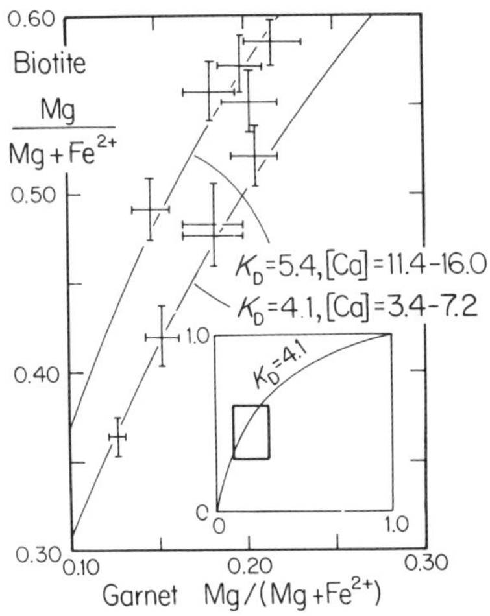

Figure 3.46 Distribution of Mg and Fe^{2+} between biotite and garnet in upper amphibolite facies gneisses from Quebec, showing displacement of distribution coefficient (K_D) with increasing atomic % of Ca in garnet ([Ca]). Data from Kretz (1990). Reprinted by permission of Blackwell Scientific Publications, Inc.

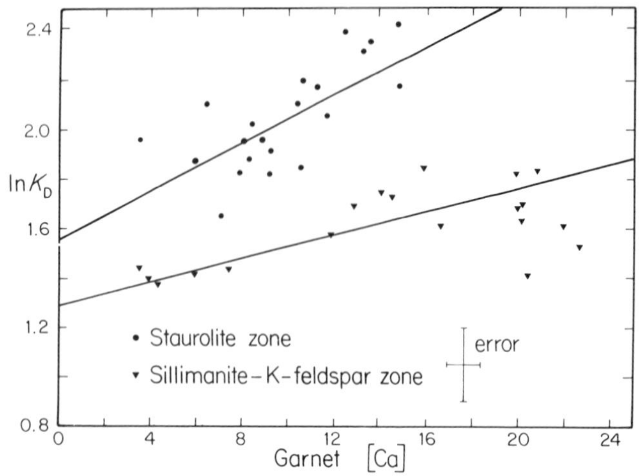

Figure 3.47 Displacement of the distribution coefficient, K_D (Mg-Fe^{2+}, biotite/garnet) with atomic % Ca in garnet ([Ca] = 100 Ca/ (Mg + Fe + Ca + Mn); staurolite zone data from Phinney (1963); sillimanite-K-feldspar zone data from Kretz (1990). Reprinted by permission of Blackwell Scientific Publications, Inc.

	K_D	[Ca]	[Mn]
1	6.4	15.3	2.8
12	6.3	20.0	3.2
13	6.0	23.5	2.5
14	5.5	12.4	10.5
18	5.7	14.0	11.9
5	6.2	19.1	9.5

The distribution coefficient in both groups is clearly larger than in specimens in which garnet is low in both Ca and Mn (i.e., $K_D = 4.1$; Figure 3.46), and the increase in K_D has been attributed to an increase in [Ca]. If Mn exerts an additional influence, this is not apparent because the higher Mn content in the second group (by a factor of 3 to 4) has not caused K_D to be noticeably larger or smaller compared to K_D of the first group. Within the encountered concentration range, spessartine is therefore viewed as an 'ideal dilutent' (Koziol 1990) in Mg-Fe-Ca garnet.

Equation (132) may now be examined in relation to the Ganguly-Kennedy equation (128) and mixing properties of garnet. We have from Sec. 2.11 (above), combined with some additional information, the following approximations regarding the binary mixtures:

almandine-pyrope:
    nearly ideal; $W_{\text{alm-prp}} \sim 0$
almandine-grossular:
    nearly ideal at $X_{\text{grs}} < 0.3$, $W_{\text{alm-grs}} = 0$

pyrope–grossular:
non-ideal; $W_{\text{prp-grs}} = 9.1 \pm 1.4$ kJ/mol (1000 °C); Wood (1988)

grossular–spessartine:
ideal; $W_{\text{grs-sps}} = 0$

almandine–spessartine:
nearly ideal (Pownceby, Wall and O'Neill 1987); $W_{\text{alm-sps}} \sim 0$

pyrope–spessartine:
$W_{\text{prp-sps}}$ not known; deduced ∼0 by Hodges and Spear (1982)

The Otter Lake data did not reveal a relation between K_D and X_{alm}, as noted above, and $W_{\text{alm-prp}}$ in the Ganguly–Kennedy equation (128) is taken as zero. A relation between K_D and X_{sps} could not be detected, and ($W_{\text{alm-sps}} - W_{\text{prp-sps}}$) is also taken as zero. Given, from the above summary, that $W_{\text{alm-grs}}$ ∼0, we have

$$\ln K_D \approx \ln K_a - \ln(\gamma_{\text{phl}}/\gamma_{\text{ann}}) + W_{\text{prp-grs}} X_{\text{grs}}/RT. \tag{133}$$

Mueller (1972) and Schulien (1980) have presented evidence that mixing of phlogopite and annite is nearly ideal and we shall assume, therefore, that the second term on the right is zero. From our proposal that $W_{\text{alm-prp}} = 0$, K_a is equal to K_D when [Ca] = 0, i.e., $\ln K_a = 1.30$ (equation (132)). Thus for the Otter Lake terrane, where the peak temperature was approximately 700 °C (970 K),

$$\ln K_D \approx 1.30 + W_{\text{prp-grs}} X_{\text{grs}}/RT, \tag{134}$$

and $W_{\text{prp-grs}} \approx 20$ kJ/mol. This approximate value may be compared with 9 kJ/mol at 1270 K, determined experimentally by Wood (1988).

The biotite crystals in exchange equilibrium with garnet are somewhat variable in Al^{VI}, Ti, Fe^{3+}, F, and vacancies, but an influence of one or more of these variables upon the Mg–Fe^{2+} distribution coefficient could not be detected.

The results from Otter Lake may be compared with those from other terranes. K_D values in amphibolite and granulite-facies gneisses of Western Australia (Stephenson 1979) average at 3.7 (Fe^{2+} in biotite determined; low Ca and Mn in garnet), similar to values at Otter Lake (mean 4.1). Relations between K_D and [Ca] were detected by Sen and Chakraborty (1968), Lyons and Morse (1970), and Hoinkes (1986), but not by Baltatzis (1979) and Schreurs (1985). A relation between K_D and [Mn] was detected by Albee (1965), Dougan (1974), and Williams and Grambling (1990) but no clear relation is apparent in the data of Sen and Chakraborty (1968) and Baltatzis (1979). Changes in K_D related to extraneous elements in biotite were described by Dahl (1969) and Dallmeyer (1974). A relation between K_D and the Ti content of biotite is not apparent in the data of Schreurs (1985).

An unmistakable increase in K_D with increasing Ca in garnet is found in certain staurolite-zone schists from Nova Scotia, which were studied by Phinney (1963). The Mg–Fe^{2+} distribution coefficient is well defined, based as it is on 22 samples and on Fe^{2+} determinations for biotite. The relation, as shown in Figure 3.47, is

$$\ln K_D = 1.57 + 0.050[\text{Ca}]. \tag{135}$$

This equation and that for the rocks of higher grade (equation (132)) are of the form

$$\ln K_D = \ln K_D^\circ + \theta[\text{Ca}], \tag{136}$$

where K_D° is the distribution coefficient at $[\text{Ca}] = 0$ and θ is the slope of the lines in Figure 3.47. Assuming a linear relation between $\ln K_D^\circ$ and θ,

$$\ln K_D^\circ = 10.8\theta + 1.03,$$

or

$$\theta = (\ln K_D^\circ - 1.03)/10.8. \tag{137}$$

By substitution of (137) in (136),

$$\ln K_D = \ln K_D^\circ + ((\ln K_D^\circ - 1.03)/10.8)([\text{Ca}])). \tag{138}$$

Now by choosing equation (131) as the preferred relation between K_D° and T, i.e.,

$$\ln K_D^\circ = (2780/T) - 1.51,$$

this equation is combined with (138) to give

$$T = \frac{2780 + 258\ [\text{Ca}]}{\ln K_D + 0.235\ [\text{Ca}] + 1.51}. \tag{139}$$

Equation (139) is proposed as a general geothermometer, giving absolute temperature (T) in terms of the Mg–Fe^{2+} distribution coefficient (K_D). Note that X in the distribution quotient stands for $\text{Mg}/(\text{Mg} + \text{Fe}^{2+})$ and $[\text{Ca}] = 100\text{Ca}/(\text{Mg} + \text{Fe}^{2+} + \text{Ca} + \text{Mn})$.

3.7.2 The distribution of Mg and Fe^{2+} between orthopyroxene and Ca pyroxene

Orthopyroxene and Ca pyroxene occur together commonly in felsic, mafic, and ultramafic rocks of the granulite facies. These minerals are of special interest because (1) numerous high-quality analyses are available, including Fe^{2+} determinations, (2) a large variation in Mg:Fe^{2+} ratio is found, covering nearly the entire composition spectrum, and (3) the mineral association provides an opportunity of examining the distribution of Fe^{2+} and Mg between two minerals in relation to the distribution of these elements between non-equivalent sites (M1, M2) within each mineral.

The distribution of Mg and Fe^{2+} between orthopyroxene and Ca pyroxene was discussed briefly in Sec. 3.5.1, where the exchange reaction was expressed in terms of simplified formulas, as follows,

$$\underset{\text{en in Opx}}{MgSiO_3} + \underset{\text{hd in Cpx}}{CaFeSi_2O_6} = \underset{\text{fs in Opx}}{FeSiO_3} + \underset{\text{di in Cpx}}{CaMgSi_2O_6}, \tag{a}$$

which defines the components enstatite (en), ferrosilite (fs), diopside (di), and hedenbergite (hd). The energy-balance equation is

$$\mu_{en} + \mu_{hd} = \mu_{fs} + \mu_{di}.$$

By substitution of four equations of the kind

$$\mu_{en} = G^{En} + RT \ln \gamma_{en} X_{en},$$

we obtain

$$\left(\frac{X_{fs}}{1 - X_{fs}} \cdot \frac{1 - X_{hd}}{X_{hd}}\right)\left(\frac{\gamma_{fs}\gamma_{di}}{\gamma_{en}\gamma_{hd}}\right) = \exp(-\Delta G_a/RT), \tag{140}$$

which is as equation (121) in Sec. 3.7.1. Note that here X is defined with the Fe component on the numerator, e.g., X_{fs} = fs/(en + fs) = $Fe^{2+}/(Mg + Fe^{2+})$ in orthopyroxene.

Data on the distribution of Mg and Fe^{2+} between associated orthopyroxene and Ca pyroxene in three granulite-facies terranes, one in Australia, one in India, and one in North America are plotted in Figure 3.48. The ratio $Fe^{2+}/(Fe^{2+} + Fe^{3+})$ was determined in all of these minerals with the exception of four that appear near the Fe terminal, where values of 0.98 (orthopyroxene) and 0.91 (Ca pyroxene) were assumed, based on analyses of Fe-rich minerals from the same region by Jen (1975). These results provide compelling evidence for the attainment of Fe^{2+}–Mg exchange equilibrium in the three widespread terranes.

In general, the distribution can be expressed well by the equation

$$\frac{X_{fs}}{1 - X_{fs}} \cdot \frac{1 - X_{hd}}{X_{hd}} = 1.80, \tag{141}$$

which is plotted in Figure 3.48. This was first demonstrated by Mueller (1960) who pointed out that the finding is consistent with ideal or near-ideal mixing in both minerals, i.e., with all activity coefficients in (140) close to 1.0. Near-ideal mixing in orthopyroxene at 700 °C to 1000 °C (despite a conspicuous ordering of Mg and Fe atoms, as described in Sec. 3.3.3) was confirmed by Chatillon-Colinet et al. (1983), in their thermochemical study of this mineral. The data of Figure 3.48 then imply near-ideal mixing in Ca pyroxene as well.

Although there is no doubt that for Mg-rich mineral pairs ($X^{Opx} < 0.6$), the distribution can be described exactly by the expression K_D = 1.80, with increasing Fe ($X^{Opx} > 0.6$) the distribution coefficient appears to decrease slightly, implying a small-magnitude departure from ideal mixing in one or both minerals. This relation, which was first detected by Binns (1962), is shown more clearly in Figure 3.49. An interpretation of this relation in terms of activity coefficients was presented by Saxena (1971) and Froese and Gordon (1974).

The observed inter-crystalline element distribution may be viewed as a consequence of the distribution of Fe^{2+} and Mg among the four sites, M1 in orthopyroxene (OpxM1), M2 in orthopyroxene (OpxM2), M1 in Ca pyroxene (CpxM1),

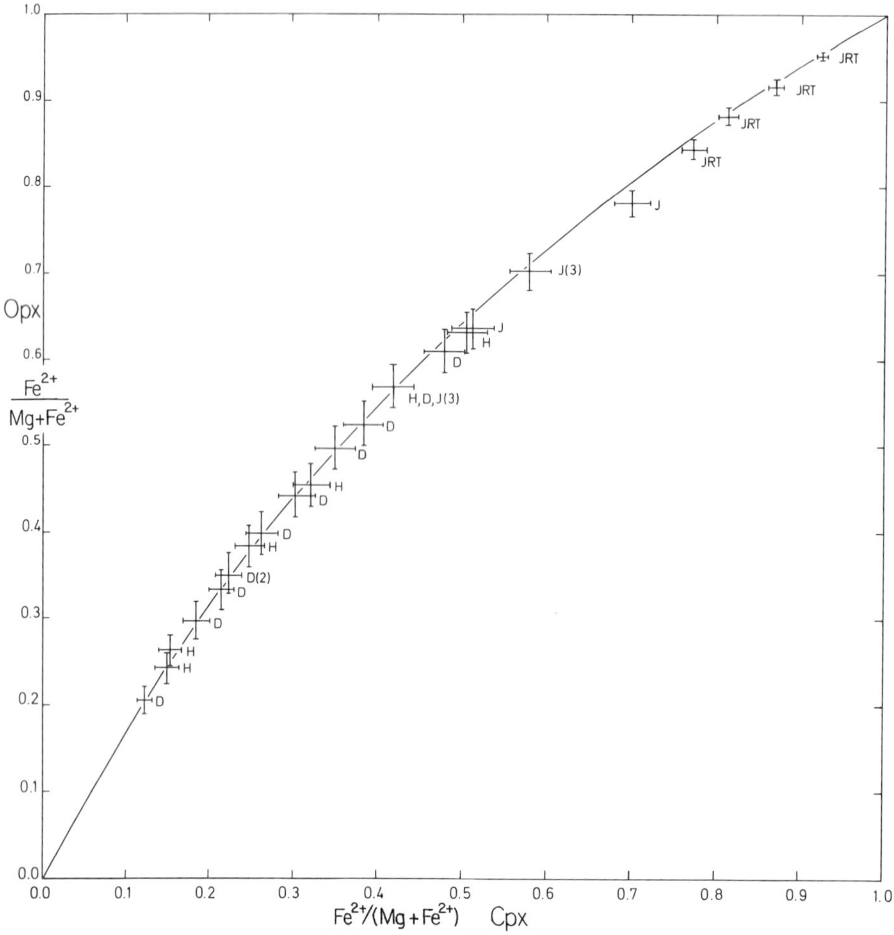

Figure 3.48 The distribution of Mg and Fe^{2+} between orthopyroxene and Ca pyroxene in granulite-facies terranes in Australia (D, Davidson 1968), India (H, Howie 1955), and in North America (J, Jen 1975; JRT, Jaffe, Robinson and Tracy 1978). Number of coincident data points (2), (3) are indicated; error bars represent estimates of 0.95 confidence limits, analytical error. The curve is for $K_D = 1.80$. From Jen and Kretz (1981)

and M2 in Ca pyroxene (CpxM2). The four between-site exchange reactions, which are illustrated in Figure 3.50, are as follows:
OpxM1–OpxM2:

$$\mathrm{Fe}^{\mathrm{OpxM2}} + \mathrm{Mg}^{\mathrm{OpxM1}} = \mathrm{Mg}^{\mathrm{OpxM2}} + \mathrm{Fe}^{\mathrm{OpxM1}},$$

$$k_{\mathrm{D}}^{\mathrm{Opx}} = \frac{x^{\mathrm{OpxM2}}}{1 - x^{\mathrm{OpxM2}}} \cdot \frac{1 - x^{\mathrm{OpxM1}}}{x^{\mathrm{OpxM1}}}, \tag{142}$$

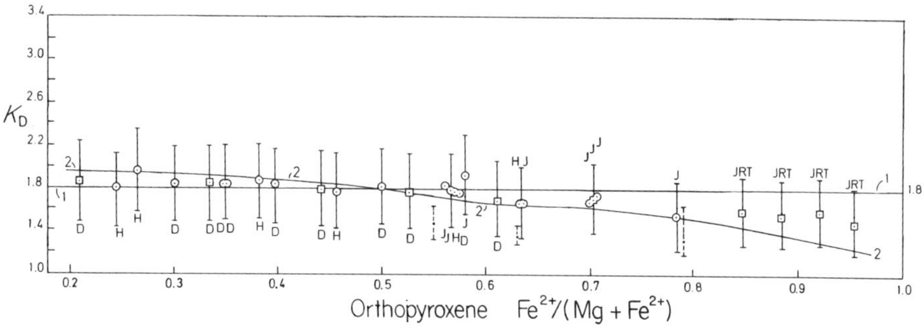

Figure 3.49 The Mg-Fe distribution coefficient for orthopyroxene-Ca-pyroxene (Figure 3.48) plotted against $Fe^{2+}/(Mg + Fe^{2+})$ in orthopyroxene; squares identify mineral pairs selected for calculation. The curves represent (1) $K_D = 1.80 =$ constant, and (2) model calculated from experimental within-crystal distribution data at 800 °C, and identical $Mg:Fe^{2+}$ ratios on the M1 sites of the two minerals. From Kretz (1981)

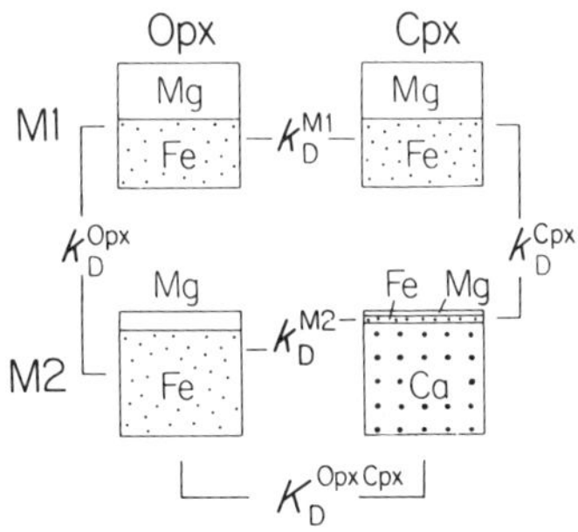

Figure 3.50 Representation of four between-site exchange reactions and distribution coefficients (k_D) and the between-crystal distribution coefficient (K_D) for associated orthopyroxene (Opx) and Ca pyroxene (Cpx), both of which have equal proportions of M1 and M2 structural sites

CpxM1-CpxM2:

$$Fe^{CpxM2} + Mg^{CpxM1} = Mg^{CpxM2} + Fe^{CpxM1},$$

$$k_D^{Cpx} = \frac{x^{CpxM2}}{1 - x^{CpxM2}} \cdot \frac{1 - x^{CpxM1}}{x^{CpxM1}}, \tag{143}$$

OpxM1-CpxM1:

$$Fe^{CpxM1} + Mg^{OpxM1} = Mg^{CpxM1} + Fe^{OpxM1},$$

$$k_D^{M1} = \frac{x^{CpxM1}}{1 - x^{CpxM1}} \cdot \frac{1 - x^{OpxM1}}{x^{OpxM1}}, \tag{144}$$

OpxM2–CpxM2:

$$\mathrm{Fe}^{\mathrm{CpxM2}} + \mathrm{Mg}^{\mathrm{OpxM2}} = \mathrm{Mg}^{\mathrm{CpxM2}} + \mathrm{Fe}^{\mathrm{OpxM2}},$$

$$k_{\mathrm{D}}^{\mathrm{M2}} = \frac{x^{\mathrm{CpxM2}}}{1 - x^{\mathrm{CpxM2}}} \cdot \frac{1 - x^{\mathrm{OpxM2}}}{x^{\mathrm{OpxM2}}}. \tag{145}$$

In (142) to (145), x refers to the on-site atomic ratio, $\mathrm{Fe}^{2+}/(\mathrm{Mg} + \mathrm{Fe}^{2+})$, and k_{D} stands for the between-site distribution coefficient, which in (142) and (143) is the disorder parameter, as defined in Sec. 3.3.3. The four distribution coefficients in (142) to (145) are not independently variable; for example,

$$k_{\mathrm{D}}^{\mathrm{M2}} = (k_{\mathrm{D}}^{\mathrm{Cpx}} k_{\mathrm{D}}^{\mathrm{M1}})/k_{\mathrm{D}}^{\mathrm{Opx}}, \tag{146}$$

and attention is here focused only on $k_{\mathrm{D}}^{\mathrm{Opx}}$, $k_{\mathrm{D}}^{\mathrm{Cpx}}$, and $k_{\mathrm{D}}^{\mathrm{M1}}$.

The ratio $X = \mathrm{Fe}^{2+}/(\mathrm{Mg} + \mathrm{Fe}^{2+})$ in a crystal as a whole is determined by this ratio on each site (x), and on the number of sites (relative to six oxygen atoms) which are occupied by ($\mathrm{Mg} + \mathrm{Fe}^{2+}$) atoms, denoted n:

$$X^{\mathrm{Opx}} = (n^{\mathrm{OpxM1}} x^{\mathrm{OpxM1}} + n^{\mathrm{OpxM2}} x^{\mathrm{OpxM2}})/(n^{\mathrm{OpxM1}} + n^{\mathrm{OpxM2}}), \tag{147}$$

$$X^{\mathrm{Cpx}} = (n^{\mathrm{CpxM1}} x^{\mathrm{CpxM1}} + n^{\mathrm{CpxM2}} x^{\mathrm{CpxM2}})/(n^{\mathrm{CpxM1}} + n^{\mathrm{CpxM2}}). \tag{148}$$

Thus K_{D} as defined in equation (140),

$$K_{\mathrm{D}} = \frac{X^{\mathrm{Opx}}}{1 - X^{\mathrm{Opx}}} \cdot \frac{1 - X^{\mathrm{Cpx}}}{X^{\mathrm{Cpx}}}, \tag{149}$$

can be expressed in terms of the four on-site atomic fractions and the four terms in n. One could therefore derive an equation for K_{D} in terms of $k_{\mathrm{D}}^{\mathrm{Opx}}$, $k_{\mathrm{D}}^{\mathrm{Cpx}}$, and $k_{\mathrm{D}}^{\mathrm{M1}}$. In an equation of this kind, n^{OpxM1} and n^{OpxM2} could be taken as constants (slightly smaller than 1.0 to allow for minor Ca and Na in M2 and $\mathrm{Al}^{\mathrm{VI}}$, Fe^{3+}, and Ti in M1), and similarly for n^{CpxM1}, but, as pointed out in Sec. 3.5.1, the number of Ca atoms in M2 of Ca pyroxene decreases with increasing X^{Cpx}, and therefore n^{CpxM2} must be expressed as a function of X^{Cpx}.

Rather than attempting a general solution to the problem, the writer (Kretz 1981) proceeded by calculating K_{D} for each of the ten selected mineral pairs which are identified in Figure 3.49, and by comparing the calculated and observed results. For each mineral pair, the n numbers were estimated by assigning Ca, Na, and K to M2, and $\mathrm{Al}^{\mathrm{VI}}$, Fe^{3+}, and Ti to M1. A value of $k_{\mathrm{D}}^{\mathrm{Cpx}} = 19$ (800 °C) was obtained from an extrapolation of the experimental (X-ray) results of McCallister, Finger and Ohashi (1976) and was assumed to remain independent of X^{Cpx}. For each Ca pyroxene, x^{CpxM1} was then calculated by an iterative solution of the equation

$$k_{\mathrm{D}} n_{\mathrm{r}} = [(n_{\mathrm{r}} + 1)X - x^{\mathrm{M1}}][(1/x^{\mathrm{M1}}) + (k_{\mathrm{D}} - 1)], \tag{150}$$

which is obtained by substitution of (143) in (148); here $n_{\mathrm{r}} = n^{\mathrm{M2}}/n^{\mathrm{M1}}$. Next, the Mueller (1962) proposal, $x^{\mathrm{OpxM1}} = x^{\mathrm{CpxM1}}$, was adopted, and by use of the Saxena and Ghose (1971) experimental (Mössbauer) results for $k_D^{Opx}(800\,^\circ C)$, x^{OpxM1} and hence X^{Opx} (calculated) provided values of K_D (calculated) for each mineral pair. The K_D (calculated) values are shown as line 2 in Figure 3.49 where good agreement with K_D (observed) is demonstrated, especially within the intermediate range of composition. Moreover, the model successfully predicts a slight decrease in K_D with increasing X^{Opx}, as observed. Thus K_D (observed) can be understood in terms of $k_D^{Opx} = 7.3$, decreasing to 2.5, $k_D^{Cpx} = 19$, and $k_D^{M1} = 1.0$.

The above discussion draws attention to the role of structural sites in determining inter-crystalline distribution coefficients. The same considerations apply to more complex associations of Mg–Fe minerals. For example, in the association garnet–Ca-pyroxene–hornblende–biotite in high-grade gneisses (Kretz 1978), there are six between-crystal distribution coefficients (three are independent) and 15 between-site distribution coefficients (viewing M1 and M2 in biotite and M1 and M3 in hornblende as single sites), and the following calculated sequence of increasing Mg/(Mg + Fe^{2+}) was determined:

$$\mathrm{Grt}\{X\} = \mathrm{Cpx\ M2} \overset{5.0}{>} \mathrm{Hbl\ M1\ M3} \overset{1.2}{>} \mathrm{Bt\ M1\ M2} \overset{2}{\gtrsim} \mathrm{Cpx\ M1} = \mathrm{Hbl\ M2},$$

where numbers are between-site distribution coefficients. These results agree well with those predicted from crystal-field theory, which takes into consideration crystal-field stabilization energy for Fe^{2+} and site configurations, as discussed by Burns (1970). Additional variations on this theme were composed by Matsui and Banno (1965), Ghose (1962), Grover and Orville (1969), Blander (1972), and Fleet (1974).

3.7.3 The distribution of Mg and Fe^{2+} between olivine and spinel

Olivine, $(Mg,Fe^{2+})Si_{1/2}O_2$, and spinel, $(Mg,Fe^{2+})(Cr,Al,Fe^{3+})_2O_4$, occur in bodies of alpine metaperidotite as described by Medaris (1975), in the Bushveld Complex (Cameron 1975) and other layered intrusions, in ultramafic nodules of mantle origin (Basu and MacGregor 1975) and in meteorites (Bunch and Olsen 1975). The problem of the distribution of Mg and Fe^{2+} between olivine and spinel in these rocks, and the displacement of the exchange equilibrium by temperature and the Cr:Al:Fe^{3+} ratios of spinel was first examined in detail by Irvine (1965). This example of Mg–Fe^{2+} distribution provides an opportunity of bringing together experimental and natural data to define the equilibrium relations, particularly the influence of Cr:Al in spinel on the Mg–Fe^{2+} distribution coefficient.

In common spinel, as defined above, Mg and Fe^{2+} atoms occupy tetrahedral sites while the trivalent atoms occupy octahedral sites. The principal end-member minerals are:

$MgAl_2O_4$	MgAl spinel
$Fe^{2+}Al_2O_4$	FeAl spinel (hercynite)
$MgCr_2O_4$	MgCr spinel (picrochromite)
$Fe^{2+}Cr_2O_4$	FeCr spinel (chromite)

We are here concerned mainly with occurrences of metaperidotite, where the Fe^{3+} content of spinel is normally low, with $Fe^{3+}/(Cr + Al + Fe^{3+}) < 0.10$ (Medaris 1975). Thus the simplified formula is $(Mg,Fe^{2+})(Cr,Al)_2O_4$.

Consider first the distribution of Mg and Fe^{2+} between olivine and Al spinel, as expressed by the exchange equation

$$FeSi_{1/2}O_2 + MgAl_2O_4 = MgSi_{1/2}O_2 + FeAl_2O_4. \tag{a}$$

The corresponding energy-balance equation is obtained, as in Sec. 3.7.1,

$$RT \ln\left[\left(\frac{X_{fo}X_{FeAl}}{X_{fa}X_{MgAl}}\right)\left(\frac{\gamma_{fo}\gamma_{FeAl}}{\gamma_{fa}\gamma_{MgAl}}\right)\right] = -[(G^{Fo} - G^{Fa}) + (G^{FeAl} - G^{MgAl})], \tag{151}$$

where MgAl and FeAl identify spinel components (subscripts) and phases (superscripts). Consider now the distribution of Mg and Fe^{2+} between olivine and Cr-bearing spinel of composition $(Mg,Fe^{2+})(Cr,Al)_2O_4$, expressed by the exchange equation

$$FeSi_{1/2}O_2 + Mg(Cr_\alpha Al_\beta)_2O_4 = MgSi_{1/2}O_2 + Fe(Cr_\alpha Al_\beta)_2O_4. \tag{b}$$

The equation is written in this form (following Roeder, Campbell and Jamieson 1979) because in any experimental or natural system, the Cr:Al ratio remains fixed as the Mg–Fe^{2+} exchange reaction takes place, provided no other reactions occur simultaneously, i.e., α and β are constants. In other systems these constants can, of course, assume different values. The energy-balance equation for (b) is

$$\mu_{fa} + \mu_{MgCr_\alpha Al_\beta} = \mu_{fo} + \mu_{FeCr_\alpha Al_\beta}, \tag{152}$$

where

$$\mu_{MgCr_\alpha Al_\beta} = G^{MgCr_\alpha Al_\beta} + RT \ln a_{MgCr_\alpha Al_\beta}. \tag{153}$$

Thus the standard state for each of $Mg(Cr_\alpha Al_\beta)$ spinel and $Fe(Cr_\alpha Al_\beta)$ spinel is a crystal that is variable in composition with respect to Cr and Al but with composition (α, β) defined for every system in which the mineral occurs. By substitution in (152), as in Sec. 3.7.1, we obtain

$$RT \ln\left[\left(\frac{X_{fo}X_{FeCr_\alpha Al_\beta}}{X_{fa}X_{MgCr_\alpha Al_\beta}}\right)\left(\frac{\gamma_{fo}\gamma_{FeCr_\alpha Al_\beta}}{\gamma_{fa}\gamma_{MgCr_\alpha Al_\beta}}\right)\right] = -[(G^{Fo} - G^{Fa}) + (G^{FeCr_\alpha Al_\beta} - G^{MgCr_\alpha Al_\beta})]. \tag{154}$$

At a given temperature and pressure, the second bracketed term on the right will take on different values in different systems, depending upon the ambient Cr:Al ratio of the spinel phase, and thus the distribution coefficient in (154) is in general expected to vary with temperature, pressure, and the Cr:Al ratio in spinel.

The term $G^{FeCr_\alpha Al_\beta}$ in (154) refers to the molar Gibbs energy of the mixture ($FeCr_2O_4$, $FeAl_2O_4$) at specified α and β. By allowing α and β $(= 1 - \alpha)$ to vary, we have from Sec. 2.4,

$$G^{Fe(Cr,Al)} = X_{FeCr}G^{FeCr} + X_{FeAl}G^{FeAl} + X_{FeCr}RT\ln X_{FeCr} + X_{FeAl}RT\ln X_{FeAl}. \quad (155)$$

(assuming ideal mixing), and a similar equation for $G^{Mg(Cr,Al)}$. By entering these equations in (154) (letting Y_{Cr} stand for the mole fraction Cr/(Cr + Al) $= \alpha/(\alpha+\beta)$) we obtain through cancellation of $RT\ln X$ terms the following equation which holds for any Cr:Al ratio in spinel:

$$RT\ln\left[\left(\frac{X_{fo}X_{Fe(Cr,Al)}}{X_{fa}X_{Mg(Cr,Al)}}\right)\left(\frac{\gamma_{fo}\gamma_{Fe(Cr,Al)}}{\gamma_{fa}\gamma_{Mg(Cr,Al)}}\right)\right] = -[(G^{Fo} - G^{Fa}) + (G^{FeAl} - G^{MgAl}) + Y_{Cr}(+G^{FeCr} - G^{MgCr} - G^{FeAl} + G^{MgAl})]. \quad (156)$$

This is equivalent to the Irvine (1965) equation. The first two bracketed terms on the right are the same as those in (151) and define the exchange energy when $Y_{Cr} = 0$. The third bracketed term is constant at a given temperature and pressure, and hence a linear relation between $\ln K_D$ and Y_{Cr} or Y_{Al} is expected, provided the activity-coefficient quotient on the left of (156) remains constant. As Y_{Cr} increases to 1.0, G^{FeAl} and G^{MgAl} drop out of this equation, leaving, in square brackets, the exchange energy for Mg–Fe^{2+} exchange between olivine and Cr spinel. Similar equations may be derived for Mg–Fe^{2+} exchange between olivine and spinel in which Fe^{3+} is present with or in place of Al and Cr.

We shall now sketch the Mg–Fe exchange relations, drawing on certain experimental and natural data. Thermodynamic considerations, as expressed above and in Sec. 3.7.1 provide expectations concerning the nature of these relations, i.e., $K_D \approx$ constant (ideal or near-ideal mixing of Mg and Fe^{2+} components in both minerals), a linear change of $\ln K_D$ with $(1/T)$, a small or negligible effect of pressure, and a linear or near-linear relation between $\ln K_D$ and Cr/(Cr + Al) in spinel. The Gibbs energy of formation of the spinel end-members (equation (156)) is not known with sufficient accuracy to provide a satisfactory thermochemical solution to this problem.

Experimental data on Mg–Fe^{2+} exchange between olivine and Al spinel and between olivine and Fe^{3+} spinel were obtained by Jamieson and Roeder (1984) who equilibrated synthetic crystals, mounted in a wire loop, within a furnace tube at 1300 °C. Oxygen fugacity was controlled by a mixture of CO_2 and H_2 or by CO_2. A close approach to equilibrium, after 24 hours, was demonstrated by reversals of the exchange reaction. The results are shown in Figure 3.51.

Experimental data on the distribution of Mg and Fe^{2+} between olivine and (Cr,Al) spinel were obtained by Engi (1983) who subjected finely ground synthetic crystals in a (Mg,Fe) chloride flux to temperatures of 700 to 900 °C in a cold-seal pressure vessel which was held (in most experiments) at 1.0 kbar pressure. A

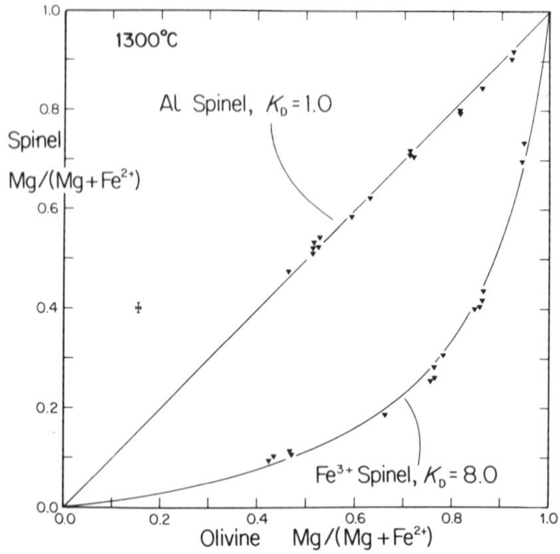

Figure 3.51 Distribution of Mg and Fe^{2+} between olivine and Al spinel and between olivine and Fe^{3+} spinel at 1300 °C, 1 bar. Data from Jamieson and Roeder (1984)

close approach to exchange equilibrium was achieved in many of the experiments, accompanied by recrystallization, which caused mean particle size to increase from 1 μm to about 20 μm. Some of the results are shown in Figure 3.52.

In all of the above experiments, concentrations of Mg and Fe in reaction products were determined by microprobe analysis, and variation in these determinations for each mineral in each reaction product was propagated to X (= Mg/(Mg + Fe^{2+})) and reported as a variation in X. It is not possible, however, to obtain meaningful estimates of standard deviation from only a few measurements (most of these will be either smaller or larger than an estimate based on 100 or 200 measurements) and that may be why the reported errors vary by factors as large as 8 (Jamieson and Roeder 1984) and 85 (Engi 1983). Therefore, average values of reported errors were here used in the construction of error bars (0.95 confidence) which are indicated in Figures 3.51 and 3.52.

We see immediately from Figure 3.51 that at 1300 °C Mg and Fe^{2+} are very nearly equally distributed between olivine and Al spinel ($K_D = 1.0$) and that Fe^{3+} spinel shows a strong preference for Fe^{2+}($K_D = 8.0$). A small variation in both distribution coefficients was interpreted by Jamieson and Roeder (1984) in terms of small-magnitude departures from ideal mixing in spinel. Suppose that mixing in both minerals (olivine and spinel) can be described by use of the symmetrical Margules parameters, W_{Ol} and W_{Spl} (Sec. 2.10), as follows:

$$RT \ln \gamma_{fo} = X_{fa}^2 W_{Ol},$$

$$RT \ln \gamma_{fa} = X_{fo}^2 W_{Ol},$$

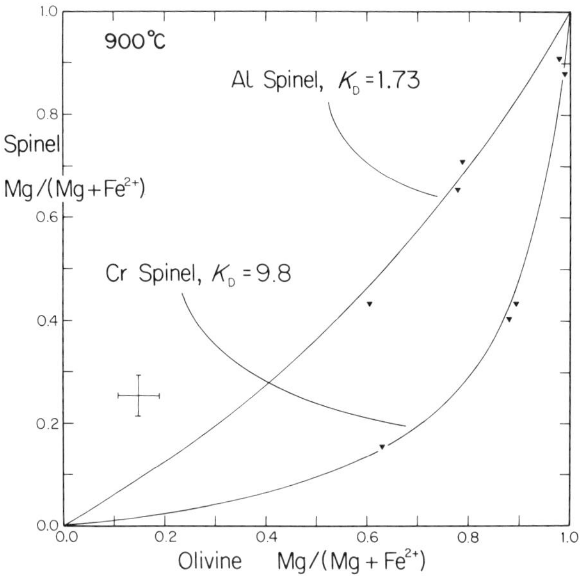

Figure 3.52 Distribution of Mg and Fe^{2+} between olivine and Al spinel and between olivine and Cr spinel at 900 °C, 1.0 kbar. Data from Engi (1983)

$$RT \ln \gamma_{\mathrm{FeAl}} = X^2_{\mathrm{MgAl}} W_{\mathrm{Spl}},$$

$$RT \ln \gamma_{\mathrm{MgAl}} = X^2_{\mathrm{FeAl}} W_{\mathrm{Spl}}.$$

Then, with reference to equation (151),

$$RT \ln \left(\frac{\gamma_{\mathrm{fo}} \gamma_{\mathrm{FeAl}}}{\gamma_{\mathrm{fa}} \gamma_{\mathrm{MgAl}}} \right) = W_{\mathrm{Ol}}(2X_{\mathrm{fa}} - 1) - W_{\mathrm{Spl}}(2X_{\mathrm{FeAl}} - 1). \tag{157}$$

By assuming W_{Ol} is equal to 0 or a few kJ/mol (Sec. 2.11, Figure 2.16), Jamieson and Roeder found that in either case, $W_{\mathrm{Spl}} = -3.1 \pm 0.6$ kJ/mol.

The experimental data also reveal a conspicuous change in K_D (Mg, Fe^{2+}; olivine/Al spinel) from 1.0 (Figure 3.51) at 1300 °C, to 1.73 (Figure 3.52) at 900 °C. At 900 °C, Cr spinel shows a strong preference for Fe^{2+}, with $K_D = 9.8$ (Figure 3.52). At this temperature, as at 1300 °C, the exchange isotherms can be described by a distribution coefficient that is independent of the Mg:Fe^{2+} ratio, consistent with only minor departure from ideal mixing in both minerals.

Numerous natural data are available on the variation of K_D (Mg, Fe^{2+}; olivine/(Cr, Al, Fe^{3+}) spinel) with variation in the Cr:Al:Fe^{3+} content of spinel. For example, the relation found by Paktunç (1983, 1984) in upper amphibolite facies ultramafic rocks in Manitoba (Figure 3.53) is

$$\ln K_D = -2.81 Y_{\mathrm{Al}} + 3.56. \tag{158}$$

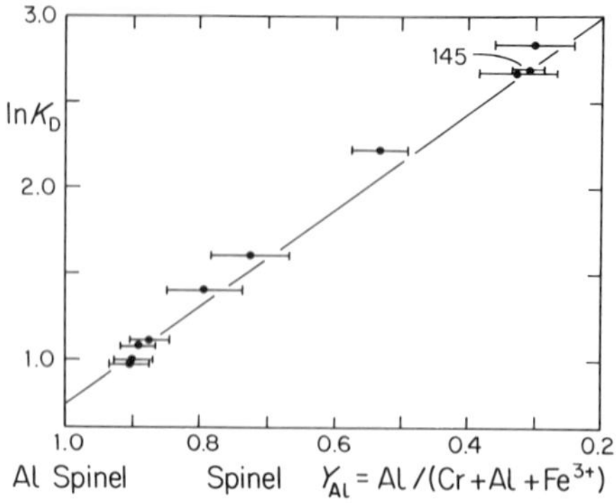

Figure 3.53 Variation in the distribution coefficient K_D (Mg–Fe^{2+}; olivine/spinel) with $Y_{Al} = Al/(Cr + Al + Fe^{3+})$ in spinel, in ultramafic rocks (upper amphibolite facies) from Manitoba. Error bars to Y_{Al} are approximate (based on 10 spot analyses in specimen 145) and to $\ln K_D$ are smaller. Data from Paktunç (1983, 1984)

Here $Y_{Al} = Al/(Cr + Al + Fe^{3+})$ in spinel in which the Fe^{3+} content is very low ($Y_{Fe} < 0.01$). This is similar to relations found by Evans and Frost (1975) and Medaris (1975) in amphibolite-facies and granulite-facies metaperidotite from the Alps and elsewhere.

The composition of olivine and spinel has been measured in some igneous rocks for which the temperature of crystallization can be determined with confidence. Thus Evans and Wright (1972) have measured K_D in a sample of Kilauea Iki pumice (Hawaii) which has crystallized at 1200–1225 °C, and this information is also useful in describing the effect of Cr:Al:Fe^{3+} on the distribution coefficient.

The relation of the Mg–Fe^{2+} distribution coefficient to temperature and Cr:Al:Fe^{3+} of spinel can now be determined approximately by use of the above data. First, a relation with temperature is defined for olivine–Al-spinel (Figure 3.54) based on $K_D = 1.0$ at 1300 °C (Figure 3.51), $K_D = 1.53$ at 800 °C (Engi 1983; mean of 1.40, 1.54, 1.65, omitting 1.09; these are used in preference to the 900 °C data because of longer run times), and $K_D = 2.13$ from Figure 3.53 at an assumed temperature of 600 °C. The equation is

$$\ln K_D = (1470/T) - 0.935. \tag{159}$$

The relation of K_D to Y_{Al} (Figure 3.55) is then defined by constructing isotherms; one of these is the 600 °C isotherm of Figure 3.53 assumed to be linear, and another is the 1200 °C isotherm passing through the Evans and Wright (1972) Kilauea Iki data point (sample S-20; $n = 15$; K_D (mean) $= 5.10$, $s = 0.23$; Y_{Al} (mean) $=$ 0.282, $s = 0.014$), also assumed to be linear. Others are located from a linear $\ln K_D$ versus $(1/T)$ variation between the 600 and 1200 °C isotherms at $Y_{Al} = 0.28$.

The construction agrees well with the 800 °C results of Engi (1983) at $Y_{Al} = 0.5$, i.e. $K_D = 5.14$ (mean of 4.01, 5.01, 5.71, 5.81), and moderately well with the 900 °C

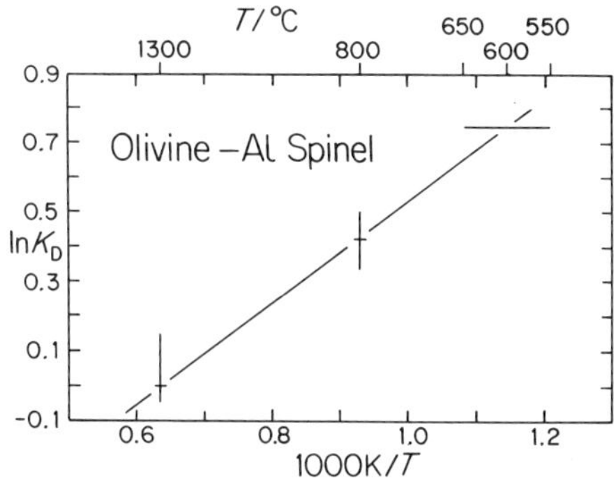

Figure 3.54 Proposal for a relation between ln K_D (Mg–Fe^{2+}; olivine/Al spinel) and absolute temperature, based on the results of Figures 3.51 and 3.53, and experimental results of Engi (1983) at 800 °C

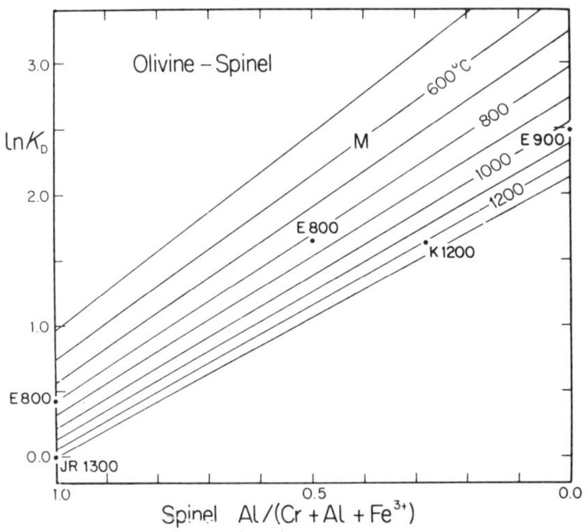

Figure 3.55 Proposal for a variation of K_D (Mg–Fe^{2+}; olivine/Al spinel) with Al/(Cr + Al + Fe^{3+}) in spinel and temperature (°C), based on experimental data of Jamieson and Roeder (1984, point JR) and Engi (1983, points E), combined with natural data from Manitoba reported by Paktunç (1983, line M) and from Kilauea Iki reported by Evans and Wright (1972, point K)

results at $Y_{Cr} = 1.0$, i.e., $K_D = 12.0$ (mean of 9.2, 10.8, 11.3, 16.6, omitting two at 5.5); these data points are shown in Figure 3.55.

While plotting the natural data, the ratio $Y_{Al} = Al/(Cr + Al + Fe^{3+})$ was used as a measure of the composition of the octahedral sites of spinel. This ratio is preferred over $Cr/(Cr + Al + Fe^{3+})$ because increasing concentrations of Cr and Fe^{3+} displace K_D (Mg, Fe^{2+}) by nearly the same amount (Figures 3.51, 3.52), i.e. by a maximum of $8K_D$ units.

An equation for the model of Figure 3.55 is obtained as follows. The equation for an isotherm is

$$\ln K_{\mathrm{D}} = \ln K_{\mathrm{D}}^{\circ} + (1 - Y_{\mathrm{Al}})\theta, \tag{160}$$

where K_{D}° is the distribution coefficient at $Y_{\mathrm{Al}} = 1.0$ and θ is the slope, which is related to K_{D}° by

$$\theta = 0.935 \ln K_{\mathrm{D}}^{\circ} + 2.13. \tag{161}$$

By substitution of (159), which refers to $Y_{\mathrm{Al}} = 1.0$, i.e.,

$$\ln K_{\mathrm{D}}^{\circ} = (1470/T) - 0.935, \tag{162}$$

and (161) into (160) we obtain

$$T = \frac{1470 + 1374(1 - Y_{\mathrm{Al}})}{\ln K_{\mathrm{D}} - 1.256(1 - Y_{\mathrm{Al}}) + 0.935}, \tag{163}$$

where T is absolute temperature, $Y_{\mathrm{Al}} = \mathrm{Al}/(\mathrm{Cr} + \mathrm{Al} + \mathrm{Fe}^{3+})$ in spinel and

$$K_{\mathrm{D}} = \left(\frac{X_{\mathrm{Mg}}}{1 - X_{\mathrm{Mg}}}\right)^{\mathrm{Ol}} \left(\frac{1 - X_{\mathrm{Mg}}}{X_{\mathrm{Mg}}}\right)^{\mathrm{Spl}},$$

in which $X = \mathrm{Mg}/(\mathrm{Mg} + \mathrm{Fe}^{2+})$.

The isotherms of Figure 3.55 differ slightly from those constructed by Fabriès (1979) which are based on fewer experimental controls and on an assignment of 700 °C (rather than 600 °C) to data from high-grade metaperidotite. The justification for a choice of 600 °C is found in Figure 3.54 and in the possibility that some Mg–Fe exchange occurred during cooling, to give an exchange temperature slightly lower than the peak metamorphic temperature.

Engi (1983) constructed isotherms based on a selection of his experimental data and on thermodynamic data and mixing models. The 600 °C isotherm agrees fairly well with that shown in Figure 3.55, and the 1300 °C isotherm originates at $\ln K_{\mathrm{D}} = 0$, as in Figure 3.55, but all isotherms are curved, and they prescribe a temperature of ca. 1500 °C for the Kilauea crystals, which is too high.

The isotherms of Roeder, Campbell and Jamieson (1979) are not supported by data from metaperidotite (Engi and Evans 1980, Paktunç 1984).

Ozawa (1983) has compiled and compared the various olivine–spinel geothermometers that have been proposed, and has pointed out some of the difficulties that can be encountered in attempting to obtain a temperature estimate. The present proposal is Figure 3.55 and equation (163).

3.8 THE DISTRIBUTION OF MINOR AND TRACE ELEMENTS BETWEEN MINERALS

Early research on the distribution of minor and trace elements between minerals in plutonic rocks was carried out by Bray (1942) and Nockolds and Mitchell (1948).

This was followed by several studies which produced a large amount of information on the distribution of Sc, V, Cr, Mn, Co, Ni, Cu, Zn, Sr, Y, Zr, Mo, Sn, Ba, Pb, and other elements in metamorphic assemblages, including, for example, orthopyroxene and Ca pyroxene (Howie 1955, Leelanandam 1967), amphibole and biotite (De Vore 1955, Nickel 1955, Moxham 1965, Annersten and Ekström 1971), and garnet and biotite (Turekian and Phinney 1962, Albee 1965).

Early trace-element analyses were obtained principally by emission spectroscopy, a method which has been replaced by several others: X-ray fluorescence spectroscopy (Johnson and King 1987), atomic absorption analysis (Angino and Billings 1972), electron microprobe analysis (Merlet and Bodinier 1990), instrumental neutron-activation analysis (Baedecker and McKown 1987), inductively coupled plasma spectroscopy (Montaser and Golightly 1987, Reed 1990), ion microprobe analysis (Hinton 1990), and proton microprobe analysis (Johansson 1989, Fraser 1990); the references may be examined for information on the given method.

Numerous experimental data are available on the distribution of a trace element between a mineral and an associated liquid or vapor. For example, the distribution of Rb and Cs between NaCl and vapor was investigated by Reichert (1966) and Schock and Puchelt (1971), and the distribution of Rb between sanidine and vapor and between phlogopite and vapor was investigated by Chelishchev (1967) and by Beswick (1973). A review of early results was prepared by McIntire (1963). Information of this kind can be used to calculate the distribution of a trace element between two minerals, for example, the distribution of Rb between sanidine and phlogopite (Beswick 1973). Experimental results on the distribution of trace elements between crystals and a melt phase were reviewed by Irving (1978).

The thermodynamics of trace-element distribution are fundamentally the same as those applied above in dealing with transfer reactions (Sec. 3.5) and exchange reactions (Sec. 3.7). Numerous problems remain, however, in the crystal chemistry of trace-element behavior, which has been discussed in general by Goldschmidt (1954) and later in terms of crystal-field theory (Schwarcz 1967, Matsui and Banno 1970, Burns 1973), electronegativity (Ramberg 1952b), lattice strain (Iiyama and Volfinger 1976), ion-site dimensions (Volfinger and Robert 1980), and adsorption (De Vore 1955).

We shall now examine some examples of minor and trace-element distribution, referring in each case to natural data, and where possible, to experimental results.

3.8.1 The distribution of strontium between K-feldspar and plagioclase

Common felsic and mafic rocks contain 200 to 500 ppm Sr (Wehmiller 1972), and this element shows a preference for the feldspar minerals over others, including biotite and garnet. In a review of the distribution of Sr between feldspar minerals, Heier (1962) found that the Sr content of K-feldspar (Kf) relative to that of associated plagioclase (Pl) is somewhat variable, with most ratios falling between

0.5 and 2.0. Barth (1961) proposed that this variation is related to temperature, approximately as

$$\ln K_{\mathrm{D}} = -(4190/T) + 5.8. \tag{164}$$

Here the distribution coefficient is defined as

$$K_{\mathrm{D}} = \frac{\text{mol fraction } \mathrm{SrAl_2Si_2O_8} \text{ in Kf}}{\text{mol fraction } \mathrm{SrAl_2Si_2O_8} \text{ in Pl}}, \tag{165}$$

and T is absolute temperature. Thus $K_{\mathrm{D}} \sim 1.0$ at 450 °C, increasing with increasing temperature.

The reaction that governs the distribution of Sr between the two feldspar phases could be viewed as an exchange reaction: for example, the migration of one Sr atom from K-feldspar to plagioclase in exchange for one Ca atom, or the migration of one Sr atom plus one vacancy from plagioclase to K-feldspar in exchange for two Na atoms. It seems more likely, however, as proposed by Barth (1961), that the process is a transfer reaction, i.e., the transfer of n moles of $SrAl_2Si_2O_8$ from one phase to the other, entirely analogous to the transfer of $NaAlSi_3O_8$, as discussed in Sec. 3.5.2. Then equilibrium demands an equality in the chemical potential of components albite, orthoclase, anorthite, *and* $SrAl_2Si_2O_8$ (Sr-fsp):

$$\mu^{\mathrm{Kf}}_{\text{Sr-fsp}} = \mu^{\mathrm{Pl}}_{\text{Sr-fsp}}. \tag{166}$$

This leads to the following equation, which is analogous to equation (60) of Sec. 3.5.2,

$$X^{\mathrm{Kf}}_{\text{Sr-fsp}}/X^{\mathrm{Pl}}_{\text{Sr-fsp}} = \left(\gamma^{\mathrm{Pl}}_{\text{Sr-fsp}}/\gamma^{\mathrm{Kf}}_{\text{Sr-fsp}}\right) \exp\left[-\left(G^{\mathrm{Kf}}_{\text{Sr-fsp}} - G^{\mathrm{Pl}}_{\text{Sr-fsp}}\right)/RT\right]. \tag{167}$$

Here the symbols G stand for the Gibbs energy of pure $SrAl_2Si_2O_8$ in the (imaginary) form of K-feldspar and plagioclase. If concentrations of Sr are sufficiently low, the activity coefficients (γ) are independent of Sr concentration (Henry's rule) but may vary with ab:an in plagioclase and ab:or in K-feldspar. Thus the ratio on the left (K_{D}) is expected to vary with temperature, pressure, and the larger compositional changes in both phases.

Iiyama (1968) has measured the distribution of small concentrations of Sr ($<$ 10 000 ppm) between K-feldspar and plagioclase at 600 °C and 1.0 kbar. Gels of prescribed composition were crystallized in separate compartments in contact with a vapor which contained Sr tagged with ^{85}Sr, and concentration changes were measured radiometrically. In one set of experiments which involved K-feldspar (or 75, ab 25) and plagioclase (ab 70, an 30), a distribution coefficient (equation (165)) of 0.80 was observed, and this was independent of Sr concentration, as expected. In other experiments, with large-scale differences in feldspar composition, K_{D} was found to change only slightly, to 0.85.

Virgo (1968) has examined the distribution of Sr in several terranes, mainly in Australia. The two feldspar minerals were separated and analyzed by X-ray fluorescence spectroscopy. In most areas, an increase in K_D was observed, with increasing grade, as is shown in Figure 3.56.

Scotford (1973) has investigated the distribution of Sr in muscovite- and biotite-bearing schist and gneiss and associated metamorphosed quartz-feldspar veins in a metamorphic terrane in New Hampshire, USA. The feldspar minerals were separated and analyzed by atomic absorption spectroscopy. Results from the staurolite–kyanite zone are shown in Figure 3.56. On entering the sillimanite zone, and then the sillimanite–K-feldspar zone, average values of the distribution coefficient increase progressively.

A new proposal for a relation between the Sr distribution coefficient and temperature is shown in Figure 3.57. It is based on the following information:

(1) Virgo (1968) data for the granulite facies (Musgrave $K_D = 1.48$ ($n = 6$), Broken Hill $K_D = 1.25$ ($n = 6$), Sri Lanka $K_D = 1.38$ ($n = 5$), omitting Langöy $K_D = 1.73$ ($n = 5$)); K_D (mean) $= 1.37$ at an estimated temperature of 770 °C (between 740 and 800 °C); point V1.

(2) Virgo (1968) data for the upper amphibolite facies (Musgrave $K_D = 1.04$ ($n = 3$), Broken Hill $K_D = 1.00$ ($n = 2$), Langöy $K_D = 0.91$ ($n = 2$), Pewsey Vale $K_D = 0.94$ ($n = 2$), Springton $K_D = 1.04$ ($n = 4$), omitting Sri Lanka $K_D = 1.38$ ($n = 4$)), K_D (mean) $= 0.99$ at an estimated temperature of 650 °C (between 620 and 680 °C); point V2.

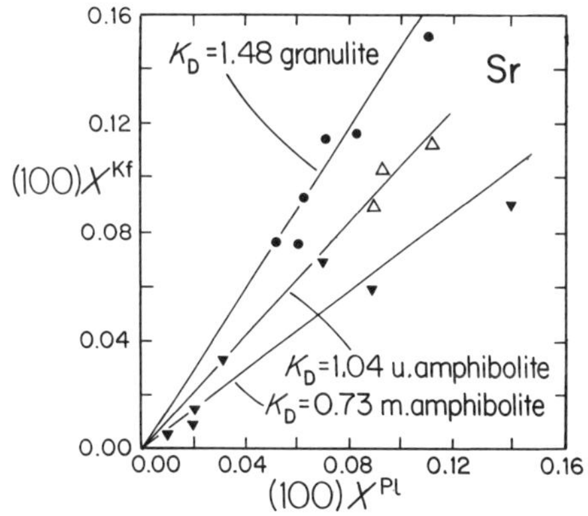

Figure 3.56 Examples of the distribution of strontium ($SrAl_2Si_2O_8$) between K-feldspar (Kf) and plagioclase (Pl) in the middle amphibolite facies (New Hampshire, $K_D = 0.73$; Scotford 1973), and in the upper amphibolite and granulite facies (Musgrove Ranges, Australia, $K_D = 1.04, 1.48$; Virgo 1968); X denotes mole fraction $SrAl_2Si_2O_8$, and $K_D = X^{Kf}/X^{Pl}$. Virgo data © University of Chicago Press, reproduced by permission

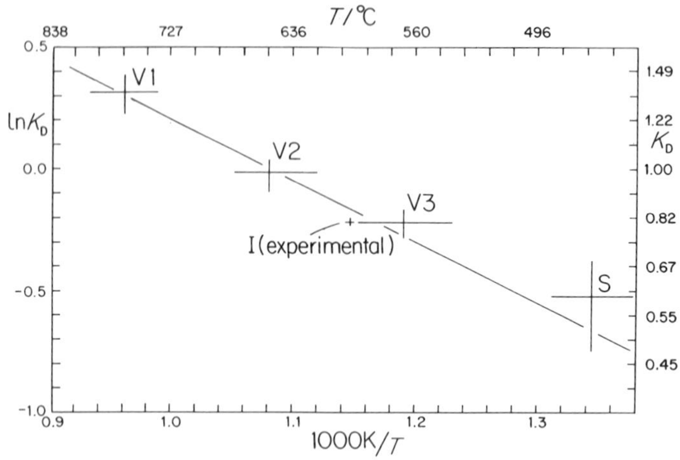

Figure 3.57 Proposed relation between the Sr distribution coefficient, K_D (Sr; Kf/Pl) and absolute temperature (T), based on data of Virgo (1968, points V1, V2, V3), Scotford (1973, point S), and experimental results of Iiyama (1968, point I)

(3) Virgo (1968) data for the mid-amphibolite facies (Palmer $K_D = 0.84$ ($n = 2$), Springton $K_D = 0.75$ ($n = 2$)), K_D (mean) $= 0.80$ at an estimated temperature of 570 °C (between 540 and 600 °C); point V3.

(4) Scotford (1973) data for the staurolite–kyanite zone, New Hampshire ($K_D = 0.40$ to 0.69 ($n = 5$), omitting two extremes (096, 1.16)), K_D (mean) $= 0.60$ at a temperature estimated at 470 °C (between 450 and 490 °C) based on the distribution of albite between the two feldspar phases (Sec. 3.4.1); point S.

(5) The Iiyama (1968) experimental data referred to above, i.e., $K_D = 0.80$ at 600 °C; point I (Figure 3.57).

The relation, which is written

$$\ln K_D(\text{Sr, Kf/Pl}) = -(2500/T) + 2.7, \qquad (168)$$

refers to feldspar phases with common ab:an:or proportions; the effect of pressure is not considered.

From a study of Donegal granite (Ireland), Hall (1967) reported distribution coefficients which are generally less than 1.0 and several are near 0.60. According to equation (168), these numbers imply temperatures in the vicinity of 500 °C.

From a study of ignimbrites in Italy, Dupuy (1972) reported K_D (mean) values from four areas of 0.96, 0.87, 0.61, 0.47, much smaller than expected. Dupuy considered that the high Sr concentrations of plagioclase (relative to K-feldspar) may result in part from fractional crystallization. Equilibrium with regard to Sr was evidently not achieved in these volcanic rocks.

3.8.2 The distribution of Mn between garnet and biotite and between Ca pyroxene and hornblende

Manganese is a minor constituent of most occurrences of ferromagnesian minerals but in garnet which has crystallized in low- to medium-grade meta-shale, it may be present in abundance. As an illustration of the distribution of Mn among ferromagnesian minerals we shall examine the distribution of this element in the high amphibolite facies gneisses and amphibolites that provided the study material for Secs. 3.6.1 and 3.7.1, above.

In spessartine-rich garnet, Mn obviously occurs as a divalent ion which resides principally or entirely in the (Mg,Fe^{2+}) sites. In diopside, Ghose and Schindler (1969) have detected Mn^{2+} in both the Mg and Ca sites, but in some crystals most or all of the ions were present in the Mg site. In the Ca pyroxene crystals of the present study, an estimated 2 to 7% of the Ca sites are occupied by (Mg,Fe) atoms, and Mn is assumed to be present on both sites, substituting for (Mg,Fe^{2+}). Similarly, in biotite and hornblende, Mn is assumed to substitute for (Mg,Fe) and the Mn concentration in all minerals is expressed as $Mn/(Mg + Fe^{2+} + Mn)$.

Consider first the Mn content of garnet relative to that of biotite, with simplified formulas written as

$$\begin{aligned}\text{garnet}: &\quad (Mg_{\alpha 1}Fe_{\beta 1}Mn_{\gamma 1})Al_{2/3}SiO_4\\ \text{biotite}: &\quad K_{1/3}(Mg_{\alpha 2}Fe_{\beta 2}Mn_{\gamma 2})Al_{1/3}SiO_{10/3}(OH)_{2/3}\end{aligned}$$

where subscripts $\alpha 1$, etc. refer to atomic fractions.

In 6 of the 17 study samples to be examined below, biotite and garnet are associated with sillimanite, K-feldspar, and quartz, and for these samples we have the following possible transfer equilibrium:

$$\underset{\text{biotite}}{K_{1/3}(Mg_{\alpha 2}Fe_{\beta 2}Mn_{\gamma 2})Al_{1/3}SiO_{10/3}(OH)_{2/3}} + \underset{\text{sillimanite}}{(1/3)Al_2SiO_5} + \underset{\text{quartz}}{(2/3)SiO_2}$$

$$= \underset{\text{garnet}}{(Mg_{\alpha 1}Fe_{\beta 1}Mn_{\gamma 1})Al_{2/3}SiO_4} + \underset{\text{K-feldspar}}{(1/3)KAlSi_3O_8} + (1/3)H_2O, \tag{a}$$

which is the equation at the beginning of Sec. 3.6.1 but with Mn added. We noted in Sec. 3.6.1 that an increase in temperature causes both garnet and biotite to become enriched in Mg, as the composition points move up the van Laar curves. With the presence of Mn, the van Laar curves become surfaces (Seltz 1935) which, although curved, in general slope upwards while moving from the Mn corner of the Mn–Fe–Mg composition triangle to the Fe corner, and from there to the Mg corner. At a given temperature, horizontal tie lines from points on the lower biotite surface to the upper garnet surface determine the Mg, Fe, and Mn content (α, β, γ) of both minerals for a given bulk composition; one of these tie lines at temperature T_1 is shown in Figure 3.58. A set of tie lines for T_1 represents changes in the Mn content of garnet and biotite which are sympathetic. Similar conditions hold at a higher temperature, T_2, or for a different activity of H_2O. In any case, in assemblages with

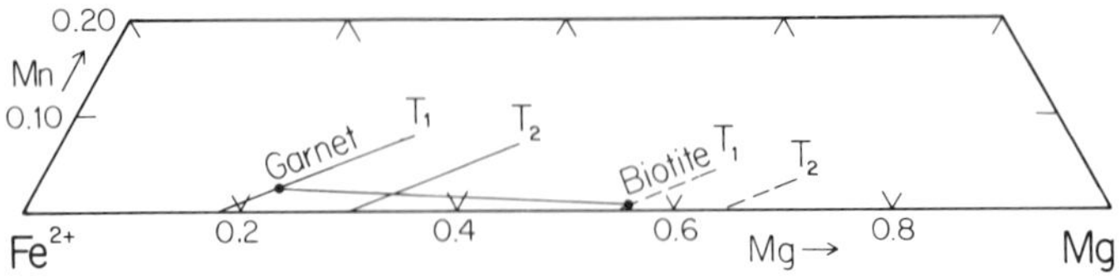

Figure 3.58 Schematic illustration of isotherms on the upper (garnet) and lower (biotite) van Laar surfaces at temperatures T_1 and T_2, and one garnet–biotite tie line; the assemblage is garnet–biotite–quartz–sillimanite–K-feldspar–H_2O, with the activity of H_2O constant

sillimanite, K-feldspar, and quartz, the Mn content of garnet and of biotite are not fixed at a given temperature but are expected to vary sympathetically.

More common is the condition where the two ferromagnesian minerals are not interrelated by a transfer equilibrium, and 11 of the garnet–biotite study samples fall in this category. Here the Mn content of garnet relative to that of biotite is evidently governed entirely by an exchange reaction. Imagine, for example, a biotite crystal that contains Mn, and an associated garnet crystal that does not, i.e., in equation (a), $\gamma 1 = 0$. In the ensuing exchange reaction, Mn atoms move from biotite to garnet in exchange for an equal number of (Mg, Fe^{2+}) atoms, i.e., the following reaction equation is displaced to the right:

$$\underset{\text{MgFe grt in Grt}}{(\mathrm{Mg}_{\alpha 1}\mathrm{Fe}_{\beta 1})\mathrm{Al}_{2/3}\mathrm{SiO}_4} + \underset{\text{Mn bt in Bt}}{\mathrm{K}_{1/3}\mathrm{MnAl}_{1/3}\mathrm{SiO}_{10/3}(\mathrm{OH})_{2/3}}$$

$$= \underset{\text{Mn grt in Grt}}{\mathrm{MnAl}_{2/3}\mathrm{SiO}_4} + \underset{\text{MgFe bt in Bt}}{\mathrm{K}_{1/3}(\mathrm{Mg}_{\alpha 2}\mathrm{Fe}_{\beta 2})\mathrm{Al}_{1/3}\mathrm{SiO}_{10/3}(\mathrm{OH})_{2/3}} \tag{b}$$

A calculation will show that this reaction must produce a change in the $\alpha 1 : \beta 1$ ratio of the garnet crystals and in the $\alpha 2 : \beta 2$ ratio of the biotite crystals, provided Mg–Fe^{2+} exchange equilibrium (Sec. 3.7.1) is preserved. Very near equilibrium, however, where all changes are infinitesimal, all of these quantities, including $\gamma 1$ and $\gamma 2$, become determined for a particular system, and, by the procedure of Sec. 3.7.1,

$$RT \ln \left(\frac{X_{\mathrm{Mn\ grt}}}{X_{\mathrm{Mg}_{\alpha 1}\mathrm{Fe}_{\beta 1}\mathrm{grt}}} \cdot \frac{X_{\mathrm{Mg}_{\alpha 2}\mathrm{Fe}_{\beta 2}\mathrm{bt}}}{X_{\mathrm{Mn\ bt}}} \right) \left(\frac{\gamma_{\mathrm{Mn\ grt}} \gamma_{\mathrm{Mg}_{\alpha 2}\mathrm{Fe}_{\beta 2}\mathrm{bt}}}{\gamma_{\mathrm{Mg}_{\alpha 1}\mathrm{Fe}_{\beta 1}\mathrm{grt}} \gamma_{\mathrm{Mn\ bt}}} \right)$$

$$= -(G^{\mathrm{MnGrt}} + G^{\mathrm{Mg}_{\alpha 2}\mathrm{Fe}_{\beta 2}\mathrm{Bt}} - G^{\mathrm{Mg}_{\alpha 1}\mathrm{Fe}_{\beta 1}\mathrm{Grt}} - G^{\mathrm{MnBt}}). \tag{169}$$

The right-hand side of this equation is the Gibbs energy change for reaction (b), which is a function of temperature, pressure, and the Mg:Fe ratio of garnet ($\alpha 1, \beta 1$) and biotite ($\alpha 2, \beta 2$), these being related by an Fe^{2+}–Mg exchange equilibrium. Each of the four activity coefficients in (169) is also a function of Mg:Fe and Mn/(Mg + Fe + Mn), but if the Mg:Fe ratios remain constant and the Mn concentrations are sufficiently low, then by the Henry and Raoult rules, all activity coefficients and hence the entire quotient remain independent of $X_{\mathrm{Mn\ grt}}$ and $X_{\mathrm{Mn\ bt}}$.

Finally, with the two Mn mole fractions as small numbers, the mole fractions of $Mg_{\alpha 1}$ $Fe_{\beta 1}$ grt in garnet and of $Mg_{\alpha 2}$ $Fe_{\beta 2}$ bt in biotite are only slightly less than 1.0; thus the distribution coefficient reduces to a simple ratio,

$$K_D = X_{Mn\ grt}/X_{Mn\ bt}, \tag{170}$$

which is expected to vary with Mg:Fe, temperature, and pressure. Equation (170) is one form of the Nernst distribution equation.

Allowing now for a variation in the composition of the garnet and biotite crystals greater than that implied by the above simplified formulas, K_D could be displaced by the entry of Ca to garnet, and of Fe^{3+}, Ti, F, and additional Al to biotite. Similar compositional effects are expected when examining the distribution of Mn between hornblende and Ca pyroxene.

The distribution of Mn between garnet and biotite in high amphibolite facies gneisses is shown in Figure 3.59a. Mineral analyses were obtained by emission spectroscopy (Mn) and by microprobe analyses; Fe^{2+} in some biotite samples was determined by titration and in others was estimated, based on these determinations. In garnet, formula calculations showed the presence of very minor Fe^{3+}.

The six mineral pairs that contain the lowest Mn concentrations (Figure 3.59a) are those that occur with sillimanite, K-feldspar, and quartz, and in these samples the Mn content of the minerals was presumably determined by a transfer equilibrium, as noted above. The Mn variation as found here can be understood as resulting partially from a variation in bulk composition and partially from a variation in temperature or H_2O activity; the two concentration terms vary sympathetically, as expected, and the relation is evidently linear.

The remaining results (Figure 3.59a) define the relation

$$K_D(\text{Mn} - (\text{Mg,Fe}), \text{Grt/Bt}) = X^{\text{Grt}}_{\text{Mn}}/X^{\text{Bt}}_{\text{Mn}} = 27,$$

which was presumably determined entirely by Mn–(Mg,Fe^{2+}) exchange equilibrium.

The distribution of Mn between Ca pyroxene and hornblende in associated amphibolite is shown in Figure 3.59b. Atomic absorption spectroscopy was used to obtain analyses of highly pure mineral separates; Fe^{2+} was measured for some of the minerals, and from these results was estimated in the remaining minerals. The Mn distribution coefficient, from Figure 3.59b, is

$$K_D(\text{Mn} - (\text{Mg,Fe}), \text{Cpx/Hbl}) = X^{\text{Cpx}}_{\text{Mn}}/X^{\text{Hbl}}_{\text{Mn}} = 1.44.$$

This is identical to the result obtained by Mueller (1960) for Ca-pyroxene–actinolite in a Quebec iron formation.

In both examples of Mn distribution the relation is linear, as expected, and a variation in this coefficient with larger compositional changes (e.g., Mg/(Mg + Fe^{2+}) which in hornblende ranges from 0.50 to 0.75) cannot be detected.

When the above results are combined with

$$K_D(\text{Mn} - (\text{Mg,Fe}), \text{Grt/Hbl}) = 9.0,$$

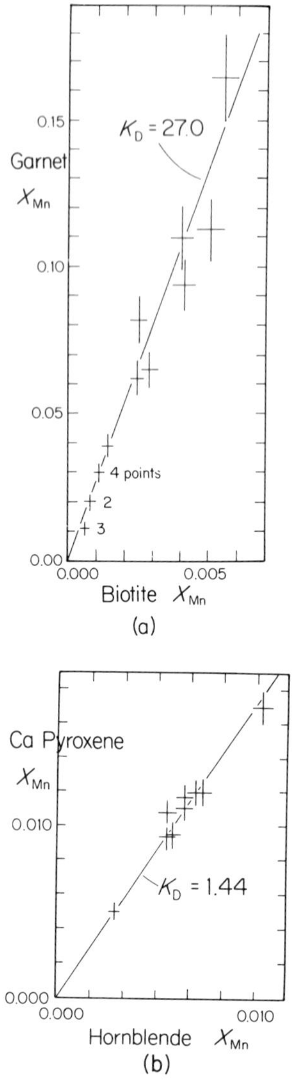

Figure 3.59 Distribution of Mn between (a) garnet and biotite and (b) Ca pyroxene and hornblende, in high amphibolite facies gneiss and amphibolite; $X_{Mn} = Mn/(Mg + Fe^{2+} + Mn)$. Data from Kretz (1959, 1978). © University of Chicago Press, reproduced by permission

from Kretz (1978), we obtain the following sequence of decreasing $Mn/(Mg + Fe^{2+} + Mn)$:

$$\text{garnet} \overset{6.4}{>} \text{Ca pyroxene} \overset{1.44}{>} \text{hornblende} \overset{3.0}{>} \text{biotite},$$

where the numbers are distribution coefficients. Theoretically, these coefficients are a function of temperature, but at present little information is available to define the relationships.

3.8.3 The distribution of vanadium among hornblende, biotite, and garnet

The mean vanadium concentration of basaltic rocks is 250 ppm and of shale, 130 ppm. During metamorphism this element becomes concentrated in magnetite, hornblende, and biotite; concentrations in the feldspar minerals are very low. In magnetite and rare occurrences of V-rich mica, the trivalent form (V^{3+}) dominates, and where it occupies six-coordinated positions, the ionic radius is 0.74 Å (Burkart-Baumann 1972). But for vanadium, as for many trace elements, the sites occupied by the atoms in mineral structures are not known with confidence, and the mass fraction, 'parts per million' (ppm), is normally used as a measure of concentration (denoted C), in place of an atomic or molecular fraction (X). The concentration of an element in mineral $\alpha(C^{\alpha})$ relative to that in mineral $\beta(C^{\beta})$ is then expressed as a distribution ratio,

$$R_{\mathrm{D}} = C^{\alpha}/C^{\beta}.$$

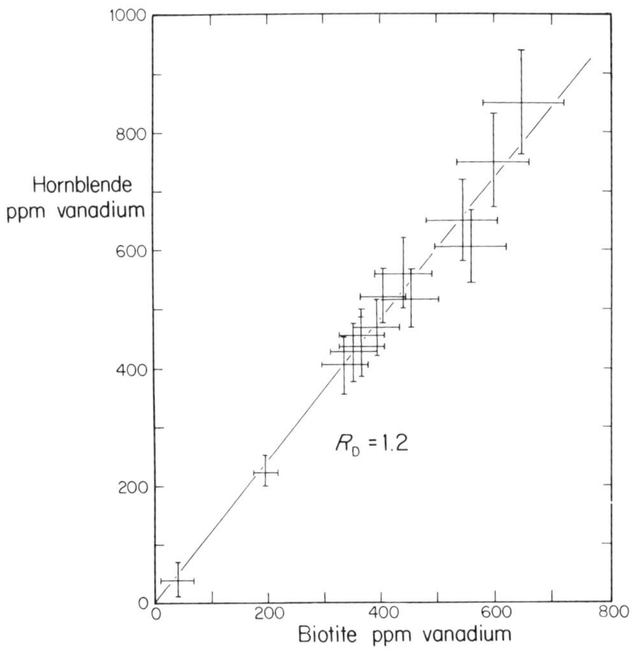

Figure 3.60 The distribution of vanadium between hornblende and biotite in high amphibolite facies gneiss and amphibolite; error bars represent 0.95 confidence limits, analytical error. Data from Kretz (1959). © University of Chicago Press, reproduced by permission

The distribution of vanadium between hornblende and biotite in the high amphibolite facies gneisses and amphibolites referred to in the previous section is illustrated in Figure 3.60. Emission spectroscopy was used to determine vanadium in highly pure mineral separates. The distribution ratio is

$$R_{\mathrm{D}} = C_{\mathrm{V}}^{\mathrm{Hbl}}/C_{\mathrm{V}}^{\mathrm{Bt}} = 1.20.$$

R_{D} is related to the distribution coefficient by a factor, and in general, the thermodynamic considerations expressed in Sec. 3.8.2 apply here as well. Thus we may propose, from the data of Figure 3.60, that exchange equilibrium was established closely in the study specimens, and that the equilibrium was not perturbed noticeably by large compositional changes (especially the Mg:Fe^{2+} ratios) in the hornblende and biotite crystals. Also, linear activity–concentration relations must hold in both crystals at $C < 1000$ ppm V, where activity refers to that of a vanadium end-member component, not yet defined.

Some of the above study specimens contain garnet and in these and in accompanying gneisses, the distribution of vanadium between garnet and biotite crystal varies greatly; thus the distribution ratio, defined as

$$R_{\mathrm{D}} = C_{\mathrm{V}}^{\mathrm{Grt}}/C_{\mathrm{V}}^{\mathrm{Bt}},$$

ranges from 0.09 to 0.52. This variation is related to the Ca content of garnet, as

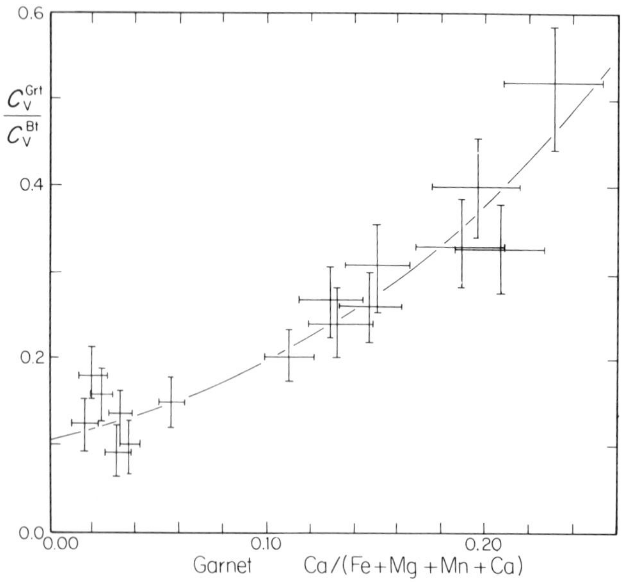

Figure 3.61 Relation between the vanadium distribution ratio, R_{D} (V; garnet/biotite) and the atomic fraction of Ca in garnet. Data from Kretz (1959). © University of Chicago Press, reproduced by permission

is shown in Figure 3.61, through the equation

$$\ln R_D(\mathrm{V, Grt/Bt}) = 6.49X_{\mathrm{Ca}} - 2.27,$$

where X_{Ca} is the atomic fraction $\mathrm{Ca/(Mg{+}Fe^{2+}{+}Mn{+}Ca)}$ in garnet, i.e., the fraction of cubic sites occupied by Ca atoms. Thus, as X_{Ca} increases from zero to 0.25, the number of vanadium atoms accepted by a garnet crystal *relative* to the number in an associated biotite crystal, increases from 0.1 to 0.5. The major-element compositional effect, which was expressed in Sec. 3.8.2 as a thermodynamic expectation, is here clearly demonstrated.

3.8.4 The distribution of Sr, Mn, and Fe between calcite and dolomite

The distribution of small concentrations of Sr (< 1600 ppm), Mn^{2+} (< 1100 ppm), and Fe^{2+} (< 500 ppm) between crystals of calcite and dolomite in high amphibolite facies marble provides a further example of trace-element distribution patterns. This example is of special interest because it is possible to obtain, from crystal-chemical considerations, numerical estimates for distribution coefficients, which can then be compared with those that are measured.

Within the Otter Lake terrane (Grenville Province), common marble consists mainly of calcite, < 30% dolomite, and minor phlogopite and graphite; tremolite, diopside, and forsterite are locally present. The marble is interlayered with garnet–biotite gneiss and amphibolite, which provided the study material for Secs. 3.6.1, 3.7.1, and 3.8.3, above.

The formulas of associated calcite and dolomite may be written as

$$\text{calcite:} \quad \mathrm{(Ca,Mg,Sr,Mn,Fe)^{CaCal}CO_3}$$
$$\text{dolomite:} \quad \mathrm{(Ca,Mg,Sr,Mn,Fe)_{0.5}^{CaDol}(Mg,Sr,Mn,Fe)_{0.5}^{MgDol}CO_3}$$

where superscripts identify the Ca site of calcite and the Ca and Mg sites of dolomite. Some Mg is present in CaCal, as demanded by the calcite–dolomite solvus equilibrium (Sec. 3.4.2); and in the formulas, as written, Sr, Mn, and Fe are not entirely excluded from each site.

The between-crystal exchange reaction for each trace element (e.g., Fe) is

$$\underset{\text{in dolomite}}{\mathrm{FeCO_3}} + \underset{\text{in calcite}}{\mathrm{MCO_3}} = \underset{\text{in calcite}}{\mathrm{FeCO_3}} + \underset{\text{in dolomite}}{\mathrm{MCO_3}}, \tag{a}$$

where M stands for the exchangeable cations (Ca,Mg,Sr,Mn). The distribution coefficient is

$$K_D(\mathrm{Fe, Cal/Dol}) = X_{\mathrm{Fe}}^{\mathrm{Cal}}/X_{\mathrm{Fe}}^{\mathrm{Dol}} \tag{171}$$

where X_{Fe} is the atomic ratio Fe/(Ca + Mg + Sr + Mn + Fe).

The between-crystal distribution of Sr, Mn, and Fe is shown in Figure 3.62. The data were obtained by atomic absorption analysis of calcite and dolomite separated from small volumes of rock, about 5 cm in diameter. The purity of the separates was

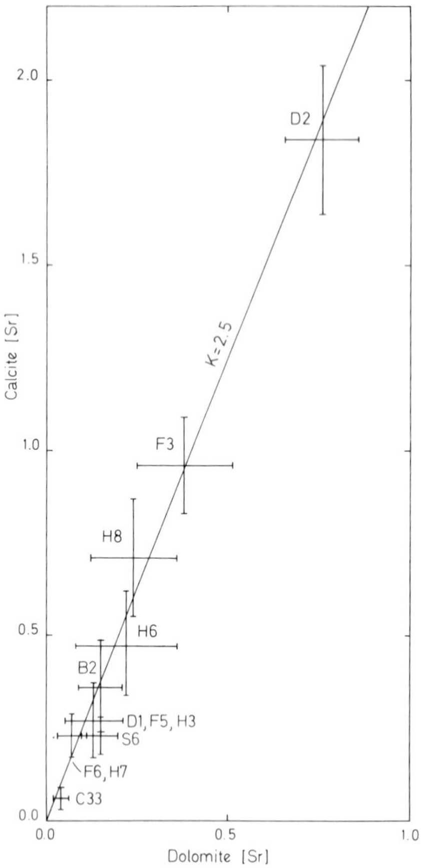
Calcite [Sr]
Dolomite [Sr]
K=2.5
D2
F3
H8
H6
B2
D1, F5, H3
S6
F6, H7
C33
0.0
0.5
1.0
1.5
2.0

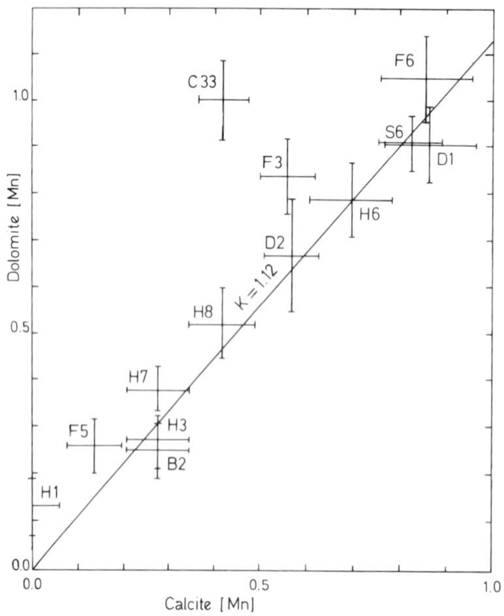
Dolomite [Mn]
Calcite [Mn]
K=1.12
C33
F6
S6
D1
F3
H6
D2
H8
H7
H3
F5
B2
H1
0.0
0.5
1.0

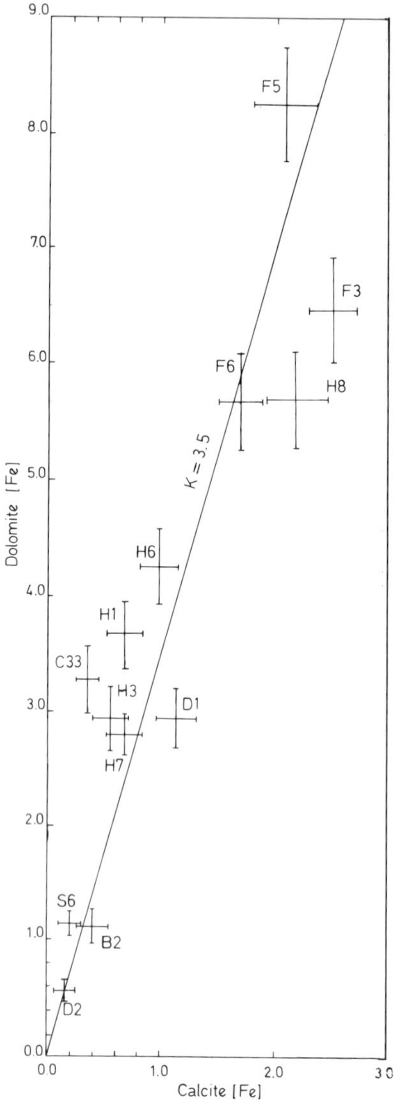
Dolomite [Fe]
Calcite [Fe]
K = 3.5
F5
F3
F6
H8
H6
H1
C33
H3
D1
H7
S6
B2
D2
0.0
1.0
2.0
3.0
4.0
5.0
6.0
7.0
8.0
9.0

monitored by applying a calcite stain to portions of about 200 particles. The procedure was designed to prevent loss, from the calcite concentrate, of small dolomite lamellae and rods which exsolved during cooling from metamorphic temperatures; however, some loss may have occurred.

Apparently linear plots (Figure 3.62) are obtained, as expected. In specimen C 33, Mn and Fe are anomalous and Sr is low; post-metamorphic loss of Sr may have disturbed the Mn and Fe distribution. Some of the scatter in the Fe diagram may be viewed as departure from exchange equilibrium, or the result of a loss of exsolved dolomite during separation of calcite crystals for analysis, giving Fe determinations for some of the samples that are too low. The determined distribution coefficients are:

$$\begin{aligned} K_\mathrm{D}(\mathrm{Sr, Cal/Dol}) &= 2.3 \\ K_\mathrm{D}(\mathrm{Mn, Cal/Dol}) &= 0.85 \\ K_\mathrm{D}(\mathrm{Fe, Cal/Dol}) &= 0.28 \end{aligned} \tag{172}$$

Thus the sequence Sr, Mn, Fe, which is one of decreasing ionic radius (1.33, 0.91, 0.86 Å, Appendix A7) is a sequence of increasing rejection by the calcite structure and increasing acceptance by the dolomite structure.

The above results may be used to obtain estimates of trace-element distribution between the Ca and Mg sites of dolomite at the time and temperature of metamorphism. The between-crystal distribution of each element is determined by three between-site exchange reactions. With regard to Fe, they are as follows:

$$\begin{aligned} \mathrm{Fe^{CaDol}} + \mathrm{M^{MgDol}} &= \mathrm{Fe^{MgDol}} + \mathrm{M^{CaDol}} \\ \mathrm{Fe^{CaDol}} + \mathrm{M^{CaCal}} &= \mathrm{Fe^{CaCal}} + \mathrm{M^{CaDol}} \\ \mathrm{Fe^{MgDol}} + \mathrm{M^{CaCal}} &= \mathrm{Fe^{CaCal}} + \mathrm{M^{MgDol}} \end{aligned}$$

The corresponding between-site distribution coefficients are:

$$k_\mathrm{D}(\mathrm{Fe, MgDol/CaDol}) = x_\mathrm{Fe}^\mathrm{MgDol}/x_\mathrm{Fe}^\mathrm{CaDol} \tag{173}$$

$$k_\mathrm{D}(\mathrm{Fe, CaCal/CaDol}) = x_\mathrm{Fe}^\mathrm{CaCal}/x_\mathrm{Fe}^\mathrm{CaDol} \tag{174}$$

$$k_\mathrm{D}(\mathrm{Fe, CaCal/MgDol}) = x_\mathrm{Fe}^\mathrm{CaCal}/x_\mathrm{Fe}^\mathrm{MgDol} \tag{175}$$

where x_Fe is the atomic fraction of Fe on the designated site.

We now assume that Fe atoms are equally distributed between the Ca site of calcite and the Ca site of dolomite, i.e.,

$$x_\mathrm{Fe}^\mathrm{CaCal} = x_\mathrm{Fe}^\mathrm{CaDol} = X_\mathrm{Fe}^\mathrm{Cal} \tag{176}$$

$$k_\mathrm{D}(\mathrm{Fe, CaCal/CaDol}) = 1.0 \tag{177}$$

Figure 3.62 Distribution of Sr, Mn, and Fe between calcite and dolomite of high amphibolite facies marble from the Grenville Province (Canadian Shield); [] denotes atomic fraction ×1000, e.g., [Fe] = Fe/(Ca + Mg + Sr + Mn + Fe) × 1000; error bars are 0.95 confidence limits, analytical error; K stands for K_D. From Kretz (1980). Reproduced by permission of Oxford University Press

By substitution of (176) in (171),

$$K_{\mathrm{D}}(\mathrm{Fe, Cal/Dol}) = x_{\mathrm{Fe}}^{\mathrm{CaDol}} / X_{\mathrm{Fe}}^{\mathrm{Dol}} \tag{178}$$

In dolomite, the atomic fraction of Fe in the crystal as a whole is related to that on the two sites by

$$X_{\mathrm{Fe}}^{\mathrm{Dol}} = (x_{\mathrm{Fe}}^{\mathrm{CaDol}} + x_{\mathrm{Fe}}^{\mathrm{MgDol}})/2 \tag{179}$$

By combining (173), (178), and (179) we obtain

$$k_{\mathrm{D}}(\mathrm{Fe, MgDol/CaDol}) = (2/K_{\mathrm{D}}(\mathrm{Fe, Cal/Dol})) - 1 \tag{180}$$

The *calculated* within-dolomite distribution coefficients (from (172) and (180)) are

$$k_{\mathrm{D}}(\mathrm{Sr, MgDol/CaDol}) \sim 0$$

$$k_{\mathrm{D}}(\mathrm{Mn, MgDol/CaDol}) = 1.35$$

$$k_{\mathrm{D}}(\mathrm{Fe, MgDol/CaDol}) = 6.1$$

These calculations imply that all of the Sr in dolomite resides in the Ca site (in agreement with Jacobson and Usdowski (1976)), and that concentrations of both Mn and Fe on the Mg site are greater than on the Ca site by factors of 1.35 and 6.1 respectively.

One could now attempt a prediction of the within-dolomite distribution coefficient for Mn and Fe, based on ionic size. Equation (180) could then be used to obtain numerical estimates of calcite–dolomite distribution coefficients.

The model (Kretz 1982b) supposes that in the dolomite structure, trace-element ions with radius larger than that of Ca (1.08 Å) will be found almost entirely in the Ca site, ions with radius smaller than that of Mg (0.80 Å) will be found almost entirely in the Mg site, and ions of intermediate size (e.g., Fe, 0.86 Å) will be distributed between the two sites as follows:

$$k_{\mathrm{D}}(\mathrm{Fe, MgDol/CaDol}) = \frac{r_{\mathrm{Ca}} - r_{\mathrm{Fe}}}{r_{\mathrm{Fe}} - r_{\mathrm{Mg}}} \tag{181}$$

where r is ionic radius. Thus ions approaching Ca in size will be increasingly concentrated on the Ca site, and ions approaching Mg in size will be increasingly concentrated on the Mg site. By use of ionic radii (listed in Appendix A7) and equation (181) we obtain the theoretical within-dolomite distribution coefficients for Fe, Mn, and Sr listed in Table 3.7. These numbers and equation (180) then produce the theoretical calcite–dolomite distribution coefficients, also listed in this table. They refer to conditions that existed during metamorphism, i.e. 650–700 °C, 7 kbar. The theory agrees moderately well with observation.

Several studies have shown that in crystals of sedimentary dolomite, concentrations of trace Mn on the Mg site, as determined by electron paramagnetic resonance, are greater than on the Ca site by a factor that ranges from 2 to 500 (Vinokurov, Zaripov and Stepanov 1961, Schindler and Ghose 1970, Wildeman 1970, Prissok

Table 3.7 Distribution of trace amounts of Fe, Mn, and Sr between calcite and dolomite and within dolomite crystals

	Fe	Mn	Sr
Calcite-dolomite, K_D (Cal/Dol)			
From observation in marble, Fig. 3.62	0.28	0.85	2.3
From ion-size model (below)	0.43	0.79	2.0
Within dolomite, k_D (MgDol/CaDol)			
From K_D (Cal/Dol), Fig. 3.62, assuming $x^{CaDol} = x^{CaCal}$	5.9	1.24	~0
From ion-size model	3.7	1.54	0.0
From measurement, contact aureole, Texas (Lloyd, Lumsden and Gregg 1985)	—	9 to 19	—

and Lehmann 1986), and most of these crystals evidently do not represent equilibrium with regard to Mn ordering. Data on dolomite from a metamorphic aureole were reported by Lloyd, Lumsden and Gregg (1985). At points beyond 100 m from a syenite–gabbro pluton in west Texas, k_D (Mn; MgDol/CaDol) ratios were found to range from 21 to 110 in an irregular pattern, but nearer the pluton, where the marble microstructure showed evidence of recrystallization, the above ratio was reduced to a mean value of 13 (Table 3.7). This evidently represents an approach to equilibrium with regard to Mn ordering, but it is not certain if equilibrium was approached closely, and if so, whether or not it was preserved during cooling.

From the above discussions, it is clear that very many opportunities exist for the further study of the behavior of atoms of trace elements during metamorphic crystallization.

3.9 OXIDATION–REDUCTION EQUILIBRIA

Reactions that bring about a transfer of electrons are referred to as 'oxidation–reduction' (redox) reactions. In the most familiar of these reactions, electrons are transferred from iron atoms to oxygen atoms or vice versa, but many other kinds of atoms can act as electron donors or acceptors.

Although oxygen is extremely abundant within the Crust and Mantle, the chemical activity of this element there is extremely small. In the atmosphere, oxygen is 'free' to oxidize ferrous iron, manganese, and other atoms, and does so during weathering, but within the Earth, where atmospheric gases are not present, the oxygen atoms form strong bonds with Si and other cations, and hypothetical oxidizing equilibria of the kind

$$SiO_2 + 2Fe_3O_4 = 3Fe_2O_3 + SiO$$

are displaced far to the left.

Some general questions regarding redox equilibrium will be considered as we examine, first, the magnetite–hematite equilibrium. Next we shall review the magnetite–ilmenite equilibrium, for which abundant experimental data are available, and finally some examples of silicate–oxide equilibrium will be described. For a general introduction to the subject, a reading of Eugster (1959, 1972), Mueller (1961b), Miyashiro (1964), and Ernst (1976, pp. 217–229) will be very helpful.

3.9.1 The magnetite–hematite equilibrium

Hematite and magnetite occur together in metamorphosed iron formation, quartzite, and some low-grade meta-shale, but in nearly all medium-to high-grade rocks of different composition, magnetite is commonly present but not hematite. The two minerals are interrelated by the univariant reaction

$$\underset{\text{magnetite (Mt)}}{4Fe^{2+}Fe_2^{3+}O_4} + O_2 = \underset{\text{hematite (Hem)}}{6Fe_2^{3+}O_3} \qquad \text{(a)}$$

With oxygen as a pure gas, this equilibrium is displaced to the right, and the widespread presence of magnetite can only be understood by supposing that the activity of oxygen is reduced, by being present at very low concentrations.

At equilibrium, assuming that magnetite and hematite are free of extraneous elements, then from Sec. 3.2.1,

$$RT \ln f\,(O_2) = \Delta G_{a(T,1)} + \Delta V_{\text{solids}}(P - 1),$$

where ΔG_a is the Gibbs energy change of reaction (a) (with oxygen as a pure phase) at 1 bar and the temperature of interest, and $f\,(O_2)$ is the fugacity of oxygen in the vapor phase.

For equilibria of this kind, experimental and thermochemical data are used to calculate a relation between $f\,(O_2)$ and temperature. The equilibrium equation for (a), from Eugster and Wones (1962), is

$$\log_{10} f\,(O_2) = -(24\,912/T) + 14.41 + 0.019(P - 1)/T,$$

which shows that the effect of confining pressure (P) is very small or negligible. We see from this equation and its graphical expression (Figure 3.63) that at 600 °C, for example, an extremely low fugacity of oxygen of 10^{-14} bar is required for the equilibrium association of the two oxide minerals. It is possible therefore that the implied very sparse oxygen molecules are dissolved in the magnetite and hematite crystals themselves, and a vapor phase is not needed.

The magnetite–hematite association is one of the several buffers used in experimental work (see Sec. 3.2.3) to control oxygen fugacity. Another is

$$\underset{\text{wüstite}}{3FeO} + (1/2)O_2 = \underset{\text{magnetite}}{Fe_3O_4}, \qquad \text{(b)}$$

which is also plotted in Figure 3.63, thereby defining the stability field of pure magnetite.

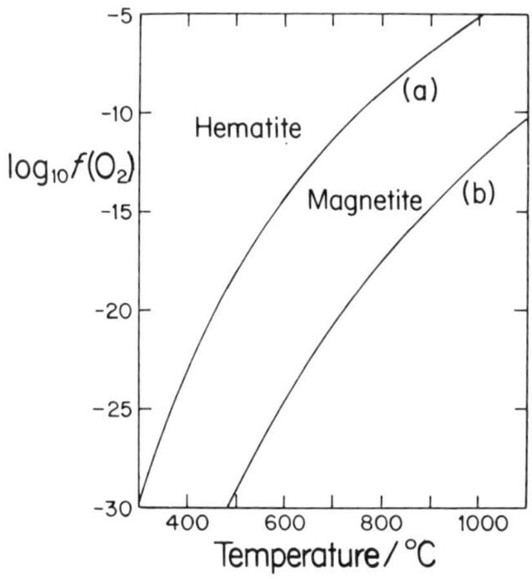

Figure 3.63 Boundary curves for the equilibria, magnetite + oxygen = hematite (a), and wüstite + oxygen = magnetite (b); f denotes fugacity. From Eugster and Wones (1962). Reproduced by permission of Oxford University Press

Miyashiro (1964) considered the case of an aqueous vapor in equilibrium with magnetite and hematite. In water, the dissociation equilibrium,

$$H_2O = H_2 + (1/2)O_2, \tag{c}$$

is defined by the equation

$$K_c = f\ (H_2)[f\ (O_2)]^{1/2}/f\ (H_2O),$$

which permits f (H_2) and f (O_2) to vary without changing the equilibrium constant, K_c. Miyashiro showed that pure water or water that contains some dissolved oxygen (e.g., ground water) is in equilibrium with hematite (not magnetite), but in the presence of graphite, f (O_2) is sufficiently diminished to fall in the magnetite field of Figure 3.63.

Provided that concentrations of Ti, Mn, and other elements are very small, the magnetite–hematite equilibrium at constant confining pressure and temperature is invariant, i.e., f (O_2) is defined. The presence of hematite alone or magnetite alone then implies f (O_2) values greater than or less than the defined value. Interlayered strata containing different oxide assemblages, as described by Mueller (1960), Rumble (1973), and others therefore represent a *variation* of f (O_2) which is difficult to reconcile with the presence, during metamorphism, of a mobile vapor phase. This problem was recognized by Eugster (1959), Mueller (1961b), and Miyashiro (1964) who proposed that the Fe^{2+}:Fe^{3+} composition of metamorphic rocks was in general determined prior to metamorphism, and was preserved. The implied immobility of oxygen (present at extremely low concentrations) can be understood in terms of buffering of oxygen by the solid phases (Eugster 1959), a small diffusive flux resulting from minute concentrations (Mueller 1961b), and the presence or absence of graphite (French 1966).

3.9.2 The magnetite–ilmenite equilibrium

Magnetite and ilmenite are very common minor minerals in schist, gneiss, amphibolite, granulite, and other crystalline rocks. For example, occurrences from the Adirondack region of New York, were described by Buddington and Lindsley (1964).

Disregarding the presence of small amounts of Mn, V, and other elements, magnetite (Mt) in these occurrences is a mixture of components $Fe^{2+}Fe_2^{3+}O_4$ (magnetite) and $Fe_2^{2+}TiO_4$ (ulvöspinel), and ilmenite (Ilm) is a mixture of components $Fe^{2+}TiO_3$ (ilmenite) and $Fe_2^{3+}O_3$ (hematite). These two solution series appear in the TiO_2–$Fe^{2+}O$–$Fe^{3+}O_{3/2}$ composition triangle shown in Figure 3.64. In high-grade gneisses, ilmenite normally contains a small proportion of hematite while magnetite contains a moderate amount of ulvöspinel, to give tie lines similar to those shown in Figure 3.64.

The mineral formulas may be written as

$$\text{ilmenite:} \quad (Fe^{2+}TiO_3, Fe_2^{3+}O_3),$$
$$\text{magnetite:} \quad (Fe^{2+}TiO_3, Fe_2^{3+}O_3)Fe^{2+}O,$$

to emphasize that $Fe^{2+}TiO_3$ and $Fe_2^{3+}O_3$ are shared by both. The magnetite–ilmenite equilibrium may therefore be viewed, in part, as an exchange equilibrium (Rumble 1970),

$$\underset{\substack{\text{ilm}\\ \text{in ilmenite}}}{Fe^{2+}TiO_3} + \underset{\substack{\text{mt}\\ \text{in magnetite}}}{(Fe_2^{3+}O_3)FeO} = \underset{\substack{\text{hem}\\ \text{in ilmenite}}}{Fe_2^{3+}O_3} + \underset{\substack{\text{usp}\\ \text{in magnetite}}}{(Fe^{2+}TiO_3)FeO}, \qquad \text{(a)}$$

$$RT\ln\left(\left(\frac{X_{\text{usp}}}{1-X_{\text{usp}}}\right)\left(\frac{1-X_{\text{ilm}}}{X_{\text{ilm}}}\right)\left(\frac{\gamma_{\text{usp}}\gamma_{\text{hem}}}{\gamma_{\text{mt}}\gamma_{\text{ilm}}}\right)\right) = -\Delta G_{\text{a}}, \qquad (182)$$

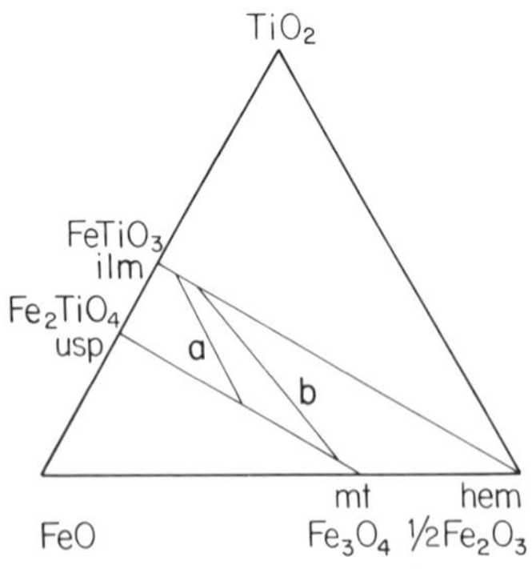

Figure 3.64 Ilmenite–hematite (ilm–hem) solid-solution series, and ulvöspinel–magnetite (usp–mt) solid-solution series in the TiO_2–$Fe^{2+}O$–$(1/2)Fe_2^{3+}O_3$ composition triangle. The composition of associated minerals at 700 °C is illustrated by tie lines (a) and (b) for an ambient oxygen fugacity represented by $\log_{10}(O_2)$ of −18 and −15 respectively. Data from Spencer and Lindsley (1981)

and in part by an oxidation equilibrium (Rumble 1970, Spencer and Lindsley 1981),

$$\underset{\substack{\text{mt} \\ \text{in magnetite}}}{Fe_4^{2+}Fe_8^{3+}O_{16}} + O_2 = \underset{\substack{\text{hem} \\ \text{in ilmenite}}}{Fe_{12}^{3+}O_{18}}, \tag{b}$$

$$RT \ln\left(\left(\frac{X_{\text{hem}}}{X_{\text{mt}}}\frac{\gamma_{\text{hem}}}{\gamma_{\text{mt}}}\right)\left(\frac{1}{f\,(O_2)}\right)\right) = -\Delta G_{\text{b}}. \tag{183}$$

Numerous experimental results on the composition of magnetite and ilmenite have been reported, chiefly by D.H. Lindsley and co-workers (see Andersen and Lindsley 1988). Most of the results were obtained by use of a hydrothermal apparatus, with oxygen fugacity controlled by a buffer assemblage, and phase composition determined by microprobe analysis. Andersen and Lindsley (1988) used these data together with thermodynamic and crystal-chemical considerations to obtain the phase relations shown in Figure 3.65. This figure expresses the composition of magnetite (X_{usp}) and associated ilmenite (X_{ilm}) as a function of temperature, and for given X_{usp}, X_{ilm} and T, provides the ambient fugacity of oxygen, $f\,(O_2)$, *relative* to that of the fayalite–magnetite–quartz (FMQ) buffer at the same temperature. Thus the vertical axis is

$$\Delta = \log_{10}(f\,(O_2)/f^{\,\text{FMQ}}(O_2)),$$

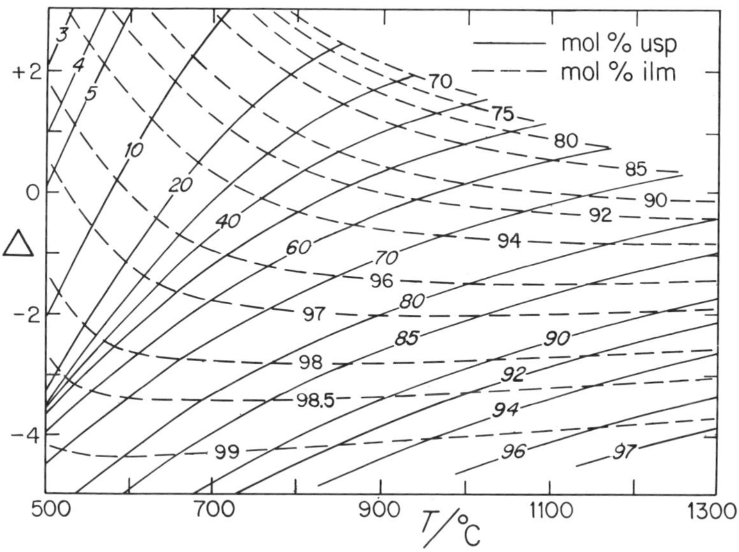

Figure 3.65 Composition of associated magnetite (mole % ulvöspinel (usp)) and ilmenite (mole % ilmenite (ilm)) as a function of temperature and oxygen fugacity (f). $\Delta = \log_{10}[f\,(O_2)/f^{\,\text{FMQ}}(O_2)]$ where $f^{\,\text{FMQ}}(O_2)$ refers to the fayalite–magnetite–quartz equilibrium (equation (184)). From Andersen and Lindsley (1988)

where $f^{\,FMQ}(O_2)$, from Myers and Eugster (1983), is

$$\log_{10} f^{\,FMQ}(O_2) \pm 0.167 = -(24\,441.9/T) + 8.290, \qquad (184)$$

(T is absolute temperature), which is valid at 1.0 kbar and 600 to 1140 °C. Thus an estimate of temperature and oxygen fugacity for naturally occurring magnetite and ilmenite of determined composition may be obtained from Figure 3.65 and equation (184), or by use of a program available from Andersen and Lindsley (1988).

In crystalline rocks, including gabbroic rocks, crystals of magnetite commonly contain oriented lamellae of ilmenite and small attached grains of ilmenite. This ilmenite was evidently extracted from magnetite as the activity of oxygen increased during cooling,

$$\underset{\substack{\text{usp}\\ \text{from magnetite}}}{3Fe_2TiO_4} + (1/2)O_2 = \underset{\substack{\text{ilmenite}\\ \text{lamellae}}}{3FeTiO_3} + \underset{\substack{\text{mt}\\ \text{into magnetite}}}{Fe_3O_4}.$$

Bohlen and Essene (1977) have analyzed magnetite and ilmenite in granulites of the Adirondack Highlands in an attempt to recover peak metamorphic temperatures. Some loss of Ti from magnetite occurred by the above reaction, but by measuring the volume of the ilmenite microcrystals, estimates of the original magnetite composition were obtained. The resulting temperature estimates are shown in Figure 3.39 (p. 203) and agree well with those determined from the distribution of albite between plagioclase and K-feldspar in the same terrane. Calculated oxygen fugacities fall in the range of 10^{-20} to 10^{-16} bar.

3.9.3 Oxide–silicate equilibria

In the several ferromagnesian minerals that occur in low- to high-grade metamorphic terranes, the iron present is dominantly ferrous iron, and the question of the influence of ambient oxygen activity in determining the composition of these crystals is of interest. Studies of silicate-bearing iron formations, in which hematite and magnetite provide some information on oxygen activity, have been particularly informative in this regard. But before examining some of these results we shall describe briefly the information obtained from an experimental study of the stability of Fe–Mg biotite.

The stability of Fe biotite (annite) was investigated experimentally by Eugster and Wones (1962) and that of Fe–Mg biotite by Wones and Eugster (1965). This is a difficult problem because the breakdown of Fe–Mg biotite depends on temperature, pressure, H_2O activity, O_2 activity, Fe:Mg ratio, and the presence of minor components. In the experiments, synthetic crystals were subjected to pressures of 1.0 and 2.1 kbar in the temperature range of 400 to 900 °C and in the presence of H_2O and an oxygen-buffer assemblage.

Some of the results are shown in Figure 3.66, where a–b–c–d–e delimits the stability field of pure annite at 2.1 kbar fluid pressure. Line b–c is the boundary

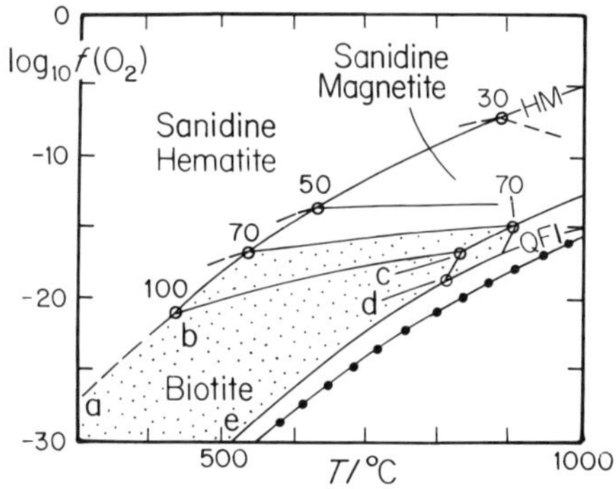

Figure 3.66 Stability of biotite in presence of H_2O and quartz; numbers 100, 70, 50, 30 refer to the mole fraction Fe component (annite) at 2.07 kbar H_2O pressure and with oxygen fugacity (f) determined by the HM and QFI buffers; curve with dots is the upper stability limit of pure Mg biotite. Note decreasing Fe content of biotite (line b–c) with increasing oxygen fugacity. Data from Wones and Eugster (1965); figure (simplified) from Ernst (1976)

for the reaction

$$\underset{\text{annite}}{KFe_3^{2+}AlSi_3O_{10}(OH)_2} + (1/2)O_2 = \underset{\text{K-feldspar}}{KAlSi_3O_8} + \underset{\text{magnetite}}{Fe^{2+}Fe_2^{3+}O_4} + H_2O. \quad \text{(a)}$$

With increasing concentrations of the Mg biotite (phlogopite) component in the biotite crystals, the chemical potential of the annite component is diminished, and a higher fugacity of oxygen is needed to compensate for this effect; hence the boundary shifts upwards, as shown. This shift expresses a relationship between the fugacity of oxygen and the Mg:Fe composition of biotite at constant temperature and pressure.

A relationship of this kind has been discovered in several natural assemblages. Mueller (1960) in a study of a Quebec iron formation found that the $Fe^{2+}/(Mg + Fe^{2+})$ ratio of actinolite in assemblages with talc and quartz was generally large and variable wherever magnetite was the only Fe oxide present, and small and restricted in the presence of magnetite *and* hematite (Figure 3.67a). All samples were taken from a small area of presumably constant temperature and pressure, and the control on actinolite composition was therefore interpreted in terms of the following redox equilibrium,

$$\underset{\text{Fe actinolite}}{(1/5)Ca_2Fe_5Si_8O_{22}(OH)_2} + \underset{\text{talc}}{(1/3)Mg_3Si_4O_{10}(OH)_2} + (1/6)O_2$$

$$= \underset{\text{Mg actinolite}}{(1/5)Ca_2Mg_5Si_8O_{22}(OH)_2} + \underset{\textit{magnetite}}{(1/3)Fe_3O_4} + \underset{\text{quartz}}{(4/3)SiO_2} + (1/3)H_2O. \quad \text{(b)}$$

An increase in the chemical potential of oxygen will drive this equilibrium to the *right*, decreasing the Fe content of actinolite, until hematite appears, and both the

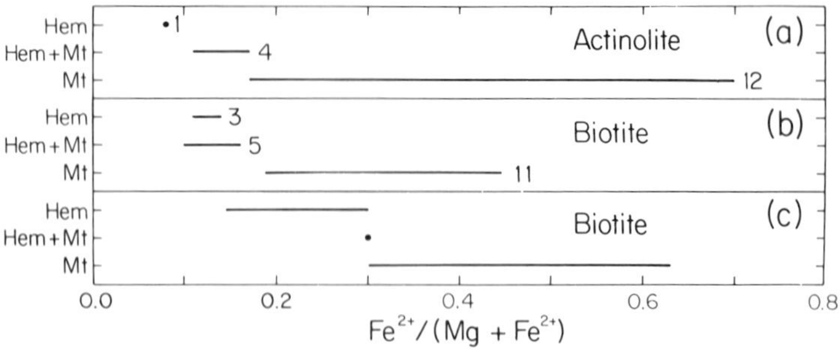

Figure 3.67 $Fe^{2+}/(Mg+Fe^{2+})$ of actinolite or biotite related to presence of hematite (Hem) and/or magnetite (Mt) in (a) iron formation in Quebec (Mueller 1960), (b) iron formation in Sweden (Annersten 1968), and (c) experimental system (Figure 3.66, 870 °C); numbers in (a) and (b) refer to number of rock specimens examined

chemical potential of oxygen and the Fe:Mg ratio of actinolite become fixed,

$$\underset{\text{Fe actinolite}}{(1/5)Ca_2Fe_5Si_8O_{22}(OH)_2} + \underset{\text{talc}}{(1/3)Mg_3Si_4O_{10}(OH)_2} + (1/5)O_2$$

$$= \underset{\text{Mg actinolite}}{(1/5)Ca_2Mg_5Si_8O_{22}(OH)_2} + \underset{\textit{magnetite}}{(1/5)Fe_3O_4} + \underset{\textit{hematite}}{(1/5)Fe_2O_3}$$

$$+ \underset{\text{quartz}}{(4/3)SiO_2} + (1/3)H_2O. \qquad \text{(c)}$$

A further illustration of the control of oxygen activity is provided by an iron formation in Sweden, investigated by Annersten (1968). Here too, the Mg:Fe ratio of silicates can be related to oxygen activity, as signalled by the appearance of hematite. Thus X_{Fe} of biotite in assemblages with K-feldspar, quartz, and magnetite was found to be variable, but where hematite appeared, X_{Fe} was small (nearly pure phlogopite) and restricted, as shown in Figure 3.67b. The reaction here was evidently the same as that investigated experimentally by Wones and Eugster (1965), as described above (equation (a)). Thus from Figure 3.66 (at 880 °C) we see that in the presence of magnetite, biotite crystals with $X_{Fe} > 0.30$ are stable; in the presence of magnetite *and* hematite, $X_{Fe} = 0.30$, and in the presence of hematite (without magnetite), $X_{Fe} < 0.30$, as shown in Figure 3.67c. In the iron formation, temperatures were less than 880 °C, presumably because of a low activity of H_2O.

Similar demonstrations of the importance of oxidation–reduction equilibria in metamorphism were reported by James (1955), Chinner (1960), Kranck (1961), Condie (1967), Butler (1969), and Klein (1966). An example of extreme reduction (to produce metallic iron) during metasomatism was described by Frost (1985).

4 Chemical Kinetics

All chemical changes, including phase transformations, may be viewed as changes in the arrangement of atoms and molecules. It is well known that the rate at which these changes proceed can vary enormously: compare, for example, the rapid (explosive) combination of oxygen and hydrogen to form water, with the much slower oxidation of iron in air. A complete understanding of reaction rates obviously requires some knowledge of the mechanisms of atomic and molecular rearrangements. The study of reaction rate and mechanism is referred to as 'chemical kinetics'.

Some of the earliest experiments in chemical kinetics (c. 1865) were carried out by A.V. Harcourt and W. Esson, who found that the rate at which certain chemicals react in solution depends on the concentration of the reactant species. Later, S. Arrhenius (c. 1890) defined a general relation between reaction rate and temperature, resulting in the concept of an energy of activation which must be overcome as reactants change to products (Laidler 1987, Back and Laidler 1967). In the decades that followed, the understanding of chemical reactions has advanced rapidly; a survey of the subject is presented by Steinfeld, Francisco and Hase (1989).

Mineral reactions that occur within the solid Earth must also proceed at certain rates and by certain mechanisms. Aspects of the kinetics of metamorphic reactions were examined by Fyfe, Turner and Verhoogen (1958), Gray (1969), Carmichael (1969), Walther and Wood (1984), Rubie and Thompson (1985), and others, and the subject has received much attention recently. Although metamorphic changes cannot be observed, some progress can nevertheless be achieved by applying various methods. For example, mineral reactions can be studied experimentally to obtain plausible rates for natural reactions, and certain rock microstructures can be examined in detail to obtain information on crystallization mechanisms. In the present chapter, some of the principles of chemical kinetics as they relate to crystallization are reviewed, together with some experimental results. In the following chapter, certain aspects of microstructure and microcomposition of crustal samples will be examined in terms of reaction kinetics.

4.1 THE REACTION RATE

A chemical change in which b moles of B react with a moles of A to produce c moles of C may be expressed by use of a stoichiometric equation,

$$aA + bB = cC, \tag{1}$$

where a, b, c are referred to as 'stoichiometric coefficients'. An example is the homogeneous (single-phase) reaction

$$2H_2 + 1O_2 \rightarrow 2H_2O. \tag{2}$$

It is well known, however, (Steinfeld, Francisco and Hase 1989) that some reactions proceed in two or more steps, not in one step, as is implied by equation (2). With regard to (2), one of the intermediate steps is

$$O + H_2 \rightarrow OH + H. \tag{3}$$

Despite these complications, the stoichiometric equation may be used to describe the overall reaction, and the rate or velocity (v) may be defined as the rate of disappearance of either A or B (equation (1)), or the rate of appearance of C. Note that where a and b are not equal, as in (2), the rates of disappearance of A and B are not equal. Thus, in general,

$$v = -\frac{1}{a}\frac{d[A]}{dt} = -\frac{1}{b}\frac{d[B]}{dt} = +\frac{1}{c}\frac{d[C]}{dt}, \tag{4}$$

where square brackets denote concentration in moles dm^{-3} (moles $liter^{-1}$) or moles cm^{-3}, and t is time, for example in seconds (s).

In the experimental study of reaction kinetics, a chemical system is initially placed at some distance from equilibrium, and the disappearance of reactants or the appearance of products is then measured as a function of elapsed time. In gas and liquid mixtures, the rate can often be expressed by use of the equation

$$v = k[A]^n, \tag{5}$$

where n (the order of the reaction) is zero or a simple integer or fraction, and k is a constant referred to as the 'rate constant' (Steinfeld, Francisco and Hase 1989). Equations for reactions of order zero ($n = 0$) and one ($n = 1$) are here derived to illustrate this approach to reaction kinetics.

A zero-order reaction is one that can be described by the equation

$$v = -\frac{d[A]}{dt} = k[A]^0 = k, \tag{6}$$

and in this case, the rate of disappearance of A is independent of the concentration of A. Note that the rate constant (k) is in units mol $V^{-1}\ t^{-1}$, where V is volume. An expression for the concentration of A at elapsed time (t), denoted $[A]_t$, as a function of t is obtained by integration:

$$\int_{[A]_0}^{[A]_t} d[A] = -k\int_{t_1=0}^{t_2=t} dt, \tag{7}$$

$$[A]_t = [A]_0 - kt, \tag{8}$$

where $[A]_0$ is the concentration of A at $t = 0$. This relation produces a linear $[A]_t$ versus t plot.

A first-order reaction (i.e., first-order with respect to species A) is described by the equation

$$v = -\frac{d[A]}{dt} = k[A]^1. \tag{9}$$

As above, the change in $[A]_t$ with time is obtained by integration,

$$\int_{[A]_0}^{[A]_t} \frac{d[A]}{[A]} = -k \int_{t_1=0}^{t_2=t} dt, \tag{10}$$

$$\ln([A]_t/[A]_0) = -kt, \tag{11}$$

or

$$[A]_t = [A]_0 e^{-kt}, \tag{12}$$

and $[A]_t$ decreases exponentially with time. For a first-order reaction, the rate constant (k) is in units of t^{-1}. Notice that as t approaches $(1/k)$, the concentration of A decreases by a factor of 1/e, or approximately $\frac{1}{3}$.

Similarly, equations for reactions of order $\frac{1}{2}$, $\frac{3}{2}$, 2, and 3 may be derived; these are presented by Laidler (1987, p. 22).

In many investigations of the effect of temperature (T) on reaction rate, the results can be expressed by use of the equation

$$k = A e^{-E_a/RT}, \tag{13}$$

which is referred to as the 'Arrhenius equation'. Here E_a (the activation energy) can be equated to the height of the potential-energy barrier, and the pre-exponential factor (A) is in some way determined by the frequency of molecular collisions. The exact meanings of E_a and A for particular systems are often not fully understood (Laidler 1969).

Several theoretical treatments of reaction rate exist, but that which has been most successful is the activated-complex theory, which was introduced in 1935 by H. Eyring and by M.G. Evans and M. Polanyi (Laidler 1969). This theory considers reaction in terms of molecular collisions, and postulates the presence of a particular transient combination and arrangement of atoms which is referred to as an 'activated complex'. The energy of the complex, compared with that of the reactants, is identified as the potential-energy barrier for the reaction (Figure 4.1). The following brief description of the theory is from Laidler (1969).

Consider a reaction,

$$A + B \rightarrow X^{\ddagger} \rightarrow C, \tag{14}$$

where A, B, C are molecules, and $X^{\ddagger}$ is an activated complex. Suppose the reaction reaches equilibrium,

$$A + B \leftrightharpoons X^{\ddagger} \leftrightharpoons C, \tag{15}$$

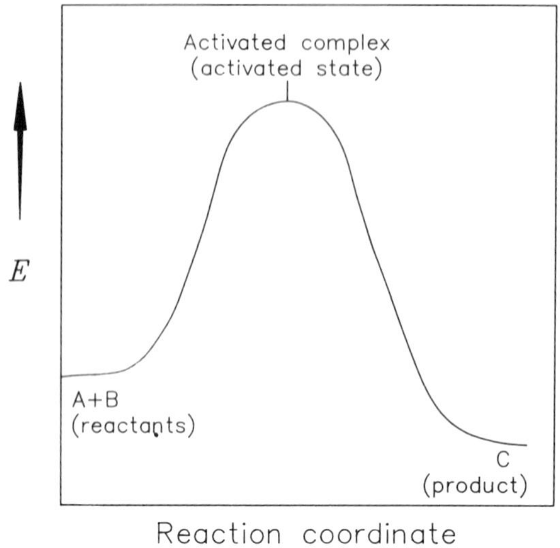

Figure 4.1 Potential energy (E) profile for reaction A + B → C. From Laidler (1969, *Theory of Chemical Reaction Rates*). Reproduced by permission of McGraw-Hill

i.e., a point at which the activated complex is at equilibrium with both reactants *and* products. The reaction may now be viewed as moving to the right and to the left at equal rates, and thus the rates of production of activated complexes from (A + B) and from C are equal.

The theory assumes that as the reaction is again caused to move to the right, a condition of 'equilibrium' is maintained between the activated complexes (those created from A and B, not from C), and the reactants, A + B. In general, the activated complexes are viewed as having a very short life-span, and once formed from A + B must necessarily change to C, not back to A + B.

The 'equilibrium' between the activated-complex state and the reactants may be defined by the equation

$$\frac{[X^{\ddagger}]}{[A][B]} = K^{\ddagger}, \tag{16}$$

where the quotient on the left is actually an activity quotient, but is here written as a concentration quotient, and $K^{\ddagger}$ is the 'equilibrium' constant. Now this constant may be expressed in terms of partition functions, denoted Q (Laidler 1969, p. 47) which are determined by the particular translational, rotational, and vibrational properties of the molecules A, B, and $X^{\ddagger}$, and the hypothetical energy of activation at absolute zero (E_0),

$$\frac{X^{\ddagger}}{[A][B]} = \frac{Q^{\ddagger}}{Q_A Q_B} e^{-E_0/RT}, \tag{17}$$

where R is the gas constant. Moreover, $Q^{\ddagger}$ is expressible as

$$Q^{\ddagger} = Q_{\ddagger}(kT/h\nu), \tag{18}$$

where $Q_{\ddagger}$ refers to a limited vibrational freedom for the activated complex, k is the Boltzmann constant, h is the Planck constant, and ν is the frequency of vibration of the activated complexes in the degree of freedom corresponding to their decomposition, i.e., ν is the frequency of decomposition. By substitution of (18) in (17) we obtain

$$\nu[X]^{\ddagger} = [\mathrm{A}][\mathrm{B}]\left\{\frac{kT}{h}\frac{Q_{\ddagger}}{Q_{\mathrm{A}}Q_{\mathrm{B}}}\mathrm{e}^{-E_0/RT}\right\}. \tag{19}$$

The term on the left is the frequency of decomposition of activated complexes within the system, i.e., the reaction rate, and the term in large brackets is the rate constant (k). From information on the properties of the reactant molecules, and a decision regarding the structure of the activated molecules, it is thus possible, by use of equation (19), to calculate rate constants at given temperatures. For many reactions, agreement between experimental results and theory has been satisfactory.

Reactions involving solids obviously must differ in many respects from those that occur in gas or liquid solutions. Consider, for example, a decomposition reaction,

$$\mathrm{A} \rightarrow \mathrm{B} + \mathrm{C},$$

where A and B are crystalline solids and C is a gas. As crystals of A are consumed (dissipate), new crystals of B are created. The new crystals will increase in size (grow) at rates that may be determined by the removal, rearrangement, or attachment of atoms at crystal surfaces (surface reaction) of by the transport of atoms and molecules to or from reaction sites (diffusion). Moreover, the geometry of reactant crystals, (i.e., crystal size, dislocations, grain boundaries), may influence reaction kinetics. The reactions are further complicated when metastable intermediary phases are formed.

From the above considerations, one must suppose that most reactions involving solids must proceed in two or more steps, the most obvious being nucleation followed by growth. It is commonly assumed (Kingery 1959) that under given conditions, each step, which may be identified as a microprocess, is capable of proceeding at a particular rate, and that the slowest of these steps governs the rate of the entire reaction. For example, the oxidation of a metal to produce a surface layer of oxide crystals is commonly observed to proceed at a parabolic rate (volume transformed proportional to time to the power $\frac{1}{2}$), a relation to be expected when the diffusion of oxygen through the surface layer governs the rate of reaction. On the other hand, the rate of oxidation of graphite, at certain temperatures, appears to be controlled by the rate of reaction at the graphite surface. These two kinds of reactions are referred to as 'diffusion-controlled' and 'surface-reaction-controlled' respectively. Burke (1965) has pointed out, however, that this distinction is meaningful only if one of the microprocesses is much slower than the others, and in some cases, two or more microprocesses, in cooperation, may govern the overall reaction rate.

In dealing with solid reactions, activated-complex theory is not always applied (Young 1966), but activated complexes may nevertheless be present and govern the rates of certain microprocesses. For example, the theory has been applied to diffusion in crystals, by viewing atoms, while they are moving from one site to another, as occupying an activated state.

What kind of reaction-rate equations can be anticipated for different kinds of solid-state reactions? Consider first a simple, homogeneous transformation, $\alpha \rightarrow \beta$, in which the product (β) forms domains within a reactant crystal (α). The ordering of Al and Si atoms in K-feldspar is an example of this kind of transformation. Suppose that the probability of transformation everywhere in α is the same; then the volume transformed in a short time interval is proportional to the volume of α remaining at that time (Christian 1975), i.e.,

$$\frac{\mathrm{d}V^{\beta}}{\mathrm{d}t} = k(V - V^{\beta}), \tag{20}$$

where V is the initial volume of α, and k is a constant. Upon integration,

$$(V^{\beta}/V) = 1 - \exp(-kt). \tag{21}$$

The symbol ζ (zeta) is now introduced to stand for the fraction transformed; thus,

$$\zeta = 1 - \mathrm{e}^{-kt}. \tag{22}$$

This is an exponential (first-order) relation, which, when plotted, produces a curve of the kind shown in Figure 4.2.

Most solid reactions of the kind $\alpha \rightarrow \beta$ proceed via the nucleation and growth of the product crystals (i.e., they are heterogeneous reactions) and a well-defined interface is formed between them. An example is the transformation of aragonite to calcite, which was investigated by Carlson and Rosenfeld (1981). Suppose that the product (β) nucleates at random within the reactant crystal (α) and that each nucleus grows at a constant linear rate. Suppose, further, that nucleation continues to occur as reaction progresses. Then the rate of reaction, initially, can be expressed in terms of the nucleation and growth rates, and obviously, numerous possibilities exist for both. As reaction progresses, more and more product crystals will impinge on each other, causing an overall decline in reaction rate. Avrami (1941) has suggested that all reactions of this kind (provided that the nucleation rate is constant or decreases with time) can be expressed approximately by the following equation (Christian 1975):

$$\zeta = 1 - \exp(-kt^{n}), \tag{23}$$

where k is a constant and $n > 1$. (With $n = 1$, equation (23) is the same as (22).) With n = 3, a plot of equation (23) produces the sigmoidal curve, shown

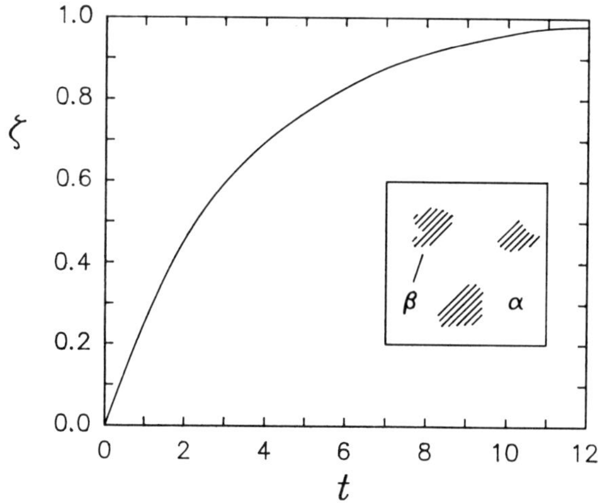

Figure 4.2 Fraction transformed (ζ) versus time (t) relation for a homogeneous transformation, $\alpha \rightarrow \beta$ (equation (22))

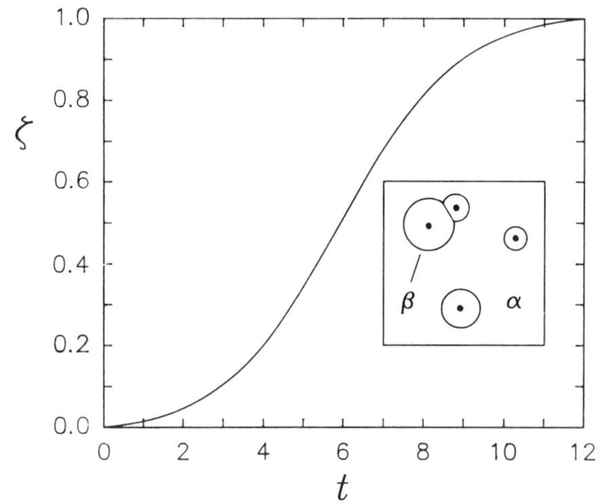

Figure 4.3 Fraction transformed (ζ) versus time (t) relation for a nucleation-and-growth transformation, $\alpha \rightarrow \beta$, with impingement (equation (23), with $n = 3$)

in Figure 4.3. The rate of reaction in terms of rates of nucleation and growth is examined in more detail below (Sec. 4.6).

Consider now a dehydration reaction,

$$\alpha \rightarrow \beta + H_2O,$$

or a similar reaction in which CO_2 or other gaseous molecules are released. Sources to some of the many reactions of this kind that have been studied experimentally are listed by Young (1966). In general, these are also nucleation-and-growth reactions, and sigmoidal reaction curves are expected, and have been verified. Much variation exists, however, and the initial acceleratory part of the idealized reaction is frequently missing; rather, the rate initially is constant, and then declines. Compared with the simpler phase transformations, described above, the transfer of H_2O away from reaction sites and the adsorption of H_2O on product crystals form additional microprocesses. The large body of theory that exists for reactions of this kind is reviewed by Young (1966).

4.2 EXPERIMENTAL STUDY OF MINERAL REACTIONS

Some experimental data on three devolatilization reactions are now presented to illustrate the kind of kinetic information that is obtainable, and to highlight some of the conclusions that may be drawn regarding the natural counterpart reactions. Compilations of experimental results, including general discussions of the kinetics of mineral reactions, were presented by Rubie and Thompson (1985) and Kerrick, Lasaga and Raeburn (1991).

4.2.1 Dehydration of brucite

A detailed study of the reaction

$$Mg(OH)_2 \rightarrow MgO + H_2O$$

was carried out by Anderson and Horlock (1962). Experiments were performed *in vacuo,* under conditions that do not occur naturally, but are most likely to provide information on reaction kinetics.

The starting material was synthetic $Mg(OH)_2$ as hexagonal discs, ~1000 Å in diameter, and natural brucite, as crystals several millimeters in diameter. Samples were heated for 1 h at 200 °C before increasing the temperature in less than 4 min to decomposition temperatures which were chosen between 247 and 302 °C. Reaction progress was determined by measuring the loss of weight as H_2O was released. Both reactants and products were examined by optical and scanning-electron microscopy, and by X-ray diffraction. Representative results are presented in Figure 4.4, where $\Delta W/W_0$ is relative weight loss, which must increase to 0.31 for complete decomposition. For values of $\Delta W/W_0$ extending to approximately 0.15, the reaction can be described by a linear (zero-order) equation,

$$(\Delta W/W) = kt. \tag{24}$$

A representative Arrhenius plot is presented in Figure 4.5. The Arrhenius equation,

$$k = A \exp[-E_a/RT], \tag{25}$$

is satisfied; for the synthetic crystals, $A = 0.5 \times 10^{-10}\ s^{-1}$ and $E_a = 27.6$ kcal/mol.

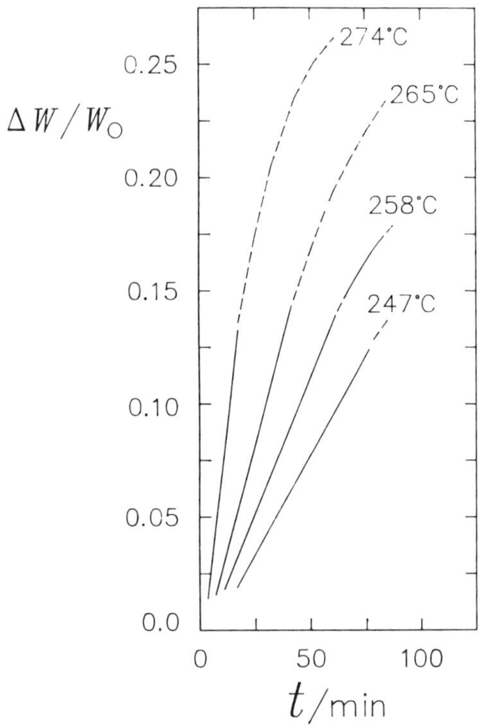

Figure 4.4 Typical decomposition curves for $Mg(OH)_2$ powder, in vacuo, and at the indicated temperatures; $\Delta W/W_0$ is the fraction of weight loss, resulting from release of H_2O. From Anderson and Horlock (1962)

From an examination of partially decomposed crystals, it was evident that reaction began at the 'edge' of the crystals, and progressed inwards, parallel to the basal plane, as shown in Figure 4.6. The product (MgO) crystallized as aggregates of very small crystals (~100 Å) which formed behind the reaction front, thus preserving the disc-like shape of the reactant crystals. H_2O produced at the interface migrated outwards.

For a disc-shaped crystal of brucite of height h (Figure 4.6), the initial volume of brucite ($t = 0$) is

$$V_0 = \pi r_0^2 h, \tag{26}$$

and the fraction of brucite decomposed at time t (denoted α) is

$$\alpha = 1 - (r^2/r_0^2), \tag{27}$$

thus,

$$(1-\alpha)^{1/2} = r/r_0. \tag{28}$$

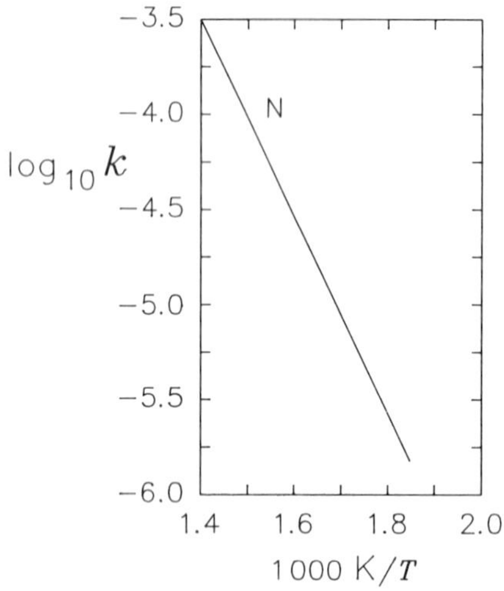

Figure 4.5 Arrhenius relation for natural brucite (N); k is the slope of linear plots as in Figure 4.4, and T is temperature. From Anderson and Horlock (1962)

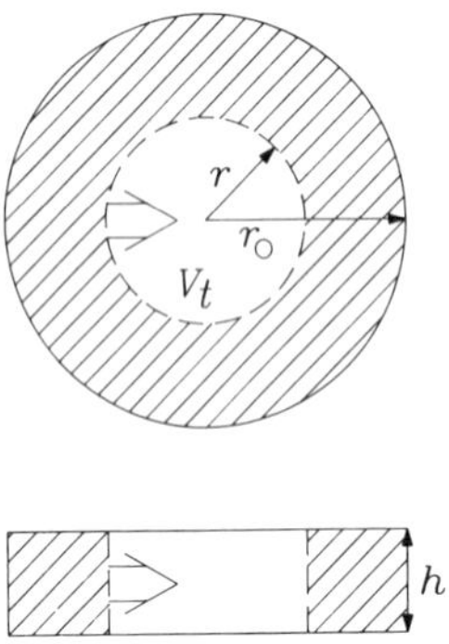

Figure 4.6 Idealized brucite crystal, showing position of reaction front, as modelled by Anderson and Horlock (1962). V_t is the volume of brucite remaining at time t.

The authors had observed microscopically that the rate of advance of the reaction interface (i.e., the rate of decrease of r) was very nearly constant, and they defined a 'growth'-rate constant, k', as

$$-\mathrm{d}r/\mathrm{d}t = k'. \tag{29}$$

By integration,

$$\int_{r_0}^{r} \mathrm{d}r = -k' \int_0^t \mathrm{d}t, \tag{30}$$

$$r - r_0 = -k't. \tag{31}$$

Dividing both sides by r_0, and by substitution of (28) in (31),

$$(1-\alpha)^{1/2} = 1 - (k't/r_0). \tag{32}$$

By this model, $(1-\alpha)^{1/2}$ bears a linear relation to t, and this was confirmed by the experimental data.

From the above and other results, Anderson and Horlock interpreted the decomposition of brucite as occurring in three steps: (1) during the initial stage, the brucite structure became somewhat perturbed, resulting possibly from some OH^- ions reacting to O^{2-} + H_2O; (2) removal of H_2O, which evidently occurred by diffusion from the reaction front (outward), parallel to (0001) of the partially destroyed brucite structure; and (3) nucleation and growth of MgO, lagging somewhat behind H_2O removal. The adsorption of H_2O on the surface of MgO crystals may explain the observed retardation in reaction rate during the second half of the reaction duration. Although the reaction is dominantly a nucleation-and-growth process, an early acceleration of rate (cf. Figure 4.3) was not detected, and the authors proposed that the removal of H_2O from the reaction sites, rather than the crystallization of MgO, was the rate-determining step.

More recent studies by Chaix-Pluchery et al. (1984) have clarified the nature of the precursor step (1), above. From conductivity and other measurements, it is apparent that an increase in H^+ (proton) mobility occurs nearly 100 K below the dehydration temperature, resulting in an increase in the fraction of $(OH)^-$ sites occupied by O^{2-} ions and H_2O molecules,

$$(OH)^- + (OH)^- = H_2O + O^{2-}. \tag{33}$$

From 75 to 15 K below the dehydration temperature, H_2O and O^{2-} become increasingly ordered in the hexagonal plane, followed by the outward diffusion of H_2O and the crystallization of MgO, as described above.

4.2.2 Calcite + quartz → wollastonite + CO_2

The important metamorphic reaction

$$CaCO_3 + SiO_2 \rightarrow CaSiO_3 + CO_2$$

was investigated by Gordon (1971), Kridelbaugh (1973), and Tanner, Kerrick and Lasaga (1985). In all of the experiments, particles of natural calcite and quartz were thoroughly mixed, and together with CO_2 (and in some cases H_2O) were placed in a cold-seal pressure vessel, and taken to combinations of temperature and pressure well within the stability field of wollastonite + CO_2. In general, the extent of reaction was determined by measuring the loss of weight of the reactants, i.e., the release of CO_2.

Examples of the results obtained by Kridelbaugh (1973) at 1.04 kbar CO_2 pressure and 800, 850, 900 °C (i.e., 100, 150, 200 °C above the equilibrium temperature),

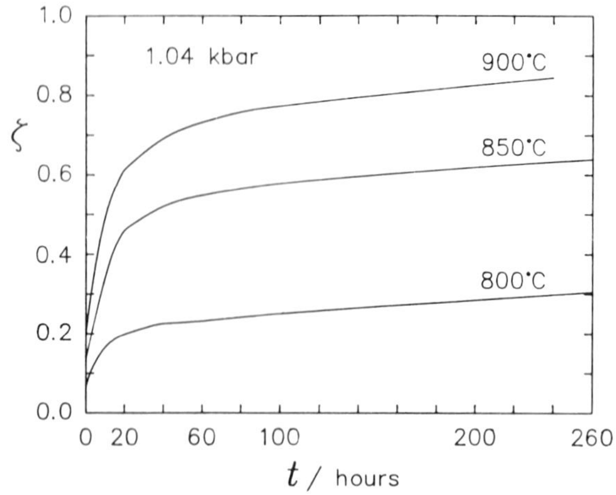

Figure 4.7 Reaction, calcite + quartz → wollastonite + CO_2; fraction transformed (ζ) as a function of time, at the indicated temperatures; initial size of reactant particles <10 μm. From Kridelbaugh (1973). Reprinted by permission of *American Journal of Science*

are reproduced in Figure 4.7. Particle size of the starting material was <10 μm. At each temperature, the reaction rate was found to be rapid, initially, followed by a decrease. The equation

$$\zeta = kt^n, \tag{34}$$

with $k = 0.1$ to 0.5 and $n = 0.30$ to 0.15 provides a satisfactory description of the data. This is a parabolic equation, but with $n < 1$, it describes a deceleration rather than an acceleration, which is expected for nucleation-and-growth reactions (Figure 4.3).

Microscopic examinations of the reaction products revealed the presence of wollastonite rims about particles of quartz. Accordingly, Kridelbaugh interpreted the rate results in terms of an initial rapid reaction at calcite–quartz contacts, followed by a decrease in rate, resulting from the need for a transfer of reactants through the wollastonite layer. A diffusion of Ca, Si, and possibly oxygen through this layer was proposed as the rate-determining microprocess.

A few experiments were carried out in the presence of a trace of H_2O, with the interesting result that wollastonite nucleated on calcite, not on quartz, thus confirming the earlier findings of Gordon (1971). These results may be viewed as a clear demonstration of the selective nature of the nucleation process.

Experimental data similar to those shown in Figure 4.7 were obtained by Tanner, Kerrick and Lasaga (1985), who attempted to fit their data to the equation

$$\zeta = kt^{1/2}, \tag{35}$$

which is the relation to be expected for a reaction controlled by one-dimensional diffusion. The fit is good except that the plotted curves do not pass through the origin and do not model adequately the rapid deceleration that occurs in the first 48 hours of reaction.

Tanner, Kerrick and Lasaga (1985) also obtained information on the reverse (carbonation) reaction. The Arrhenius relation calls for an increase in reaction rate with increasing temperature, and it is of interest to ask what happens when a system is taken away from equilibrium in the direction of decreasing temperature. Tanner et al. found that at 2 kbar and 3 kbar, a maximum in the rate of transformation of wollastonite to calcite + quartz occurred at about 100 degrees below the boundary curve. This was interpreted as resulting from a decrease in atomic mobility, which at low temperatures overshadows the increase in driving energy, ΔG. The results quantify the phenomenon of quenching (Christian 1975).

The Tanner et al. experiments were carried out in a CO_2 gas with approximately 4 mole % H_2O. In the prograde reaction, wollastonite was observed to nucleate on calcite (in agreement with the earlier results, described above). By the authors' interpretation, the diffusion of Si, possibly as H_4SiO_4, through an intergranular network in the wollastonite layer, is the rate-determining mechanism in the wollastonite-forming reaction.

4.2.3 Serpentine + brucite $\rightleftharpoons$ forsterite + H_2O

Rates of displacement of the equilibrium

$$Mg_3Si_2O_5(OH)_4 + Mg(OH)_2 = 2Mg_2SiO_4 + 3H_2O$$

were determined by Martin and Fyfe (1970) and Wegner and Ernst (1983). The hydration of olivine is a widespread, retrograde reaction in dunite, peridotite, and marble, and the dehydration of serpentine and brucite occurs during the prograde metamorphism of ultramafic rocks. The rate of displacement of the above equilibrium in both directions is therefore of interest.

The experiments of Wegner and Ernst were carried out by use of synthetic serpentine, brucite, and forsterite, of four particle-size fractions, all less than 150 μm. A cold-seal pressure vessel was used, in which mineral powders, together with H_2O, were taken to temperatures between 188 and 500 °C and pressures of 1, 2, and 3 kbar. These temperatures extend to approximately 100 degrees above and below the equilibrium boundary curve, which at 3 kbar is located at 400 °C. Extent of reaction was determined from the change in weight in the solid fraction.

The observed rates of dehydration are comparable to the rates of decarbonation found by Kridelbaugh, as described above. Thus for both dehydration and decarbonation at 100 °C above the corresponding boundary curves ($P = 1$ kbar, particle size <10 μm) 20% of reaction occurred in approximately 30 h.

With regard to the hydration reaction, examples of results obtained by Wegner and Ernst are reproduced in Figure 4.8, which shows that an increase in H_2O pressure from 1 to 3 kbar increases the reaction rate by a factor of 2.5. For a given pressure, the relation between fraction transformed and time is deceleratory, and Wegner and Ernst have chosen the Avrami 'nucleation-and-growth' equation

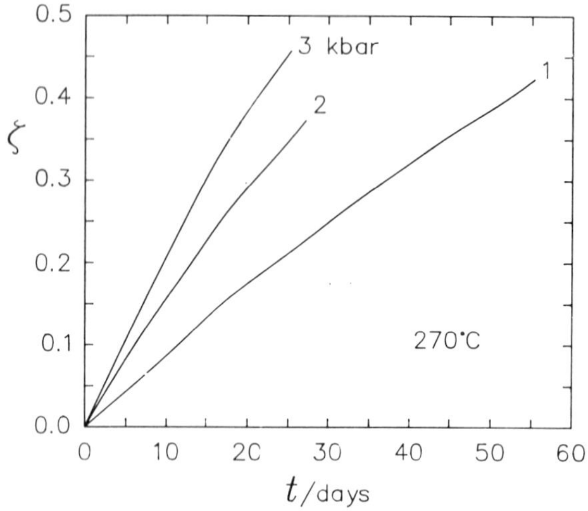

Figure 4.8 Reaction, 2 forsterite+3 $H_2O \rightarrow$ 1 serpentine+1 brucite; fraction transformed (ζ) as a function of time (t) at 270 °C and the indicated pressures; initial size of reactant particles, 105 to 149 μm. From Wegner and Ernst (1983). Reprinted by permission of *American Journal of Science*

(equation (23); Sec. 4.1) to describe the data:

$$\zeta = 1 - \exp(-kt^n), \tag{36}$$

with $n = 1$ (Figure 4.8). This equation is identical to (22) in Sec. 4.1, above, which describes a homogeneous reaction. Obviously, as was emphasized by Anderson and Horlock (1962) and Christian (1975), difficulties exist in identifying reaction mechanisms from empirical rate relations.

Reaction rates observed in all of the above experiments are remarkably rapid in relation to most geological events. Wegner and Ernst have calculated from their results that, in the presence of excess H_2O, a body of dunite of grain size 1 mm, resting under a pressure of 1 kbar at 310 °C, will be almost completely altered to serpentine and brucite in less than ten years. Evidently then, the presence of unaltered forsterite and wollastonite in surface exposures must be understood not in terms of slow rates of hydration and carbonation, but in terms of the absence of H_2O and CO_2 from the environment of these crystals.

4.3 DIFFUSION

4.3.1 Introduction to diffusion

The rearrangement of atoms and molecules that must occur in virtually all metamorphic processes implies a net displacement of these particles from one site to another across distances that range from a fraction of 1 mm to a few centimeters. Ramberg (1952a), in particular, has emphasized the importance of diffusion in bringing about these translations; more recent discussions are by Mueller (1967b), Fisher (1973), Dowty (1980), Brady (1983), and Joesten (1991a). Interest in diffusion has grown recently, as diffusional creep is considered increasingly as a mechanism for plastic flow within the Crust and Mantle. Much of our knowledge of diffusion in crystals comes from metallurgy, where the process is moderately well understood (Shewmon 1963, Borg and Dienes 1988).

In 1855, A. Fick suggested that the migration of atoms or molecules in a concentration gradient is analogous to the transfer of heat in a temperature gradient (as described by the Fourier equation), and he proposed the following relation which is here referred to as the 'first equation of diffusion':

$$J\,(\mathrm{g\ cm^{-2}\ s^{-1}}) = -D\,(\mathrm{cm^2\ s^{-1}})\frac{\mathrm{d}C\,(\mathrm{g\ cm^{-3}})}{\mathrm{d}x\,(\mathrm{cm})}. \tag{37}$$

This equation states that the flux (J) of mass in one direction (x) through unit cross-sectional area is proportional to the concentration gradient ($-\mathrm{d}C/\mathrm{d}x$); the proportionality constant (D) is referred to as the 'diffusion coefficient'. Commonly used units are given in parentheses. The first equation presents a definition of diffusion in terms of a concentration gradient; a more general definition is given below.

Diffusion is a common, everyday process, occurring for example in plants and soils, and during the oxidation of iron and the weathering of rock. The process, as it occurs in liquids can be demonstrated by placing a crystal of potassium permanganate or sodium cobaltinitrite on a sheet of wet filter paper, and observing the outward migration of the color-producing ions.

It is important to distinguish between diffusion and fluid flow (Jost 1960). Diffusion is a chemical process, driven by chemical-potential gradients, whereas the flow of a gas through a permeable rock is a physical process, occurring in response to a pressure gradient. Rates of diffusion for different molecules can vary greatly and the rates are sensitive to temperature. By comparison, the rate of permeation of different gases through a given solid does not vary greatly, and the rates are only slightly dependent on temperature. In a diffusion process, the movement of molecules is largely random, and different molecular species may experience a net translation in opposite directions; in fluid flow this is not possible. Although the term 'flow' is used in describing diffusion processes, the term 'drift', as used by Le Claire (1976), would appear to be more appropriate.

How do atoms and molecules move within and through crystals? From indirect evidence (Le Claire 1976) transfer is facilitated by the presence of vacancies, including vacant lattice sites and 'interstitial sites'. Edge dislocations and crystal boundaries (Sec. 1.4) provide paths along which diffusion may occur with relative ease.

An interpretation of the diffusion coefficient (D) as it appears in the first equation (37) may be obtained in terms of atomic movements. Following Shewmon (1963), consider a negative concentration gradient of atoms A within a crystal, in the $+x$ direction (Figure 4.9). The number of atoms of A (per unit area) on plane 1 is N_1 and on plane 2 is N_2. Atoms A could be neutral atoms occupying interstitial

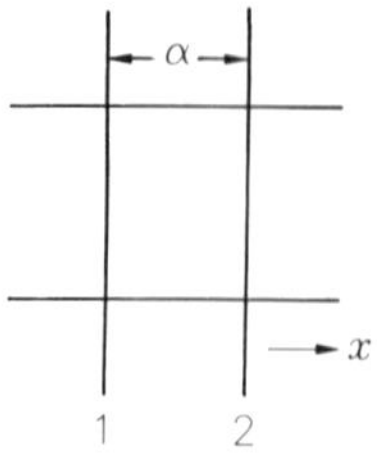

Figure 4.9 Arrangement for discussion of diffusion in terms of random jumps of atoms from one plane to another.

positions, e.g., carbon in iron. Let Γ be the frequency of jumps of atoms A from one plane to the next, and suppose the jumps occur randomly, i.e., jumps to the left and right have the same frequency. Then half of the atomic jumps (per second) from plane 1 are to the right, and half of the atomic jumps from plane 2 are to the left, and the net flux is

$$J = \left(\frac{1}{2}\right) N_1\Gamma - \left(\frac{1}{2}\right) N_2\Gamma = \left(\frac{1}{2}\right) \Gamma(N_1 - N_2). \tag{38}$$

The concentration (number of atoms per unit volume) of A at plane 1 is $C_1 = N_1/\alpha$ and at plane 2 is $C_2 = N_2/\alpha$, where α is the distance between planes of atoms. By substitution,

$$J = \left(\frac{1}{2}\right) (C_1 - C_2)\alpha\Gamma. \tag{39}$$

Assuming steady-state conditions,

$$C_1 - C_2 = -\alpha\frac{\partial C}{\partial x}, \tag{40}$$

and, therefore,

$$J = -\left(\frac{1}{2}\right) \alpha^2\Gamma\frac{\partial C}{\partial x}. \tag{41}$$

Thus by comparison of (41) with (37), the diffusion coefficient (D) may be expressed in terms of atomic jump frequency (Γ) and jump distance (α),

$$D = \left(\frac{1}{2}\right) \alpha^2\Gamma. \tag{42}$$

Notice that atomic movement in this model is random, i.e., atoms of A do not show a preference for movement in the down-concentration direction. The observed net flow to the right is simply the result of more atoms of A residing on planes to the left than on planes to the right (Figure 4.9).

At 900 °C, D (carbon in Fe) $\sim 10^{-6}$ cm^2 s^{-1}. Taking α as $\sim 10^{-8}$ cm, then from equation (42), $\Gamma \sim 10^{10}$ s^{-1}, i.e., each carbon atom changes place about 10^{10} times per second. This number, although large, is considerably less than the Debye vibrational frequency (10^{12} to 10^{13} s^{-1}), and the carbon atoms do occupy their interstitial sites most of the time.

The following five examples will serve to illustrate the many possibilities that exist for diffusion in metamorphic terranes.

In a crystal of graphite, each carbon atom is in a state of constant thermal agitation; hence, provided the temperature is sufficiently high, each atom migrates from point to point within the crystal. This random walk (Le Claire 1976) which occurs in the absence of a concentration gradient, is referred to as 'self-diffusion'. Synthetic isotopes are now widely used to measure rates of self-diffusion.

The introduction of H_2O molecules to channels in cordierite (Figure 1.1) is an example of the diffusion of particles of neutral charge in one direction, down a concentration gradient. Pre-existing 'vacant' sites are available in the channels, and sites need not be created by the displacement of other particles.

The homogenization of a zoned crystal of (Fe, Mn) garnet is an example of exchange diffusion. The atoms presumably move as ions, and charge-balance requirements demand an equal flux of Fe and Mn atoms in opposite directions; the concentration of vacancies is assumed to remain uniform. Although both Fe and Mn atoms move down their respective concentration gradients, it is more rewarding to view the process in terms of chemical potential gradients. A levelling of these gradients results in a decrease in the Gibbs energy of the crystal, by an amount determined by the energy of mixing.

Exchange diffusion also occurs during exsolution. For example, during the exsolution of dolomite from magnesian calcite, Mg ions presumably diffuse through the calcite crystal to the calcite–dolomite interface, while Ca ions migrate in the opposite direction. Meanwhile the calcite–dolomite interface migrates into calcite. Although Mg atoms migrate toward the Mg-rich phase, dolomite, they do in fact move down a concentration gradient in the surrounding calcite; similarly, Ca ions move down a Ca concentration gradient. The process stops when the chemical potential (not concentration) of $Ca_{0.5}Mg_{0.5}CO_3$ is the same in both phases, or when falling temperature greatly reduces ionic mobility.

A complex rearrangement of atoms must occur during reactions such as chlorite + quartz = garnet + H_2O. Billions of atoms of different kinds (including trace elements) are relocated across distances up to a few centimeters. Although some geologists (e.g., Walther and Wood 1984) have argued that circulating aqueous fluids, acting as solvents, can bring about these relocations, it is possible that all of them, including the removal of H_2O from reaction sites, are effected predominantly by a diffusion mechanism.

4.3.2 Solutions to the first equation of diffusion

Although the first equation, as written above, i.e.,

$$J = -D\frac{\mathrm{d}C}{\mathrm{d}x}, \tag{43}$$

appears not to be directly applicable to diffusion in silicates and other largely ionic crystals, it does nevertheless lead to some equations that are generally useful. In

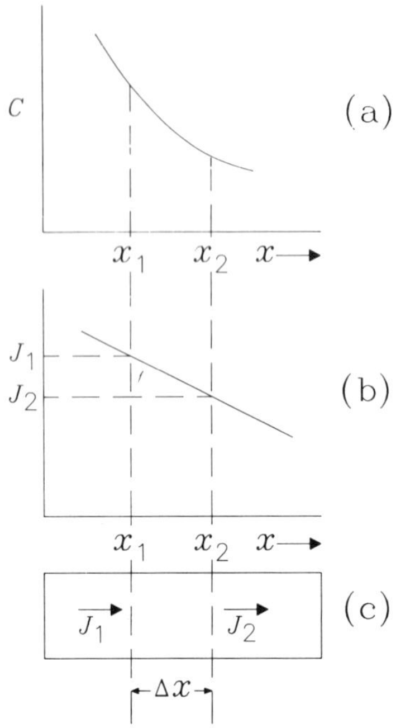

Figure 4.10 (a) A concentration (C) gradient that varies with distance (x); (b) the flux (J) at x_1 and x_2; (c) flux J_1 entering and J_2 leaving an element of volume. From Shewmon (1963)

deriving these equations, one must assume the somewhat unsatisfactory position of viewing crystals as a continuum.

Equation (43) could be used to measure D for a system in which the concentration gradient is the same at all points along x. But in many diffusion experiments, concentration gradients (at a given time) are a function of x, as shown in Figure 4.10. This implies a change in flux along the x direction. Suppose that at x_1 the flux is J_1 and at x_2 it is J_2. Provided that $\Delta x (= x_2 - x_1)$ is small,

$$-\frac{\mathrm{d}J}{\mathrm{d}x} = -\frac{(J_2 - J_1)}{\Delta x} = \frac{J_1 - J_2}{\Delta x}, \tag{44}$$

or

$$J_1 - J_2 = -\Delta x \frac{\mathrm{d}J}{\mathrm{d}x}. \tag{45}$$

The net increase in the mass of the diffusing particles within a volume delineated by unit cross-sections at x_1 and x_2, separated by a distance of Δx is $1(\Delta x)(\mathrm{d}C/\mathrm{d}t)$, i.e.,

$$J_1 - J_2 = \Delta x \frac{\mathrm{d}C}{\mathrm{d}t}, \tag{46}$$

or

$$\frac{\mathrm{d}C}{\mathrm{d}t} = -\frac{\mathrm{d}J}{\mathrm{d}x}. \tag{47}$$

Now by differentiation of (43), assuming D remains constant,

$$\frac{\mathrm{d}J}{\mathrm{d}x} = -D\frac{\mathrm{d}^2C}{\mathrm{d}x^2}, \tag{48}$$

and by substitution of (47) in (48),

$$\frac{\mathrm{d}C}{\mathrm{d}t} = D\frac{\mathrm{d}^2C}{\mathrm{d}x^2}. \tag{49}$$

This equation, which is more useful than (43), is referred to as the 'second equation of diffusion'; the derivation here is from Shewmon (1963). The equation can be rewritten as

$$\frac{\mathrm{d}C}{\mathrm{d}t} = \frac{\mathrm{d}}{\mathrm{d}x}\left\{D\frac{\mathrm{d}C}{\mathrm{d}x}\right\}, \tag{50}$$

to allow for a variation in D.

The above considerations may be extended to radial diffusion in a plane or to three-dimensional diffusion in a crystal. For diffusion in three-dimensional space, defined by coordinates x, y, z, equation (49) becomes

$$\frac{\mathrm{d}C}{\mathrm{d}t} = D\left\{\frac{\partial^2C}{\partial x^2} + \frac{\partial^2C}{\partial y^2} + \frac{\partial^2C}{\partial z^2}\right\}, \tag{51}$$

assuming D is constant. Much of the mathematics of diffusion, as presented by Crank (1975), is devoted to a solution of the second equation (49), or analogous equations such as (51), for different physical systems and different boundary conditions.

Consider, for example, the one-directional diffusion of a component from a vapor phase (α) in which the concentration is maintained at C_1 (e.g., by convection) into a solid phase (β) in which the initial concentration is zero. The concentration (C) versus distance (x) profile in phase β after a duration of time (t) is obtained by solution of equation (49) (Shewmon 1963), to give

$$C = C_1\left\{1 - \operatorname{erf}\left(\frac{x}{2(Dt)^{1/2}}\right)\right\}. \tag{52}$$

Here D is the diffusion coefficient (assumed constant) and erf is the error function; values of erf z for increments of z are listed in Table 4.1. In Figure 4.11, equation (52) is plotted for three arbitrary values of t. In experimental studies, concentration profiles of this kind are detected by chemical analysis, and the appropriate diffusion equation (e.g., equation (52)) is used to obtain values for the diffusion coefficient.

The above examplc brings out an important attribute of diffusion, viz. that the distance travelled by the diffusing atoms or molecules varies greatly. Thus, after a given interval of time, some atoms will have travelled very far from the α–β interface, while others will have penetrated the β phase by only a very small distance (Figure 4.11). It is therefore convenient to define a mean displacement ($\overline{x}$)

Table 4.1 The error function

z	erf(z)	z	erf(z)
0	0	0.85	0.7707
0.025	0.0282	0.90	0.7970
0.05	0.0564	0.95	0.8209
0.10	0.1125	1.0	0.8427
0.15	0.1680	1.1	0.8802
0.20	0.2227	1.2	0.9103
0.25	0.2763	1.3	0.9340
0.30	0.3286	1.4	0.9523
0.35	0.3794	1.5	0.9661
0.40	0.4284	1.6	0.9763
0.45	0.4755	1.7	0.9838
0.50	0.5205	1.8	0.9891
0.55	0.5633	1.9	0.9928
0.60	0.6039	2.0	0.9953
0.65	0.6420	2.2	0.9981
0.70	0.6778	2.4	0.9993
0.75	0.7112	2.6	0.9998
0.80	0.7421	2.8	0.9999

From Shewmon (1963) where sources are given. The error function is defined as

$$\mathrm{erf}(z) = \frac{2}{\pi^{1/2}} \int_0^z \mathrm{e}^{-n^2}\, \mathrm{d}n.$$

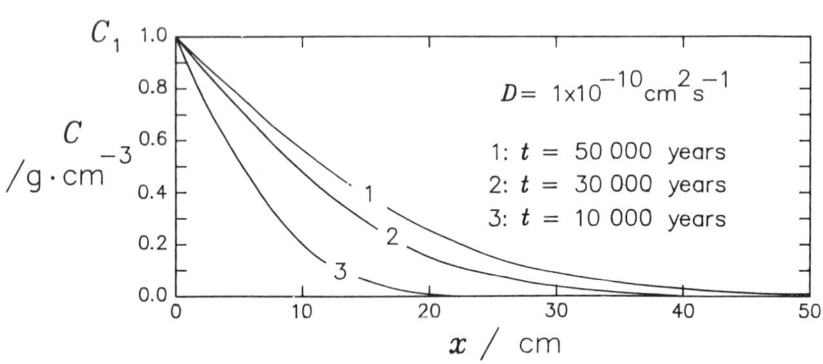

Figure 4.11 Plot of the diffusion equation, $C = C_1[1 - \mathrm{erf}(x/2(Dt)^{1/2})]$ for the indicated durations of time (t), and a diffusion coefficient (D) of 1×10^{-10} cm^2 s^{-1}; C is concentration

corresponding to the distance (x) from the α–β interface at which the concentration of the diffusing atoms has increased from zero to $C_1/2$. At this point (Figure 4.11),

$$(C_1/2) = C_1 \left\{ 1 - \mathrm{erf}\left(\frac{\overline{x}}{2(Dt)^{1/2}} \right) \right\}, \tag{53}$$

and

$$\frac{1}{2} = \text{erf}\left\{\frac{\bar{x}}{2(Dt)^{1/2}}\right\}. \tag{54}$$

From Table 4.1, $(\frac{1}{2}) \sim \text{erf}(0.5)$, and therefore

$$\frac{1}{2} \sim \frac{\bar{x}}{2(Dt)^{1/2}}, \tag{55}$$

and

$$\bar{x} \sim (Dt)^{1/2}. \tag{56}$$

Thus, penetration, as measured by $\bar{x}$, is proportional to $t^{1/2}$ and the rate of penetration (obtained by differentiation of (56), is proportional to $1/t^{1/2}$, i.e., it decreases with time. In Figure 4.12, equation (56) is plotted for increments of D.

Values of D for diffusion in crystals at temperatures of medium- to high-grade metamorphism range from 10^{-6} to 10^{-20} $cm^2\ s^{-1}$. Data for several minerals have been compiled by Freer (1981). Diffusion in liquids occurs with greater ease.

Different values of D for different crystallographic directions (D_x, D_y, D_z) in non-isotropic crystals can be accommodated by the above theory. Consider, for example, a concentration gradient in an arbitrary direction r within an orthorhombic crystal. A diffusion coefficient, D_r, for diffusion parallel to r is defined as

$$J_r = -D_r(\partial C/\partial r). \tag{57}$$

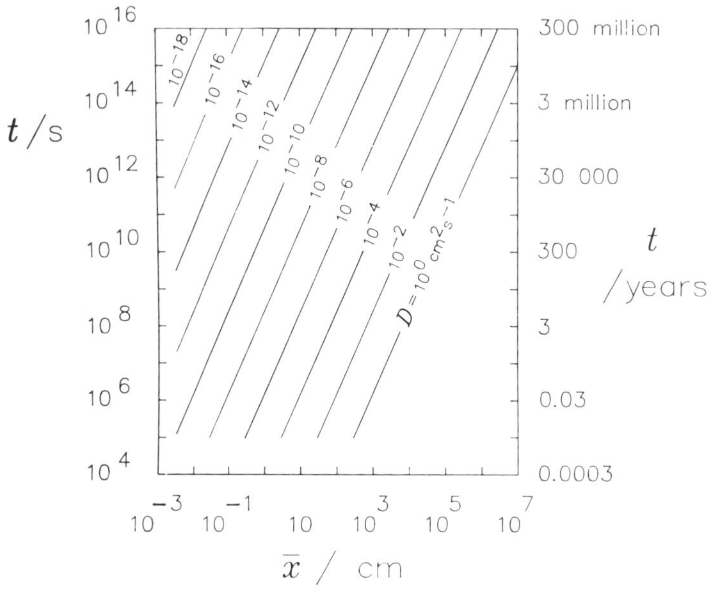

Figure 4.12 Plot of equation $\bar{x} = (Dt)^{1/2}$ for indicated values of the diffusion coefficient (D); $\bar{x}$ is mean displacement, i.e., x at $C = C_1/2$ (see Figure 4.11). From the author's unpublished notes distributed at a conference in 1974, with modifications

Now from Le Claire (1976),

$$D_r = l^2 D_x + m^2 D_y + n^2 D_z \tag{58}$$

where l, m, n are direction cosines of r with regard to x, y, z. Thus D for any direction can be expressed in terms of D_x, D_y, D_z.

From experimental studies it is clear that diffusion coefficients increase rapidly with an increase in temperature (T) and the relation can, in general, be expressed by use of the Arrhenius equation,

$$D = D_0 \exp -(E/RT). \tag{59}$$

Here D_0 and E are constants, characteristic of the system, and R is the gas constant. E, which is identified as the activation energy for diffusion, is obtained experimentally from the slope of the linear plot of ln D against $1/T$.

The time needed for concentration gradients in a crystal to 'level off', i.e., for the crystal to become homogeneous, is a question of importance, particularly with regard to crystals of garnet, plagioclase, and other minerals in which chemical zoning is produced during crystal growth. This process of diffusional homogenization is known as 'relaxation'. Consider, for example, an initial ($t = 0$) concentration profile as shown in Figure 4.13, which is a graph of the equation

$$C(-d/2 \leq x \leq 3d/2, t = 0) = \overline{C} + C' \sin(\pi x/\mathrm{d}). \tag{60}$$

We wish to determine, for a given diffusion coefficient (D), the time needed for the gradients to disappear, resulting in a uniform concentration, $C = \overline{C}$ (Figure 4.13). In this case, according to Shewmon (1969), the solution to the second equation of diffusion is

$$C(x,t) = \overline{C} + C' \sin\left\{\frac{\pi x}{d}\right\} \exp\left\{\frac{-Dt\pi^2}{d^2}\right\}. \tag{61}$$

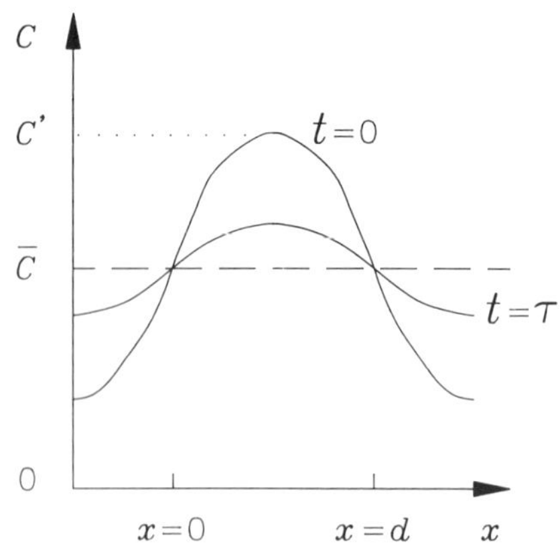

Figure 4.13 Diffusional relaxation of a concentration (C) versus distance (x) profile; full lines are initial profile ($t = 0$) and profile at time τ ($\tau = d^2/(D\pi^2)$). From Shewmon (1969). Note that C' is displacement of concentration from $\overline{C}$ at $x = d/2$

The change in concentration at $x = d/2$ is

$$C(d/2, t) - \overline{C} = C' \exp(-Dt\pi^2)/d^2. \quad (62)$$

The relaxation time is defined as $\tau = d^2/(D\pi^2)$. After a duration of τ seconds,

$$C(d/2, \tau) - \overline{C} = C' \exp(-1/1) = C'(0.368), \quad (63)$$

and, in general,

$$C(x, \tau) - \overline{C} = C'(x, t = 0)(0.368), \quad (64)$$

to give the concentration profile shown as the $t = \tau$ line in Figure 4.13. For example, with $D = 10^{-16}$ cm^2 s^{-1} and $d/2 = 0.1$ cm, the relaxation time is 10^{13} s, or 3×10^5 years. A period of time of approximately 6τ or 20×10^5 years is needed to make this crystal homogeneous to the extent that a variation in composition would be difficult to detect by microprobe analysis. If conditions are not isothermal, these estimates of time would be smaller or larger, depending on whether the temperature rises (increase in D) or falls (decrease in D).

4.3.3 Steady-state theory applied to diffusion problems

A general theory for changes in systems removed from equilibrium was developed by Onsager (1931). This theory deals with mass diffusion in terms of chemical potential, rather than concentration gradients, and accommodates the simultaneous flow of heat and mass, and the influence of one on the other. The theory is, in general, more satisfying than that presented above, and, incidentally, explains some known occurrences of diffusion up (rather than down) a concentration gradient. The theory is not, however, without difficulties; these include the somewhat vague (though mathematically precise) definition of chemical potential, as presented in Sec. 2.4, the difficulty of assigning numbers to this potential, and the finding that in minerals (unlike metals), diffusing atoms are not, in general, solution components. Aspects of the theory were first introduced to the geological literature by Fisher (1973); the following treatment is from Denbigh (1951).

The first equation of diffusion can be rewritten as

$$\mathbf{J} = L\mathbf{X}, \quad (65)$$

where $\mathbf{J}$ is a velocity vector and $\mathbf{X}$ is a chemical-potential gradient (the thermodynamic force); L is a scalar quantity analogous to the diffusion coefficient.

Consider a system in which flows of certain atoms ($\mathbf{J}_2$) and of heat ($\mathbf{J}_1$) are occurring simultaneously, and assume that $\mathbf{J}_2$ and $\mathbf{J}_1$ do not change, i.e., steady-state conditions are realized. Then,

$$\mathbf{J}_2 = L_{22}\mathbf{X}_2 + L_{21}\mathbf{X}_1, \quad (66)$$

where the second term on the right has been added to allow for the effect of the flow of heat on the flow of atoms. Similarly,

$$\mathbf{J}_1 = L_{11}\mathbf{X}_1 + L_{12}\mathbf{X}_2, \quad (67)$$

where $\mathbf{J}_1$ is the flux of heat, $\mathbf{X}_1$ is the force causing the flow of heat (i.e., the temperature gradient), and the second term on the right allows for the effect of the flow of atoms on the flow of heat. Thus,

$$\begin{aligned} \text{heat:}\quad \mathbf{J}_1 &= L_{11}\mathbf{X}_1 + L_{12}\mathbf{X}_2, \\ \text{mass:}\quad \mathbf{J}_2 &= L_{21}\mathbf{X}_1 + L_{22}\mathbf{X}_2. \end{aligned} \tag{68}$$

Where two kinds of atoms (2, 3) are diffusing simultaneously, together with a flow of heat, we have, in place of (68),

$$\begin{aligned} \text{heat:}\quad \mathbf{J}_1 &= L_{11}\mathbf{X}_1 + L_{12}\mathbf{X}_2 + L_{13}\mathbf{X}_3, \\ \text{mass:}\quad \mathbf{J}_2 &= L_{21}\mathbf{X}_1 + L_{22}\mathbf{X}_2 + L_{23}\mathbf{X}_3, \\ \text{mass:}\quad \mathbf{J}_3 &= L_{31}\mathbf{X}_1 + L_{32}\mathbf{X}_2 + L_{33}\mathbf{X}_3, \end{aligned} \tag{69}$$

and, in general,

$$\mathbf{J}_i(i = 1, 2, \ldots, k) = \sum_k L_{ik}\mathbf{X}_k. \tag{70}$$

The presence of cross-coefficients (e.g., L_{12}, L_{21}) in these equations implies a coupling of heat flow and diffusion. It follows that a temperature gradient is expected to cause diffusion (the Soret effect) and that diffusion is expected to cause a small temperature gradient (the Dufour effect). The cross-coefficients (L_{23}, L_{32}) allow for an influence of the flow of atoms of kind 3 on the flow of the others of kind 2, and vice versa.

An important part of the Onsager theory is the assumption that the cross-coefficients are equal. Thus,

$$\begin{aligned} L_{12} &= L_{21}, \\ L_{13} &= L_{31}, \\ L_{32} &= L_{23}. \end{aligned} \tag{71}$$

This introduces a symmetry to the equations, e.g.,

$$\left(\frac{\partial \mathbf{J}_1}{\partial \mathbf{X}_2}\right)_{\mathbf{X}_1} = \left(\frac{\partial \mathbf{J}_2}{\partial \mathbf{X}_1}\right)_{\mathbf{X}_2}. \tag{72}$$

In dealing with changes that occur at some distance from equilibrium, the rate of production of entropy is commonly used to describe these changes. Onsager therefore introduced the proposal that the thermodynamic forces ($\mathbf{X}_1$, $\mathbf{X}_2$) be chosen to satisfy the equation

$$T\Theta = \mathbf{J}_1\mathbf{X}_1 + \mathbf{J}_2\mathbf{X}_2, \tag{73}$$

Where Θ is the rate of creation of entropy per unit volume, and T is absolute temperature. Thus, in general,

$$\Theta = (1/T)\sum_i \mathbf{J}_i\mathbf{X}_i, \tag{74}$$

and for isothermal diffusion,

$$\Theta = -(1/T)\sum_i \mathbf{J}_i \frac{\partial \mu_i}{\partial x}. \tag{75}$$

Denbigh (1951) has shown that this expression follows naturally from the equations of equilibrium thermodynamics.

The exsolution of alkali feldspar by a nucleation-and-growth mechanism is an example of a diffusion problem that can be examined in terms of chemical-potential gradients. At equilibrium, the chemical potential of $NaAlSi_3O_8$ (not the concentration) is equal in the two exsolved phases, albite and K-feldspar, and similarly for component $KAlSi_3O_8$. As temperature decreases and exsolution progresses, K must diffuse into K-feldspar, in exchange for Na, and Na must diffuse into albite in exchange for K. All ions diffuse down concentration gradients, but one could view gradients in chemical potential as the ultimate cause of diffusion. Moreover, in a detailed study, one could examine the influence of a flow of heat (which accompanies solution and exsolution processes), on the diffusion of atoms.

4.3.4 Exchange and self-diffusion

The rate of exchange diffusion (also known as 'interdiffusion' or 'chemical diffusion') compared to the rates of self-diffusion in crystals of variable composition is a problem of some interest. With regard to a metallic mixture, (A, B), where A and B are different kinds of atoms, Darken (1948) derived the following equation:

$$\tilde{D} = (D_A^* X_B + D_B^* X_A)\left\{\frac{d\ln a_A}{d\ln X_A}\right\}. \tag{76}$$

Here $\tilde{D}$ is the diffusion coefficient for A–B exchange diffusion in a crystal of composition X_A (= mole fraction A/(A + B)), and D_A^* and D_B^* are diffusion coefficients for the self-diffusion of A and B in pure A and pure B respectively; a_A is the activity of component A. In deriving this equation, cross-coefficients were assumed to be zero and the vacancy concentration was assumed to remain constant. For ideal or dilute solutions ($\gamma = 1.0$ or a constant, respectively) the term $d\ln a/d\ln X$, drops out of the equation.

A test of equation (76) is provided by the following experimental data (at 1000 °C) on (Ag, Au) at $X_{Ag} = 0.50 : \tilde{D} = 9.7 \times 10^{-9}, D_{Ag}^* = 8.2 \times 10^{-9}, D_{Au}^* = 3.2 \times 10^{-9}$ cm^2 s^{-1}; using equation (76), $\tilde{D}$ (calculated) = 9.3×10^{-9} cm^2 s^{-1}, slightly less than 9.7×10^{-9} cm^2 s^{-1} obtained experimentally (Shewmon 1963).

With regard to ionic crystals, e.g., (A, B) M, in which the charge on ion A is Z_A and on B is Z_B, Barrer and Rees (1964) derived the following equation for A–B exchange diffusion within the crystals:

$$\tilde{D} = D_A^* D_B^* \left(\frac{C_B Z_B^2 \dfrac{\partial \ln a_A}{\partial \ln C_A} + C_A Z_A^2 \dfrac{\partial \ln a_B}{\partial \ln C_B}}{D_A^* C_A Z_A^2 + D_B^* C_B Z_B^2} \right), \tag{77}$$

where C is concentration. Provided that the charge on ions A and B is equal ($Z_A = Z_B$), and that

$$\frac{\partial \ln a_A}{\partial \ln C_A} = \frac{\partial \ln a_B}{\partial \ln C_B},$$

equation (77) becomes the Manning (1968) equation,

$$\tilde{D} = \frac{D_A^* D_B^*}{D_A^* X_A + D_B^* X_B} \left\{ \frac{\partial \ln a_A}{\partial \ln X_A} \right\}, \tag{78}$$

where X_A and X_B are ionic or molar fractions. Equation (78) is similar to (76) and becomes identical to (76) when $D_A^* = D_B^*$. In view of uncertainties in D^* and activity, as determined experimentally, equation (78) can provide only a rough estimate of $\tilde{D}$. With regard to olivine, good agreement between $\tilde{D}$ (measured) and $\tilde{D}$ (calculated) was, however, obtained by Hermeling and Schmalzried (1984).

4.3.5 Crystal-boundary diffusion

In most models of crystal boundaries (Sec. 1.5) the structure of the boundary is relatively open, and diffusion through the boundary is therefore expected to be more rapid than through the adjoining crystals. This expectation was confirmed by Hoffman and Turnbull (1951) who used the experimental arrangement shown in Figure 4.14 to study the self-diffusion of Ag along Ag–Ag boundaries.

Le Claire (1963) and Kaur and Gust (1989) have compared the different procedures that have been used to obtain estimates of boundary-diffusion coefficients (D') from penetration profiles. In general, the boundary is viewed as a slab of

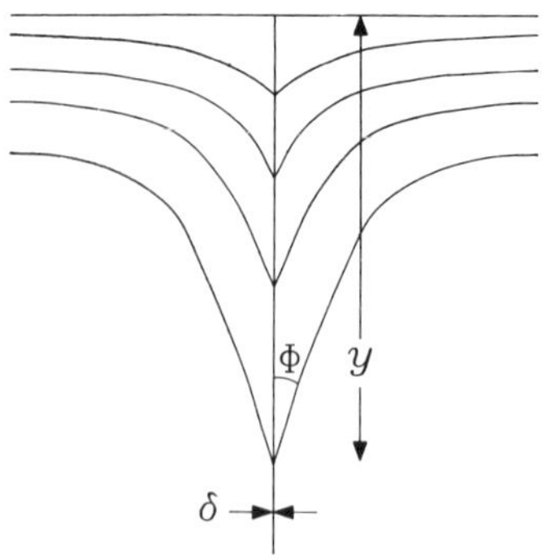

Figure 4.14 Typical concentration profiles resulting from enhanced diffusion on a crystal boundary; δ denotes boundary width. From Le Claire (1963)

uniform width, δ, and it is assumed that atoms diffuse into the adjacent crystals from the slab and from the surface on which the diffusant was deposited. Thus the diffusion equation is

$$\frac{\partial C}{\partial t} = D\left\{\frac{\partial^2 C}{\partial y^2} + \frac{\partial^2 C}{\partial x^2}\right\}, \tag{79}$$

where D is the coefficient for volume diffusion, assumed constant. The solution to this equation is facilitated by introducing the following dimensionless quantities:

$$\eta = \frac{y}{(Dt)^{1/2}}, \quad \beta = \frac{D'}{D}\frac{(1/2)\delta}{(Dt)^{1/2}}.$$

It is of interest to examine, first, the relation between ϕ (Figure 4.14) and β. From an approximate analysis by Fisher (1951),

$$\cot\phi = \pi^{-1/4}\beta^{1/2}. \tag{80}$$

Thus, deep penetration on the boundary (ϕ small) is obtained when β is small, but the cusp becomes less pronounced with time, i.e., increasing t causes a decrease in β. In experimental studies, β should ideally be greater than 10.

In one solution to the diffusion equation (Le Claire 1963), the mean concentration (boundary slab plus adjacent crystals), denoted $\overline{C}$, as a function of depth, y, is used to evaluate the product, $D'\delta$:

$$D'\delta = \left\{\frac{\partial \ln \overline{C}}{\partial y^{6/5}}\right\}^{-5/3}\left\{\frac{4D}{t}\right\}^{1/2}\left[\left\{\frac{\partial \ln \overline{C}}{\partial(\eta\beta^{-1/2})^{6/5}}\right\}^{5/3}\right], \tag{81}$$

where D is the coefficient for volume diffusion and t is time. Empirically, the term in square brackets is found to be nearly constant, at 0.661, provided β is larger than 10; thus $D'\delta$ is obtained from a linear plot of ln $\overline{C}$ against $y^{6/5}$. In experimental studies, δ is rarely known with confidence, but estimates of δ (e.g., 7 Å for NiO–NiO boundaries, reported by Atkinson and Taylor (1986)) provide estimates for D'.

In metal and oxide systems D' generally exceeds D by a large factor. For example, at 500 °C, grain-boundary diffusion of Ag on Ag–Ag boundaries exceeds volume diffusion by a factor of approximately 10^5 (Hoffman and Turnbull 1951). Thus the activation energy for boundary diffusion is commonly less than for volume diffusion; for example, in the diffusion of Ni, Co, and Cr on NiO–NiO boundaries, the factor is 0.68 to 0.79 (Atkinson and Taylor 1986). The few data that exist at present on boundary diffusion in silicate systems were compiled by Joesten (1991a); an example of diffusion on feldspar boundaries is presented below.

The problem of crystal-boundary diffusion under natural conditions was recently discussed by Joesten (1991a), who emphasized that much variation must be expected. One possibility is that diffusion within and through a volume of rock occurs almost entirely along the boundaries, with little or no penetration of the adjacent crystals. In this case the second equation of diffusion may be used in place of (81) above. In view of different models of crystal-boundary structure and composition, as reviewed in Sec. 1.5, it is possible that values of D' can be larger than those obtained experimentally. If so, a penetration greater than a few centimeters (Joesten 1991a) could occur in 1 million years at metamorphic temperatures.

4.3.6 Diffusion and chemical reaction

In many of the possible natural examples of diffusion that one could propose, diffusion is accompanied by chemical reaction, which both mobilizes and immobilizes diffusing particles. The partial immobilization of H_2O during hydration reactions, as H_2O diffuses through small volumes of rock, and locally to crystal interiors, provides examples of both diffusion and reaction occurring simultaneously. With regard to the rate of diffusion relative to the rate of reaction, one can visualize the extreme case of the reaction rate being very slow (diffusion without reaction, as discussed above) and the other extreme of the reaction rate being fast (diffusion-controlled reaction). Between these extremes there are many possibilities for diffusion modified by reaction (partial immobilization), some of which may lead to steady-state conditions.

Following Crank (1975), consider the unidirectional flow of atoms or molecules through a medium, and suppose that the concentration (s) of the immobilized particles and the concentration (C) of the particles that are free to move are related as

$$s = RC^n, \tag{82}$$

where R and n are constants. This equation implies near-equilibrium conditions for the immobilization reaction, and it implies that the reaction is diffusion-controlled. Consider the relatively simple case of $n = 1$. The second equation of diffusion is now written

$$\frac{\partial C}{\partial t} = D\frac{\partial^2 C}{\partial x^2} - \frac{\partial s}{\partial t}, \tag{83}$$

where D is constant and the second term on the right allows for partial immobilization. By differentiation of (82) with respect to x, followed by substitution in (83), we obtain

$$\frac{\partial C}{\partial t} = \left\{\frac{D}{R+1}\right\}\frac{\partial^2 C}{\partial x^2}. \tag{84}$$

This is the diffusion equation with $D/(R+1)$ in place of D. The result is a decrease in the rate of diffusion by a factor of $1/(R+1)$. For example, if $R = 9$, the rate of diffusion is diminished by a factor of $\frac{1}{10}$.

Suppose that n is greater or less than 1, and that $\partial s/\partial t$ is sufficiently large compared to $\partial C/\partial t$ that $\partial C/\partial t$ in (83) may be neglected. Then,

$$\frac{\partial s}{\partial t} = \frac{\partial}{\partial x}\left\{D\frac{\partial C}{\partial x}\right\}. \tag{85}$$

Again, by differentiation of (82) with respect to x, followed by substitution in (85),

$$\frac{\partial s}{\partial t} = \frac{\partial}{\partial x}\left\{\frac{D}{n}\left\{\frac{1}{R}\right\}^{1/n} s^{(1-n)/n}\frac{\partial s}{\partial x}\right\}. \tag{86}$$

Thus, the general form of the diffusion equation is preserved, but with D variable; under the more general conditions expressed by (86), diffusion is a function of both s and C.

In general, mathematical solutions to the diffusion equation when immobilization (through reaction or adsorption) accompanies diffusion, lead to complex expressions, as presented by Crank (1975, Ch. 14). These solutions lead to the conclusion that (a) when the reaction rate is very slow (less than one-thousandth that of diffusion), behavior is the same as in the absence of reaction, and (b) when rates of reaction and diffusion are similar, behavior is nearly the same as for a reaction rate that is extremely rapid relative to diffusion.

4.3.7 Experimental data on diffusion in minerals

Results on diffusion in silicate and oxide minerals were compiled by Freer (1980, 1981), and numerous additional data have become available recently. The following examination of some representative results will provide information on experimental methods that are currently in use, and will convey an impression of the magnitude of atomic and molecular migrations in different silicate crystals in a wide range of temperature.

Chabazite

Diffusion in zeolite minerals, as represented by chabazite, is of interest because the Al silicate framework is open and rigid, allowing the migration of H_2O and cations to occur within the crystals with relative ease. Also, as noted in Sec. 1.5, these crystals may provide a rough analogue of crystal-boundary structures in portions of the solid Earth.

Chabazite, with simplified formula, $CaAl_2Si_4O_{12}{\cdot}6H_2O$, is a framework structure in which (Si, Al)O_4 tetrahedra are linked to form 4-, 6-, and 8-member rings. The structure may be viewed as consisting of two kinds of cages, a distorted hexagonal prism, and a larger ellipsoidal cage, bounded by twelve 4-member rings, two 6-member rings, and six 8-member rings. Crystals typically contain 22 wt % H_2O, and the water molecules, as well as ions of Ca and Na (where present) are located within the cages. The 'hole' in the 8-member rings is about 4.0 Å in diameter, and hence water molecules and various cations can presumably pass through the crystal with ease. Moreover, the framework remains virtually unchanged as water molecules are removed by heating (Smith 1962).

Barrer and Fender (1961) studied the rate of intrinsic diffusion of H_2O into slightly dehydrated chabazite by placing crystals in a loose-mesh fiberglass basket, and by measuring the gain of weight as the crystals were bathed in a stream of N_2–H_2O vapor. The results obtained at 31 and 75 °C are shown in Figure 4.15a,

Table 4.2 Data on diffusion in minerals

Diffusion of	Diffusion in	Note	°C	D $cm^2\ s^{-1}$	D_0 $cm^2\ s^{-1}$	E $kJ\ mol^{-1}$	Ref.
H_2O	chabazite	intrinsic	45	4.8×10^{-6}	1.2×10^{-3}	28	1
H_2O	chabazite	intrinsic T-extrapolation	500	1.6×10^{-5}			
H_2O	chabazite	self	45	1.3×10^{-7}	1.2×10^{-1}	36	1
H_2O	chabazite	self T-extrapolation	500	4.4×10^{-4}			
H_2O	liquid water	self	45	3.9×10^{-5}			1
H_2O	ice	self	−2	1.0×10^{-10}			1
Ca	chabazite	self	60	5.9×10^{-14}	12.0×10^{-2}	77	2
Ca	chabazite	self T-extrapolation	500	7.3×10^{-7}			
Ni–Mg	olivine (⊥ (001))	exchange X_{fo} = 0.2	1350	1.4×10^{-11}			3
		0.4	1350	1.5×10^{-11}			
		0.6	1350	1.4×10^{-11}			
		0.8	1350	1.2×10^{-11}			
		1.0 (X-extrapolation)	1350	1.2×10^{-11}	2.23×10^{2}	414	
		1.0 (T-extrapolation)	600	3.9×10^{-23}			
Mn–Mg	olivine (⊥ (001))	exchange X_{fo} = 0.2	1200	13.9×10^{-11}			
		0.4	1200	5.4×10^{-11}			
		0.6	1200	2.4×10^{-11}			
		0.8	1200	1.4×10^{-11}			
		1.0 (X-extrapolation)	1200	0.59×10^{-11}	9.69×10^{3}	429	
		1.0 (T-extrapolation)	600	2.1×10^{-22}			
Fe–Mg	olivine (⊥ (001))	exchange X_{fo} = 1.0 (X-extrapolation)	1100	4×10^{-12}	1.08×10^{-2}	252	4
		1.0 (T-extrapolation)	600	8.9×10^{-18}			
O	Mg-olivine	self	1400	$\sim 1 \times 10^{-14}$	2.3×10^{-6}	290	5
Si	Mg-olivine	self	1400	$\sim 1 \times 10^{-18}$	1.5×10^{-6}	380	6
Na–K	feldspar (⊥ (001))	exchange X_{fo} = 0.4	900	1.0×10^{-13}			7
		= 0.4	1000	1.0×10^{-12}			
		= 0.4 (T-extrapolation)	600	$\sim 1.0 \times 10^{-17}$			

O	Ab-Kf	volume	600	$\sim 1.0 \times 10^{-14}$	1.5×10^{-8}	109	8
		boundary; δ (assumed) $\sim 10^{-7}$ cm	600	$\sim 1.0 \times 10^{-9}$	4.7×10^{-6}	180	

1, Barrer and Fender (1961);
2, Kretz (unpublished);
3, Morioka (1981);
4, Misener (1974);
5, Jaoul, Houlier and Abel (1983);
6, Jaoul et al., (1981);
7, Christoffersen, Yund and Tullis (1983);
8, Nagy and Giletti (1986).

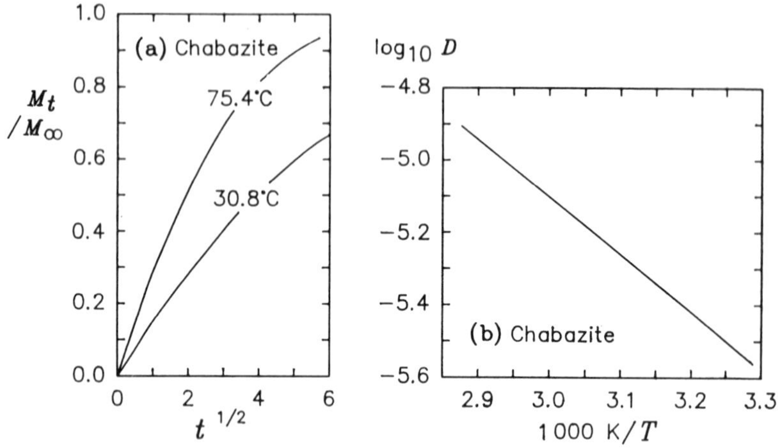

Figure 4.15 (a) Intrinsic diffusion of H_2O in chabazite; M_t/M_∞ is fraction gain in weight and t is time in minutes; (b) Arrhenius plot of data; D is the diffusion coefficient. From Barrer and Fender (1961, *J. Phys. Chem. Solids*). Reproduced by permission of Pergamon Press Ltd.

where the increase in weight (expressed as a fraction of the total small increase that was recorded at the end of the experiment) is plotted as a function of $t^{1/2}$. By use of the appropriate diffusion equation, the diffusion coefficient was determined. For example, at 45 °C, D (intrinsic) = 4.8×10^{-6} cm^2 s^{-1}. The increase of D with temperature obeys well the Arrhenius equation,

$$D = D_0 \exp(-E/RT), \tag{87}$$

as shown in Figure 4.15b; values of D_0 and E are listed in Table 4.2.

The self-diffusion of H_2O in chabazite was studied by beginning with crystals which contain D_2O, and by measuring the loss of weight as D_2O molecules diffused from the crystals, in exchange for H_2O. The value of D^* for self-diffusion at 45 °C was found to be slightly smaller than that for intrinsic diffusion, and it lies approximately midway between the coefficient for self-diffusion in water at the same temperature and that for ice, just below the melting point (Table 4.2). Barrer and Fender estimated an uncertainty in the intrinsic-diffusion and self-diffusion coefficients of 30%.

By use of the Arrhenius equation (87), diffusion coefficients for intrinsic- and self-diffusion of H_2O in chabazite have been extrapolated to 500 °C (Table 4.2) to give values of $\sim 1 \times 10^{-5}$ cm^2 s^{-1}. These numbers are intended only as a very rough estimate of diffusion rates for H_2O molecules in open silicate structures, as they might exist at crystal boundaries or particular regions along crystal boundaries during metamorphism. From Figure 4.12, D $\sim 1 \times 10^{-5}$ corresponds to a mean displacement of approximately 40 m in 30 000 years.

The self-diffusion of Ca in hydrous Ca chabazite was investigated by Barrer, Bartholomew and Rees (1963) and others. The results obtained by the writer at

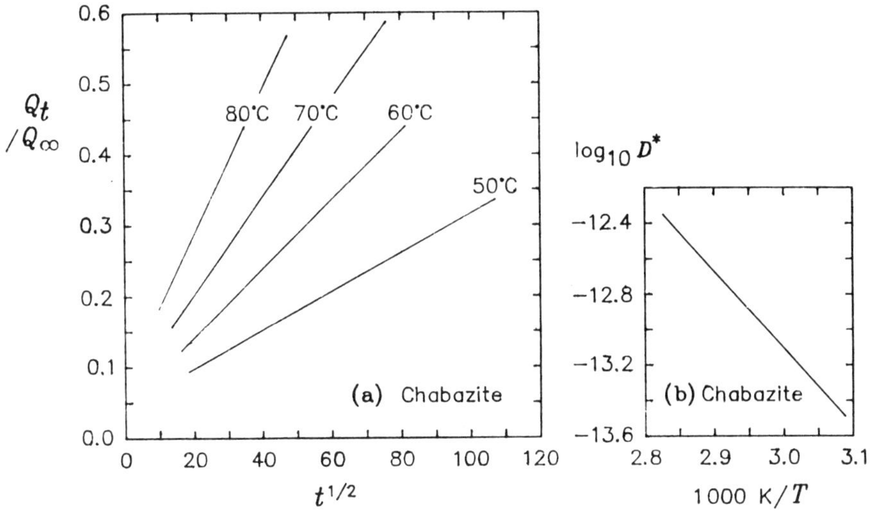

Figure 4.16 (a) Self-diffusion of Ca in Ca chabazite; Q_t/Q_∞ is fraction of exchange of ^{40}Ca, and t is time in minutes; (b) Arrhenius plot for the data of (a); D^* is the diffusion coefficient. From unpublished results obtained by author in laboratory of R.M. Barrer

50 to 80 °C are presented in Figure 4.16a. These data were obtained by immersing and stirring crystals in a solution of $CaCl_2$ which was tagged with ^{40}Ca, and by measuring the loss of ^{40}Ca from the solution (i.e., the uptake of ^{40}Ca by the crystals) radiometrically. By use of the appropriate diffusion equation (Barrer, Bartholomew and Rees 1963), i.e.,

$$\frac{Q_t}{Q_\infty} = \frac{Q_0}{Q_0 - Q_\infty} \cdot \frac{2a}{v} \left\{ \frac{D^* t}{\pi} \right\}^{1/2}, \tag{88}$$

values of D^* are obtained from the data of Figure 4.16a. In equation (88), Q_0 is the amount of radioactive cation in the solution initially, and Q_t and Q_∞ are the amounts of exchange at time t and at equilibrium, in crystals of volume v and surface area a. For example, at 60 °C, $D^* = 5.9 \times 10^{-14}$ cm^2 s^{-1}. A major source of error lies in the estimate of surface area, which here was obtained from measurements of particle size. The Arrhenius equation is well obeyed (Figure 4.16b) to produce estimates of D_0 and E (Table 4.2). An extrapolation to 500 °C gives $D^* \sim 1 \times 10^{-6}$, smaller, by a factor of 1/10, than the estimate for D^* (H_2O) at the same temperature. From Figure 4.12, $D^* \sim 1 \times 10^{-6}$ cm^2 s^{-1} translates to a mean displacement of approximately 10 m in 30 000 years.

Olivine

Ni–Mg and Mn–Mg exchange diffusion in olivine was investigated by Morioka (1981). Although the oxygen atoms in olivine are closely packed, half of the octahedral 'sites' are vacant (Morioka and Nagasawa 1991) and these, together with additional vacancies, present numerous opportunities for cation exchange to occur within the crystal.

Morioka (1981) obtained synthetic crystals of pure Mg_2SiO_4 and Ni_2SiO_4 which were cut into cubes (3 mm^3 in volume) with sides perpendicular to crystal axes. Diffusion couples were then formed by bringing into contact polished surfaces (normal to the *c* axis) of Ni olivine and Mg olivine, and tying the crystals together with Pt wire. The diffusion couples were placed in quartz crucibles and heated in air at temperatures between 1100 and 1400 °C for durations of 10 to 300 hours. Upon removal from the furnace, each bicrystal was cut normal to the interface, and the Mg and Ni concentration as a function of distance normal to the interface was determined by microprobe analysis. The same method was used to study Mn–Mg exchange diffusion. Representative results for Ni–Mg and Mn–Mg exchange are presented by the concentration profiles of Figure 4.17a and b.

In general, the diffusion coefficient for exchange diffusion is expected to vary with composition. Let C denote the concentration (in g cm^{-3}) of an element of interest (e.g., Ni), and let C_1 denote the concentration at any value between zero and the initial concentration, C_∞. Assume that the change of volume resulting from a change in composition is negligible, i.e., the interface of the diffusion couple remains stationary, and let x denote distance from the interface, as in Figure 4.18.

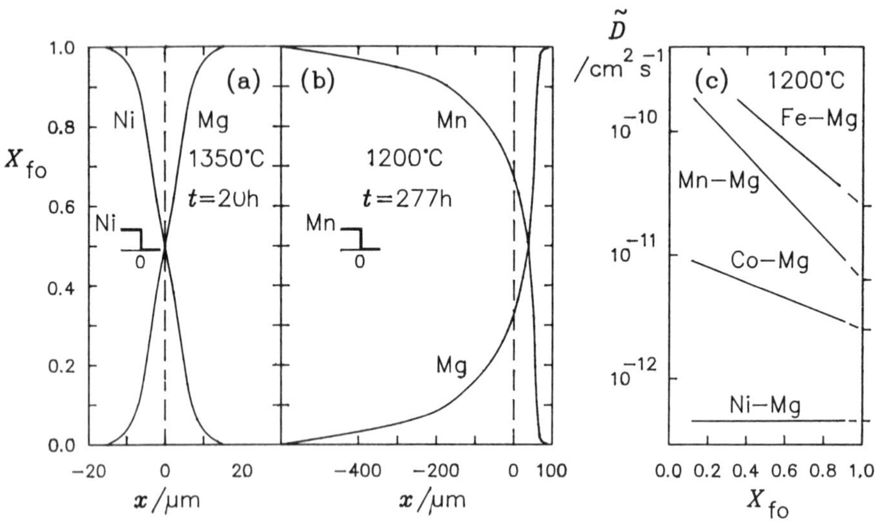

Figure 4.17 Data on diffusion in olivine; (a) concentration (X_{fo} = mole fraction Mg_2SiO_4) versus distance (x) profiles for Ni–Mg exchange diffusion, and (b) for Mn–Mg exchange diffusion; (c) Arrhenius plots for four different exchange couples at 1200 °C; $\tilde{D}$ is the diffusion coefficient. From Morioka (1981), reproduced by permission of Pergamon Press Ltd and Morioka and Nagasawa (1991), reproduced by permission of Springer-Verlag

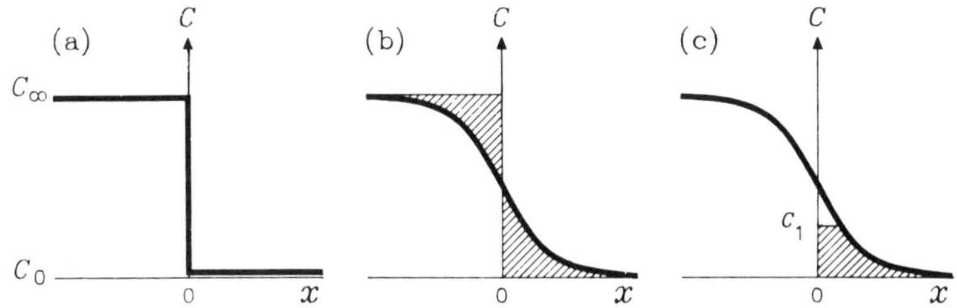

Figure 4.18 Evaluation of the diffusion coefficient ($\tilde{D}$) from a concentration (C) versus distance (x) profile; (a) profile at time $t = 0$; (b) profile after a period of elapsed time (t); shaded areas must be equal; (c) calculation of $\tilde{D}$ at $C = C_1$, using the slope of the curve and area under the curve, as described in text. In this example, volume changes are negligible, and the original interface position (in (a)) remains stationary. From Crank (1975). Reproduced by permission of Oxford University Press

Then the appropriate diffusion equation, as given by Crank (1975), is

$$\tilde{D}_{C_1} = -\frac{1}{2t}\left\{\frac{\mathrm{d}x}{\mathrm{d}C}\right\}\left\{\int_0^{C_1} x\,\mathrm{d}C\right\}, \tag{89}$$

where t is time.

Morioka (1981) used equation (89) to obtain values of $\tilde{D}$ at concentration increments. Following a common procedure (Crank 1975, p. 231), x is measured from a place (the Matano interface) which is located from a consideration of mass balance (Figure 4.18b); in the case of olivine, this is the original interface, shown in Figure 4.18a. In equation (89) $\mathrm{d}x/\mathrm{d}C$ is the slope of the profile at C_1, and the integral is the area under the curve (Figure 4.18c). Representative results for $\tilde{D}$ (Ni–Mg) and $\tilde{D}$ (Mn–Mg) are given in Table 4.2. Note that $\tilde{D}$ (Mn–Mg) at 1200 °C varies greatly with composition, resulting in highly asymmetrical concentration profiles (Figure 4.17b); the dependence of $\tilde{D}$ on concentration ($X_{\mathrm{fo}} = \mathrm{Mg}/(\mathrm{Mg} + \mathrm{Mn})$) was found to be exponential (Figure 4.17c).

There is a rather large uncertainty of 20 to 30% in these results. Thus the use of equations which accommodate a change in volume would appear at present to be unwarranted. Experimental reproducibility, in general, is discussed by Rothman (1984).

Diffusion involving Fe (Fe–Mg exchange) in olivine was investigated by Buening and Buseck (1973), and the results for diffusion parallel to [001] at 1200 °C (Figure 4.17c) indicate that Fe–Mg diffusion is more rapid than that involving Ni and Mn. Similar results were obtained by Misener (1974) who showed, in addition, that an increase in pressure from 1 to 25 000 bar decreases the diffusion coefficient by a factor of $\frac{1}{10}$. Both of the above studies showed clearly that Fe–Mg diffusion is noticeably anisotropic, with $\tilde{D}$ ([001]) greater than $\tilde{D}$ ([100], [010]) by a factor that lies between 3 and 10.

For the purpose of comparing different rates of exchange, the linear X versus $\log \tilde{D}$ plots shown in Figure 4.17c are frequently extrapolated to zero concentration for the ion of interest, i.e., to $X_{\text{fo}} = 1.0$. The resulting numbers for Ni–Mg, Mn–Mg, and Fe–Mg are listed in Table 4.2 together with D_0 and E which define the Arrhenius change in diffusion coefficients with temperature. Extrapolation to a metamorphic temperature of 600 °C yields corresponding diffusion coefficients of 3.9×10^{-23}, 2.1×10^{-22}, and 8.9×10^{-18} cm^2 s^{-1}. The last number (for Fe–Mg exchange) translates to a mean displacement (Figure 4.12) of 0.1 mm in 300 000 years.

Data on the self-diffusion of oxygen and silicon in Mg olivine were obtained by Jaoul et al. (1981, 1983) (Table 4.2). These diffusion coefficients are much smaller than those for divalent cation exchange. For example, at 1400 °C, $\tilde{D}$ (Fe–Mg) $\sim 10^{-10}$, D^* (oxygen) $\sim 10^{-14}$, D^* (Si) $\sim 10^{-18}$ cm^2 s^{-1}.

Alkali feldspar

Na–K exchange diffusion in alkali feldspar was investigated by Christoffersen, Yund and Tullis (1983), using a procedure similar to that described above for olivine. Diffusion couples were made from natural crystals of nearly pure K-feldspar and albite. Most of the experiments were carried out at 15 kbar and at 900 or 1000 °C. Results for diffusion ⊥ (001) at different values of $X_{\text{or}} = \text{K}/(\text{Na} + \text{K})$ are presented in Figure 4.19 and Table 4.2. Repeated experiments gave estimates of $\tilde{D}$ that differed by a factor as large as 2.

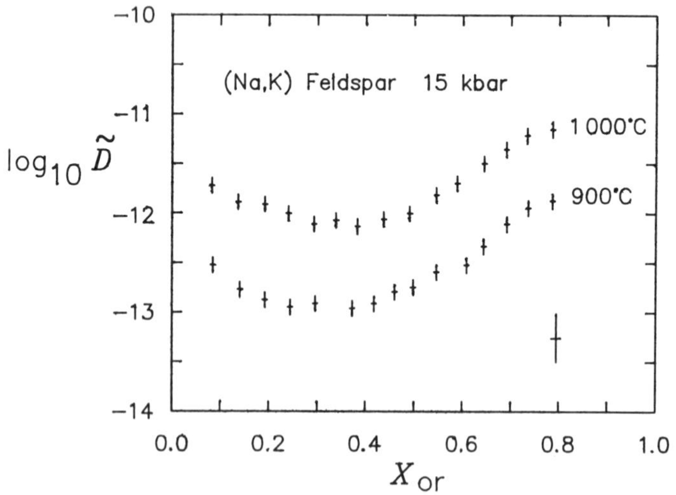

Figure 4.19 Data on Na–K exchange diffusion, ⊥ (001), in alkali feldspar; $\tilde{D}$ denotes diffusion coefficient, and X_{or} is mole fraction $KAlSi_3O_8$. Small and large error bars refer to analytical and experimental error respectively. From Christofferson, Yund and Tullis (1983)

At 1000 °C and 15 kbar, diffusion was found to be anisotropic, with $\tilde{D}$ ($\perp$ (001)) > $\tilde{D}$ ([011]) > $\tilde{D}$ ($\perp$ (010)), and $\tilde{D}$ ($\perp$ (001)/$\tilde{D}$ ($\perp$ (010)) ~ 12. An increase in pressure from 5 to 15 kbar did not produce a detectable change in $\tilde{D}$ ($\perp$ (001)), nor was $\tilde{D}$ found to be sensitive to the presence or absence of a trace amount of H_2O.

An estimate of $\tilde{D}$ (Na–K; $\perp$ (001); $X_{or} = 0.4$) at 600 °C is obtained from an Arrhenius extrapolation of the data in Figure 4.19 to give $\tilde{D} \sim 1 \times 10^{-17}$ cm^2 s^{-1}. This is similar to the estimated $\tilde{D}$ (Fe–Mg) in olivine at $X_{fo} = 1.0$ and 600 °C, as noted above, and similarly translates (Figure 4.12) to a mean displacement of 0.1 mm in 300 000 years. Additional information on volume diffusion in feldspar is presented by Yund (1983b).

The diffusion of oxygen on low-angle boundaries between K-feldspar and Na-feldspar in natural perthite was investigated by Nagy and Giletti (1986). Perthite fragments, together with H_2O enriched in ^{18}O, were sealed in gold tubes, placed in a cold-seal pressure vessel, and taken to temperatures between 500 and 800 °C at a pressure of 1.0 kbar and for periods of 1×10^4 to 5×10^6 seconds. An ion microprobe was then used to measure the concentration of ^{18}O within the crystals, and to locate concentration contours, which showed a preferred penetration along grain boundaries. Equation (81) in Sec. 4.3.5 was used to derive from these measurements, the following Arrhenius equations for volume and grain-boundary diffusion:

$$D^* = 1.5^{+72}_{-1.3} \times 10^{-8} \exp\left\{-\frac{109 \pm 12}{RT}\right\}, \tag{90}$$

$$\delta D' = 4.7^{+240}_{-4.6} \times 10^{-6} \exp\left\{-\frac{180 \pm 29}{RT}\right\}. \tag{91}$$

Here the activation energy is in kJ/mol and the effective crystal-boundary width (δ) is in centimeters. Assuming $\delta \sim 10$ Å, D' (600 °C, 1 kbar) $\sim 10^{-9}$ cm^2 s^{-1}, which exceeds the coefficient for volume diffusion by a factor of $\sim 10^5$ (Table 4.2), similar to the factor found in metals. These diffusion coefficients evidently refer to oxygen self-diffusion, but the authors have suggested that oxygen may have migrated as OH^- or H_2O. The relatively high activation energy for boundary diffusion (equation (91)) is not understood; in oxides the activation energy for boundary diffusion is in general less than that for volume diffusion.

In conclusion, available experimental results on diffusion in common silicate crystals confirm the long-held view that in the middle range of crustal temperature (500–600 °C) rates of diffusion are low in relation to crystal size, even on a scale of 1 million years. The rates are not, however, infinitesimal, and are certainly sufficiently rapid to enable exsolution and other within-crystal processes to occur under favorable temperature–time conditions.

A very large increase in cation mobility is found in the more open silicate structures (e.g., in chabazite, compared to feldspar), where diffusion coefficients, extrapolated to 500 to 600 °C, are larger by a factor of approximately 10^{10}. Assuming that crystal-boundary structures in the Earth are rather open as they are in metals (where the factor is $\sim 10^5$), or that zeolite crystals do in fact form an analogue, then rates of diffusion on boundaries are more than adequate to bring about the rearrangement of atoms needed for metamorphic crystallization and differentiation.

4.4 CRYSTAL NUCLEATION

The solid Earth consists of very approximately 10^{30} crystals, most of which were created by processes of nucleation and growth. Nucleation is the first step in the formation of a crystal, when a number of atoms assemble in a structure which is similar to that of the completed crystal. Nucleation is followed by growth, which takes place as atoms are added and the crystal increases in size.

Within the Earth, crystal nucleation occurs in a vapor, liquid, or solid medium; at the surface it occurs predominantly in the formation of carbonate and other minerals in sea water and saline lakes, and in the formation of ice and snow. Discussions on nucleation in natural systems were presented by Wager (1959), Spry (1969), Knight (1971), Hornbogen (1973), and Kirkpatrick (1981).

Nuclei are extremely small and have a life-span of only a fraction of a second; hence nucleus formation is not easily studied and is not well understood. Some information on nucleation sites may be obtained from the distribution of fully grown crystals. For example, crystals of ice on window panes in northern winters are often observed to be concentrated on certain invisible lines, which shows that nucleation is a very selective process. Scanning tunneling microscopy, which is capable of detecting particles of extremely small size, is now being used to study the nucleation of metals and other solids (Neddermeyer 1990). The classical theory of nucleation and the theory of spinodal decomposition are landmarks in the study of nucleation.

Our concern at present is nucleation within a medium that is largely or entirely solid, and which lies little or moderately removed from equilibrium. The process will be examined in three kinds of reaction: simple phase change, exsolution, and the production of new minerals in multi-phase crystal assemblies.

4.4.1 Nucleation in a simple phase transformation

When a pure metal is heated very slowly at a given pressure, melting begins at a specific temperature, referred to as the 'melting point'. Upon further heating, followed by slow cooling, crystallization does not occur at the melting point, but at a point several degrees below this temperature. The classical theory of nucleation provides an explanation for the barrier to nucleation that exists for metals and other kinds of crystals.

The classical theory was proposed by Gibbs (1878), Volmer and Weber (1926), and Becker and Döring (1935), and was initially applied to condensation and other transformations which do not require a change in composition. The following summary is from Burke (1965) and Raghaven and Cohen (1975).

Consider a perfect crystal of unit volume which consists of a single substance (α), and suppose that the crystal is taken to a temperature above the point at which an $\alpha \rightarrow \beta$ solid-state phase transformation is expected to occur. The classical theory of nucleation states that as a result of thermal fluctuations, small domains of atoms with the structure of phase β are continually forming and disappearing. These sub-microscopic domains, which for the present are assumed to be spherical and of radius r, are referred to as 'embryos'.

Let ΔG_e refer to the energy of formation of an embryo, according to the equation

$$\Delta G_e = \left(\frac{4}{3}\right)\pi r^3 \Delta G_v + 4\pi r^2 \sigma, \tag{92}$$

where $\Delta G_v (< 0)$ is the Gibbs energy change for the $\alpha \rightarrow \beta$ phase transformation per unit volume of β, and $\sigma (> 0)$ is the energy of the α–β interface per unit area of interface. At small values of r, the second term on the right dominates, and at larger values of r, the first term dominates. Consequently, with increasing size, ΔG_e passes through a maximum, as shown in Figure 4.20.

The theory postulates an 'equilibrium' size distribution for the embryos; thus thermal fluctuations can cause a particular embryo to increase or decrease in size, with a certain probability that it will increase to the critical size, denoted r_c (Figure 4.20), corresponding to a maximum in ΔG_e. At this point the embryo becomes a nucleus. A small further increase in size results in a decrease in ΔG_e and the nucleus is therefore expected to enlarge spontaneously to become a crystal. Thus, interfacial energy is identified as the sought-for barrier to crystallization.

Specific interfacial energy is the energy per unit area of the boundary that separates any two domains of like or unlike phase. In metal systems, boundaries are classified as coherent, semi-coherent, and non-coherent, as shown in Figure 4.21,

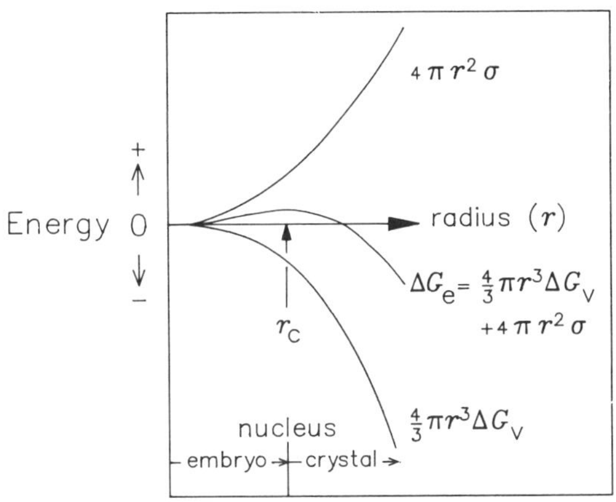

Figure 4.20 Contributions of volume (Gibbs) energy (bottom curve), and interfacial energy (top curve) to the energy of formation of a spherical embryo ($r < r_c$), nucleus ($r = r_c$), and crystal ($r > r_c$)

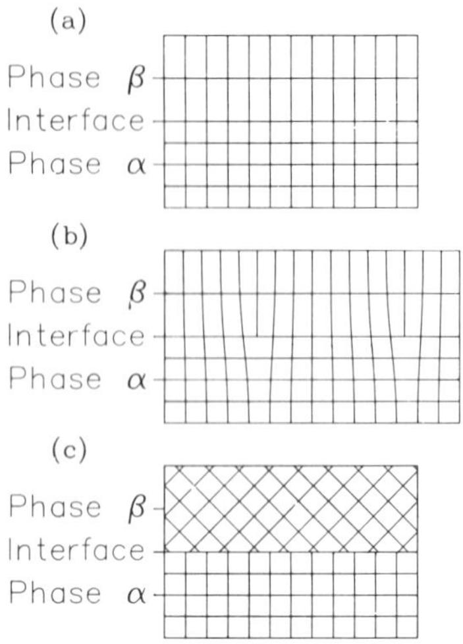

Figure 4.21 Illustration of (a) coherent, (b) semi-coherent, and (c) non-coherent interfaces between two crystals. From Raghaven and Cohen (1975)

with energies (σ) of approximately 10 to 50, 100 to 300, and 500 to 1500 erg cm^{-2} respectively (Raghaven and Cohen 1975). Thus, the energy is sufficiently small (1 erg $= 10^{-7}$ J) that in discussions of phase equilibrium it is normally disregarded. But for certain processes, especially those that involve very small particles, this form of energy can be important.

In the classical theory, the embryo–matrix interface is assumed to resemble the interface that exists between macroscopic crystals of α and β. This assumption has been criticized by Hoare and Pal (1972), Christian (1975), and others, and in the following further development of the theory, this assumption must be kept in mind.

The energy of formation of a *nucleus* in terms of ΔG_v and σ is obtained as follows. At $r = r_c$ (Figure 4.20),

$$\frac{\partial(\Delta G_e)}{\partial r} = 0, \tag{93}$$

and by differentiation of (92), followed by a substitution in (93),

$$r_c = -2\sigma/\Delta G_v. \tag{94}$$

The same result is obtained for a cubic embryo in which r_c is the distance from center to face. By substitution of (94) in (92),

$$\Delta G_c = \frac{16\pi}{3}\frac{\sigma^3}{(\Delta G_v)^2}, \tag{95}$$

where ΔG_c refers to the energy of formation of a particle of critical size, i.e., a nucleus.

In nucleation as pictured above, all atoms in the crystal α represent points at which nucleation may occur with equal probability. An expression for the rate of nucleation of β in α is obtained as follows. Let N_N stand for the very many atomic sites in crystal α at which embryos can form, and let N_r denote the number of embryos present, with radius r. Then an 'equilibrium' size distribution of embryos requires that at a given temperature and for a given size, r,

$$\frac{N_r}{N_N} = K_r, \tag{96}$$

where K_r is an 'equilibrium' constant. This may be related to ΔG_e (the energy of embryo formation) by

$$K_r = e^{-\Delta G_e/kT}, \tag{97}$$

where k is the Boltzmann constant and T is absolute temperature. Thus the 'equilibrium' number of embryos of critical size (N_c), i.e., the number of nuclei, is

$$N_c = N_N e^{-\Delta G_c/kT}. \tag{98}$$

Although the movement of atoms needed to cause an embryo to expand is very small, an energy of activation (E_a) associated with the transfer of atoms across the embryo–matrix interface must be taken into consideration. At the critical size, the addition of a single atom changes an embryo to a nucleus, and the frequency of this event may be approximated by the term

$$(n_s \nu p) e^{-E_a/kT}, \tag{99}$$

n_s is the number of atoms of α at the interface, ν is the frequency of vibration of these atoms p is the probability that the vibration direction is toward the embryo, and they are multiplied by a factor which acknowledges that attachment on the β particle can occur only at certain sites. Expression (99), i.e., the frequency (s^{-1}) of the step in changing embryos to nuclei, when multiplied by the number of embryos which are just below critical size (i.e., N_c in equation (98)) gives the steady-state rate of nucleation (I),

$$I = N_N (n_s \nu p) e^{-E_a/kT} e^{-\Delta G_c/kT}, \tag{100}$$

or

$$I = A e^{-(E_a + \Delta G_c)/kT}. \tag{101}$$

Initially, and for isothermal nucleation, the pre-exponential term, denoted A, is nearly constant, and for metals is very approximately 10^{36} cm^{-3} s^{-1}. Thus, $(E_a + \Delta G_c)/kT$ must be about 70 for a measurable rate of nucleation (Raghaven and Cohen 1975). The term E_a is approximately constant, but ΔG_c (through ΔG_v) is dependent on temperature; a small decrease in ΔG_c (increase in ΔG_v) will cause a very large increase in nucleation rate. For example, at 450 °C (i.e., $kT = 10^{20}$ J) and for $A = 10^{36}$ cm^{-3} s^{-1}, the theory predicts that when $(E_a + \Delta G_c) = 10^{-18}$ J, the nucleation rate is 10^{-8} cm^{-3} s^{-1}, i.e., 1 nucleus per cm^3 per year; then, decreasing $(E_a + \Delta G_c)$ by a factor of $\frac{1}{10}$ causes an increase in the nucleation rate by the astonishing factor of 10^{40}. But with $(E_a + \Delta G_c) < kT$, and $I \sim A$, then I remains nearly independent of changes in $(E_a + \Delta G_c)$.

Symbol I in equation (101) stands for the rate of nucleation for unit volume of α, which initially contains N_N possible nucleation sites. As crystallization progresses, some nucleation sites will be swept out by the advancing α–β interface, and consequently I must decrease with time. If virtually all nucleation occurs in the first small interval of time, this decrease may be neglected.

The above theory provides a partial understanding of nucleus formation in a melt, where nucleation may be homogeneous. Just below the equilibrium temperature, ΔG_v is small, ΔG_c is large (equation (95)), $\exp(-\Delta G_c/kT)$ is very small and, therefore, I is small. With a further decrease in temperature, ΔG_c becomes smaller, but now E_a increases as atomic mobility decreases. Thus I passes through a maximum, i.e., the greatest nucleation rate occurs at a particular temperature below the melting point, in agreement with observation.

The preferred nucleation of a phase at a crystal boundary is commonly observed (Shewmon 1969), and this behavior, which is referred to as 'heterogeneous nucleation', can be understood in an extension of the classical theory. From Figure 4.20, and especially from equation (95), it is clear that a reduction in interfacial energy will decrease the critical size and therefore facilitate nucleation. Consider, for example, the formation of a spherical embryo of β on an α–α grain boundary. Interfacial energy of $4\pi r^2\sigma^{\alpha-\beta}$ units is needed (equation (92)), but some of this energy can be obtained from the α–α boundary that is being destroyed, i.e., $\pi r^2\sigma^{\alpha-\alpha}$ units. Consequently the interfacial energy term that appears in equation (92) is

$$4\pi r^2\sigma^{\alpha-\beta} - \pi r^2\sigma^{\alpha-\alpha}, \tag{102}$$

or

$$\pi r^2(4\sigma^{\alpha-\beta} - \sigma^{\alpha-\alpha}), \tag{103}$$

which represents a decrease (compared with homogeneous nucleation), by a factor of $(4\sigma^{\alpha-\beta} - \sigma^{\alpha-\alpha})/4\sigma^{\alpha-\beta}$, or, assuming that $\sigma^{\alpha-\alpha} \sim \sigma^{\alpha-\beta}$, by a factor of approximately 0.75. For a cubic embryo, the factor is 0.83. But because ΔG_c is a function of σ^3 (equation (95)), the reduction in ΔG_c is by a factor of approximately $\frac{1}{2}$. Clemm and Fisher (1955) have shown that a further reduction in ΔG_c results from nucleation on grain edges (where three grains meet at a line) and at grain corners (where four grains meet at a point). $\sigma^{\alpha-\alpha}$ and $\sigma^{\alpha-\beta}$ are, however, dependent, in general, on the relative crystallographic orientation of the adjoining crystals and presumably those α–α boundaries of relatively high energy will be chosen preferentially.

The rate of nucleation on grain boundaries can be expressed by equation (100), with N redefined as the number of possible nucleation sites on the boundaries. Although N is smaller, compared with homogeneous nucleation, the nucleation rate (I) will be larger, because of the overriding effect of a decrease in ΔG_c.

Again, I for the entire system may decrease with time as nucleation sites are swept out by growing crystals of β.

In the above example of homogeneous nucleation within a perfect crystal, the possibility of a volume change in the $\alpha \rightarrow \beta$ transformation was implicitly assumed to be negligible. In general, a change in volume during phase transformations does occur, and the nucleus and immediately surrounding parts of the α crystal presumably become somewhat strained elastically. One approach to the problem is to assume that strain energy increases linearly with embryo volume, and to combine strain energy (> 0) with $\Delta G_{\rm v}$ (<0). The result is an increase in $\Delta G_{\rm c}$ (equation (95)) and a decrease in nucleation rate (equation (101)), as described by Porter and Easterling (1981). Nabarro (1940) has calculated that elastic strain energy for a given volume of crystal (β) is minimized when the particle occurs as a disc rather than a needle or sphere.

4.4.2 Nucleation and spinodal decomposition during exsolution

Exsolution, also known as 'precipitation' or 'phase separation', occurs in many metal and mineral crystals during cooling. Consider, for example, a crystal of unit volume and of composition $X_L = \mathrm{B}/(\mathrm{A}+\mathrm{B})$, which is cooled from temperature T_0 (Figure 4.22) to a temperature below the solvus. Compared with a simple phase transition, as described above, the formation of particles of a new phase during exsolution requires a change in composition.

Within all crystals of variable composition, local fluctuations in composition are expected to occur as a result of random movements of atoms. The change in Gibbs energy, resulting from a local change in composition from C_0 to $C_0 + \Delta C$ and $C_0 - \Delta C$ (disregarding other possible flows of energy), is

$$\Delta G = G_{C_0+\Delta C} + G_{C_0-\Delta C} - 2G_{C_0}. \tag{104}$$

Expanding each of the first two terms on the right (Shewmon 1969, p. 291), gives

$$G_{C_0+\Delta C} = G_{C_0} + \left(\frac{\partial G}{\partial C}\right)_{C_0} \Delta C + \left(\frac{\partial^2 G}{\partial C^2}\right)_{C_0} \frac{(\Delta C)^2}{2} + \ldots, \tag{105}$$

$$G_{C_0-\Delta C} = G_{C_0} - \left(\frac{\partial G}{\partial C}\right)_{C_0} \Delta C + \left(\frac{\partial^2 G}{\partial C^2}\right)_{C_0} \left(\frac{\Delta C}{2}\right)^2 + \ldots \tag{106}$$

By substitution in (104),

$$\Delta G = \left(\frac{\partial^2 G}{\partial C^2}\right)_{C_0} (\Delta C)^2. \tag{107}$$

For the solvus of Figure 4.22a, the Gibbs energy (G) versus concentration (C) curve at temperature T_1 is shown schematically in Figure 4.22b, immediately below. The common tangent to this curve gives the equilibrium composition of the two phases

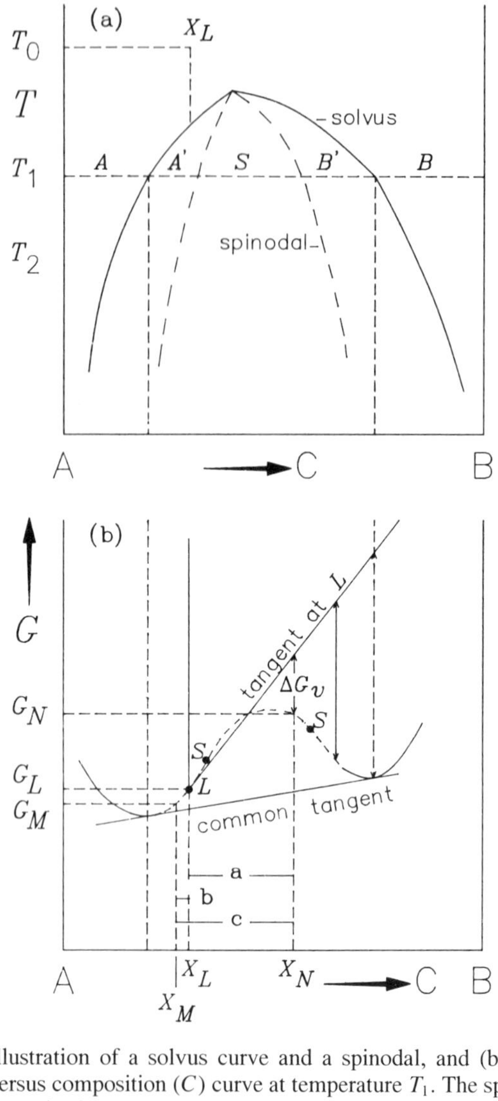

Figure 4.22 (a) Illustration of a solvus curve and a spinodal, and (b) the corresponding Gibbs energy (G) versus composition (C) curve at temperature T_1. The spinodal is delineated by inflection points (S) in the G–C curves. From Burke (1965)

for an overall composition point that lies within the solvus. Two other points are of interest, namely the inflection points (s) on the metastable (broken line) part of the energy curve; at these points, $(\partial^2 G/\partial C^2)_{C_0} = 0$, and from equation (107), $\Delta G = 0$. The locus of these points in the phase diagram (above) is referred to as the 'spinodal'.

It is now possible to recognize five composition intervals, A, A', S, B', B (Figure 4.22a). From an inspection of the energy curve (Figure 4.22b) or by use of equation (107) we note that in segments A, A', B, and B', all small fluctuations in composition result in an increase in energy (G). On the other hand, all small fluctuations in segment S (where $(\partial^2 G/\partial C^2)$ is negative) result in a decrease in energy. A crystal with a composition point that lies within the spinodal should therefore exsolve more readily than other crystals with composition points in adjacent regions A' and B', where *large* fluctuations in composition are needed to produce an overall decrease in G.

Consider first an exsolution process as it might occur within the spinodal, e.g., at temperature T_1, given a crystal of initial composition X_L (Figure 4.22a). The theory of spinodal decomposition, as proposed by Cahn (1961), differs from the classical theory of nucleation in that less importance is placed on interfacial energy, and more importance is given to fluctuations in composition, which are viewed as occurring and growing in magnitude, as in Figure 4.23. Changes in concentration of the two components, A and B, is by diffusion, which occurs up, rather than down, a concentration gradient.

Following Shewmon (1969, p. 294), recall the first equation of diffusion,

$$J = -D\frac{\partial C}{\partial x}, \tag{108}$$

which may be compared with an analogous equation,

$$J = -MC\frac{\partial \mu}{\partial x}, \tag{109}$$

where μ is chemical potential of component A or B, and M is the mobility of the component. Combining these equations,

$$-D\frac{\partial C}{\partial x} = -MC\frac{\partial \mu}{\partial x}, \tag{110}$$

$$D = MC\frac{\partial \mu}{\partial x}\cdot\frac{\partial x}{\partial C}, \tag{111}$$

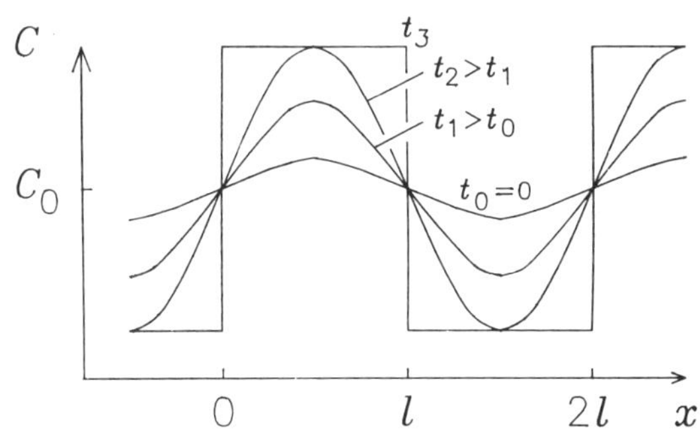

Figure 4.23 Illustration of fluctuations in composition (C), resulting in exsolution by the spinodal mechanism. The curves are concentration profiles for one of the two components (e.g., A), with increasing time, from t_0 to t_3. Notice that atoms of A migrate in the direction of increasing concentration. From Shewmon (1969)

$$D = MC\frac{\partial \mu}{\partial C}. \tag{112}$$

Now, from Darken and Gurry (1953),

$$\frac{\partial \mu}{\partial C} = (1 - C)\frac{\partial^2 G}{\partial C^2} \tag{113}$$

and, therefore,

$$D = MC(1 - C)\left(\frac{\partial^2 G}{\partial C^2}\right). \tag{114}$$

The solution to this diffusion equation is

$$C_{x,t} = C_{x,0}\exp\left(\frac{-\pi^2 Dt}{\lambda^2}\right) \tag{115}$$

(cf. Sec 4.3.2, Figure 4.13), where λ is the wavelength of the sinuous fluctuations (Figure 4.23). With diffusion occurring up-concentration, $\partial^2 C/\partial x^2$ is negative, and therefore, from the second equation of diffusion, D is negative. Notice that the time exponent in (115) is proportional to $1/\lambda^2$; thus fluctuation with small wavelength is expected to grow most rapidly.

But obviously λ cannot approach zero, and equation (115) is therefore incomplete. Two other sources of energy must be considered — interface and strain energy. An energy analogous to interfacial energy of the classical theory arises from the many A–B nearest neighbors (of high energy compared to A–A and B–B nearest neighbors) that occur in the boundary regions (Figure 4.23). This energy has been defined as $\alpha(\Delta C/\lambda)^2$ where α is a constant dependent on the A–A, B–B, and A–B interactions. Adding this term to equation (107) gives

$$\Delta G = (\Delta C)^2\left[\left(\frac{\partial^2 G}{\partial C^2}\right)_{C_0} + \alpha/\lambda^2)\right]. \tag{116}$$

Moreover, a quantity of strain energy may arise at the newly created phase boundaries, which are assumed to be coherent or nearly so. The strain energy is

$$\Delta G_\epsilon = \eta^2(\Delta C)^2(E/(2(1 - \nu)), \tag{117}$$

where η is the change in lattice spacing, $\mathrm{d}a_0/\mathrm{d}C$, E is the Young's modulus, and ν is the Poisson ratio. Adding this term to (116) gives, finally,

$$\Delta G = (\Delta C)^2\left[\left(\frac{\partial^2 G}{\partial C^2}\right)_{C_0} + (\alpha/\lambda^2) + \eta^2 E/(2 - 2\nu)\right]. \tag{118}$$

The last term in the square brackets is important in determining the geometry of the two-phase aggregate. If η is very small, strain energy is nearly zero, and the particles of exsolved phase are expected to be nearly equidimensional. If η is not negligible, and E varies with orientation, the exsolved phase will tend to form oriented lamellae. Such lamellae are observed in Al–Cu, alkali feldspar, and other crystals. A large value of η for near-coherent domains translates to a large ΔG_ϵ which will effectively displace the equilibrium (non-coherent) solvus. In some metal systems, the coherent solvus lies several tens of degrees below the non-coherent (thermodynamic) solvus, and a similar displacement is expected in silicate and other mineral systems (Yund and McCallister 1970).

Consider now an exsolution process in a crystal with composition represented by a point outside of the spinodal, for example, a crystal of composition X_L at temperature T_1 (Figure 4.22). As noted above, a rather large fluctuation in composition is needed to bring about a reduction in Gibbs energy, leading to completed exsolution. Following R. Becker, as quoted by Burke (1965), suppose that fluctuations have created an embryo of composition X_N (Figure 4.22b), resulting in a very small change in matrix composition, to X_M (exaggerated, for clarity, in Figure 4.22b). The Gibbs energy of formation of the embryo, per unit volume of embryo, is

$$\Delta G_{\rm v} = \left(\frac{b}{c}G_N + \frac{a}{c}G_M - G_L\right)/(b/c). \tag{119}$$

Here $a = (X_N - X_L)$, $b = (X_L - X_M)$ and $c = a + b$, i.e., b/c and a/c are the volume fractions of embryo and matrix, as determined by the lever rule (Figure 4.22b). G_N, G_M, and G_L represent the Gibbs energy per unit volume of embryo, matrix, and original crystal respectively. Simplifying (119),

$$\Delta G_{\rm v} = G_N + (a/b)G_M - (c/b)G_L. \tag{120}$$

But $a = -b + c$. Thus,

$$\Delta G_{\rm v} = G_N - G_M - (c/b)(G_L - G_M). \tag{121}$$

G_M differs only slightly from G_L, and X_M from X_L, and therefore,

$$\frac{G_L - G_M}{X_L - X_M} \sim \left(\frac{\mathrm{d}G}{\mathrm{d}X}\right)_{X_L}. \tag{122}$$

Also, $c \sim a$, and thus,

$$\Delta G_{\rm v} \sim \left[G_N - G_M - a\left(\frac{\mathrm{d}G}{\mathrm{d}X}\right)_{X_L}\right]. \tag{123}$$

In Figure 4.22b we see, from geometry, that the term in square brackets is the distance from the energy curve at X_N to a tangent drawn at L. Notice that $\Delta G_{\rm v}$ will vary depending on embryo composition; in the chosen example (X_N) and for all composition points lying to the right of the intersection of the energy curve by the tangent, embryo formation results in an overall decrease in energy. Whether or not an embryo of composition X_N becomes a nucleus will depend on whether or not critical size can be attained, taking interfacial energy into consideration.

The energy of embryo formation will also depend on the initial composition of the crystal, which in the above example is X_L. From Figure 4.22b we see that as X_L increases and point L approaches the spinodal (s), the slope of the tangent drawn at L increases, and for a given embryo composition, ΔG_v increases in the negative sense. Thus the rate of nucleation is expected to increase, in agreement with some data from metals. The same behavior is expected to result from a decrease in temperature, which moves the initial composition point, L, toward the spinodal.

Christie (1968) suggested that exsolution in alkali feldspar may occur by spinodal decomposition, and, since then, delicate lamellar intergrowths have been detected which have almost certainly formed by this mechanism. The subject is examined in detail by Putnis and McConnell (1980) who present additional mineralogical examples.

4.4.3 Nucleation in a multi-phase crystal assembly

A process analogous to nucleation, as defined above, must occur in the more complex systems in which staurolite, garnet, sillimanite, and other minerals crystallize within the Earth. But these complex systems are unique because in general, an immense variation must exist in the kind of possible nucleation sites that are present when certain mineral-forming reactions are in progress. For example, where andalusite changes to sillimanite, the classical theory would appear to predict the nucleation of sillimanite within andalusite (small E_a) or possibly on andalusite–quartz boundaries (where the increase in interfacial energy associated with embryo formation is small). But if the rate of reaction is slow compared with atomic mobility, various foreign boundaries (e.g., quartz–biotite) could become 'available' for nucleation, and this may explain why sillimanite crystals do in fact commonly occur several micrometers from andalusite. Additional complexity must arise where volumes of rock are deformed plastically while nucleation is in progress.

Four possibilities for nucleation kinetics in a multi-phase crystal assembly will now be examined. Imagine a fine-grained aggregate of crystals, in which a reaction has produced 100 crystals of a new phase, which occupies about 10% of the total volume. Assume that the very many points on crystal boundaries in the matrix were possible nucleation sites. The duration of the reaction is taken as 1 unit of time, and the rate of crystal growth ($\mathrm{d}r/\mathrm{d}t$) is assumed to have been constant. Now for isothermal nucleation ($\Delta G_v \sim$ constant), equation (101) provides one possibility (model 1) for the rate of nucleation,

$$\frac{\mathrm{d}N}{\mathrm{d}t} = A\,\exp\{-[(E_a + \Delta G_c)/kT]\}. \tag{124}$$

Upon integration,

$$N = A\{\mathrm{e}^{-[(E_a+\Delta G_c)/kT]}\}t, \tag{125}$$

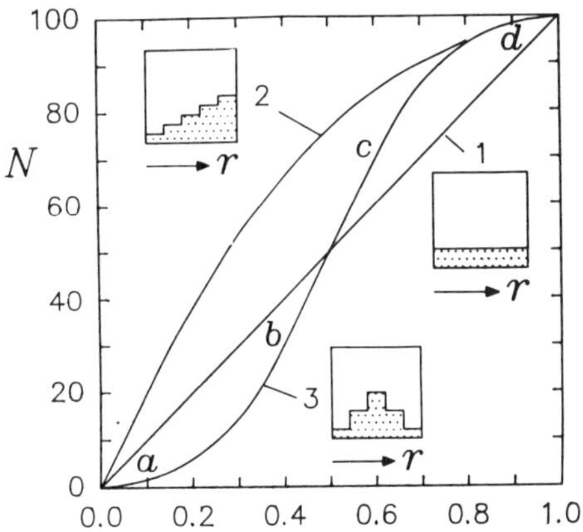

Figure 4.24 Three models (1, 2, 3) for the rate of nucleation of product crystals in a multiphase crystal assembly; N denotes number of crystals per unit volume, and t is time; inset histograms are corresponding crystal-size distributions (r = radius)

which is plotted as curve 1 in Figure 4.24. The nucleation rate, being constant, results in ten new crystals in each of ten increments of time, to produce a mesa-like histogram for the crystal-size distribution.

A decline in nucleation rate has been observed for some dehydration reactions (Hannay 1976, p. 166), resulting evidently from a progressive depletion of a relatively few (N_F) favorable nucleation sites. Suppose (model 2), that the decrease can be expressed as

$$\frac{dN}{dt} = k_N(N_F - N), \tag{126}$$

where k_N is a constant. Upon integration,

$$N = N_F(1 - e^{-k_N t}), \tag{127}$$

which is plotted as curve 2 in Figure 4.24. The resulting crystal-size distribution is positively skewed.

As a third possibility (model 3), consider a system in which a reaction moved to completion in a temperature increment sufficiently small to cause negligible changes in E_a and kT. Although nucleation occurred continuously three stages may be recognized. In the first stage (a to b, Figure 4.24, curve 3) ΔG_v increased (in the negative sense) with increasing temperature and departure from equilibrium, which, according to the equation

$$\Delta G_c = \frac{16\pi}{3}\frac{\sigma^3}{(\Delta G_v)^2} \tag{128}$$

(equation (95) of Sec. 4.4.1), resulted in a decrease in the energy of nucleus formation (ΔG_c), and hence (from equation (124)) an increase in nucleation rate. Assume that the decrease in ΔG_c with time (t) was linear,

$$\Delta G_c = (\Delta G_c)_{t=0} - mt, \tag{129}$$

where time is measured from the beginning of nucleation, and m is a constant. By substitution in (124),

$$\frac{dN}{dt} = A\,e^{-E_a/kT} \cdot e^{-(\Delta G_c)_{t=0}/kT} \cdot e^{mt/kT}, \tag{130}$$

and,

$$\frac{dN}{dt} = [(dN/dt)_{t=0}] \cdot e^{mt/kT} \quad 0 < t < t_b. \tag{131}$$

Upon integration,

$$N = [(dN/dt)_{t=0}(kT/m)] \cdot (e^{(m/kT)t} - 1), \tag{132}$$

or

$$\ln N = \ln[(dN/dt)_{t=0}(kT/m)] + (m/kT)t \quad 0 < t < t_b. \tag{133}$$

The result is an exponential increase in number of crystals, as illustrated by segment a–b of curve 3 (Figure 4.24.)

During the second stage, b to c, ΔG_v arrived at a constant value, and ΔG_c also remained constant, which implies (equation (128)) that nucleation sites were still abundantly present, and interfacial energy associated with nucleus formation (σ) remained uniform at the sites selected for nucleation. With ΔG_c constant, the nucleation rate in time interval b to c remained constant (equation (125)):

$$N = A\,\exp\{-[(E_a + \Delta G_c)/kT]\}t, \quad t_b < t < t_c, \tag{134}$$

yielding equal numbers of new crystals in equal increments of time, as illustrated by linear segment b–c of curve 3 (Figure 4.24).

In the final stage, c to d, there was a decline in nucleation rate resulting possibly from a small increase in σ (equation (128)) and a larger increase in ΔG_c, related to the most favorable nucleation sites being taken. This decline is expressed by segment c–d of curve 3 (Figure 4.24)

Model 3 (in three stages) results in a sigmoidal nucleation curve (curve 3) and a normal or slightly skewed size distribution, as commonly found for garnet and other minerals. Sigmoidal nucleation curves may be described by use of the Avrami equation, which in Sec 4.1 was derived for the rate of a reaction. The equation, adapted to nucleation, is

$$N = 100(1 - e^{-k_N t^n}), \tag{135}$$

where N extends from zero to 100 crystals, and t from zero to 1.0 time units; k_N and n are empirical constants. Curve 3 of Figure 4.24 is a plot of the equation for $k_N = 5.6$ and $n = 3$.

Carlson (1989) proposed a nucleation model (model 4) in which a decline in nucleation rate in the final stage of crystallization is attributed to a decrease in volume of matrix available to nucleus formation, resulting from the presence of diffusion haloes about growing crystals, haloes in which nucleation is not likely to occur. The equation is

$$\frac{dN}{dt} = \left(\frac{dN}{dt}\right)_{t=0} (1 - V_u)\exp(\alpha t), \tag{136}$$

where V_u is the fraction of 'undepleted' matrix ($V_u = 1$ at $t = 0$) and α is a constant. The equation when plotted produces a sigmoidal curve similar to curve 3 in Figure 4.24. Various unimodal size distributions, whether normal or skewed, can be described by adjusting $(dN/dt)_{t=0}$ and α to fit the data.

Additional models could be constructed but what is now urgently needed is information on the nature and abundance of favorable nucleation sites that exist and are actually chosen while nucleation is in progress. Where nucleation occurs within a crystal, these sites are almost certainly different kinds of imperfections. Where nucleation occurs in an assembly of foreign crystals, the sites may lie on certain boundaries between certain crystals, where interfacial energy is relatively high or where embryo formation results in a small amount of strain energy (e.g., points of low pressure). But it is important not to lose sight of the element of atomic-scale randomness which is fundamental to the classical theory. Locally within the Earth, where temperature changes rapidly relative to atomic mobility, randomness may be demonstrated in the selection of nucleation sites *near* the reactant crystals — sites that are both favorable and less favorable — while atomic immobility makes inaccessible other sites further removed. With a very *slow* change in temperature, sufficient time may be available for activated atoms to search out the most favorable sites, not ignoring those that lie at some distance from the reactant crystals. But whether or not a particular site is chosen remains a matter of probability. In this sense, very slow reactions that occur within the solid Earth may be unique, and impossible to reproduce experimentally.

4.5 CRYSTAL GROWTH

With the addition of atoms to the surface of a solid nucleus, the nucleus becomes a crystal. Compared with nucleation, crystal growth may be readily observed, and is more amenable to study. Indeed, countless experiments have been carried out to measure the rate of growth of crystals in a vapor, aqueous solution, or melt; many of the early results are reviewed by Buckley (1951). General principles of growth are covered by Strickland-Constable (1968), Hartman (1973), Kirkpatrick (1975), Sunagawa (1984), and Chernov (1989). The growth of dendritic crystals of snow, metals, silicates, and other phases has attracted much attention recently, and this problem is discussed, for example, by Langer (1980, 1989) and Fowler (1990). In crystallization from a vapor or melt, removal of the heat of crystallization is commonly important in determining growth rate and crystal morphology.

Following a brief review of growth in a fluid phase, our attention will turn to crystal growth in a solid medium and the role of surface processes and diffusion in governing the rate of growth.

4.5.1 Growth in a fluid environment

Frequently, crystals that grow in a fluid medium are bounded by planar or near-planar surfaces (faces), and two principal mechanisms by which these surfaces advance have been proposed. In the Volmer model (Volmer 1939), a two-dimensional nucleus forms on the surface and spreads to the face edges, and the process is repeated. One difficulty with this model is that high supersaturation is needed for growth to occur (Burton, Cabrera and Frank 1951), whereas growth has been observed frequently when the supersaturation is extremely small. In the Frank model (Frank 1949), the spreading of layers is facilitated by emerging screw dislocations (Sec. 1.4) as shown in Figure 4.25, and surface nucleation is not necessary. The advance of a step created by a screw dislocation should result in the step forming a spiral pattern when viewed from above; such a pattern was first observed by Griffin (1950) on a face of natural beryl. Subsequently, spiral steps have been detected on biotite (Amelinckx 1952; Figure 4.25e), garnet (Endo and Sunagawa 1968), kaolinite (Sunagawa and Koshino 1975), and other natural and synthetic crystals.

Recent studies have revealed much complexity in the structure and topography of crystal faces, and these results raise new questions regarding growth mechanisms. In general, several micro-mechanisms may be involved, including the diffusion of atoms across faces and through surface layers, and the stimulation or inhibition of growth by crystal imperfections and foreign particles. The computer simulation of crystal growth (e.g., Gilmer 1980) represents a relatively new approach to the problem.

4.5.2 Growth in a solid environment

Our concern at present lies principally in the growth of crystals in a medium which is entirely or predominantly solid. Growth of a crystal that is produced by a simple phase transformation will be examined as an example of growth determined by an interface process, and that resulting from exsolution will be examined

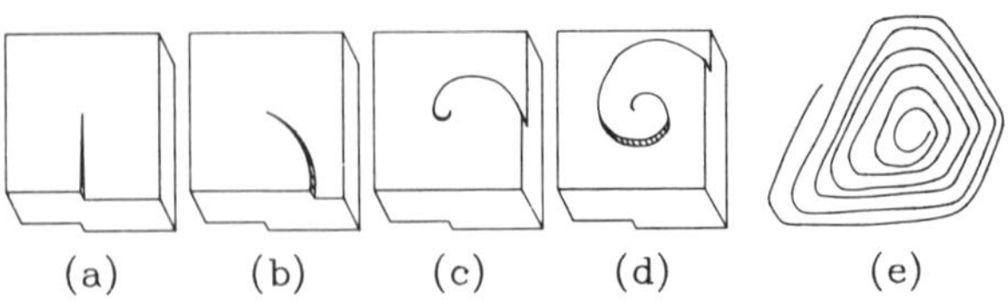

Figure 4.25 (a) to (d), development of a crystal growth spiral, centered at an emerging screw dislocation (the Frank model); (e) growth spiral on (001) of phlogopite from Vesuvius (Amelinckx 1952). Figure (e) reproduced by permission of Gauthier-Villars Publishers

as an example of growth determined by the diffusion of atoms to and from the interface.

Growth determined by interface processes

Consider a crystal (α) of a single substance, which is heated to a fixed temperature (T) above an $\alpha \rightarrow \beta$ transformation temperature, and imagine the nucleation and growth of β at a point in α. The rate of growth of β is defined as the rate of advance of the α–β interface, $\mathrm{d}r^{\beta}/\mathrm{d}t$, where r^{β} is distance normal to the interface. In the interface model, as described by Christian (1975), the rate of growth is determined by the Gibbs energy change for the reaction (Δg, per atom), the activation energy associated with the transfer of atoms across the interface ($\Delta g^{\ddagger}$, per atom), and the characteristic frequency (ν) which is a function of the vibrational frequency of the atoms, and in the activated complex theory, is equated to kT/h. (k and h are the Boltzmann and Planck constants.) Now the frequency of transfer of individual atoms from α to β is

$$\nu \exp[-\Delta g^{\ddagger}/kT], \tag{137}$$

and the frequency of transfer of individual atoms from β to α (Figure 4.26) is

$$\nu \exp[-(\Delta g^{\ddagger} + \Delta g)/kT], \tag{138}$$

and therefore the net frequency of transfer of atoms from α to β is (137) – (138), i.e.,

$$\nu\{\exp[-\Delta g^{\ddagger}/kT] - \exp[-(\Delta g^{\ddagger} + \Delta g)/kT]\}, \tag{139}$$

or, by rearrangement of terms,

$$\nu\{\exp[-\Delta g^{\ddagger}/kT](1 - \exp[-\Delta g/kT])\}. \tag{140}$$

This expression, multiplied by distance across the interface (i.e., δ in Figure 4.26) is the velocity of advance of the α–β interface, i.e., the growth rate,

$$U = \frac{\mathrm{d}r^{\beta}}{\mathrm{d}t} = \nu\delta\{\exp[-\Delta g^{\ddagger}/kT](1 - \exp[-\Delta g/kT])\}. \tag{141}$$

At equilibrium ($\Delta g = 0$), $\exp[-\Delta g/kT] = 1.0$, and thus the growth rate $(U) = 0$. At small departures from equilibrium ($\Delta g \ll kT$),

$$1 - \exp[-\Delta g/kT] = \Delta g/kT, \tag{142}$$

and

$$U = \nu\delta\{(\Delta g/kT)\exp[-\Delta g^{\ddagger}/kT]\}. \tag{143}$$

Thus at small departures from equilibrium, the growth rate is a near-linear function of Δg, and at a fixed temperature, the growth rate is constant. For a sphere or cube, a constant growth rate implies an increase in volume, proportional to t^3.

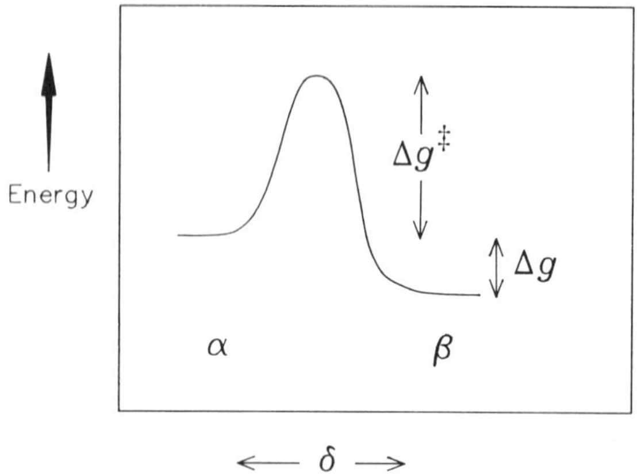

Figure 4.26 Energy barrier ($\Delta g^{\ddagger}$) to transfer of atoms from phase α to β during the growth of crystal β; Gibbs energy change for the $\alpha \rightarrow \beta$ reaction is Δg; distance across interface is δ

For the reaction $\alpha \rightarrow \beta$ which occurs with a rise in temperature, the growth rate, by this theory, will increase in response to an increase in Δg and a decrease in $\Delta g^{\ddagger}$. For the reverse reaction, i.e., upon cooling, the increase in Δg is eventually surpassed by an *increase* in $\Delta g^{\ddagger}$ (decrease in atomic mobility), and U passes through a maximum.

An example of growth in a system similar to that described above, is the growth of calcite from aragonite, which was investigated by Carlson and Rosenfeld (1981). Oriented crystals of natural aragonite were placed in a heating stage on a microscope, and growth of calcite was observed directly. The temperature was first taken to about 455 °C, where calcite crystals nucleated and began to grow. The temperature was then decreased rapidly to a set value between 375 and 455 °C (about 500 K from the boundary curve) where growth rates in the order of 10 μm h^{-1} were recorded by use of a graduated eyepiece.

Calcite crystals which were embedded in aragonite were observed to grow more rapidly in directions parallel to crystallographic c of aragonite than parallel to a. But for a particular direction and temperature, the growth rate was found to be constant, in agreement with the interface theory presented above. The increase in volume (ΔV for the reaction) did not appear to affect the growth rate appreciably. Growth rates did increase conspicuously with an increase in temperature (i.e., an increase in Δg) as expected. Additional data on the reaction were recently obtained by Liu and Yund (1993).

Growth determined by diffusion

In many nucleation-and-growth processes, the composition of the product crystals differs from that of the reactant crystals, and the possibility exists that the diffusion of atoms toward and away from the advancing growth surface determines entirely, or in part, the rate of growth.

Consider a crystal of phase α and formula (A, B) in which, at a given temperature, a crystal of β nucleates and grows as a product of the reaction $(A, B)^{\alpha} \rightarrow (A, B)^{\beta}$. Suppose that at equilibrium, the concentration of B (denoted C, and defined as the number of atoms of B per cm^3) is greater in β than in α, as in Figure 4.27. Also, assume that the interface process, as described above, is rapid relative to the rate of diffusion of B atoms to the interface. Then a concentration profile normal to the interface will form, as shown schematically in Figure 4.27. Assume, further, that the particle of β is spherical, that the composition of β remains constant (at C^{β}), that the composition of α at the interface is the equilibrium composition (C_e^{α}), and that the initial composition of α is C_0^{α}. We wish to obtain an expression for the rate of growth of the β crystal, dr^{β}/dt, in terms of the diffusion coefficient (D) which refers to A–B exchange diffusion in α.

Following Burke (1965), in a small advance of the interface, dr, the number of atoms of B added to β per unit area of interface is $C^{\beta}dr$; of these, $C_e^{\alpha}dr$ atoms are from α immediately at the interface, and the remainder, $(C^{\beta} - C_e^{\alpha})dr$, arrived by diffusion. The diffusive flux, therefore, is

$$(C^{\beta} - C_e^{\alpha})\frac{dr^{\beta}}{dt}, \qquad (144)$$

where r^{β} is the distance from the center of crystal β to the surface. Now, by the first equation of diffusion, this flux is equal to

$$D\left(\frac{\partial C}{\partial r}\right)_{r=r^{\beta}} \qquad (145)$$

which is positive because $\partial C/\partial r$, as drawn in Figure 4.27 is positive. Thus, by combining (144) and (145), and rearranging terms, we obtain for the growth-rate

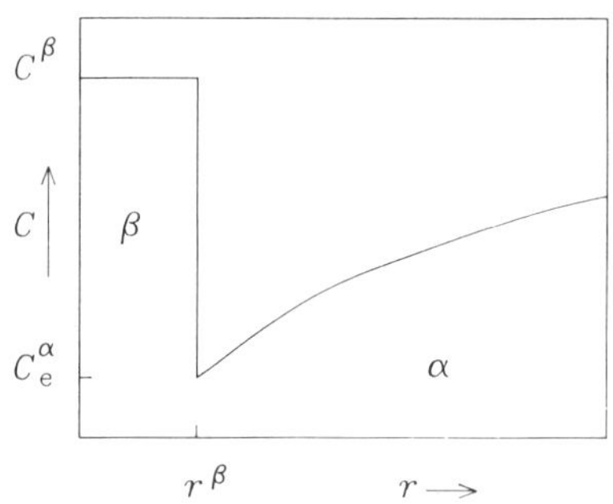

Figure 4.27 Concentration profile for the diffusion-controlled growth of crystal β (A,B) growing in crystal α (A,B); C is concentration of B per unit volume; r is distance normal to the α–β interface

equation,

$$\frac{dr^\beta}{dt} = \frac{D}{(C^\beta - C_e^\alpha)} \cdot \left(\frac{\partial C}{\partial r}\right)_{r=r^\beta}. \tag{146}$$

Provided that the width of the depleted zone is large relative to r^β, then the diffusion of B to the interface is analogous to the diffusion of B through a spherical shell of inner radius r_1 and outer radius r_2. Under these conditions (from Crank 1975),

$$\frac{\partial C}{\partial r} = \frac{C_2 - C_1}{(1/r_1) - (1/r_2)} \cdot \frac{1}{r^2}. \tag{147}$$

Taking $r_1 = r^\beta$, $r_2 \gg r^\beta$, $C_1 = C_e^\alpha$, and $C_2 = C_t^\alpha$ (where C_t^α is C far from the interface),

$$\left(\frac{\partial C}{\partial r}\right)_{r=r^\beta} = \frac{C_t^\alpha - C_e^\alpha}{(1/r^\beta) - (0)} \cdot \frac{1}{(r^\beta)^2}. \tag{148}$$

Combining (146) and (148),

$$U = \frac{dr^\beta}{dt} = D\left(\frac{C_t^\alpha - C_e^\alpha}{C^\beta - C_e^\alpha}\right) \cdot \frac{1}{r^\beta}. \tag{149}$$

Although the concentration of B far from β must slowly decrease, the quotient in brackets is initially constant, and with increasing t is approximately constant. Thus the growth rate (U) is proportional to $1/r^\beta$, and, by integration of (149),

$$\int r^\beta dr^\beta = D\left(\frac{C_t^\alpha - C_e^\alpha}{C^\beta - C_e^\alpha}\right) \int dt, \tag{150}$$

$$(r^\beta)^2 = 2D\left(\frac{C_t^\alpha - C_e^\alpha}{C^\beta - C_e^\alpha}\right) t, \tag{151}$$

i.e.,

$$r^\beta \propto t^{1/2}, \tag{152}$$

$$V^\beta \propto t^{3/2}, \tag{153}$$

where V^β is the volume of the spherical crystal of β.

In the above integration, the diffusion coefficient (D) was assumed to remain constant, but this coefficient refers to A–B exchange diffusion, which in general is concentration-dependent. Some departure from the $r^\beta \propto t^{1/2}$ relation might therefore be expected.

Normally, several crystals of β will be growing simultaneously, and after a time, will begin to compete for atoms of B. First, the depleted haloes of adjacent crystals may meet (referred to as 'soft impingement'), and then, some of these crystals themselves may meet ('hard impingement'), causing growth at the junctions to stop. Soft impingement can be accommodated approximately by multiplying the right-hand side of equation (149) by $(1 - y)$, where y is the fraction of β crystallized. Thus, during advanced crystallization, the parabolic decrease of r^{β} with time ($r^{\beta} \propto t^{1/2}$) will experience a further deceleration.

In the above theoretical model, the composition of the growing crystals was assumed to remain constant. This is a valid assumption for some mineral reactions, for example, the exsolution of dolomite from magnesian calcite. In this case, the reaction progress, defined as the volume fraction of product (dolomite), obviously cannot change from zero to 1.0, but only from zero to 0.2 or 0.3, depending on the initial composition of calcite and temperature. Thus, although soft impingement may occur, hard impingement is not a problem, and may occur only very locally, where two product crystals nucleate close together. For other exsolution reactions, the composition of the newly formed crystals is expected to change somewhat as growth progresses; an example is the growth of albite in alkali feldspar, where exsolution occurs by a nucleation-and-growth mechanism.

An opportunity for the extent of reaction to increase from zero to 1.0 exists for certain continuous reactions, such as (Mg, Fe) olivine → (Mg, Fe) spinel, but here the composition of the product crystals is expected to change significantly as growth progresses. Equation (149) must therefore by modified to allow C^{β} to change with time. Also, if the reaction is diffusion-controlled, both soft and hard impingement must occur.

In the above treatment, the shape of β crystals was, for convenience, assumed to be a sphere. With regard to exsolution reactions, however, the crystals commonly assume the shape of a plate or rod rather than a sphere. In a definitive, study of the problem, Ham (1958) arrived at the important conclusion that for widely spaced particles and small departure from equilibrium, parabolic growth is expected for plates and rods as well as for spheres, provided the growth shape remains constant. For crystal growth further from equilibrium, concentration gradients will be steep and depleted zones will be narrower than those shown in Figure 4.27, but Burke (1965) has argued that parabolic growth of isolated crystals remains theoretically plausible. A detailed discussion of all aspects of solid-state growth is presented by Christian (1975).

For an example of diffusion-determined growth, we now refer to some of the experimental results obtained by Yund, McCallister and Savin (1972), who investigated the exsolution of kalsilite from nepheline. In one group of experiments, crystals of synthetic nepheline of composition C_0 = 47.7 mole % $NaAlSiO_4$ (53.3 mole % $KAlSiO_4$) were annealed at 600 °C and 1 kbar, well within the solvus (Figure 4.28 inset), and the extent of kalsilite exsolution (ζ) was determined as a function of time. ζ is defined as,

$$\zeta = \frac{C_t - 47.7}{68.8 - 47.7},$$

where C_t is the mole % $NaAlSiO_4$ in nepheline at time t (measured by use of X-ray diffraction) and 68.8 is the equilibrium composition at 600 °C, as prescribed

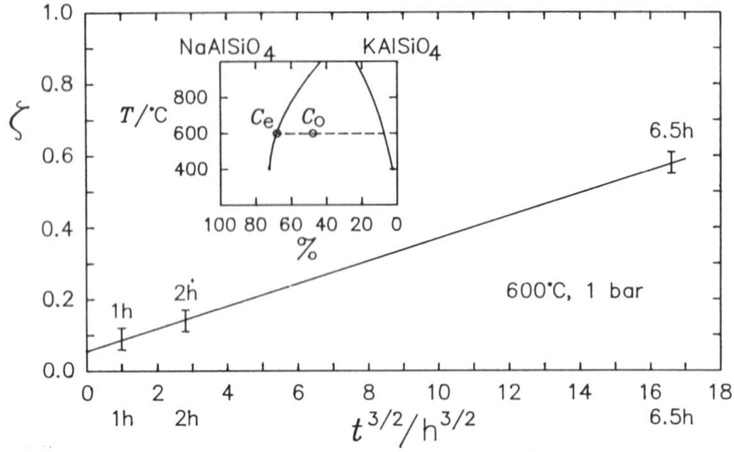

Figure 4.28 Rate of exsolution of kalsilite from nepheline (Na, K)$AlSiO_4$ (of initial composition 47.7 mole % $NaAlSiO_4$), at 600 °C, 1 bar; fraction transformed (ζ) is plotted against time (hours) to the power $\frac{3}{2}$; data from Yund, McCallister and Savin (1972). Inset shows the solvus, from Tuttle and Smith (1958), reproduced by permission of *American Journal of Science*, with location of initial composition (C_0) and equilibrium composition (C_e) at 600 °C. Nucleation occurred very rapidly, and the ζ–t relation is therefore an expression of the rate of growth of kalsilite crystals

by the solvus. Thus, as ζ changes from zero to 1.0, the volume fraction of exsolved kalsilite (as determined by the lever rule), changes from zero to 35%

The authors presented independent evidence that nucleation occurred very rapidly, and consequently the measured increase of ζ is a true measure of crystal growth. From the above theory for diffusion-controlled growth (equations (151) and (153)), ζ, which is a linear function of the volume of kalsilite crystallized, is expected to increase (initially) as $t^{3/2}$; Figure 4.28 shows that the experimental data appear to satisfy this expectation. With increasing time (i.e., for $t > 7$ h) ζ was found to increase more slowly, possibly the consequence of soft impingement. In the many additional experiments that were carried out between 400 and 700 °C, the growth rate was found to increase with temperature in accord with the Arrhenius equation.

Growth determined by interface processes and diffusion

Growth determined by the mechanisms described above represents two limiting cases. It is commonly assumed that growth is either interface- or diffusion-controlled, but Burke (1965), in particular, has emphasized that both processes may be determinative. Thus if the rates of both processes are approximately equal, then n in the expression $r^{\beta} \propto t^{n}$ will lie between 1.0 (interface control) and 0.5 (diffusion control), decreasing with impingement. Moreover, in many natural processes, the temperature changes with time, and early interface control could conceivably change to later diffusion control, or vice versa.

4.5.3 Growth in a multi-phase crystal assembly

The growth of crystals of garnet and other minerals in an assembly of reactant and passive crystals can be examined in relation to the theory described above. Consider, for example, a euhedral crystal of garnet that is growing in a matrix of chlorite and abundant quartz, and suppose that the crystal is in contact only with quartz. Assume, further, that the atoms which enter the growing crystal are released by the reaction

$$\text{chlorite} + \text{quartz} \rightarrow \text{garnet} + H_2O,$$

and that these atoms move to the growth surface along quartz–quartz grain boundaries, which terminate at the growth surface. Although some of the atoms for garnet growth may come from the immediately surrounding quartz crystals, most of these crystals must be passive in the sense that they remain unaffected by the reaction or they participate only in being displaced from the path of the advancing growth surface.

Consider first that garnet growth is principally interface-controlled. Then the processes (1) reaction of chlorite and quartz at some distance from the garnet crystal, (2) diffusion of those atoms that are required for growth to the growth surface, and (3) diffusion away from the surface of those atoms not needed for growth, must all occur at a rate that is rapid relative to the correct placement of atoms on the garnet surface. Under these conditions, the interface theory described above (Sec. 4.5.2) applies, and a linear growth rate ($r \propto t$) is expected (Figure 4.29). This requires, however, a constant driving force (Δg), which, for a continuous reaction of the kind here considered, could be maintained approximately by a suitable rise in temperature. If grain-boundary diffusion is responsible for the transport of growth atoms to the garnet surface, concentration gradients must exist within these boundaries, but the gradients could be very small — too small to influence appreciably the rate of growth.

Consider next the possibility that growth is controlled by diffusion. Assume (for convenience) that the growing garnet crystal is a sphere, which, as above, is embedded in a matrix of quartz crystals, denoted q. The garnet crystal has nucleated at a q–q boundary, and now a number of lines appear on the surface — lines which represent the intersection of the garnet surface with q–q boundaries. Assume further that growth atoms arrive by diffusion along pipe-like channels which lie within the q–q boundaries and terminate at the surface. Suppose that n_p pipes, each of cross-sectional area a_p, intersect the surface; the cross-sectional area for the flux of growth atoms is therefore $n_p a_p$. As the atoms arrive, they spread out over the surface of the garnet crystal, and become attached at a rate that is fast relative to diffusion.

Assume that, following an initial period of growth, a steady state is obtained, with cross-sectional area ($n_p a_p$), concentration gradients in channels, and the diffusion coefficients all remaining constant as growth continues. Thus, from the first equation of diffusion, the flux of garnet-forming atoms is also constant. One must imagine several kinds of atoms (Fe, Mg, Al, Si, O) diffusing through the channels toward the crystal (presumably as ions), while other atoms (Si, O) diffuse in the

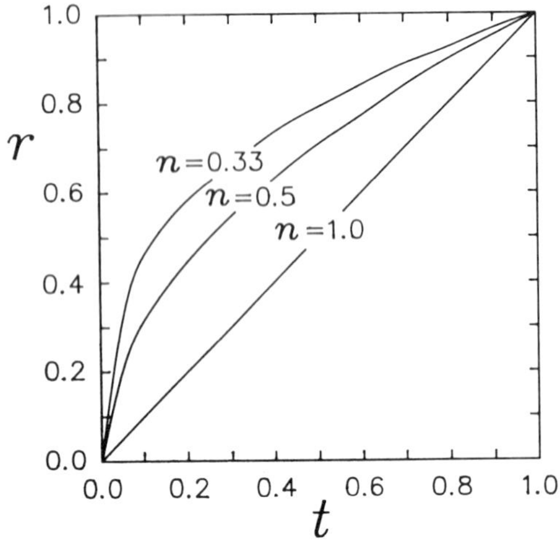

Figure 4.29 Illustration of crystal growth, described by the relation, $r \propto t^n$ for three values of n; r is crystal radius and t is time

opposite direction, maintaining a local balance of charge. The diffusion coefficient refers to this complex system of exchange diffusion, or a part of it which controls the magnitude of the flux.

Given, then, that the total flux of atoms arriving at the garnet crystal is constant, the rate of increase of crystal volume (V) is constant,

$$\frac{\mathrm{d}V}{\mathrm{d}t} = k_J, \tag{154}$$

and the rate of increase of radius (the growth rate) is

$$\frac{\mathrm{d}r}{\mathrm{d}t} = \frac{\mathrm{d}V}{\mathrm{d}t} / \frac{\mathrm{d}V}{\mathrm{d}r}. \tag{155}$$

But,

$$V = (4/3)\pi r^3, \tag{156}$$

$$\frac{\mathrm{d}V}{\mathrm{d}r} = 4\pi r^2. \tag{157}$$

By substitution of (154) and (157) in (155),

$$\frac{\mathrm{d}r}{\mathrm{d}t} = k_J/(4\pi r^2). \tag{158}$$

Integrating (158) to obtain a relation between r and t,

$$\int r^2\,\mathrm{d}r = (k_J/4\pi)\int \mathrm{d}t, \tag{159}$$

$$r = (3k_J/4\pi)^{1/3}t^{0.33}, \tag{160}$$

$$r \propto t^{0.33}. \tag{161}$$

The form of this parabolic expression is shown in Figure 4.29, where it may be compared with $r \propto t^{0.5}$ which refers to diffusion-controlled growth in a transformation, as defined in Sec. 4.5.2.

In general, as a crystal becomes larger, the length of intersection of q–q boundaries will increase, or in terms of the pipe model, the number of pipes leading to the growth surface will increase. It is intuitively obvious, and the construction of simple models will confirm, that as r exceeds the radius of the matrix crystals by a factor of 3 or more, the length of q–q boundaries intersected by the crystal increases rapidly. The effect will be an increase in the flux of growth atoms and an increase in growth rate, although this could be compensated to some extent by a decrease in concentration gradient and flux within each pipe. Thus, in the relation

$$r \propto t^n,$$

n is expected to increase with r, i.e., with time, from 0.33 to a number that is not to exceed 0.50.

4.6 REACTION RATE IN TERMS OF NUCLEATION AND GROWTH RATES

The rate of production of crystals (total volume) in a nucleation-and-growth reaction can obviously be expressed by combining, in a single expression, the rate of nucleation and the rate of growth. This problem, which was considered briefly in Sec. 4.1, is now examined more thoroughly. Three cases will be discussed: (1) initial $\alpha \rightarrow \beta$ transformation, without impingement, (2) nucleation and growth in a multi-phase crystal assembly, and (3) complete $\alpha \rightarrow \beta$ transformation, allowing for impingement. Much of what follows is from Christian (1975) who has summarized and expanded on the theoretical treatment of Johnson and Mehl (1939) and Avrami (1939, 1940, 1941).

4.6.1 Transformation without impingement

Consider a crystal of α in which islands of β nucleate at random and grow at a rate $U = \mathrm{d}r/\mathrm{d}t$ (Figure 4.30). Assume that the growth rate is constant, as it might be for isothermal, interface-controlled growth, and assume that the crystals are spherical. Then the volume of any crystal that has nucleated at time τ and is examined at time t is

$$v_\tau = (4/3)\pi[U(t-\tau)]^3. \tag{162}$$

(Recall that radius = growth rate × time.) As crystals of β increase in number and size, the total volume of β (denoted V^β) increases, and the volume of untransformed α (denoted V^α) decreases proportionately. It is important to define carefully the

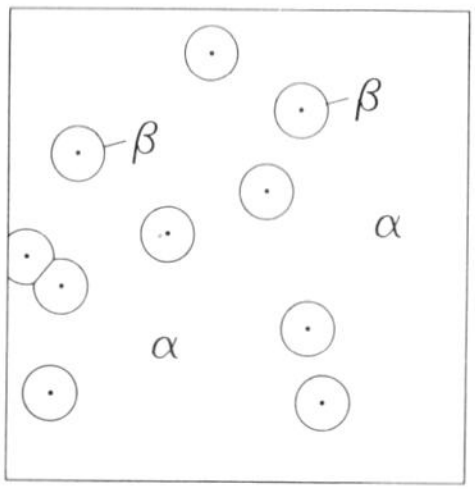

Figure 4.30 Two-dimensional representation of the random nucleation of crystals of β in a crystal of α; note minor impingement.

nucleation rate, ${}^{V}I$, which is the rate of appearance of nuclei per unit volume, e.g., 10 new nuclei per cubic centimeter per second. Then $({}^{V}I)V^{\alpha}$ is the nucleation rate (new nuclei per second) in *all* of untransformed α at a particular time, and $({}^{V}I)V^{\alpha}\,\mathrm{d}\tau$ is the number of new nuclei formed in all of untransformed α in the time interval $\mathrm{d}\tau$. Now the total volume of β produced in time interval $\tau + \mathrm{d}\tau$ is

$$V^{\beta}_{\tau+\mathrm{d}\tau} = v_{\tau}[({}^{V}I)V^{\alpha}\,\mathrm{d}\tau], \tag{163}$$

and the total volume of β produced at time t is

$$V^{\beta} = \int_{\tau=0}^{\tau=t} (4/3)\pi[U(t-\tau)]^3[({}^{V}I)V^{\alpha}]\,\mathrm{d}\tau. \tag{164}$$

This equation can be applied to the first stage of a reaction, when the volume of untransformed α is only slightly less than the initial volume of α, i.e $V^{\alpha} \sim V_0^{\alpha}$, and before appreciable impingement occurs. Different possibilities exist for ${}^{V}I$, but consider for example that ${}^{V}I$ is constant, i.e., nuclei form continuously, at a constant rate, as reaction progresses. Then equation (164) becomes

$$V^{\beta}/V_0^{\alpha} = (4/3)\pi U^3({}^{V}I)\int_{\tau=0}^{\tau=t} (t-\tau)^3\,\mathrm{d}\tau. \tag{165}$$

On the left is the volume fraction transformed, ζ, and by integration of the right-hand side,

$$\zeta = (1/3)\pi U^3({}^{V}I)t^4. \tag{166}$$

Thus $V^{\beta} \propto t^4$, and the reaction rate, obtained by differentiation, is proportional to t^3. When $({}^{V}I)$ or U or both are functions of time, these terms must be placed within the integrand. For example, with $({}^{V}I)$ constant and growth controlled by diffusion, i.e., $r = k_u t^{1/2}$, where k_u is a constant, we obtain

$$\zeta = (8/15)\pi k_u^3({}^{V}I)t^{5/2}, \tag{167}$$

and the reaction rate is proportional to $t^{3/2}$.

4.6.2 Nucleation and growth in a multi-phase crystal assembly

Consider a reaction $\alpha \rightarrow \beta$ which occurs in the presence of an assembly of passive crystals, q, as shown in Figure 4.31. In this model, crystals of β nucleate on q–q

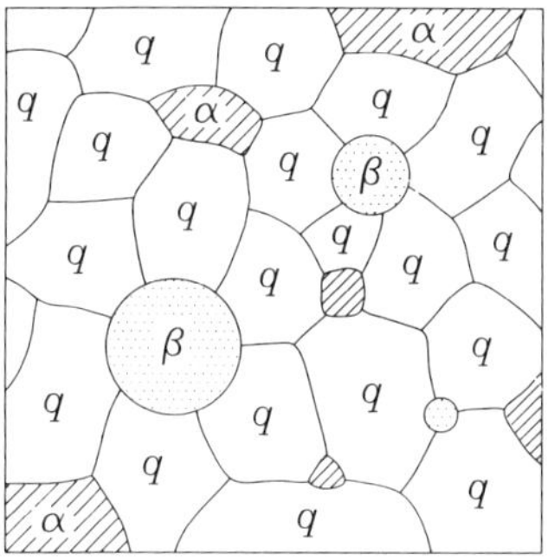

Figure 4.31 Two-dimensional representation of the nucleation and growth of crystals of β in an assembly of reactant crystals (α) and passive crystals (q)

boundaries and grow into q rather than into α, as in the case described above (Figure 4.30). Given that the final volume of β in the entire system is small (less than 10%), and provided that the spatial nucleation of β is approximately random, only minor β–β impingement will occur. Thus equation (164) can be used to describe the entire $\alpha \rightarrow \beta$ transformation. Now, however, the nucleation rate must be redefined. Let ${}^{V}I$ for the present case denote the rate of appearance of nuclei per unit volume of the crystal-assemblage system, and let V^s denote the volume of this system. Then the nucleation rate for the system (number of new nuclei in the entire system in unit time) is $I = ({}^{V}I)V^s$. A small change in volume (V^s) may occur as the $\alpha \rightarrow \beta$ reaction progresses (the volume change for the reaction, and a volume change resulting from the addition or removal of components), but assume for the present that V^s is nearly independent of time. Then equation (164) applied to this model, is

$$V^{\beta}/V^{s} = (4/3)\pi \int_{\tau=0}^{\tau=t} [U(t-\tau)]^3 [I]\,\mathrm{d}\tau. \tag{168}$$

This equation can be used for product crystals of any shape by substituting the appropriate shape factor for $(\frac{4}{3})\pi$. Students are invited to solve and plot this equation for different combinations of growth rate (U) and nucleation rate (I) (Kretz 1974).

4.6.3 Transformation taken to completion: allowance for impingement

Equation (164) was derived for the initial stage in an $\alpha \rightarrow \beta$ reaction, where crystals of β grow in α (Figure 4.30). We will now return to this model and derive a more general equation which will make allowance for the sweeping out of possible nucleation sites as crystals of β increase in size, and for the progressive termination of growth resulting from β–β impingement.

${}^{V}I$ and V^{α} are defined as in Sec. 4.6.1, and again, $({}^{V}I)V^{\alpha}\,\mathrm{d}\tau$ is the number of nuclei of β to appear in time interval $\mathrm{d}\tau$. The volume of β is V^{β}. We now define $({}^{V}I)V^{\beta}\,\mathrm{d}\tau$ as the number of nuclei of β which would have formed had the volume of α not been reduced by growth of crystals of β; Avrami (1940) has referred to these as 'phantom' nuclei. Avrami also introduced the concept of an 'extended volume' of β, denoted V_{ex}^{β} which is defined by

$$\mathrm{d}V_{ex}^{\beta} = v_{\tau}[(^{V}I)(V^{\alpha} + V^{\beta})]\,\mathrm{d}\tau, \tag{169}$$

or by (169) in integrated form,

$$V_{ex}^{\beta} = (4/3)\pi V \int_{\tau=0}^{\tau=t} U^{3}(^{V}I)(t-\tau)^{3}\,\mathrm{d}\tau, \tag{170}$$

where, by assuming a negligible change in reaction volume, $V = V^{\alpha} + V^{\beta}$. The extended volume, V_{ex}^{β} is greater than V^{β} because of the presence of 'phantom' crystals which nucleated at places already occupied by β, and because impingement was not permitted, i.e., adjacent crystals of β grew into each other, and some volumes of β were counted twice.

An expression for V_{ex}^{β} in terms of V^{β} is now needed. Consider a region of volume V which contains both α and β (as in Figure 4.30). The fraction of volume occupied by β is V^{β}/V; the fraction of volume occupied by α is $1-(V^{\beta}/V) = V-(V^{\beta}/V)$. Initially, V^{β} and V_{ex}^{β} are equal, but with time, $\mathrm{d}V_{ex}^{\beta}$ exceeds $\mathrm{d}V^{\beta}$, and the relation is written

$$\mathrm{d}V^{\beta} = (V^{\alpha}/V)\,\mathrm{d}V_{ex}^{\beta}, \tag{171}$$

or

$$\mathrm{d}V^{\beta} = [1-(V^{\beta}/V)]\,\mathrm{d}V_{ex}^{\beta}. \tag{172}$$

By integration of (172),

$$V_{ex}^{\beta} = -V\ln[1-(V^{\beta}/V)], \tag{173}$$

and by substitution of (173) in (170),

$$\ln[1-(V^{\beta}/V)] = -(4/3)\pi \int_{\tau=0}^{\tau=t} U^{3}(^{V}I)(t-\tau)^{3}\,\mathrm{d}\tau. \tag{174}$$

Provided that U and ^{V}I remain constant,

$$\ln[1-(V^{\beta}/V)] = -(\pi/3)U^{3}(^{V}I)t^{4}. \tag{175}$$

Replacing V^{β}/V by ζ and rearranging terms,

$$\zeta = 1-\exp[-(\pi/3)U^{3}(^{V}I)t^{4}]. \tag{176}$$

Avrami proposed the following semi-theoretical equation, which was shown subsequently to be capable of accommodating a variety of reaction rates,

$$\zeta = 1-\exp(-kt^{n}). \tag{177}$$

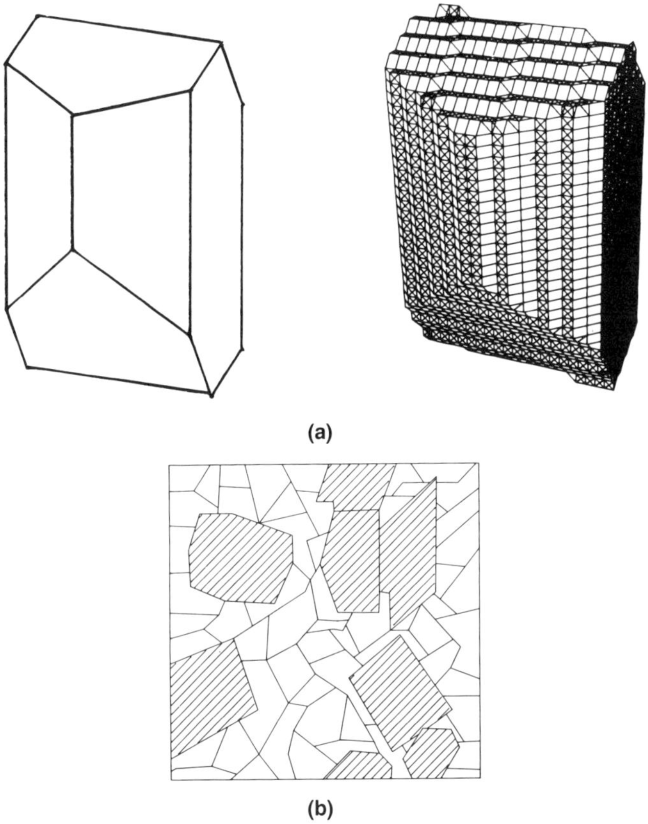

Figure 4.32 Computer simulation of crystal nucleation and growth; (a) simulated plagioclase crystal; (b) section through a three-dimensional 'space' in which nine plagioclase crystals (stippled) were nucleated instantaneously on random sites, with random orientation, and constant growth rate; these were then joined by 100 additional plagioclase crystals, treated similarly. The first group imposed their forms on the second group; impingement in the second group was mutual. Figure (a) is from Sempels and Raymond (1980, *Computers and Geosciences*), reproduced by permission of Pergamon Press, and (b) is from Sempels (1978)

Here k is a constant determined by the growth rate, the nucleation rate, and the shape of β crystals. With U and $(^V I)$ constant (equation (176)), n in (177) is 4. When $(^V I)$ decreases with time, n decreases and in general lies between 4 and 3. Equations (176) and (177) allow for impingement and should remain valid during the entire $\alpha \rightarrow \beta$ transformation, i.e., until α is completely replaced by β. The equations plot as sigmoidal curves, as shown for example in Figure 4.3 of Sec. 4.1; the decline in reaction rate results from impingement. The equations could also apply to nucleation and growth in multi-phase assemblies, as described in Sec. 4.6.2, when the volume fraction of the product crystals is sufficiently large to lead to an appreciable amount of impingement.

As an alternative to the experimental study of crystallization, Sempels (1978) has developed a computer program to simulate the nucleation and growth of crystals in three dimensions. The program permits the user to predetermine all of the following: nucleation sites (random, non-random), number of crystal phases present, nucleation rate (instantaneous, constant, variable), crystal orientation (random, preferred), and crystal shape (e.g. Figure 4.32a) including characteristic forms for various minerals (Sempels and Raymond 1980). In addition, the nature of impingement can be controlled, i.e., mutual impingement, or with crystals of one phase imposing their forms on those of another phase. A section through a three-dimensional assembly is shown, for example, in Figure 4.32b. This procedure appears capable of reproducing some of the complex microstructures found in Nature.

4.7 CONCLUDING REMARKS

In this chapter we have reviewed some of the elementary theory that exists to describe the rates of chemical reaction, diffusion, and the nucleation and growth of crystals. Some experimental results on reactions involving crystals were also examined. In most experimental studies of chemical kinetics, temperature is the chosen thermodynamic variable, and a closed system is taken to a fixed temperature and uniform pressure, where the rate of change of a process is measured.

Within the Earth, mineral reactions presumably occur frequently as temperature changes continuously. Consider, for example, the rise in temperature at a point (α) within a contact aureole, as shown in Figure 1.28 (p. 50), and suppose that a discontinuous reaction is initiated in response to an increase in temperature. Provided the reaction is endothermic, a portion of the heat that enters a small volume of rock at α will be consumed (the heat of reaction) but the temperature could continue to increase, thus increasing the driving force (ΔG) for the reaction. If the reaction is continuous, the driving force could remain approximately constant as temperature rises. Similarly, a change in pressure could provide the necessary driving force. Although rates of change are determined by activation energy, not ΔG, a change in ΔG will in general cause an increase or decrease in activation energy.

Natural systems also differ from the experimental counterparts where the systems are open to one or more components. Thus, in a dehydration reaction, the release of H_2O could escape continuously as reaction progresses, but in addition, H_2O molecules from adjacent or underlying volumes of rock must, in general, pass through the volume under study (Walther and Orville 1982). This flux of H_2O and other species could influence reaction rate and mechanisms. Moreover, the flux could, under certain conditions, provide the driving force for a reaction inasmuch as it changes the composition of the system; this possibility is examined in detail by Ferry (1991). Conspicuous examples of mineral reactions driven by an influx of matter are found in zones of contact metasomatism.

Moreover, many natural systems differ from experimental or theoretical analogues in that changes occur while a shear stress is applied, causing the systems to deform plastically. In metal, silicate, and other systems, shear stress is capable of increasing the density of dislocations, and the energy of these dislocations provides a driving force for recrystallization (Nicolas and Poirier 1976). With regard to phase transformations, few experimental data are available, but those of Dachille and Roy (1964), Newton, Goldsmith and Smith (1969), and others have demonstrated that the effect of shear stress is to increase reaction rates, and in some cases to cause metastable crystallization.

But of the various conditions that can be imposed on a natural system, the flux of heat into and out of the system, to cause a change in temperature, must be of primary importance in determining reaction rates and mechanisms. For, obviously, an appropriate decrease in heat flux could readily reduce the rate of a thermally activated reaction to zero. Given that a change in temperature has occurred, to render a system unstable, the question of how the system adjusts to this change — how the trillions of atoms rearrange themselves in an attempt to reduce the energy of the system — is the question that has occupied us so far. Some examples will be considered in the following chapter.

5 Granular Microstructures and Crystallization Mechanisms

We shall now examine several examples of granular microstructure in relation to the information that is provided on the mechanisms and rates of metamorphic processes. Some of the examples were chosen intentionally to demonstrate the benefits of combining microstructural patterns with microchemical analyses.

5.1 MECHANISM OF REACTION: ANDALUSITE → SILLIMANITE AND KYANITE → SILLIMANITE

The first appearance of sillimanite in metamorphic terranes is frequently viewed as an expression of the reaction

$$\text{andalusite} \rightarrow \text{sillimanite}$$

or

$$\text{kyanite} \rightarrow \text{sillimanite.}$$

We now examine some aspects of the mechanism of these important reactions, based on microstructural information. Rock samples that contain andalusite and sillimanite or kyanite and sillimanite are best suited for this purpose.

Pitcher (1965) and Kerrick (1990) have reviewed the numerous occurrences of associated Al silicates that have been described, and a few examples are referred to in Chapter 3 (Sec. 3.1.2). In places, sillimanite has obviously 'replaced' andalusite, but more frequently, small crystals of sillimanite are seen to have crystallized a few micrometers to a few millimeters from much larger crystals of andalusite, often in contact with quartz or biotite, as shown in Figure 5.1.

The nucleation and growth of product crystals at some distance from reactant crystals appears to be commonplace for metamorphic reactions in general (Carmichael 1969), and, where this occurs, matrix crystals must become 'involved' to some extent in the reaction. For example, where a sillimanite crystal nucleates on a muscovite–muscovite boundary, some muscovite must be 'removed' to provide space for the growing sillimanite crystal, and conversely, where atoms are dislodged from the surface of andalusite crystals, adjustments must occur, involving muscovite or other minerals, to take up the space vacated by andalusite. These rearrangements of matter were recognized by Harker (1932) and were first studied in detail by Carmichael (1969).

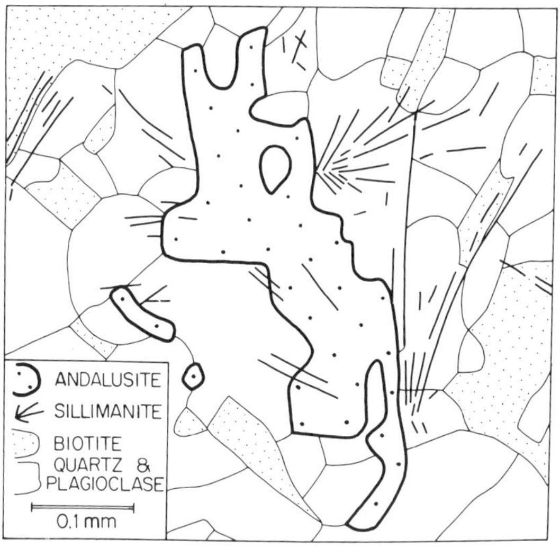

Figure 5.1 Distribution of sillimanite relative to andalusite and other minerals in schist at the north-west contact of the small Prestige pluton, 60 km north-east of Yellowknife (Figure 1.17 of Chapter 1)

One of the reactions examined by Carmichael is

$$\text{kyanite} \rightarrow \text{sillimanite},$$

which defines an isograd in the Grenville Province as noted in Chapter 1 (Sec. 1.9). Before describing the results of this study, let us imagine a volume of rock in which sillimanite crystals nucleate and grow *within* an existing crystal of kyanite, as in the simple phase transformation that was pictured in Sec. 4.6 (Figure 4.30). The reaction may be written

$$\underset{\text{kyanite (132 cm}^3)}{3Al_2SiO_5} \rightleftharpoons [6Al + 3Si + 15O] \rightleftharpoons \underset{\text{sillimanite (150 cm}^3)}{3Al_2SiO_5}, \tag{1}$$

where the square brackets enclose atoms that occur neither in kyanite nor in sillimanite, but are in the process of passing from kyanite to sillimanite, and (at a slower rate), from sillimanite to kyanite. In this case, transport distances for the atoms are obviously very small. Equation (1) may also be written as

$$\overset{\text{kyanite}}{\underset{\text{sillimanite}}{3Al_2SiO_5}} \rightleftharpoons [6Al + 3Si + 15O] \tag{2}$$

where arrows represent (before → after) changes in atomic arrangement.

Consider now a crystal of kyanite in contact with muscovite (at site A), and the nucleation and growth of a crystal of sillimanite on a muscovite–muscovite boundary (at site B) as represented in Figure 5.2. One possibility is that Al_2SiO_5 is transported from A to B, while $KAl_3Si_3O_{10}(OH)_2$ is transported from B to A:

$$\begin{array}{ccccc} \text{kyanite} & & \text{muscovite 'molecule'} & & \text{muscovite} \\ 3Al_2SiO_5 & + & [K + 3Al + 3Si + 12O + 2H] & \underset{\text{at B}}{\overset{\text{at A}}{\rightleftarrows}} & 1KAl_3Si_3O_{10}(OH)_2 \\ \text{sillimanite} & & \text{muscovite 'molecule'} & & \text{muscovite} \end{array}$$

$$\begin{array}{cc} & \text{kyanite 'molecule'} \\ + & [6Al + 3Si + 15O]. \\ & \text{sillimanite 'molecule'} \end{array} \quad (3)$$

In this case, all of the Al, Si, O atoms that enter sillimanite were derived from kyanite at A, and all of the K, Al, Si, O, H atoms released from muscovite at B enter newly crystallizing muscovite at A.

Carmichael (1969) has argued that at least some of the Al and Si atoms released from muscovite at B could enter crystallizing sillimanite, and some of the Al and Si atoms needed for muscovite crystallization at A could come from the decomposing kyanite. This is not an unreasonable expectation, provided that the elements occur as atoms or ions, for it would greatly reduce the amount of atomic traffic needed for reaction to occur. Numerous possibilities now exist; one of these is that no transport of Si occurs between A and B,

$$\begin{array}{ccccc} \text{kyanite (132)} & & & & \text{muscovite (141)} \\ 3Al_2SiO_5 & + & [K + 2H] & \underset{\text{at B}}{\overset{\text{at A}}{\rightleftarrows}} & 1KAl_3Si_3O_{10}(OH)_2 + [3Al + 3O]. \\ \text{sillimanite (150)} & & & & \text{muscovite (141)} \end{array} \quad (4)$$

By this mechanism, half of the Al atoms in crystallizing sillimanite are obtained from adjacent muscovite and half from distant kyanite.

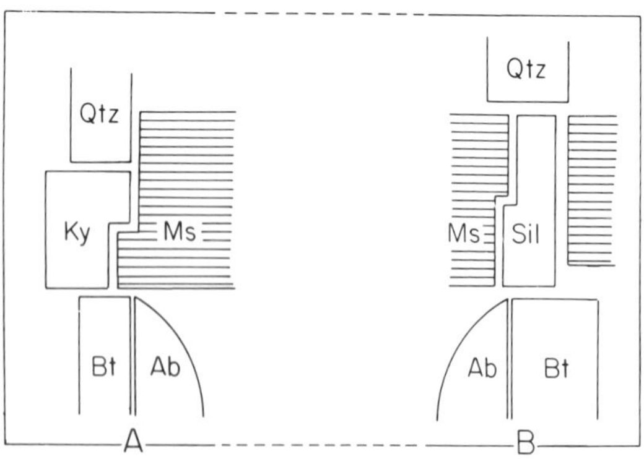

Figure 5.2 Schematic distribution of crystals at two sites, A and B, in a small volume of rock in which the reaction kyanite (Ky) → sillimanite (Sil) is in progress. Other minerals are quartz (Qtz), muscovite (Ms), biotite (Bt), and albite (Ab)

The process preferred by Carmichael (1969) is one that involves quartz as well as muscovite, and occurs with no transfer of Al from A to B,

$$\begin{array}{cccccc} \text{kyanite (132)} & \text{quartz (136)} & & \text{at A} & \text{muscovite (282)} & \\ 3Al_2SiO_5 & +\ 6SiO_2 & +\ [2K + 4H] & \rightleftarrows & 2KAl_3Si_3O_{10}(OH)_2 & +\ [3Si + 3O]. \quad (5) \\ \text{sillimanite (150)} & \text{quartz (136)} & & \text{at B} & \text{muscovite (282)} & \end{array}$$

By this mechanism, *all* of the Al released by kyanite at A enters adjoining muscovite, and *all* of the Al that enters sillimanite at B is derived from adjoining muscovite. The equation as written results in only small changes in volume at A and at B.

The association of sillimanite with biotite is well known and has been described from numerous localities. Carmichael has expanded the mechanism expressed by equation (4) to involve biotite and albite as well as muscovite and quartz. For convenience, let crystals of biotite and albite occur at both A and B, as in Figure 5.2. A possible mechanism for the kyanite-to-sillimanite reaction, as proposed by Carmichael (assuming again that Al is 'immobile'), is as follows:

$$\begin{array}{ccccccc} \text{kyanite (132)} & \text{quartz (68)} & & \text{biotite (150)} & & \\ 3Al_2SiO_5 & +\ 3SiO_2 & + & KMg_2FeAlSi_3O_{10}(OH)_2 & + & [K + Na + 2H] \\ \text{sillimanite (150)} & \text{quartz (68)} & & \text{biotite (150)} & & \end{array}$$

$$\begin{array}{cccccc} \text{at A} & \text{muscovite (282)} & & \text{albite (100)} & & \\ \rightleftarrows & 2KAl_3Si_3O_{10}(OH)_2 & + & NaAlSi_3O_8 & + & [2Mg + Fe + O]. \\ \text{at B} & \text{muscovite (282)} & & \text{albite (100)} & & \end{array} \qquad (6)$$

Notice that, by this mechanism, equal volumes of biotite and sillimanite are produced at B. Thus the frequently observed inclusions of sillimanite in biotite are explained by the simultaneous crystallization of these two minerals. This does not exclude the possibility of sillimanite nucleation on biotite–biotite or biotite–quartz boundaries.

Obviously, many possibilities exist for writing reaction equations such as (3) to (6) above, which in their most complex manifestations could involve *some* crystals of all minerals present, in addition to *all* crystals that actually define the 'net' reaction. Moreover, site A, as described in the equations and in Figure 5.2 could in fact refer to many different sites surrounding each B site, and arrows to show atomic displacements were therefore intentionally omitted from Figure 5.2. Additional complexity must arise where bodies of rock are deformed plastically, while reactions are in progress.

These considerations lead to the following view of a small volume of rock ($\sim 8\ cm^3$) in which the temperature (T) increases slowly from T_1, just above the kyanite–sillimanite boundary curve, to T_2, just below. At T_1, crystals of kyanite and other minerals occur in contact with a crystal-boundary phase, which Carmichael (1969) has assumed to be a fluid but is here viewed as a structured phase similar to that described in Sec. 1.5.2. Where sillimanite eventually nucleates on a quartz–quartz boundary (for example), some Al atoms must have been present on that boundary. It seems necessary to postulate, therefore, that the boundary phase everywhere in the volume of rock contains atoms of all of the elements present in that volume, not only those that reside in the two adjoining crystals. Now, as temperature increases and crosses the kyanite–sillimanite boundary curve, the concentration of Si, Al, and O within the crystal-boundary phase increases, and sites for nucleation are selected which are not necessarily adjacent to or within crystals of kyanite.

The reluctance of sillimanite to nucleate within kyanite may lie in the large increase in volume (equation (1)), resulting in a large strain energy, a large energy of nucleus formation, and therefore a large critical size. But the nucleation of sillimanite in andalusite (where the volume change is negative) is also uncommon, and another explanation must be sought. Oriented inclusion of sillimanite (fibrolite) in biotite was described by Chinner (1961) who suggested that epitaxial nucleation of sillimanite had occurred on or within biotite crystals. But sillimanite also occurs on quartz–quartz and other boundaries, as described by Spry (1969) and Vernon and Flood (1977), and evidently various kinds of sites on various kinds of crystal boundaries are capable of providing energetically favorable locations for sillimanite nucleation.

Equations such as (5) do not appear capable of providing the high activity of Al that is needed for sillimanite nucleation and it seems necessary to postulate that, at least initially, some Al migrated from kyanite to the nucleation sites, i.e., the presence of andalusite was ‘felt’ at these sites. Later, these equations may describe well the rearrangement of atoms that occurs during sillimanite growth.

There are, however, some difficulties with this model, wherein elements are viewed as occurring as atoms or ions. Where Al, Si, and O atoms are released from kyanite, they obviously become indistinguishable from Al, Si, and O atoms also present in the crystal-boundary phase, but derived from muscovite and quartz. The model therefore appears to be in violation of the concept of a chemical reaction in which one atomic configuration changes to another, via an activated complex (Sec. 4.1). It is possible therefore that equations of the kinds (5) and (6) represent an oversimplification, and that various *molecular* species (rather than ions) are present on the kyanite and sillimanite surfaces and in the crystal-boundary phase, as the kyanite → sillimanite reaction is in progress. A characterization of these species would greatly improve our understanding of metamorphic reactions in general.

5.2 KINETICS OF THE CRYSTALLIZATION OF GARNET

Several properties of garnet make this mineral particularly suitable for a study of the kinetics of natural metamorphic reactions. These properties include a broad stability field, a tendency to form relatively large and equidimensional crystals, and the common presence of compositional zoning. Consequently, many studies have been carried out on garnet crystallization, based on information on crystal size, crystal distribution, and compositional zoning, for example, by Galwey and Jones (1963), Hollister (1966), Jones and Galwey (1966), Edmunds and Atherton (1971), Ono (1975), Finlay and Kerr (1987), Cashman and Ferry (1988), and Carlson (1989). The following summary of a recent study (Kretz 1993) will serve to demonstrate the kind of information that can be obtained from the microstructure and micro-composition of a single garnet population.

A sample was collected from the cordierite zone of a broad, low-pressure thermal dome in the southern Slave Province of the Canadian Shield; the location is 0.8 km north of a tadpole-shaped granite pluton, 39 km northeast of Yellowknife (Figure 1.17, p. 30). The sample is homogeneous schist composed of garnet (1.5%), biotite (28%), plagioclase (48%), quartz (22%), and minor ilmenite, apatite, and secondary chlorite. Garnet was possibly produced from Al-rich chlorite (equation (7)), or from muscovite and a more common chlorite (equation (8)), while the needed Ca was possibly obtained from plagioclase (equation (9)):

$$\underset{\text{chlorite}}{2[(Fe,Mg,Mn)_{4.5}Al_{1.5}][Al_{1.5}Si_{2.5}]O_{10}(OH)_8} + \underset{\text{quartz}}{4SiO_2}$$

$$= \underset{\text{garnet}}{3(Fe,Mg,Mn)_3Al_2Si_3O_{12}} + 8H_2O, \qquad (7)$$

$$\underset{\text{chlorite}}{3[(Fe,Mg,Mn)_5Al][AlSi_3]O_{10}(OH)_8} + \underset{\text{muscovite}}{KAl_3Si_3O_{10}(OH)_2} + \underset{\text{quartz}}{3SiO_2}$$

$$= \underset{\text{garnet}}{4(Fe,Mg,Mn)_3Al_2Si_3O_{12}} + \underset{\text{biotite}}{1K(Fe,Mg,Mn)_3AlSi_3O_{10}(OH)_2} + 12H_2O, \qquad (8)$$

$$\underset{\text{chlorite}}{[(Fe,Mg,Mn)_5Al][AlSi_3]O_{10}(OH)_8} + \underset{\text{anorthite}}{CaAl_2Si_2O_8} + \underset{\text{quartz}}{SiO_2}$$

$$= \underset{\text{garnet}}{[(Fe,Mg,Mn)_5Ca]Al_4Si_6O_{24}} + 4H_2O. \qquad (9)$$

An estimate for the reaction pressure, based on the coexistence of andalusite and sillimanite in neighboring samples is 2 to 4 kbar, and a temperature estimate, based on the distribution of Mg and Fe between biotite and the margins of garnet crystals (Sec. 3.7.1, equation (139)) is 550 °C.

The garnet crystals, both large and small, are bounded by dodecahedral {110} and trapezohedral {211} faces, and when examined by SEM, the topography of the faces was found to be variable and locally, extremely complex (Figure 5.3).

The spatial distribution of the garnet crystals was investigated by dissecting a portion of the sample measuring $1.5 \times 1.4 \times 1.1$ cm (volume B) and by measuring the x, y, z coordinates of all of the 91 crystals that were encountered. The distribution in a volume $1.1 \times 1.0 \times 0.9$ cm (volume B′) which is centrally located in volume B is illustrated in Figure 5.4. The concept of randomness in the spatial distribution of garnet and other crystals which occupy a small proportion of a rock volume may

Figure 5.3 SEM image of garnet crystal, 0.36 mm in diameter, showing four-sided dodecahedral faces and six-sided trapezohedral faces, and (below) microtopography (ridges and bumps) on trapezohedral faces. From Kretz (1993), reproduced by permission of Blackwell Scientific Publications, Inc.

be defined by supposing that the volume is subdivided into a large number of cells and by specifying that the probability that a cell will receive a garnet nucleus is the same for each cell. Then the N sites that are chosen will be randomly distributed. With increasing crystal density and size, nucleation sites are increasingly destroyed and the remaining ones become less uniformly distributed; but for the study sample, where crystal density is 39 crystals/cm^3 and garnet volume is only 1.5%, the above concept of randomness is applicable.

Chandrasekhar (1943) derived the following equation for a random distribution of points in space:

$$s_e = 0.554N^{-0.333}, \tag{10}$$

Figure 5.4 Model of the distribution and size of 39 garnet crystals in 1 cm^3 of schist. From Kretz (1993), reproduced by permission of Blackwell Scientific Publications, Inc.

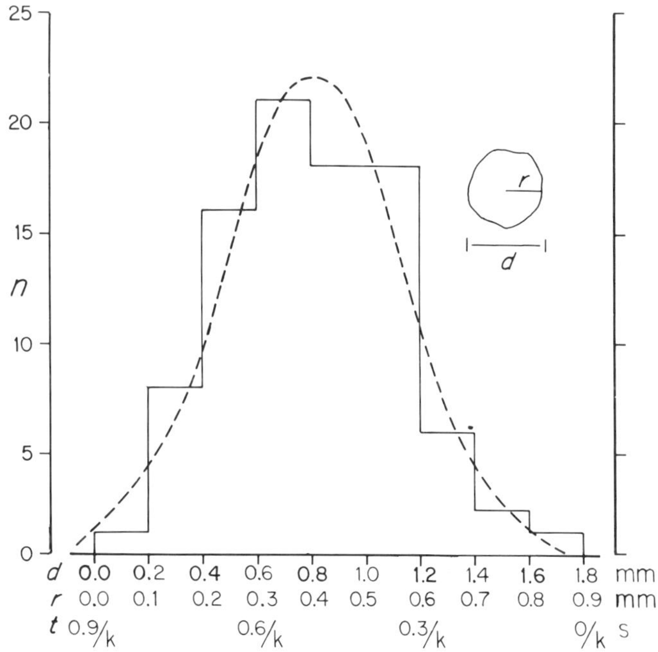

Figure 5.5 Garnet crystal-size distribution (n = number of crystals), using true diameter (d) or radius (r) as a measure of size; broken line is the fitted normal curve; time scale (t) is explained in text. From Kretz (1993), reproduced by permission of Blackwell Scientific Publications, Inc.

where N is the number of actual points/cm^3 and s_e is the mean of all of the distances measured from each of the N points to its nearest neighbor. Applying this equation to the 39 crystals in volume B′ (volume = 0.99 cm^3; $N = 39.4$ crystals/cm^3), we note that s_e for a random distribution is 0.163 cm. The measured mean distance to nearest neighbor in B′ is $s = 0.177$ cm, which is sufficiently close to 0.163 cm to demonstrate that the observed distribution is nearly random. A clustering of the crystals would decrease s to a very small number, while an arrangement that is most widely spaced would increase s to a maximum of approximately 0.30 cm.

Within the population of 91 garnet crystals in volume B, the diameters of the smallest and largest crystals are 0.19 and 1.67 mm, and the size distribution, as presented in Figure 5.5, is normal, with mean 0.81 mm and standard deviation 0.32 mm. Crystals that in three dimensions are relatively isolated (Figure 5.4) are not necessarily relatively large, and, in fact, one of the most isolated crystals is one of the smallest of the population. Thus, a relation between size and spatial distribution could not be detected.

Microprobe analyses were obtained on sections through the centers of several crystals of different size, both from volume B and from an adjacent volume, A, of the hand specimen. The results for four crystals from volume A are presented in Figures 5.6 and 5.7. The concentration profiles for the largest crystal (a.1) are similar to those described from medium-grade meta-shale in other metamorphic terranes. The atomic percentage of Mn ([Mn] = 100Mn/(Fe + Mg + Mn + Ca)) decreases from 14.5 at the center to 7.5, and then increases in the outer margin to

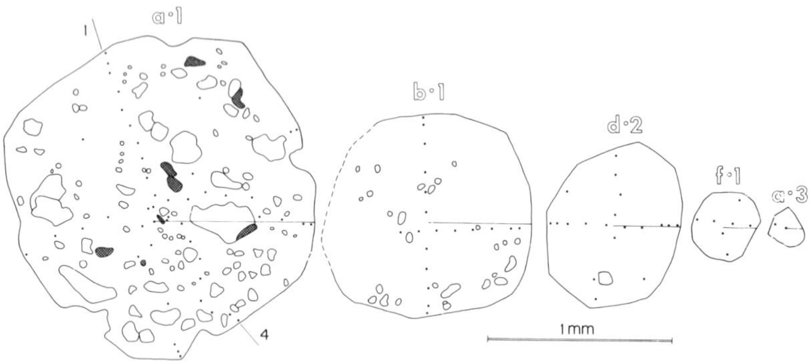

Figure 5.6 Sections through centers of garnet crystals of different sizes, showing location of spot microprobe analysis; inclusions are quartz (clear) and ilmenite (shaded); horizontal lines begin at growth centers (estimated from compositional zoning) and the length is the average distance from growth center to surface. From Kretz (1993), reproduced by permission of Blackwell Scientific Publications, Inc.

Figure 5.7 Concentration (atomic %, e.g. [Mn] = 100 Mn/(Fe + Mg + Mn + Ca)) versus distance (center to surface) profiles for the four largest crystals in Figure 5.6; arrows are growth vectors; length of vertical bars is ±2s (0.95 confidence limits), analytical precision. From Kretz (1993), reproduced by permission of Blackwell Scientific Publications, Inc.

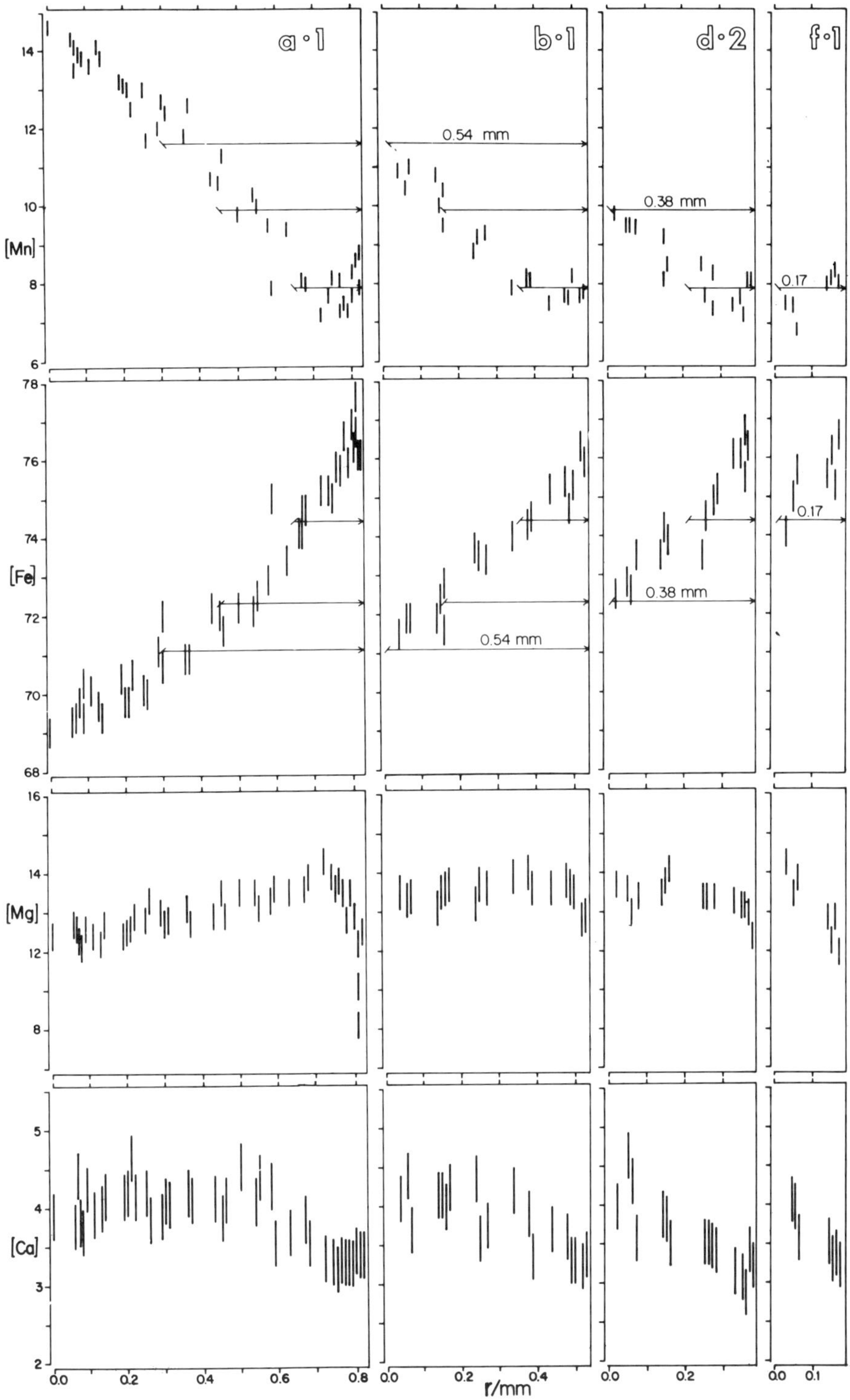
a·1
b·1
d·2
f·1
[Mn]
[Fe]
[Mg]
[Ca]
0.54 mm
0.38 mm
0.17
r/mm

8.5; [Fe] increases continuously from 67 at the center to 77 at the surface; [Mg] increases from 12.5 to 13.5 and then falls sharply to 11; [Ca] remains unchanged at 4, and then drops to 3.3. Progressively smaller crystals have progressively lower Mn and higher Fe concentrations at their centers, while marginal and surface composition of all crystals is the same (Figure 5.6).

The above data are now examined in relation to the growth rate (dr/dt), the nucleation rate (dN/dt), the change in composition of crystallizing garnet (dC/dt) and the rate of increase of garnet volume at the sample site (dV/dt).

The compositional data (Figure 5.7) lead to the proposal that growth of progressively smaller crystals was initiated at progressively later points in time, and that growth of all crystals was completed at the same time. Moreover, the Mn:Fe:Mg:Ca proportions of the atoms being deposited on the surfaces of growing garnet crystals in any brief interval of time were the same at all points of garnet growth throughout the study volume, as well as at other points where new crystals were created during that time interval. These and other atoms needed for crystal growth (Si, Al, O) were presumably concentrated in a crystal-boundary phase which provided access to the points of crystal growth.

The concept of a growth vector is now introduced to derive an expression for the rate of growth, based on compositional zoning. In Figure 5.7, growth vectors are drawn from the centers of crystals to the surfaces, and they are then placed in crystals of larger size to show that all vectors advanced at nearly the same rate, regardless of the size of crystal in which the vector is located. This implies that the same equation can be used to describe concentration–distance relations in all crystals, regardless of size. An empirical equation for the relation between [Fe] and distance measured from crystal surface toward the center (denoted r') to fit all of the data is

$$[\mathrm{Fe}] = 69 + 10.6(0.83 - r')^{1.5}. \tag{11}$$

The above results are in agreement with the growth model

$$\frac{dr}{dt} = k_U, \tag{12}$$

where r is radial distance from crystal center, t is time, and k_U is the growth constant. Now for a spherical or near-spherical crystal, the volume (v) is

$$v = (4/3)\pi r^3, \tag{13}$$

and the change of volume with radius is

$$\frac{dv}{dr} = 4\pi r^2 = a, \tag{14}$$

where a is surface area. The rate of change of volume is

$$\frac{dv}{dt} = \frac{dv}{dr} \cdot \frac{dr}{dt}, \tag{15}$$

and, by substitution of (12) and (14),

$$\frac{\mathrm{d}v}{\mathrm{d}t} = k_U a. \tag{16}$$

Thus the rate of increase of radius of the garnet crystals was evidently constant, and the rate of increase of volume was proportional to surface area. This is consistent with growth determined by the arrangement of atoms on the crystal surface, as expressed by the theoretical equation, from Sec. 4.5.2, eq. (143),

$$\frac{\mathrm{d}r}{\mathrm{d}t} = \nu\delta\{(\Delta g/kT)\exp[-\Delta g^{\ddagger}/kT]\}, \tag{17}$$

which predicts (for a simple phase transformation) a constant growth rate at constant temperature. The garnet-forming reaction is continuous and is expected to occur with increasing temperature, but provided that the Gibbs energy change (Δg) and activation energy associated with the transfer of atoms ($\Delta g^{\ddagger}$) remain nearly constant, a near-constant growth rate is theoretically possible.

The above conclusion regarding surface-controlled growth $r \propto t^1$ may be compared with the theory of diffusion-controlled growth (Secs. 4.5.2 and 4.5.3) which predicts a relation, $r \propto t^n$, where n is 0.5 or less.

The two central problems related to garnet nucleation are the nature of the nucleation sites and the rate of nucleation. Information on the nature of the sites cannot be obtained from observation, but in view of the classical theory (Sec. 4.4), certain chlorite–quartz crystal boundaries or triple junctions of relatively high interfacial energy form one possibility. In view of the random motion of atoms, upon which the classical theory is built, a high density of permissible nucleation sites (in the fine-grained chlorite–quartz matrix) should result in a distribution of crystals that is random, as found in the study sample.

The crystal-size distribution (Figure 5.5) will now be examined to provide some information on the rate of nucleation, expressed in relation to the rate of crystal growth, k_U. Assuming that the size distribution was not modified subsequent to garnet crystallization, it is apparent that the relatively few large crystals represent an initial stage in which the nucleation rate was small and increasing, the many medium-sized crystals represent an intermediate stage, of rapid nucleation, and the relatively small crystals represent a final stage, of decline. Note that there is a lower limit to size (0.19 mm diameter), which implies that crystallization continued very briefly after nucleation ceased.

Recalling the above evidence for a constant growth rate (k_U), we may now define a time scale extending from $r = 0.9$ mm (Figure 5.5), where $t = 0$, to $r = 0$, where $t = 0.9/k_U$ seconds. A relation between number of crystals (N) as a function of time is then obtained immediately by plotting the cumulative curve for the reverse histogram (Figure 5.8), i.e., the histogram with $r = 0.9$ mm on the left and $r = 0$ on the right (Figure 5.8 inset). The rate of nucleation at any time (e.g., $t = 0.6/k_U$ seconds) is the slope of the sigmoidal nucleation curve at that time.

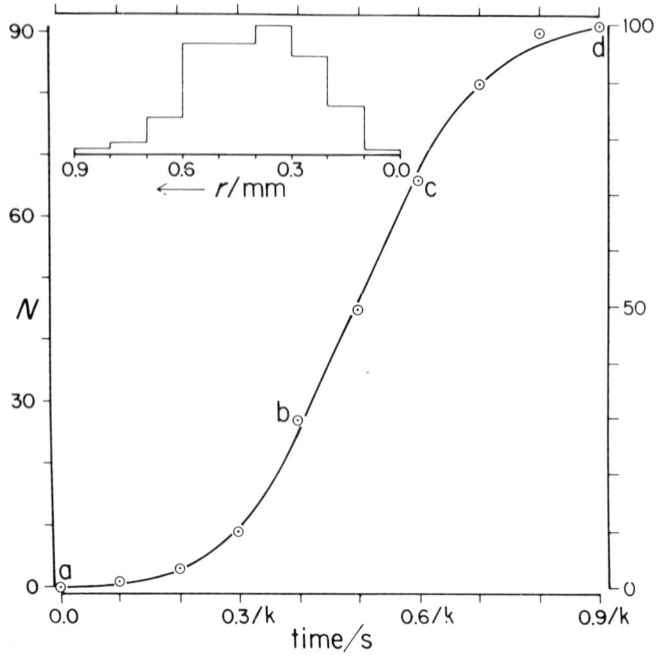

Figure 5.8 Number of crystals/cm^3 (N) as a function of time, obtained from a cumulative plot (data points) of the reverse histogram (inset, from Figure 5.5); an Avrami expression (equation (19)) is drawn to fit the data. From Kretz (1993), reproduced by permission of Blackwell Scientific Publications, Inc.

A description of the nucleation curve in Figure 5.8 may be obtained by use of the Avrami equation (Sec. 4.4.3). The variables are N, normalized to 100 crystals (Figure 5.8, right-hand vertical axis) and time (t). The equation is

$$N = 100[1 - \exp\{-(k_N t^n)\}]. \tag{18}$$

Constants k_N and n are obtained from the straight line which results from a plot of $-\ln[1 - (N/100)]$ against t on log-log paper. This produces the equation

$$N = 100[1 - \exp\{-(5.2t^{3.4})\}], \tag{19}$$

which is drawn in Figure 5.8.

Nucleation model 3 of Sec. 4.4.3 is now applied in the interpretation of the garnet nucleation curve. The curve is divided into three segments (a–b, b–c, c–d; Figure 5.8) and each is treated separately. Recall (Sec. 4.4.3) that the initial increase in nucleation rate is attributed to a decrease in the Gibbs energy of nucleus formation (ΔG_c), leading to an exponential relationship (Sec. 4.4.3, equation (133)),

$$\ln N = \ln[(\mathrm{d}N/\mathrm{d}t)_{t=0}(kT/m)] + (m/kT)t. \tag{20}$$

Segment a–b of the garnet curve (Figure 5.8) is of this form:

$$\ln N = -1.20 + 11.5k_U t. \tag{21}$$

The intermediate stage, represented by segment b–c (Figure 5.8) is a time of constant rate of nucleation, expressed theoretically by a linear equation (Sec. 4.4.3, equation (125)),

$$N = A\{\exp -[(E_a + \Delta G_c)/kT]\}t. \tag{22}$$

The equation for segment b–c of the garnet curve is of this form:

$$N = -56 + 215k_U t. \tag{23}$$

The final decline in nucleation rate in model 3 of Sec. 4.4.3 was attributed qualitatively to a depletion of favorable nucleation sites, i.e., an increase in the interfacial energy term (σ), and therefore an increase in ΔG_c. This interpretation is proposed for the decline in garnet nucleation (segment c–d, Figure 5.8), which is very approximately exponential.

The observed variation in garnet composition (Figure 5.7) is now examined in relation to both a transfer reaction, by which atoms needed for garnet crystallization were obtained principally from chlorite, and to exchange reactions, by which some Fe and Mn atoms were transported to garnet from biotite, in exchange for Mg, the two kinds of reactions occurring simultaneously.

The entire history of change in Fe/Mg/Mn/Ca atomic proportions entering the garnet structure at growth surfaces is recorded in the largest crystal of the population (a.1) in which the growth vector (from center to surface) may be regarded as a time scale. Transfer reactions are, of course, sensitive to changes in temperature, and in the following interpretation, garnet crystallization is assumed to have occurred with rising temperature; moreover, the increase is assumed to have been near-linear with time,

$$\frac{dT}{dt} = k_T. \tag{24}$$

Consequently, the growth vector in crystal a.1 may also be viewed as a temperature scale. This follows from the equation (25), in which C stands for concentration, e.g., $Fe/(Fe + Mg + Mn + Ca)$ and r_a is radial distance in crystal a.1.

$$\frac{dC}{dT} = \frac{dr_a}{dt} \cdot \frac{dt}{dT} \cdot \frac{dC}{dr_a}. \tag{25}$$

By substitution of $dr_a/dt = k_U$, and $dT/dt = k_T$,

$$\frac{dC}{dT} = \frac{k_U}{k_T}\frac{dC}{dr_a}. \tag{26}$$

The general nature of the Fe and Mn profiles in crystal a.1 can be understood by referring to a model reaction (equation (7), above, simplified to exclude Mg), which was examined from a thermodynamic viewpoint in Chapter 2, Sec. 2.5. Recall that

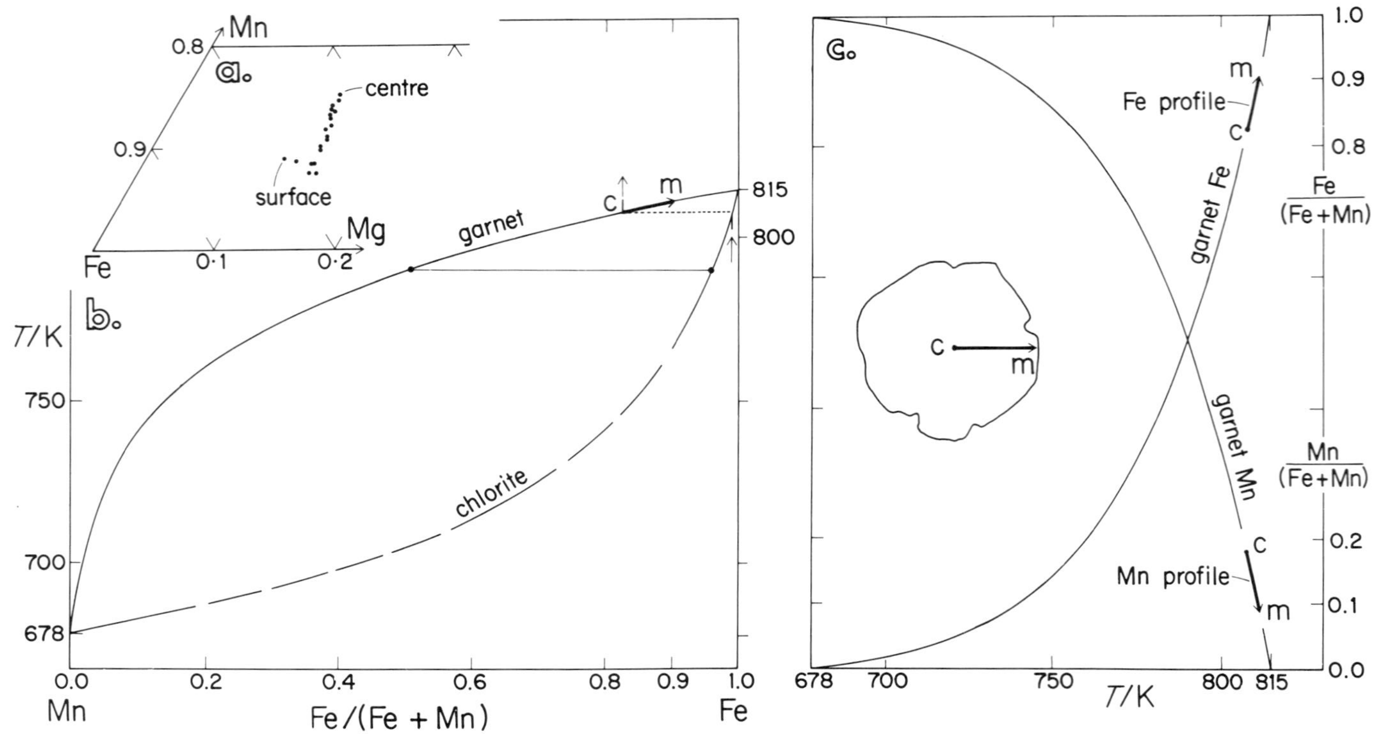

Figure 5.9 (a) Mn–Fe–Mg composition triangle showing the path followed by a point representing the Mn/Fe/Mg ratio in garnet crystal a.1; all points except the last two are presumed to lie on the garnet van Laar surface which relates composition to temperature, (b) Solution to the upper (garnet) van Laar equation in an idealized Fe–Mn system; the full garnet–chlorite tie line is from Kurata and Banno (1974); line c–m is a possible center-to-surface profile. (c) Garnet curve of (b) replotted to show profiles as they appear in Figure 5.7, assuming radial distance in the largest crystal is proportional to temperature. From Kretz (1993), reproduced by permission of Blackwell Scientific Publications, Inc.

in this model, the Mn and Fe end-member equilibria (at 2.0 kbar) are at 678 and 815 K, the heat of reaction is taken as 100 and 140 kJ/mol respectively, and both chlorite and garnet are assumed to be ideal mixtures. Application of the van Laar analysis then leads to an equilibrium phase diagram which is reproduced in Figure 5.9b. Referring to this figure, a near-equilibrium reaction of Fe-rich chlorite of composition represented by the arrow, requires that the garnet must change in composition from c to m, but in the absence of diffusive homogenization of garnet, the crystals will be zoned, with increasing Fe and decreasing Mn from center to surface, as shown by paths c–m in Figure 5.9c. This model can be enlarged to accommodate Mg; path c–m in Figure 5.9b then becomes a path on the garnet van Laar surface in the Fe–Mn–Mg phase diagram (Figure 5.9a). Both models explain qualitatively the observed Fe and Mn profiles (Figure 5.7), but they do not explain the Mn reversal, which is expressed in Figure 5.9a by a movement of the composition point off the van Laar surface.

The Mn reversal (which in terms of garnet volume is very significant), and the high mean Fe and Mn content of the garnet crystals combined (compared with the Fe and Mn content of down-grade, biotite-zone chlorite) can be understood by supposing that the chlorite-to-garnet transfer reaction was accompanied, increasingly, by Fe–Mg and Mn–Mg exchange between garnet surfaces and the otherwise passive biotite crystals. Exchange reactions of this kind are in fact expected to occur because newly formed layers of garnet on growing crystals will not necessarily be in exchange equilibrium with associated crystals of biotite.

A calculation was carried out to simulate the redistribution of atoms that occurred as the pre-garnet assemblage, chlorite (the assumed principal reactant), quartz, plagioclase, biotite changed to garnet, quartz, plagioclase, biotite, which occur in volume proportions 1.5:22:48:28. Following a conversion of the present minerals from volume (mm^3) to micromoles (μmol), reaction (9), above, was chosen to provide the Ca needed for garnet and reaction (7) to provide the principal share of Fe, Mg, Mn, Al, Si, and O. Biotite was assumed to remain unchanged, except for an adjustment in Fe:Mg:Mn proportions. These proportions in initial chlorite and biotite were chosen to make K_D (Mg–Fe; chlorite–biotite) = 1.1, and K_D (Mn–(Mg, Fe); chlorite–biotite) = 1.0, as commonly found elsewhere (Albee 1965); K_D is the distribution coefficient.

The results of the calculation are presented in Table 5.1. The chlorite composition lies between an Al-rich chlorite (equation (7)) and an Al-poor chlorite (equation (9)). The results can be used to write out the reaction equations, which include (7) and (9) above (individual equations for Fe, Mg, Mn) and the exchange reactions,

$$\text{Fe biotite} + \text{Mg garnet (surface)} = \text{Fe garnet (surface)} + \text{Mg biotite},$$

$$\text{Mn biotite} + \text{Mg garnet (surface)} = \text{Mn garnet (surface)} + \text{Mg biotite},$$

which were displaced to the right.

Table 5.1 Proposed mass-balance change in production of a garnet-bearing assemblage (B′) from a chlorite-bearing assemblage (X)

X
Chlorite: $33.2(Fe_{0.435}Mg_{0.555}Mn_{0.0095})_9Al_6Si_5O_{20}(OH)_{16}$
$+ 13.7(Fe_{0.435}Mg_{0.555}Mn_{0.0095})_5Al_2Si_3O_{10}(OH)_8$
Quartz: $9736SiO_2$
Plagioclase: $3290NaAlSi_3O_8 + 1424CaAl_2Si_2O_8$
Biotite: $1860K_{0.82}Na_{0.062}(Fe_{0.466}Mg_{0.524}Mn_{0.0096})_{2.24}Fe^{3+}_{0.16}Ti_{0.09}Al_{0.34}(Al_{1.25}Si_{2.75})O_{10}(OH)_2$
B′
Garnet: $127(Fe_{0.75}Mg_{0.13}Mn_{0.084}Ca_{0.036})_3Al_2Si_3O_{12}$
Quartz: $9600SiO_2$
Plagioclase: $3290NaAlSi_3O_8 + 1410CaAl_2Si_2O_8$
Biotite: $1860K_{0.82}Na_{0.062}(Fe_{0.437}Mg_{0.561}Mn_{0.0025})_{2.24}Fe^{3+}_{0.16}Ti_{0.09}Al_{0.34}(Al_{1.25}Si_{2.75})O_{10}(OH)_2$
H_2O: $321H_2O$

Reproduced by permission of Blackwell Scientific Publications, Inc.

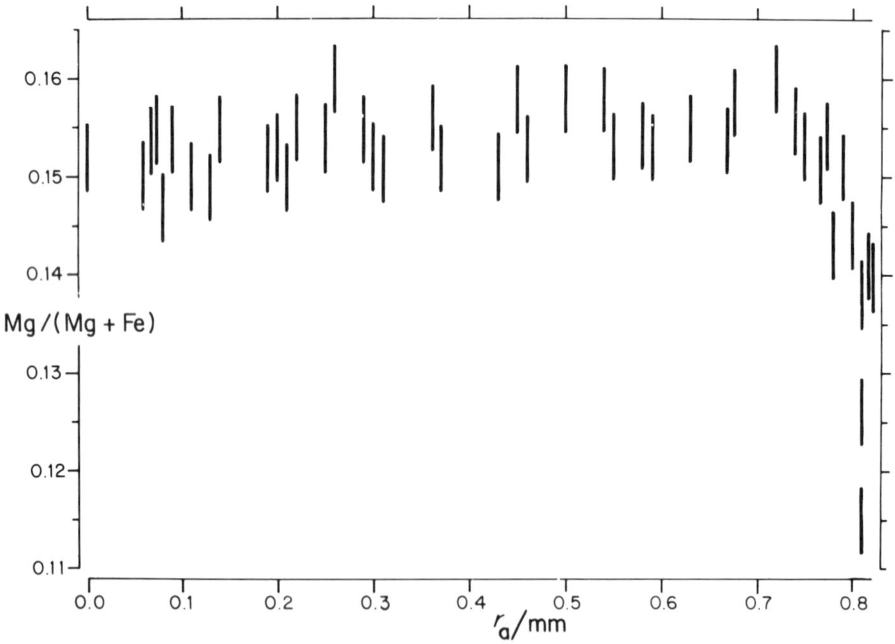

Figure 5.10 Mg/(Mg + Fe) in garnet crystal a.1; radial distance (r) and error bars as in Figure 5.7. From Kretz (1993), reproduced by permission of Blackwell Scientific Publications, Inc.

Figure 5.10 shows that $Mg/(Mg + Fe^{2+})$ of garnet remained constant at 0.155 during most of the time of garnet crystallization. The estimated ratio for biotite at the beginning of garnet crystallization is 0.529 (Table 5.1) and as presently found is measured at 0.561. These numbers are similar because the volume of biotite is much greater than that of garnet (by a factor of 19), and Mg–Fe^{2+} exchange between the two minerals will produce only a minor change in Mg:Fe of biotite. Using the estimate of 0.529 for Mg/(Mg+Fe^{2+}) in biotite at the beginning of garnet crystallization, with 0.155 for garnet, yields a distribution coefficient (K_D) of 7.1. The corresponding value for present garnet margin and present biotite is K_D = 8.9, which is consistent with exchange equilibrium at about 550 °C (Sec. 3.7.1). Thus, during most of the time of garnet crystallization (K_D, (garnet-surface–biotite) ~7.1), the garnet was apparently not in (Mg–Fe^{2+}) exchange equilibrium with the associated crystals of biotite, and only in the final stage, during crystallization of garnet rims, did Fe move from biotite to garnet, in exchange for Mg, in an attempt to achieve equilibrium, at K_D = 8.9. This explains the Mg reversal (Figure 5.7) and the decrease in Mg/(Mg + Fe) in crystal margins (Figure 5.10).

A similar interpretation extends to Mn. Initially (at the center of crystal a.1), Mn/(Mg + Fe^{2+} + Mn) = 0.15, while an estimate for this ratio in biotite at the same time (Table 5.1) is 0.0096, to give a distribution coefficient, K_D = (Mn in garnet/Mn in biotite) = 16. By comparison with the distribution of Mn in rocks of the upper amphibolite facies (Sec. 3.8.2) the expected value is 30 or more, and obviously, at the beginning of garnet crystallization, the distribution of Mn was far from equilibrium. With increasing time, the transfer reaction provided progressively less Mn to the garnet surfaces, causing even greater departure from equilibrium. Finally, however, a delayed migration of Mn to garnet from biotite, in exchange for Mg, evidently occurred, the delay resulting possibly from restricted diffusion in biotite, until at the close of crystallization, exchange equilibrium was finally established at K_D = 35, calculated from present garnet-rim–biotite analyses. This interpretation provides an explanation for Mn reversals in the margins of garnet crystals.

The proposed total redistribution of Mg, Fe, Mn, and Ca is represented in Figure 5.11, which shows that, although all of the Al, Si, O required for garnet was obtained from chlorite, most of the Mn and much of the Fe now in garnet were derived from biotite.

The total volume of garnet crystals in portion B (uncorrected for an estimated 10% of quartz and 1% of ilmenite inclusions) can be calculated from the crystal-size data and was found to be 37.4 mm^3. Given that radial distance in crystal a.1 is a linear measure of time, an expression for the rate of increase of garnet volume (i.e., the reaction rate) in portion B can be obtained from the Johnson–Mehl equation (Sec. 4.6). At time t, the volume of a spherical garnet crystal that nucleated at time τ is (4/3) $\pi k_U^3(t-\tau)^3$ mm^3, and the total volume of garnet (V) as a function of time is

$$V = \int_{\tau=0}^{\tau=t} (4/3)\pi k_U^3 (t-\tau)^3 I_{t=\tau}\, d\tau. \tag{27}$$

By use of an approximate expression for the rate of nucleation (I) which disregards the final decline in rate (Figure 5.8), i.e., $I \sim 300\ k_U^2 t$, an integration of equation (27) gives

$$V = 62.8 k_U^3 t^5 \tag{28}$$

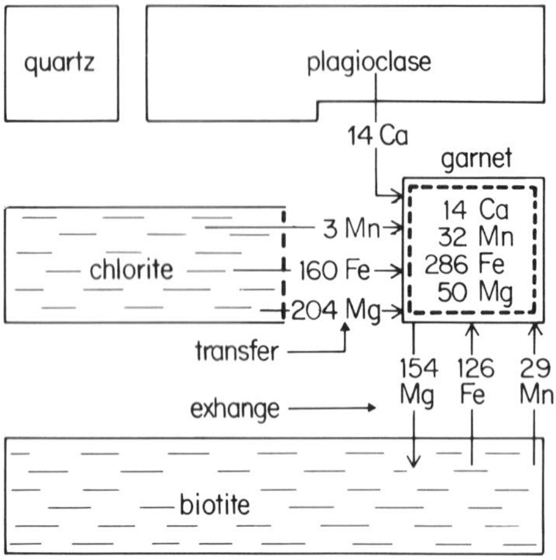

Figure 5.11 Calculated (total) rearrangement of Ca, Mn, Fe, and Mg atoms (μmol) in volume B′ (0.99 cm^3), showing the transfer and exchange parts of the garnet-forming reaction. By this interpretation, most of the Mn now in garnet was obtained from biotite. Details, including the transfer of Al and Si, are presented in Table 5.1. From Kretz (1993), reproduced by permission of Blackwell Scientific Publications, Inc.

which is plotted in Figure 5.12. Thus the volume of garnet produced increased extremely slowly during the first half of the crystallization period, and then very rapidly. The curve is similar to that obtained when the lever rule is applied to the van Laar loop shown in Figure 5.9b, and garnet volume is plotted against temperature (Figure 5.12 inset). The form of the reaction-rate equation ($V \propto t^5$) may therefore be attributed, in a large part, to the form of the van Laar equations.

In summary, we have examined here an example of microstructural and microchemical data on garnet, and an example of the interpretation of such data. The principal proposals to emerge from the study are: (1) the growth rate was constant and was controlled by surface processes rather than the diffusion of atoms to the crystals, (2) nucleation sites were randomly selected, (3) the nucleation rate increased exponentially, then remained constant, and finally declined, determined by changes in the energy of nucleus formation, (4) garnet was produced by a transfer reaction (chlorite $\rightarrow$ garnet) which was accompanied increasingly by exchange of Mg, Fe, and Mn between garnet surface layers and associated biotite, (5) the entire reaction occurred in a temperature increment of only about 20 K, and (6) the rate of reaction increased very slowly at first, and then very rapidly.

It is important to recognize, however, that garnet is an extremely common mineral that has crystallized in many different crustal environments, and that other kinds of data from other localities will call for different kinetic interpretations.

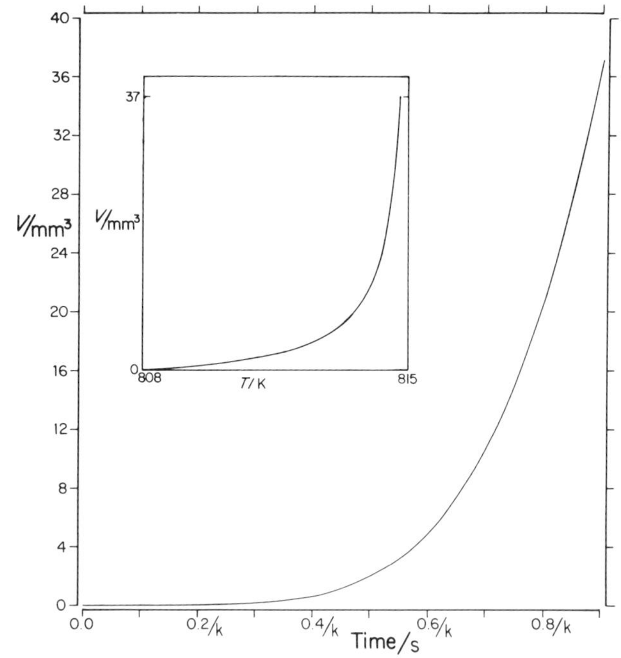

Figure 5.12 Total volume of garnet (V) in portion B plotted against time (time axis as in Figure 5.8), obtained by use of the Johnson–Mehl equation. Inset: increase of garnet volume with temperature in the idealized Mn–Fe system of Figure 5.9b. From Kretz (1993), reproduced by permission of Blackwell Scientific Publications, Inc.

5.3 MICROREACTION ZONES

Where crystals of one mineral (α) are separated from crystals of another (β) by a zone of one or more additional minerals (γ), it is clear that a reaction $\alpha + \beta \rightarrow \gamma$ was initiated but had not progressed to completion. Similarly, where an α crystal is rimmed by and invaded by another mineral (γ), irrespective of the surrounding minerals, one may consider that a reaction was initiated, $\alpha \rightarrow \gamma$, resulting from the migration of certain atoms to the reaction site. Two examples of such microstructures are now examined.

5.3.1 Corona microstructures

Zones of orthopyroxene, hornblende, garnet, and other minerals between crystals of olivine and plagioclase are moderately common in metamorphosed mafic to ultramafic rocks in high-grade metamorphic terranes. These (corona) microstructures are of special interest because the reactant crystals can be identified with confidence, and both reactant and product crystals are therefore available for study. In the example here chosen, the zones consist of orthopyroxene and a hornblende–spinel intergrowth, and the sequence is

$$\text{olivine} \mid \text{orthopyroxene} \mid \begin{matrix}\text{hornblende}\\ \text{and spinel}\end{matrix} \mid \text{plagioclase} \tag{29}$$

Reaction zones similar to those that occur between natural olivine and plagioclase have been produced experimentally and the reaction mechanisms have attracted attention in the field of solid-state chemistry. It may be helpful to begin our study with a brief reference to this literature.

Many experiments have been carried out in which particles of two different phases are mixed and then taken to a temperature where they react to yield one or more product phases. An example is the reaction calcite + quartz → wollastonite, which was described in Sec. 4.2.2. Normally, the products form layers or zones which isolate the two reactants, and further reaction occurs by transport of atoms through this layer. Although the transport almost certainly occurs by diffusion, a detailed description of the diffusion mechanism is not always tractable.

Data on the formation of spinel, as reviewed by Schmalzried (1973) and Brown, Dollimore and Galwey (1980) will serve as an illustration of the problem. Consider a crystal of MgO and one of Al_2O_3 that are initially in contact, and then react to produce a single-crystal layer of $MgAl_2O_4$, as in Figure 5.13. In mechanism (a) (Figure 5.13a), Mg^{2+} and Al^{3+} diffuse in opposite directions through the product layer, and the following reactions occur at the two interfaces:

$$\text{at 1}: \quad [2Al^{3+}] + 4MgO \rightarrow MgAl_2O_4 + [3Mg^{2+}], \tag{30}$$

$$\text{at 2}: \quad [3Mg^{2+}] + 4Al_2O_3 \rightarrow 3MgAl_2O_4 + [2Al^{3+}]. \tag{31}$$

Thus the volume of spinel produced at interface 2 exceeds that produced at interface 1 by a factor of 3. In mechanism (b) (Figure 5.13b), diffusion of two Al^{3+} occurs from interface 2 to 1, together with two O^{2-}, for a net charge of 2+, while one Mg^{2+} (of the same charge) migrates in the opposite direction, to give the following reactions at the two interfaces:

$$\text{at 1}: \quad [2Al^{3+} + 2O^{2-}] + 2MgO \rightarrow MgAl_2O_4 + [Mg^{2+}], \tag{32}$$

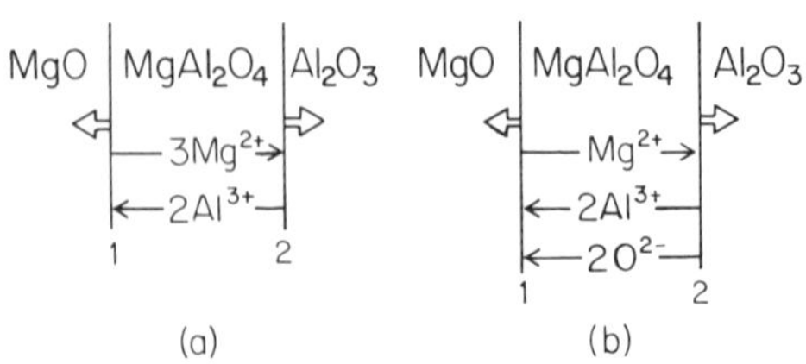

Figure 5.13 Illustration of two mechanisms for the production of $MgAl_2O_4$ spinel by the reaction, $MgO + Al_2O_3 \rightarrow MgAl_2O_4$; diffusion of ions is indicated

at 2 : $$[Mg^{2+}] + 2Al_2O_3 \rightarrow MgAl_2O_4 + [2Al^{3+} + 2O^{2-}]. \quad (33)$$

Now equal volumes of spinel are produced at the two interfaces. In (a) the original $MgO-Al_2O_3$ interface is displaced to the left by a distance $0.06w$, and in (b) to the right by a distance $0.19w$, where w is width of the $MgAl_2O_4$ layer. Other possibilities exist, but all are limited by the requirement of local charge balance.

An example of a reaction that produces two intervening layers is

$$AgCl + NaI \rightarrow AgI + NaCl, \quad (34)$$

which was discussed by Wagner (1938) and Hannay (1976). It is generally agreed that the position of the product layers is determined by the mobility of the cations relative to that of the anions. Thus high cation mobility would produce the sequence

$$AgCl|NaCl|AgI|NaI, \quad (35)$$

(as found experimentally), and high anion mobility,

$$AgCl|AgI|NaCl|NaI, \quad (36)$$

leading to the rule that phases containing the least mobile atoms become juxtaposed. This can be explained by considering the nucleation and initial growth of NaCl and AgI on the original reactant interface, as in Figure 5.14a. Given a greater mobility for Na and Ag compared with Cl and I, NaCl is expected to grow into AgCl and AgI into NaI; in fact, little or no migration of anions is needed.

An analysis of the diffusion process was presented by Wagner (1938) who proposed that because of a limited substitution of Ag for Na in NaCl, and of Na for Ag in AgI, concentration gradients can exist in these crystals. Thus a Ag^+ ion in its journey through NaCl and then through AgI will travel down a concentration gradient in both phases as in 'a bucket brigade', and similarly for Na. If the transfer

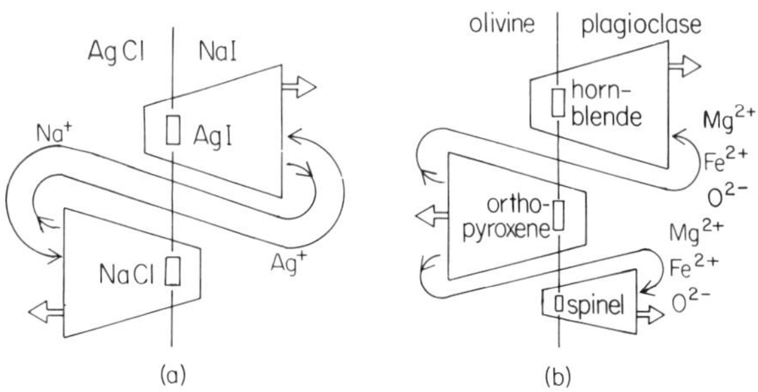

Figure 5.14 Illustration of (a) nucleation and initial growth of AgI and NaCl at a AgCl–NaI interface, for high mobility of cations relative to anions, resulting (with continued growth) of the layer sequence, AgCl|NaCl|AgI|NaI; (b) nucleation and initial growth of hornblende, orthopyroxene, and spinel at an olivine–plagioclase interface, for high mobility of Mg, Fe, and O ions, relative to Al and Si ions, resulting in the layer sequence, olivine | orthopyroxene | hornblende and spinel | plagioclase

of anions is negligible, electroneutrality demands an equal flux for Ag and Na, i.e., the process is one of exchange diffusion (Sec. 4.3.4).

Later, Wagner (1975) re-examined the problem in terms of the Onsager theory (Sec. 4.3.3) in which the flux of diffusing particles is related to chemical potential (μ) rather than concentration. But, because the particles are ions, the total force is defined as

$$\eta = \mu + zF\phi, \tag{37}$$

where z is charge number, F is the Faraday constant, and ϕ is the average electrical potential. The Onsager equations for the flux of Na^+ (J_1) and Ag^+(J_2) are therefore

$$\left.\begin{aligned} J_1 &= L_{11}\frac{d\eta_1}{dx} + L_{12}\frac{d\eta_2}{dx}, \\ J_2 &= L_{21}\frac{d\eta_1}{dx} + L_{22}\frac{d\eta_2}{dx}, \end{aligned}\right\} \tag{38}$$

where the L terms are Onsager coefficients and x is distance. Recall (Sec. 4.3.3) that the cross-coefficients, which express the influence of one diffusing particle on the other, form an essential part of the Onsager theory.

Mineral zones between crystals of olivine and plagioclase were recognized by Törnebohm (1877) and Lacroix (1889), and general information on these microstructures, and examples of recent studies, are found in papers by Nishiyama (1983), Joesten (1986), Grant (1988), Rivers and Mengel (1988), and Johnson and Carlson (1990). The example to be described here is from the Otter Lake area of the Grenville Province, where several meta-gabbro complexes occur in an upper-amphibolite to lower-granulite facies marble–gneiss–amphibolite terrane. Olivine and plagioclase occur locally in the more mafic units of these complexes, and crystals of the two minerals are invariably separated by a zone of orthopyroxene, adjoining olivine, and a zone of hornblende and spinel, adjoining plagioclase (Figure 5.15), and the reaction, in skeletal form, reads

$$\text{olivine} + \text{plagioclase} \rightarrow \text{orthopyroxene} + \text{hornblende} + \text{spinel}. \tag{39}$$

The reaction is now considered in terms of a balanced equation, atomic fluxes, crystal growth, and the change of temperature during reaction.

Using simplified formulas, it is possible to write the following reaction equation, which is fully balanced, implying that all of the atoms present in the product crystals, excluding H (in the form of H_2O) could be derived from olivine and plagioclase:

$$\underset{\text{olivine}}{7(Mg, Fe)_2SiO_4} + \underset{\text{plagioclase}}{(1NaAlSi_3O_8 + 2CaAl_2Si_2O_8)} + 1(H_2O)$$

$$\rightarrow \underset{\text{orthopyroxene}}{7(Mg, Fe)SiO_3} + \underset{\text{hornblende}}{1Ca_2Na(Mg, Fe)_5AlSi_7O_{22}(OH)_2}$$

$$+ \underset{\text{spinel}}{2(Mg, Fe)Al_2O_4}. \tag{40}$$

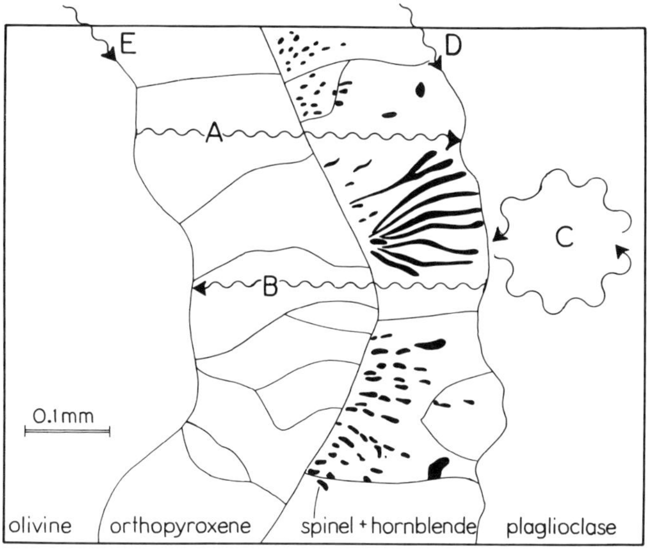

Figure 5.15 Reaction zones of orthopyroxene and a hornblende–spinel intergrowth between olivine and plagioclase; calculated migration of atoms through the reaction zones (A, B), within plagioclase (C), and from outside (D, E) is indicated in text. From Kretz, Jones and Hartree (1989)

Microprobe analyses reveal the presence of several elements in addition to those that appear in equation (40), and formulas that differ from those used in this equation, but it is still possible to obtain an equation that is balanced for the major cations. The only assumption needed is that albite and anorthite components were extracted from plagioclase independently, i.e., the composition of the plagioclase crystal changed continuously as reaction progressed. Thus in one sample (P-3) from the Picanoc Complex, mineral formulas for olivine, orthopyroxene, hornblende, and spinel were determined and are listed in Table 5.2, where stroichiometric coefficients obtained from a mass-balance calculation appear above the mineral symbols. The equation is

$$\begin{aligned}&5.12\text{ olivine} + \underbrace{0.78\text{ albite} + 1.86\text{ anorthite} + 0.006\text{ orthoclase}}_{\text{plagioclase}}\\&+(0.15\text{Ti} + 0.17\text{K} + 0.036\text{Zn} + 0.041\text{F} + 0.024\text{Cl} + 1.94\text{H})\\&\rightarrow 1.0\text{ hornblende} + 5.30\text{ orthopyroxene} + 0.86\text{ spinel}\end{aligned} \quad (41)$$

where brackets enclose atoms introduced to the reaction site. The coefficients were obtained by solving a number of simultaneous equations, one for each of Si, Al, (Mg, Fe), Ca, Na, K.

The above calculation leads to the following proposals concerning the reaction: (1) the mole fraction, an/(ab + an), of plagioclase consumed is 0.71 (this ratio in

Table 5.2 Balance sheet for reaction olivine + plagioclase → orthopyroxene + hornblende + spinel

	Reactants				Products			Sum	
	5.12 Ol	0.78 ab	1.86 an	0.006 or	1.00 Hbl	0.86 Spl	5.30 Opx	Reactants	Products
Si	5.10	2.33	3.72	0.018	5.98	—	5.19	11.17	11.17
Al	0.015	0.78	3.72	0.006	2.66	1.65	0.20	4.52	4.51
Ti	—				0.14	0.0002	0.004		0.15
Fe + Mg	10.24				4.21	0.90	5.13	10.24	10.24
Fe	3.39				1.29	0.539	1.49	3.39	3.32
Mg	6.85				2.92	0.365	3.64	6.85	6.93
Mn	0.043				0.012	0.003	0.031	0.043	0.046
Ca	0.0015		1.86		1.86	—	0.04	1.86	1.90
Na	—	0.78			0.78	—	0.002	0.78	0.78
K	—			0.006	0.17	—	—	0.006	0.17
Zn	—				—	0.036	—		0.03
F	—				0.041	—	—		0.04
Cl	—				0.024	—	—		0.02
H	—				~1.94	—	—		~1.94
O	20.5	6.2	14.9	0.05	23.9	3.5	15.9	41.6	43.3

Numbers at column headings are reaction coefficients.
(—) denotes concentration not determined and assumed to be very small.
ab and an are albite and anorthite components, assumed to have been extracted from plagioclase independently; or is orthoclase component, based on K content of plagioclase.

Fe refers to total iron; in hornblende some of this is expected to be Fe^{3+}.
For Ti, K, Zn, F, Cl, H, O, (Products) - (Reactants) = estimate of atoms introduced to reaction site relative to 1.0 hornblende.

Formulas: olivine $(Fe_{0.66}Mg_{1.34}Mn_{0.0038}Ca_{0.0003})(Al_{0.0003}Si_{0.995})O_4$
plagioclase $(Na_{0.424}Ca_{0.547}K_{0.006})(Al_{1.58}Si_{2.44})O_8$
hornblende $(Ca_{1.86}Na_{0.777}K_{0.172})(Fe_{1.29}Mg_{2.92}Mn_{0.012}Ti_{0.142}Al_{0.64})(Al_{2.02}Si_{5.98})O_{22}(F_{0.041}Cl_{0.024}OH_{1.94})$
spinel $(Fe_{0.53}Mg_{0.42}Mn_{0.003}Zn_{0.041})(Al_{1.91}Fe_{0.092}Ti_{0.0002})O_4$
orthopyroxene $(Fe_{0.28}Mg_{0.69}Mn_{0.006}Ti_{0.0007}Ca_{0.0074}Na_{0.0003}Al_{0.017})(Al_{0.021}Si_{0.98})O_3$

Data from Kretz, Jones, Hartree (1989)

the present, remaining plagioclase is 0.55), (2) all of the principal cations (Si, Al, Fe, Mg, Mn, Ca, Na) are fully accounted for (Table 5.2) and no addition or removal of atoms of these elements occurred, (3) the K content of plagioclase is very low (or/(ab + an + or) = 0.006) and nearly all of the K in hornblende was introduced to the reaction site, (4) assuming that Ti and Zn in initial olivine were very low, nearly all of the Ti in hornblende and Zn in spinel were introduced to the reaction site, (5) in addition to hydrogen, minor fluorine and chlorine were introduced to the site, (6) assuming that $\frac{1}{3}$ of the iron in hornblende is ferric iron, 15% of the total iron involved in the reaction was oxidized, (7) a small proportion of oxygen was apparently introduced, and (8) for some calculations, which used data from elsewhere in the study area or from the literature, balance was achieved for (Mg + Fe) but not for each of Mg and Fe, and here forsterite and fayalite components were evidently extracted from olivine independently, made possible by Mg–Fe exchange diffusion in olivine.

From the above information one may calculate the half-reactions (analogous to equations (30) and (31) or (32) and (33)) that occurred at the reaction interfaces: at olivine–orthopyroxene interface:

$$\begin{aligned}&[0.09\mathrm{Si} + 0.19\mathrm{Al} + 0.04\mathrm{Ca} + 0.002\mathrm{Na}] + 5.12 \text{ olivine} + (0.004\mathrm{Ti})\\ &\rightarrow 5.30 \text{ orthopyroxene} + [3.2\mathrm{Mg} + 1.9\mathrm{Fe} + 0.012\mathrm{Mn} + 4.6\mathrm{O}],\end{aligned} \tag{42}$$

at hornblende–plagioclase interface:

$$\begin{aligned}&[3.2\mathrm{Mg} + 1.9\mathrm{Fe} + 0.012\mathrm{Mn} + 4.6\mathrm{O}] + 0.78 \text{ albite} + 1.86 \text{ anorthite}\\ &+0.006 \text{ orthoclase} + (0.14\mathrm{Ti} + 0.17\mathrm{K} + 0.03\mathrm{Zn} + 0.04\mathrm{F} + 0.02\mathrm{Cl} + 1.9\mathrm{H} + 1.6\mathrm{O})\\ &\rightarrow 1.00 \text{ hornblende} + 0.86 \text{ spinel} + [0.09\mathrm{Si} + 0.19\mathrm{Al} + 0.04\mathrm{Ca} + 0.002\mathrm{Na}],\end{aligned} \tag{43}$$

where square brackets identify atoms that arrived from the opposite interface, and round brackets enclose atoms introduced from beyond the reaction zones (see Figure 5.15), relative to one formula unit of hornblende produced (equation (41)). Evidently oxygen ions as well as cations migrated through the product layers, with balance of charge being maintained locally. The partial oxidation of iron presumably occurred at the hornblende–plagioclase interface.

Provided that the atoms moved as ions and that the ions moved through the product crystals by diffusion, the diffusive fluxes (Figure 5.15) varied for different elements, being greatest for oxygen, iron, and magnesium, and least for sodium:

$$\mathrm{O}^{2-} \underset{1.4}{>} \mathrm{Mg}^{2+} \underset{1.7}{>} \mathrm{Fe}^{2+} \underset{10.0}{>} \mathrm{Al}^{3+} \underset{2.1}{>} \mathrm{Si}^{4+} \underset{2.2}{>} \mathrm{Ca}^{2+} \underset{3.3}{>} \mathrm{Mn}^{2+} \underset{6.0}{>} \mathrm{Na}, \tag{44}$$

the numbers are flux ratios.

Atoms of K, Ti, and Zn which arrived at the reaction site were possibly derived from the immediate surroundings, where other reactions involving Ca pyroxene and biotite were in progress, while H, F, and Cl were presumably derived from sources beyond the meta-gabbro complex.

Further information on atomic migration is provided by a sample from an ultramafic unit in the Grove Complex, where some crystals of olivine and plagioclase are embedded in a large crystal of Ca pyroxene. Here reaction progressed most rapidly where crystals of olivine and plagioclase happened to be in contact, as shown in Figure 5.16, but some reaction occurred even where these crystals were separated by ~0.5 mm, i.e., reaction occurred 'through' the intervening Ca pyroxene. Invariably, the product orthopyroxene nucleated on olivine and the hornblende on plagioclase (Figure 5.16, bottom left). The composition of product crystals *within* the large Ca pyroxene crystals is similar to that indicated in Table 5.2, and the atoms which were transported to the reaction sites (notably K, Ti, H) evidently migrated by volume diffusion *through* the Ca pyroxene structure. Clearly, the concept of atomic transport by dissolution and precipitation in an aqueous 'pore fluid' does not apply in this case.

It is interesting to try to reconstruct the earliest stage in the olivine–plagioclase reaction. Returning to Figure 5.15, imagine a planar interface between olivine and plagioclase on which the nucleation of orthopyroxene, hornblende, and spinel occurred, as in Figure 5.14b (above). Assuming that the flux 'order' of expression (44) is a rough measure of atomic mobility during the initial growth of nuclei, then the high mobility of Mg, Fe, and O compared with Al and Si would cause a more rapid growth of orthopyroxene *into* olivine and of hornblende and spinel *into* plagioclase, than in the opposite directions (cf. Figure 5.14a). The result is a layer of orthopyroxene adjacent to olivine and a layer of hornblende and spinel adjacent to plagioclase, rather than the inverse. The size and orientation of the product crystals were presumably determined principally during this early stage of reaction, but nucleation of spinel and possibly also of orthopyroxene and hornblende occurred while the reaction zones were increasing in width (Figure 5.15).

Using estimates of the molar volume of the minerals involved in the reaction, together with the coefficients in equation (41), volume changes may be calculated and are approximately as follows (in arbitrary units):

$$\underbrace{\underset{224}{\text{olivine}} + \underset{265}{\text{plagioclase}}}_{489} \rightarrow \underbrace{\underbrace{\underset{273}{\text{hornblende}} + \underset{34}{\text{spinel}}}_{307} + \underset{167}{\text{orthopyroxene}}}_{474} \tag{45}$$

The overall change in volume is very small, but the volume of orthopyroxene is clearly less than that of olivine and the volume of hornblende and spinel, combined, is greater than that of plagioclase. Hence, the initial olivine–plagioclase interface must have migrated toward olivine; the displacement is visible is Figure 5.16.

The transfer of atoms from reactant to product crystals (Figure 5.15) is most readily interpreted as occurring by diffusion through the product layers. In an examination of orthopyroxene–olivine interfaces, as seen for example in Figure 5.15, no evidence was found of enhanced growth of orthopyroxene where

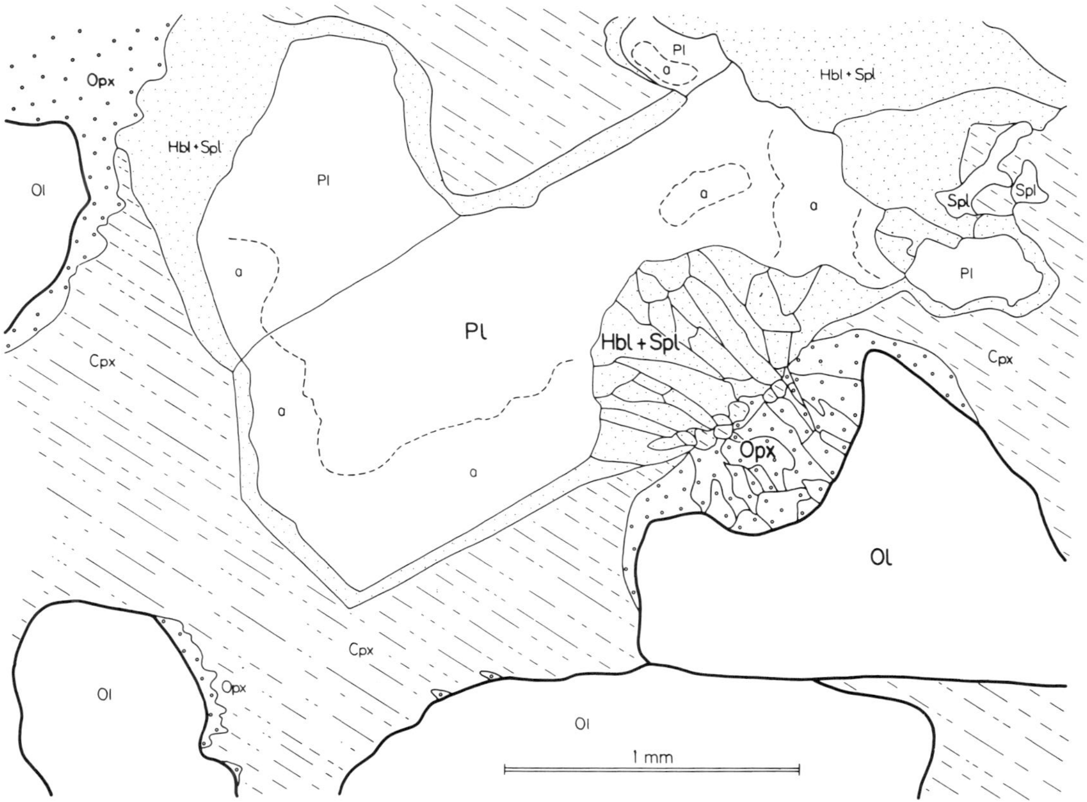

Figure 5.16 Reaction zones of orthopyroxene (Opx) and a hornblende–spinel intergrowth (Hbl + Spl) between olivine (Ol) and plagioclase (Pl), all embedded in a large crystal of Ca pyroxene (Cpx); note evidence for migration of the original Pl–Ol interface toward olivine. From Kretz, Jones and Hartree (1989)

orthopyroxene–orthopyroxene boundaries meet olivine, as might be expected if transfer occurred only or mainly on grain boundaries. The preferred interpretation, therefore, is that migration occurred largely through the orthopyroxene and hornblende crystals.

Given that the atoms moved as ions, a complex pattern of flow must have existed. With a major flow of several ions (Mg^{2+}, Fe^{2+}, O^{2-}, Al^{3+}, Si^{4+}) and a minor flow of others (Mn^{2+}, Ca^{2+}, Na^{+}) and with each element in its diffusion influencing the rate of diffusion of all others, a description of the process would require a knowledge of a large number of Onsager coefficients (equation (38)), and their variation with temperature. In a novel approach to the subject, Joesten (1977, 1986) has attempted to obtain some information on these coefficients from natural reaction zones. The method is based on the assumption that the reaction occurs by dissolution and precipitation in an aqueous fluid, that the diffusing particles are not ions but neutral units such as MgO and SiO_2, and that all of the Onsager cross-coefficients equal zero.

Because of an increase, with time, in the width of reaction zones, and hence a decrease in concentration or chemical-potential gradients, diffusion-controlled reactions of this kind follow a power equation (Jost 1960), i.e., the rate decreases sharply with time. This may explain why examples of arrested reaction are so commonly found.

Other microprocesses operated and may have governed to some extent the rate of reaction. Crystal growth occurred at both of the outer interfaces, and early growth-control may have changed to later diffusion-control (Sec. 4.5.2). The migration of H to the reaction sites may have exerted a control on the crystallization of hornblende, (and hence the rate of reaction), as proposed by Esbensen (1978), but in the sample illustrated in Figure 5.16, widths of the hornblende zone within and without Ca pyroxene crystals are equal and hydrogen appears to have been available wherever it was needed. Finally, Sørensen (1979) suggested that the reaction rate may be controlled by NaSi–CaAl exchange diffusion in plagioclase, which evidently accompanied the reaction.

The question of whether 'corona' reactions occur with rising or falling temperature has been discussed frequently. By comparison with other hydration reactions (e.g., $MgO + H_2O \rightarrow Mg(OH)_2$) and the AgCl–NaI reaction (equation (34), for which $\Delta H = -14.0$ kcal/mol at STP), the reaction under study is expected to be exothermic. In general, exothermic reactions that occur near equilibrium should occur with falling temperature, but when far from equilibrium, they could conceivably occur while temperature increases.

From experimental results on NaSi–CaAl exchange diffusion in plagioclase ($X_{\text{an}} \sim 0.80$), obtained by Grove, Baker and Kinzler (1984), i.e.,

$$\tilde{D}(\text{cm}^2\ \text{s}^{-1}) = 11.0 \exp[-123(\text{kcal mol}^{-1})/RT],$$

it is clear that the diffusion coefficient is far too small (at ~700 °C) to bring about the required change in plagioclase composition, as proposed above. But if reaction was initiated soon after solidification, say at 1100 °C, where, from the above equation, $\tilde{D} \sim 10^{-19} cm^2 s^{-1}$ (or according to Liu and Yund (1992), $\sim 10^{-15} cm^2 s^{-1}$) the necessary transfer of ions by diffusion is a possibility.

All things considered, one must agree with Griffin and Heier (1973) that corona reactions most likely occurred during very slow cooling. Then, at a point on the cooling path, increasing separation of the reactant crystals and decreasing atomic mobility brought the reaction to a premature termination.

5.3.2 Crystals of Ca pyroxene with mantles and lamellae of Ca amphibole

The alteration of Ca pyroxene to Ca amphibole occurs in a variety of rocks, especially in basalt and gabbro, as they are transformed to greenstone, amphibolite, and meta-gabbro. Compelling evidence for the reaction is found in the presence of mantles of amphibole about crystals of pyroxene, and in the apparent invasion of pyroxene crystals by lamellae of amphibole (Figure 5.17). Examples of these microstructures are here described and examined in relation to the information they contain regarding reaction mechanisms.

Consider first the presence of mantles of tremolite about crystals of diopside in marble, as in Figure 5.18. The mantle may consist of a single crystal, coaxial with the central diopside, or of several crystals of different orientation. Locally quartz and calcite are present as inclusions in the tremolite. The change implied by these microstructures can be expressed by writing a chemical equation to define the phases and components that are consumed and produced, and an interface equation (equivalent to a half-reaction), which attempts to describe atomic fluxes to and from the reaction interface.

A possible chemical equation for the pyroxene → amphibole reaction of Figure 5.18 is

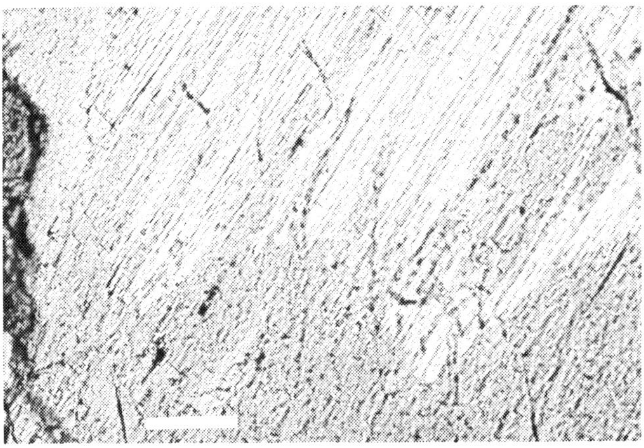

Figure 5.17 Ca pyroxene (light) with rim and lamellae of hornblende (dark) in metamorphosed ultramafic rock of the Thorne complex; Grenville Province. Bar = 0.1 mm From Kretz, Jones and Hartree (1989)

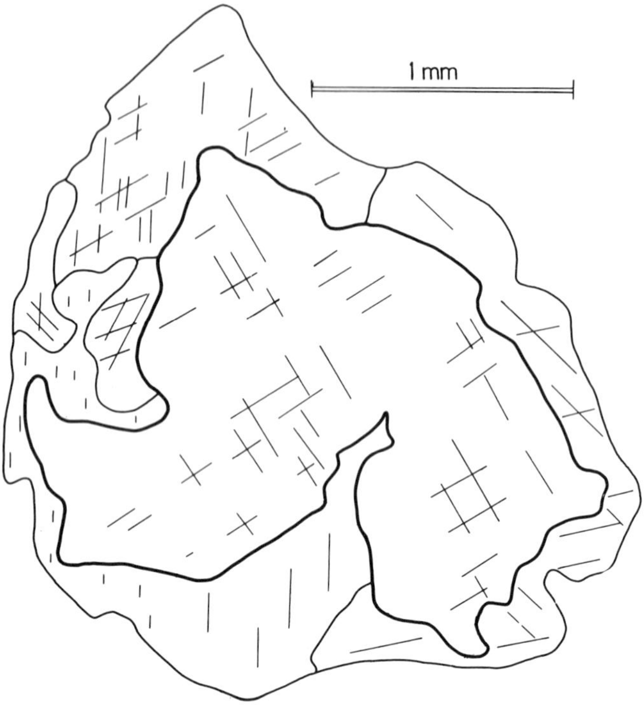

Figure 5.18 Crystal of diopside in marble, with mantle of tremolite. From Kretz (1980)

$$\underset{330}{5CaMgSi_2O_6} + (H_2O + 3CO_2)$$

$$\rightarrow \underset{273}{Ca_2Mg_5Si_8O_{22}(OH)_2} + \underset{111}{3CaCO_3} + \underset{45}{2SiO_2}, \qquad (46)$$

where numbers below the formulas are molar volumes, taken from Table 2.1 In view of the boundary curve for this reaction (Figure 3.15 of Sec. 3.2.3), the reaction could be displaced to the right by an introduction of H_2O, or a change in temperature. The increase in solid volume is by a factor of 1.3.

Different possibilities exist for the interface reaction. Consider, for example, a single-phase mantle of tremolite about a crystal of diopside, and suppose that reaction at the pyroxene–amphibole interface occurs with no local loss or gain of oxygen. Then the reaction equation, relative to 24 oxygen atoms, is

$$\underset{264}{4CaMgSi_2O_6} + [2H^+ + 1Mg^{2+}] \rightarrow \underset{273}{Ca_2Mg_5Si_8O_{22}(OH)_2} + 2\,[Ca^{2+}], \qquad (47)$$

for which the volume change is very small. By this model pyroxene single chains become amphibole double chains, as cations move to and from the interface, possibly by way of exchange diffusion between ($2H^+ + Mg^{2+}$) and ($2Ca^{2+}$) through the amphibole mantle. Equation (47) does not, of course, identify the source of the H and Mg ions, or the fate of the Ca ions.

Rims of hornblende about crystals of Ca pyroxene occur commonly in metamorphosed mafic and ultramafic rocks, and examples from a high-grade terrane are presented in Figure 5.19a and b. Figure 5.19a shows much variation in rim width, only some of which can be attributed to the cut effect. Figure 5.19b shows a group of Ca pyroxene crystals embedded in a large crystal of orthopyroxene, each with a single mantle of hornblende that is coaxial with the central Ca pyroxene, i.e., not with orthopyroxene. These observations emphasize that progress in the pyroxene → amphibole reaction has varied greatly, even on a very small scale, and that the rims have formed, regardless of whether the mineral originally in contact with Ca pyroxene was plagioclase, orthopyroxene, or another mineral of the assemblage.

The sample that was chosen in Sec. 5.3.1 (Figure 5.15) to illustrate a corona reaction (sample P-3) contains crystals of Ca pyroxene with rims of hornblende, and analyses of these phases are listed in Table 5.3. Compared with the central pyroxene, the hornblende rims carry more Al, Na, K, and H, and less Si and Ca.

Three possibilities are now considered for a chemical equation (analogous to (46)), for the production of hornblende from Ca pyroxene. First, suppose that all of the Mg and Fe in hornblende were obtained from Ca pyroxene, and the extra Al and Na from plagioclase, which by analysis is $Ca_{0.55}Na_{0.42}K_{0.006}Al_{1.58}Si_{2.44}O_8$. By use of this formula and those for Ca pyroxene and hornblende, from Table 5.3,

Table 5.3 Chemical analyses of Ca pyroxene and hornblende rim in metamorphosed ultramafic rock[1]

	Analyses			Formulas[3]	
	Ca pyroxene	Hornblende		Ca pyroxene	Hornblende
SiO_2/wt%	50.8	41.3	Si	7.52	5.93
TiO_2	0.89	1.25	Al^{iv}	0.480	2.07
Al_2O_3	4.39	16.2	Al^{vi}	0.304	0.677
[2]FeO	6.50	10.4	Ti	0.100	0.135
MgO	13.7	13.7	Fe	0.81	1.24
MnO	0.16	0.11	Mg	3.03	2.93
CaO	21.8	12.0	Mn	0.020	0.014
Na_2O	0.77	2.67	Ca	3.46	1.82
K_2O		0.92	Na	0.220	0.742
Cl		0.01	K		0.173
F		0.05	F		0.022
			Cl		0.002
			OH		~1.98

[1]From the Picanoc metagabbro complex (sample P-3) Grenville Province (Kretz, Jones and Hartree 1989a).

[2]Total iron expressed as FeO.

[3]Based on 24 oxygen atoms.

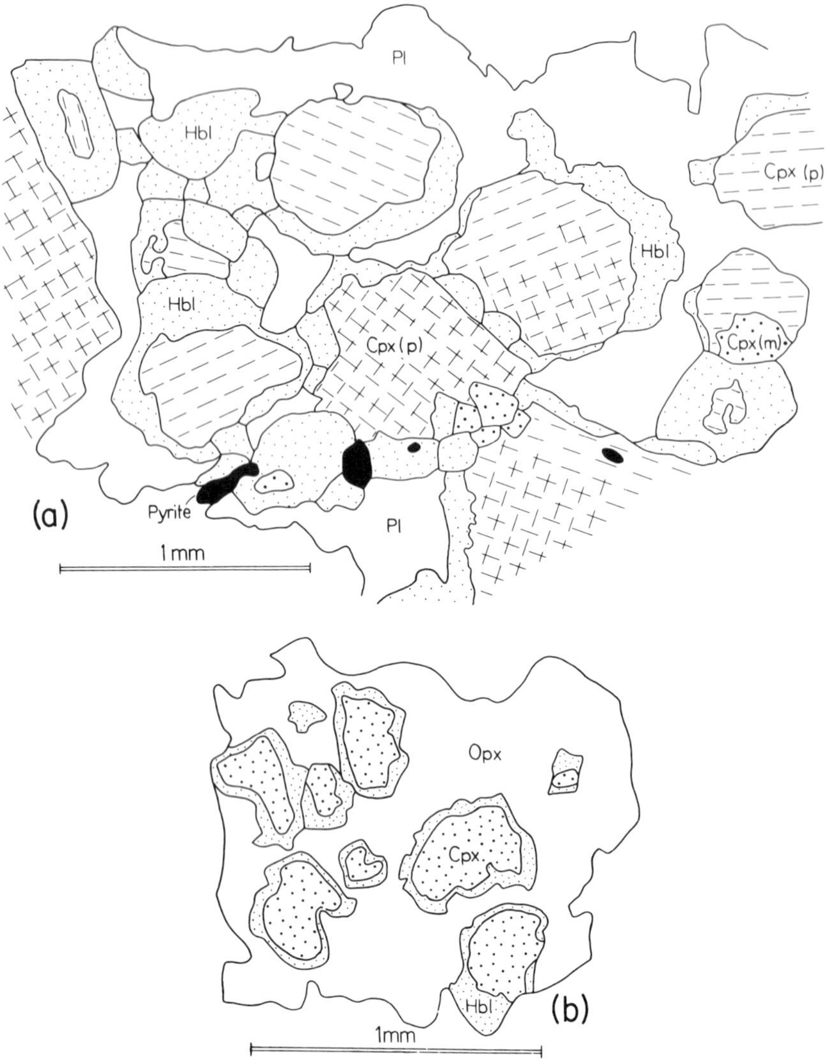

Figure 5.19 Microstructure of metagabbro. (a) Crystal of Ca pyroxene (Cpx) with rims and partial rims of hornblende (Hbl), all enclosed by a single crystal of plagioclase (Pl); symbols 'p' and 'm' identify primary and metamorphic (recrystallized) Ca pyroxene. (b) Crystals of Ca pyroxene with rims of hornblende, embedded in a single crystal of orthopyroxene (Opx); rims are single crystals, coaxial with the central Ca pyroxene. From Kretz, Jones and Hartree (1989)

we have

$$4.35 \text{ Ca pyroxene} + 1.2 \text{ plagioclase} + (0.17\text{K} + 1.98\text{H} + 0.022\text{F} + 0.002\text{Cl}) \rightarrow 1.0 \text{ hornblende} + (5.2\text{Si} + 2.6\text{Ca}). \quad (48)$$

This produces a satisfactory balance for (Mg + Fe), Al, and Na, but not for Si and Ca.

As a second attempt to write a balanced equation, assume that the extra Na and K required for hornblende were taken from plagioclase as $NaAlSi_3O_8$ (ab) and $KAlSi_3O_8$ (or) components, and that spinel

$$(Mg_{0.42}Fe_{0.53}Mn_{0.003}Al_{1.9}Fe^{3+}_{0.09}O_4)$$

and magnetite (Fe_3O_4), which are present in the sample, were additional reactants. We now have

$$\begin{aligned}&1.69 \text{ Ca pyroxene} + 0.79 \text{ spinel} + 0.60 \text{ magnetite} + 0.75 \text{ ab}\\ &+0.17 \text{ or } + (1.98\text{H} + 0.22\text{F} + 0.002\text{Cl}) \rightarrow 1.0 \text{ hornblende.}\end{aligned} \tag{49}$$

This equation gives the following numbers of atoms (on left, on right), Si (5.93, 5.93), Al (2.73, 2.74), Ti (0.04, 0.14), (Fe + Mg) (4.17, 4.17), Mn (0.01, 0.01), Ca (1.46, 1.82), Na (0.74, 0.74), K (0.17, 0.17), oxygen (23.1, 24.0), and a balance for the major elements, that is generally satisfactory. Equation (49) cannot, however, apply where spinel is absent.

The logical next step is to adopt the view that it is simply not possible to identify the source of all of the atoms that presently reside in a hornblende rim. By this we mean that if every atom in a hornblende mantle carried a label identifying the location of the atom prior to crystallization of the hornblende, we would find that although many of the atoms were derived from the Ca pyroxene that was replaced by the mantle, the others would have come from virtually all of the crystals of all of the minerals in the vicinity. In a sense, therefore, all of these crystals were reactants, and the appropriate equation is

$$\begin{aligned}&\text{Ca pyroxene} + \begin{pmatrix}\text{atoms of various elements}\\ \text{from numerous crystals in}\\ \text{the vicinity}\end{pmatrix} + \begin{pmatrix}\text{H and possibly other}\\ \text{atoms from a distant}\\ \text{source}\end{pmatrix}\\ &\rightarrow \text{hornblende} + \begin{pmatrix}\text{atoms of various elements}\\ \text{displaced to crystals in}\\ \text{the vicinity}\end{pmatrix}\end{aligned} \tag{50}$$

This is an extension of the concept of uncertainty regarding metamorphic reactions, which was discussed in Sec. 5.1.

Where hornblende crystals nucleated on the surface of a pyroxene crystal, evidently the crystals grew 'into' Ca pyroxene. In addition, some growth may have occurred in the opposite direction, similar to the growth of spinel in the reaction discussed in Sec. 5.3.1. With regard to the hornblende–Ca-pyroxene interface, an equation analogous to (47) may be proposed by again assuming no transport of oxygen. From the formulas of Table 5.3, the equation is

$$\begin{aligned}&1.0\text{Ca pyroxene} + [2.0\text{Al} + 0.4\text{Fe} + 0.5\text{Na} + 0.2\text{K} + 1.98\text{H} + 0.02\text{F} + 0.002\text{Cl}]\\ &\rightarrow 1.0 \text{ hornblende} + [1.6\text{Si} + 0.1\text{Mg} + 1.6\text{Ca}].\end{aligned} \tag{51}$$

There is here an additional assumption — that the Ca pyroxene was consumed 'as is', i.e., there was no selective extraction of certain atoms from the Ca pyroxene as the interface advanced. Equation (51) expresses rather large fluxes of atoms to and from the interface; these are relative to the rearrangement of 24 oxygen atoms from pyroxene to amphibole structures, i.e., $(Si, Al)_8O_{24} \rightarrow (Si, Al)_8O_{22} + 2O$.

Hornblende also occurs frequently as lamellae in Ca pyroxene, and such microstructures have been studied in detail by many authors. Papike, Ross and Clark (1969) proposed that the lamellae have possibly formed by simultaneous crystallization, by the replacement of one phase by the other, or by exsolution. Yamaguchi, Akai and Tomita (1978), in a study of garnet lherzolite, obtained chemical analyses of Ca pyroxene, hornblende lamellae, and depleted zones adjacent to the lamellae. The depleted zones carry less Al, Ti, Cr, Fe, and Na than pyroxene beyond; these are the very elements that are concentrated in hornblende. Assuming that the depleted zones extend one to two lamellar widths on either side of each lamella, a mass-balance calculation leads to the conclusion that the zones could have provided all of the Ca and Na that was needed, but only about half of the Al.

Where hornblende lamellae form an extension of a rim, as in Figure 5.17, it seems reasonable to suppose that the lamellae formed by the same reaction that produced the rims, the reaction occurring with relative speed along certain (110) planes. A large variation in shape and size of hornblende lamellae in the study sample is brought out particularly well in SEM images (Figure 5.20), where the microstructure is seen to be more irregular than many that are known to result from exsolution (Sec. 5.4, below). In this case, the crystallization of hornblende

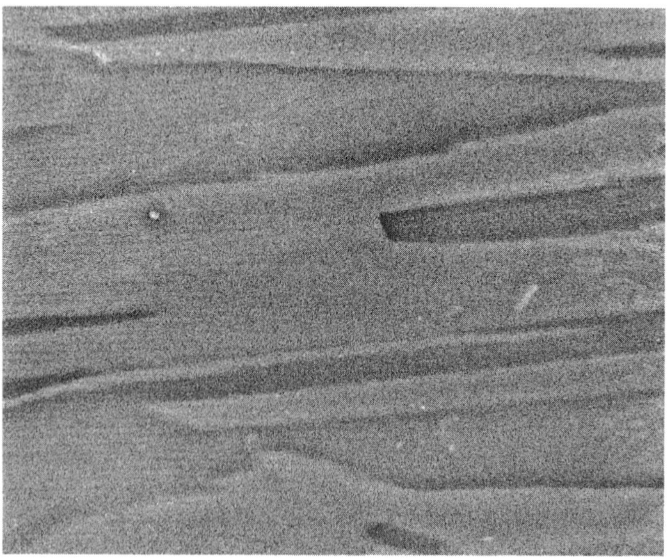

Figure 5.20 SEM image of hornblende lamellae in Ca pyroxene from metapyroxenite (Sample P-3, Table 5.3). Width of field, 30 μm

by a pyroxene → hornblende reaction implies a large flux of atoms (presumably by diffusion) either through the pyroxene structure or along pyroxene–amphibole boundaries.

TEM studies by Yamaguchi, Akai and Tomita (1978) and others have revealed amphibole lamellae as narrow as one double chain. The question of whether these and wider lamellae were produced by alteration or exsolution was discussed by Smith (1977), Yamaguchi, Akai and Tomita (1978), and Veblen and Buseck (1981).

5.4 INTERGROWTH MICROSTRUCTURES AND EXSOLUTION KINETICS

The presence of oriented lamellae or rods of one phase within a crystal of another phase is here referred to as an 'intergrowth'. A familiar example is the presence of lamellae of Na-rich feldspar in crystals of K-rich feldspar. In many natural occurrences, some uncertainty exists regarding the process that has produced the intergrowth, but frequently it is possible to state with confidence that the process was exsolution. The examples of intergrowth microstructure here chosen for study are taken from iron meteorites, and from alkali feldspar, and carbonate minerals — mineral systems in which immiscibility has been confirmed experimentally. Having accepted that particular intergrowths were indeed produced by exsolution, our attention will be drawn to the kinetics of the exsolution process.

The term 'exsolution' is here applied to the formation, on cooling, of two separate phases from one homogeneous solid solution, regardless of differences that may exist in the atomic structure of the two phases. When a new phase of different structure is produced, the process can also be viewed as a continuous phase transformation, accompanied by a change in composition.

5.4.1 Intergrowths in iron meteorites

The presence of oriented lamellae of one iron-rich phase in another (Widmanstätten microstructure) in meteorites was discovered nearly 200 years ago, and since then, various interpretations of this microstructure have been debated, including simultaneous crystallization of the two phases (Buchwald 1975). We now know that the lamellae consist of iron with < 8 wt % Ni, of a body-centered cubic structure (α iron, or kamacite) and the host is iron with $\sim$10 to 50 wt % Ni, of a face-centered cubic structure (γ phase, or taenite). Normally, four sets of lamellae are present, each parallel to an octahedral plane in the host, but in random sections (Figure 5.21a), three sets appear conspicuously. The crystallographic orientation of the lamellae (α) is related to that of the host (γ) as follows:

$$\{110\}\ \alpha \text{ parallel to } \{111\}\ \gamma,$$
$$[111]\ \alpha \text{ parallel to } [110]\ \gamma.$$

Thus planes of highest atomic density in α and β are parallel.

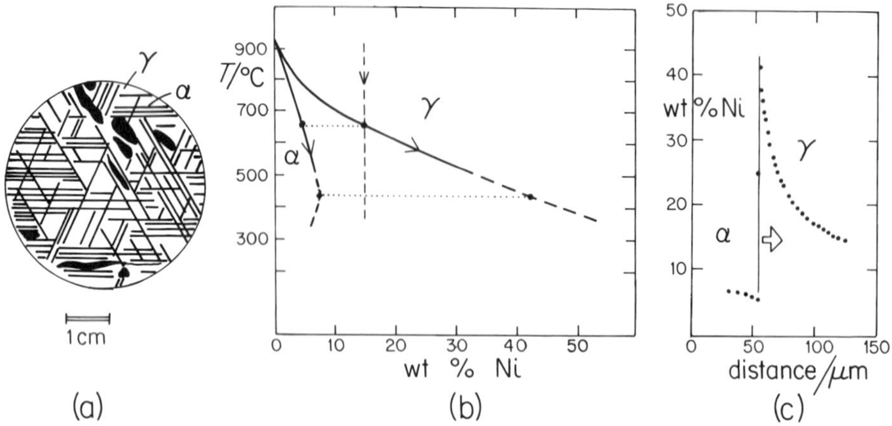

Figure 5.21 Exsolution in iron meteorites: (a) drawing of a photograph of lamellae of α-iron (kamacite) in γ-iron (taenite); four sets of lamellae are present; the fourth is nearly parallel to the section, and individuals appear as irregular areas; (b) phase diagram for the Fe–Ni system, showing the miscibility gap, and a large increase (arrow) in Ni content of γ-iron on near-equilibrium cooling; (c) Ni concentration profile across an α–γ boundary in the Carleton meteorite, a stranded diffusion gradient. (a) from Buchwald (1975), (b) and (c) from Lin, Goldstein and Williams (1977) by permission of Pergamon Press Ltd

The sub-solidus portion of the Fe–Ni phase diagram (Figure 5.21b) provides a basis for the interpretation of α–γ intergrowths as products of an exsolution process. On slow cooling of homogeneous γ iron, with 15 wt % Ni in solid solution, a separation of α iron of lower Ni content is expected to occur, beginning at approximately 650 °C. Notice (Figure 5.21b) that, as exsolution progresses, equilibrium demands of γ a rather large increase in Ni content, and provided that cooling is rapid relative to Ni–Fe exchange diffusion in γ iron, a concentration gradient will be created in γ at the γ–α interface. Stranded gradients of this kind were first detected by Agrell, Long and Ogilvie (1963), and an example is presented in Figure 5.21c. In this preliminary interpretation, the presence of minor phosphorus in both phases has been disregarded.

Narayan and Goldstein (1985) have observed exsolution in laboratory experiments. For example, in one experiment, γ iron containing 6.9 wt % Ni and 0.5 wt % P was cooled from 790 to 650 °C at a rate of 5 °C/day. When quenched to room temperature and examined microscopically, crystals of α were found to have nucleated on γ–γ boundaries, and also (as rods) within crystals of γ. The rods are small (<10 μm across) and the adjacent Ni-enriched zones are narrow (~1 μm) and (as might be expected) the Widmanstätten microstructure was not fully reproduced experimentally.

Several attempts have been made to obtain estimates of the rate of cooling of iron meteorites from information on microstructure and microcomposition. Numerical simulations of the exsolution process were constructed by Narayan and Goldstein

(1985) and Saikumar and Goldstein (1988), based on the Fe–Ni–P phase diagram and data on diffusion coefficients, with allowance made for soft impingement; an equilibrium distribution of Ni and P between γ and α at the γ–α interface was assumed. The simulations provide different concentration profiles for given initial Ni and P content and different cooling rates, which can then be matched to measured profiles of the kind shown in Figure 5.21c. This gives, for example, a cooling rate of 25 °C/ My for the meteorite Toluca. The implications these numbers have for the size of parent bodies from which the meteorites were derived, are discussed by Buchwald (1975).

In conclusion, we find in iron meteorites a relatively simple example of exsolution microstructure in which the exsolved phase forms lamellae in an orientation that was controlled by the preferred formation of {111} interfaces. These are planes of high atomic density, and therefore the α–γ interfaces lying parallel to them are almost certainly interfaces of relatively low energy. The presence of stranded concentration gradients confirms that the lamellae formed by a nucleation-and-growth process, and that growth was limited by diffusion in the host phase. But, presumably, the ultimate control on the rate of reaction was the rate of cooling, because a decrease in the rate of cooling would necessarily cause a decrease in the rate of reaction. Information on nucleation sites in meteoritic iron is not yet available, but a rather uniform width of lamellae may be interpreted as a rapid (possibly instantaneous) rate of nucleation in the initial stage of phase separation.

The chosen example also demonstrates the value of experimentation and numerical simulation, in the study of natural microstructures.

5.4.2 Exsolution microstructures in alkali feldspar

An intergrowth of Na-rich and K-rich feldspar is commonly observed microscopically, especially in various granitic and syenitic rocks, but also in gneisses and granulites. The intergrowth, which is referred to as 'perthite', may take the form of alternating lamellae or plates, as in Figure 5.22, or the Na-rich phase may form a zigzag pattern (known as 'braid perthite), or it may occur as rods or irregularly shaped crystals; these various forms have been illustrated by Smith (1974).

Intergrowths on the sub-microscopic scale were detected by Laves (1952), by use of X-rays, and later, TEM images of such microstructures were obtained by Fleet and Ribbe (1963), Lorimer and Champness (1973), Brown, Becker and Parsons (1983), and others. One variety is a lamellar microstructure very similar to that shown in Figure 5.22, but on a much finer scale; in one sample from Italy (Parsons and Brown 1991, Figure 7) the lamellae are only 20 nm (200 Å) wide. These sub-microscopic intergrowths are known as 'cryptoperthite'.

As in metals (Sec. 5.4.1) the crystallographic orientations of the two phases in most lamellar intergrowths are related, and the interfaces have a preferred orientation. Thus in microperthite examined by Laves (1952) and Smith (1961), interatomic distances along [010] and along an irrational direction which lies between [106] and [108] are equal in both phases, and the plane containing these two lines

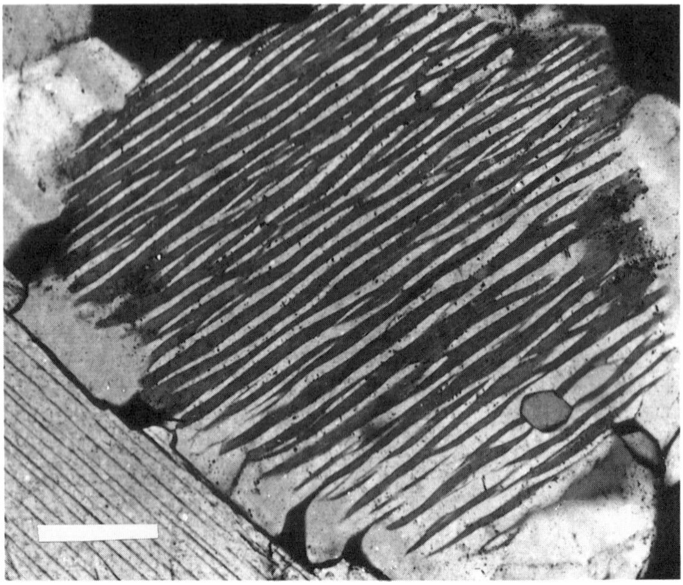

Figure 5.22 An example of microperthite (lamellae of K-feldspar in a crystal of Na-feldspar) in hornblende granulite of the Bell Mount Complex (Kretz 1980); distance across field of view is 0.5 mm. Bar = 0.1 mm

(near ($\bar{7}01$)) is the interface plane. Willaime and Brown (1974) showed that this plane, compared with other possible orientations, is energetically favorable.

Cryptoperthite was first produced experimentally by Owen and McConnell (1971), and there is now general agreement that these fine lamellar intergrowths, and possibly many that are visible microscopically are products of an exsolution reaction. Our attention here is confined to these 'exsolution microstructures'; more irregular intergrowths, some of which may be the result of complex replacement processes, will not be examined.

From the studies of Laves (1952) and Smith (1961) and more recent X-ray and TEM results, it is clear that the 'anomalous' properties of Na- and K-rich phases of cryptoperthite must be attributed to lattice strain, and that this strain is the result of coherency. The term 'coherent' as presently used, implies a *continuity* of lattice planes across the two-phase interfaces. Because lattice spacings in the sodic and potassic phases are not equal, lattices on either side of the interface are distorted (strained elasticity), as in Figure 4.21 (p. 302) and the interface is either coherent or slightly semi-coherent. As noted briefly in Sec. 4.4.2, coherency strain, where present, will cause a displacement of the equilibrium (strain-free) solvus to lower temperatures.

The coherent solvus for disordered alkali feldspar was located experimentally by Sipling and Yund (1976), and is plotted in Figure 5.23, where it may be compared with the strain-free solvus, from Sec. 3.4.1. The experimental results were obtained

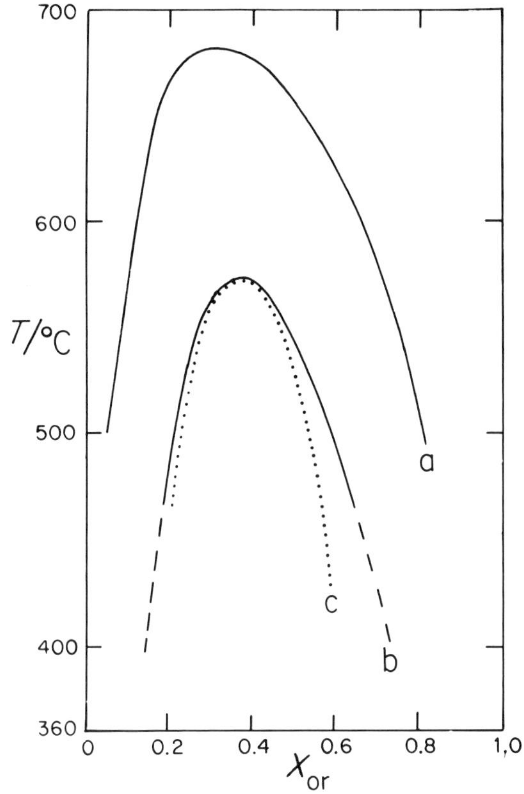

Figure 5.23 Coherent solvus (b) and coherent spinodal (c) in the alkali-feldspar system at 1 bar, as located by Sipling and Yund (1976), and, for comparison, the strain-free solvus (a) at 2.0 kbar, taken from Figure 3.26 of Sec. 3.4.1

by annealing homogeneous high sanidine of composition $X_{or} = K/(Na+K) = 0.20$ to 0.56 at temperatures between 550 and 400 °C, for durations to 1440 h. On quenching to room temperature, coherent lamellae ($\sim(\bar{8}01)$) of monoclinic sodic and potassic phases were identified by use of X-rays and TEM. The composition (X_{or}) of the two phases was determined from lattice spacings (applying the necessary correction for strain), thus obtaining, from each experiment, points on the limbs of the coherent solvus. Coherent equilibrium (metastable, because both phases remained monoclinic) was approached from both higher and lower temperatures. The solvus agrees well with results obtained earlier by Robin (1974) based on calculations of strain energy. The presence of a coherent solvus means that, for a cooling rate which is sufficiently rapid to ensure that only coherent nuclei will form and that coherency is maintained, exsolution in a crystal of intermediate composition (e.g., $X_{or} = 0.40$) cannot occur until a point well within the strain-free solvus (e.g., $\sim$570 °C) is obtained (Figure 5.23).

There is growing evidence, as reviewed by Yund (1983a), that the very fine lamellar intergrowths produced in the above and other experiments are the result of spinodal decomposition (Sec. 4.4.2). For example, the regularity in lamellar spacing is more consistent with this mechanism than with classical nucleation, which is expected to produce a more random distribution of exsolved microcrystals. A proposal for the location of the coherent spinodal was presented by Sipling and Yund (1976) from observations that in some of their experiments (described above) phase separation occurred rapidly (representing points within the spinodal), while in others it occurred more slowly (representing points outside of the spinodal); the resulting curve is shown in Figure 5.23.

An increase in lamellar width with time (known as 'coarsening') is the process by which cryptoperthite, produced by spinodal decomposition, could conceivably change to lamellar perthite, of the kind illustrated in Figure 5.22. Coarsening in alkali feldspar was studied in detail by Yund and Davidson (1978), who selected a monoclinic cryptoperthite of bulk composition, $X_{\text{or}} = 0.35$, and at temperatures of 560, 530, and 470 °C, measured the lamellar spacing (λ) as a function of time (t). For a given temperature, the results can be described by

$$\lambda = \lambda_0 + kt^{1/3}, \tag{52}$$

where λ_0 is the spacing at $t = 0$, and k is a rate constant. For example, at 530 °C, λ was found to increase from <100 Å to 314 Å in 548 days. A plot of $\ln k$ against $1/T$ produces a straight line,

$$k = 1.8 \times 10^8 \exp(-25\,000/RT), \tag{53}$$

which may be used to extrapolate k to lower temperatures. In additional experiments, the presence of H_2O under a pressure of 2 kbar was found to have no effect on coarsening kinetics.

The results could be used (with caution) to obtain an estimate of the time needed for lamellar coarsening to $\lambda \sim 10\ \mu$m (Figure 5.22). For example, at 530 °C, the determined relation between lamellar spacing (λ in Å) and time (t in years) is

$$\lambda = 74 + 210t^{1/3}, \tag{54}$$

which gives $t = 110 \times 10^6$ years to produce lamellae of width ~5 μm. But, as noted by Yund (1983a), an extrapolation of equation (52) may not be reasonable, especially in view of the loss of coherency that occurs as coarsening progresses.

In the theory of coarsening, as applied to metals and reviewed by Raghaven and Cohen (1975), the driving force for an increase in crystal size is interfacial energy, which is reduced as coarsening progresses. The theory is based on the Thomson–Freundlich equation, which expresses the concentration ($C^{\alpha(r)}$) of a component in a matrix (α) adjacent to a small spherical crystal (β) of radius r, relative to the concentration ($C^{\alpha(\infty)}$) adjacent to a much larger hypothetical crystal of radius ∞, in terms of the specific energy of the $\alpha-\beta$ interface (σ),

$$\frac{C^{\alpha(r)}}{C^{\alpha(\infty)}} = \exp\left(\frac{2V^{\beta}\sigma}{RTr}\right) \tag{55}$$

Here V^{β} is the molar volume of β, R is the gas constant, and T is temperature. Because the exponent is positive, the term on the right is > 1, and hence, $(C^{\alpha(r)}) > (C^{\alpha(\infty)})$. Thus a concentration gradient is created which may result in the diffusion of the component from the smaller to the larger crystals, causing the former to disappear and the latter to grow larger. Equation (55) applies only approximately to particles of sodic phase (Ab) in the potassic phase (Kf), or of Kf in Ab, because in the derivation of this equation, a linear relation between activity and concentration is assumed (Burke 1965), and because the particles occur as plates, i.e., σ varies with orientation.

It is helpful to construct a Gibbs energy–concentration plot for the matrix phase (e.g., Kf) and for two particles of the precipitate (Ab), one of larger size and lower energy, and one of smaller size and higher energy (Burke 1965, p. 170); the common tangents then confirm that the concentration of the segregated component, ab, ($X_{ab} = \mathrm{Na}/(\mathrm{Na}+\mathrm{K})$), in the Kf matrix is *larger* near the particle of small size. This means that the concentration of the second component (or) is *smaller* at these points.

One possible coarsening process in lamellar intergrowths is the retreat of terminations of individual plates, as discussed by Brady (1987), and illustrated in Figure 5.24. Now r in equation (55) is the radius at plate terminations, which can be viewed as curved surfaces of progressively decreasing radius, as the tips

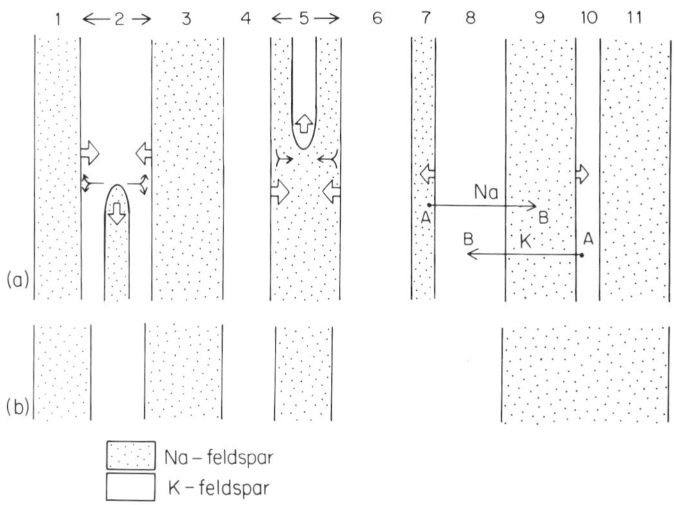

Figure 5.24 Coarsening of perthite lamellae, (a) to (b), by retreat of narrow lamellae (left half of (a) and by diffusion from narrow lamellae to nearby larger lamellae (right half of (a))

are approached. Thus gradients in the concentration are expected to extend away from the plate terminations, causing a retreat of the terminations and an advance of planar interfaces elsewhere. These gradients for the albite component (and Na) are shown as light arrows in zones 2 and 5 of Figure 5.24; identical and opposite arrows (not shown) are implied for potassium.

Where the lamellar interfaces are coherent or semi-coherent, strain is an alternative energy source for coarsening. Provided a measure of relaxation occurs in the centers of the relatively large plates (at B in plates labelled 8 and 9 in Figure 5.24), then the strain energy in the margins of small plates (at A in plates 7 and 10) will exceed that in the centers of the larger plates, and the chemical potential of both the ab and or components will be greater at A than at B. A tendency will exist, therefore, for both Na and K to migrate from points A to B, through the intervening plate of the second phase, resulting in interface migration and disappearance of the small plates.

Coarsening driven by strain energy can occur only so long as coherency persists. Eventually, the coherent boundary, in which the chemical potential of both ab and or components is high, will give way to a semi-coherent or non-coherent interface. This will be accompanied by changes in lamellar composition, as points on the coherent solvus (Figure 5.23) move toward the strain-free solvus. The result is a minor change in volume proportions and a minor displacement of phase boundaries, but no appreciable coarsening.

As noted by Yund (1983a), the conspicuous deceleration of lamellar spacing with time, and the exponential increase in rate with temperature (equations (52) and (53)) are both consistent with cation diffusion being the principal mechanism and control on the rate of coarsening, as observed experimentally. Data on Na–K exchange diffusion in (Na, K)$AlSi_3O_8$, as presented in Sec. 4.3.7, in so far as they can be extrapolated to temperatures of $\sim$530 °C, lead to the conclusion that cation mobility is adequate to bring about the rearrangement of atoms that is needed, both within the experimental systems and on a geological time scale.

As exsolution progresses, equilibrium requires of the two phases a change in Na/(Na + K) composition, defined by the coherent solvus or the strain-free solvus. Mason (1982) has obtained ion microprobe analyses of lamellae and host phases in perthite from pegmatite, and found the Na-rich phase to be nearly pure albite ($X_{or} \sim 0.01$) while the K-rich phase contains more than 92% of the orthoclase component ($X_{or} = 0.92$ to 0.96); these obviously represent points closer to the strain-free than the coherent solvus (Figure 5.23). Mason also obtained data on trace elements; for example, Ba and Rb were found to be concentrated in the K-rich phase. Provided that the perthite formed by exsolution, these results confirm that minor elements as well as Na and K migrate and participate in the exsolution process.

From the numerous natural and experimental data that are now available, as reviewed by Putnis and McConnell (1980), Yund (1983a), and Parsons and Brown (1991), it is now possible to attempt a reconstruction of the development of different exsolution microstructures, using initial composition and rate of cooling as variables.

(1) During very rapid cooling, e.g., in volcanic ejecta, alkali feldspar crystallizes as monoclinic sanidine, and exsolution does not occur.

(2) During rapid cooling (~5 °C/h), coherent phase separation in sanidine of intermediate composition occurs by spinodal decomposition, or possibly by homogeneous nucleation. Initially, both phases remain monoclinic — inversion to triclinic being prevented by coherency. Upon further cooling, the Na phase collapses to triclinic, accompanied by twinning and some loss of coherency. Then the K phase becomes triclinic, accompanied by a rotation of the lamellae to form a zigzag pattern, resulting in a decrease in interfacial energy. Only cryptoperthite is formed. For crystals rich in K, the spinodal field may not be intersected (Figure 5.23) and the Na phase must form by nucleation and growth. Possibilities for determining the rate of cooling from lamellar width, as discussed above, have been considered in detail by Yund (1983a).

(3) During slow cooling (e.g., in granite plutons and gneisses), microscopically visible perthite forms, but it is not always clear whether this has evolved through spinodal decomposition and coarsening, or through heterogeneous nucleation and growth. Following the early stage of exsolution, numerous opportunities could arise for microstructural modification by coarsening and recrystallization.

5.4.3 Calcite–dolomite intergrowths

Crystals of calcite enclosing oriented plates and rods of dolomite have been reported from several occurrences of marble and carbonatite. Coomāraswāmy (1902) detected these microstructures in marble from Sri Lanka, and Joplin (1935) proposed that similar intergrowths in Australia were produced by the exsolution of dolomite from calcite. Support for this interpretation then came from the experimental delineation of a miscibility gap in the $CaCO_3$–$CaMg(CO_3)_2$ system, as described in Sec. 3.4.2 More recent findings of dolomite inclusions in calcite are by Carpenter (1967), Nesbitt and Essene (1982), and Puhan (1984) (from marble), and by Puustinen (1974) (from carbonatite). An experimental study of carbonate exsolution was carried out by Goldsmith (1960).

As an example of calcite–dolomite intergrowths we shall examine some SEM and microchemical results obtained from high-grade marble of the Otter Lake terrane in the Grenville Province, where carbonate exsolution evidently occurred during cooling from approximately 700 °C and 7 kbar. One-quarter of 120 samples of marble from this terrane were found to contain plates or rods of dolomite in calcite, and locally both are present in the same sample (Figure 5.25). In nearly all of these samples, dolomite occurs as discrete grains, as well as inclusions. Typically, the plates lie normal to *c* of the host, while swarms of parallel rods lie in the rhombohedral plane, and both are usually but not invariably coaxial with the enclosing calcite.

An impression of the size, shape, and distribution of the dolomite microcrystals may be obtained from the SEM images in Figure 5.26, where (as a result of the

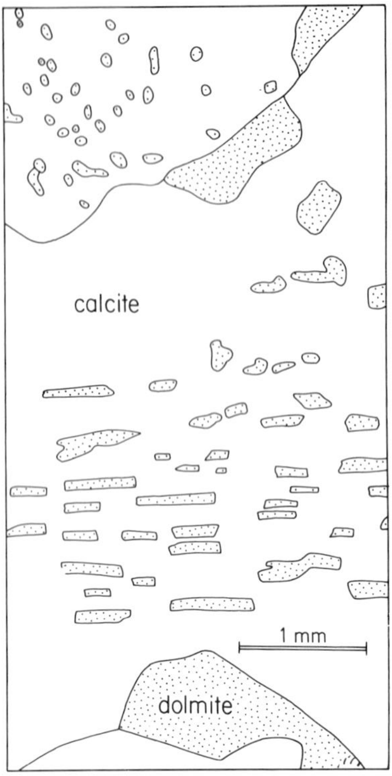

Figure 5.25 Plates (center) and rods (top) of dolomite (stippled) in calcite (clear), and larger crystals of dolomite on calcite–calcite boundaries, in high-grade Grenville marble; portions of the central calcite have recrystallized, following exsolution, to a different orientation. From Kretz (1988)

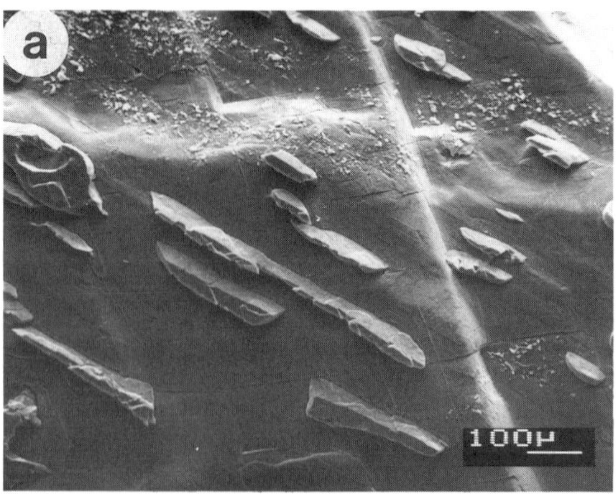

Figure 5.26 (*Continued opposite and overleaf*) SEM images of dolomite microcrystals (in relief) in calcite of high-grade Grenville marble. From Kretz (1988)

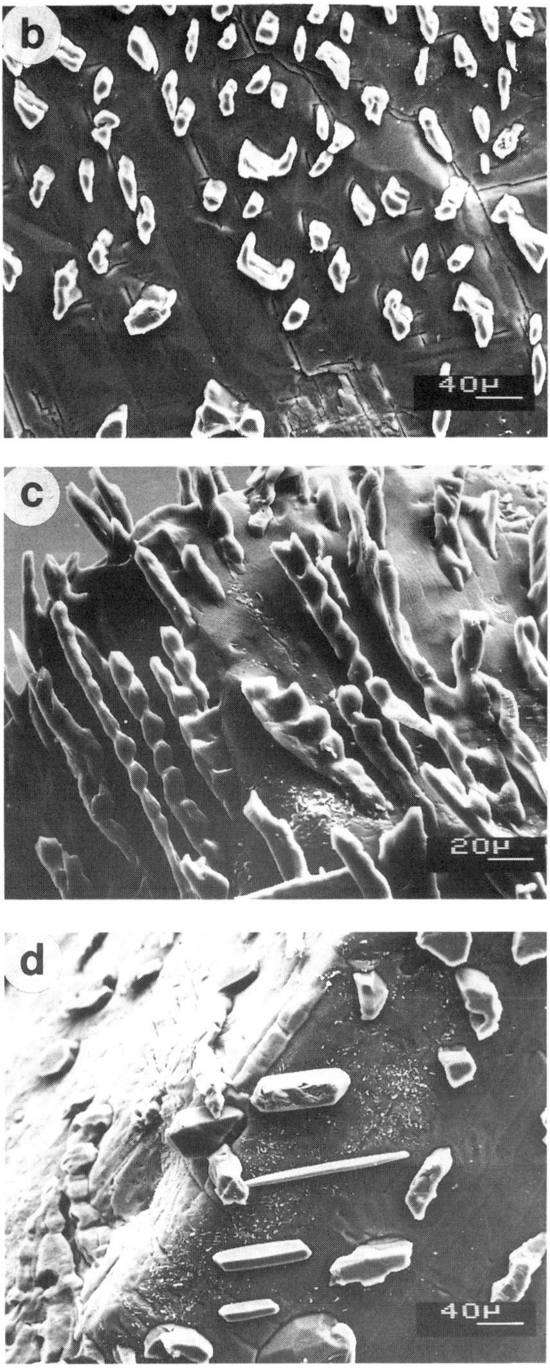
b
40μ
c
20μ
d
40μ

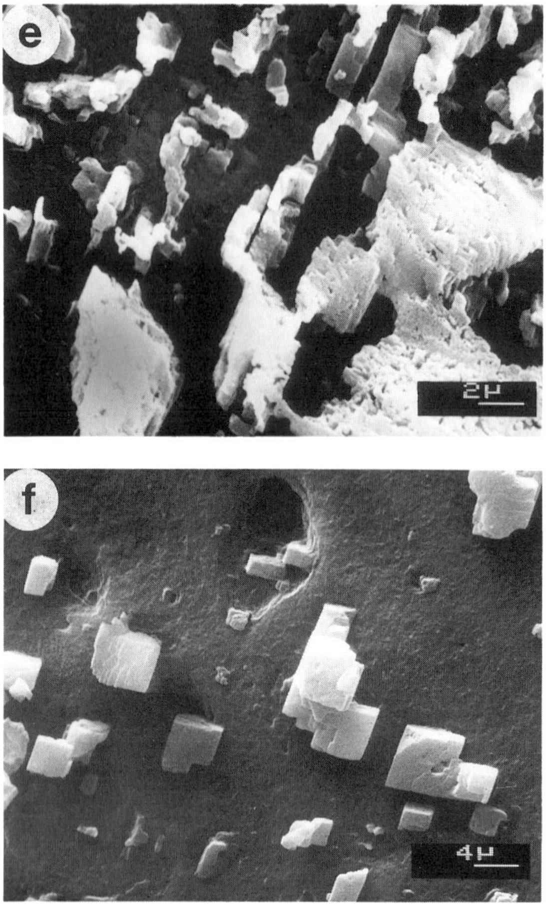

Figure 5.26 *(continued)*

selective dissolution of calcite) the dolomite particles stand in relief. Oriented plates and rods of dolomite appear in Figure 5.26a and b respectively, while a swarm of segmented rods is shown in Figure 5.26c. Figure 5.26d is of special interest because plates, rods, and very small particles are all present together; note also that the plates are terminated by rhombohedral planes. The very small particles of dolomite are viewed in Figure 5.26e, where they are interconnected, and in Figure 5.26f, where they occur as small rhombohedra, one of which (bottom right) is terminated by a basal, (0001), plane.

Figure 5.27 shows the variation in Mg concentration along a microprobe traverse that crosses a dolomite rod in one of the samples. The electron-beam diameter was 5 μm, and the spot analyses of calcite include some very small dolomite particles; hence the term *calcite* for this material. The presence of depleted zones adjacent to rods and plates confirms that the microstructures were produced by a nucleation-and-growth separation of dolomite from calcite, i.e., by exsolution.

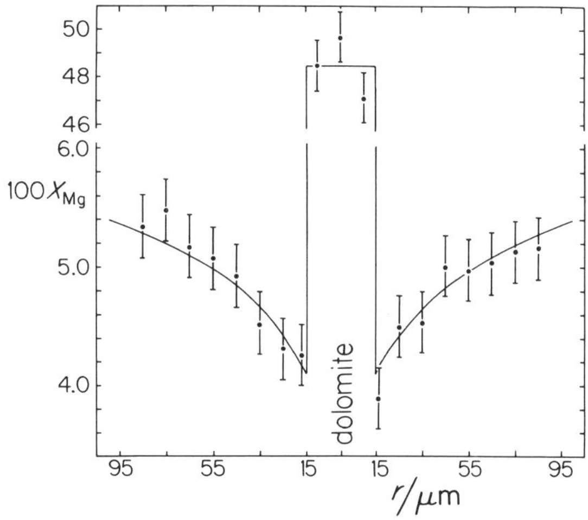

Figure 5.27 Microprobe traverse across a dolomite rod and adjacent *calcite* (i.e., calcite with 1 to 4 μm dolomite particles included in the analysis), showing variation in Mg content (100 Mg/(Ca + Mg)) with distance (r); bars represent 0.95 confidence limits, analytical error; equation used to plot the curves is given in text. From Kretz (1988)

The concentration gradient illustrated in Figure 5.27 is of the kind to be expected from a theoretical consideration of diffusion-controlled growth, as discussed in Sec. 4.5.2. A fit to the data is provided by the following diffusion equation, from Crank (1975), which applies to steady-state radial diffusion toward a long cylinder, and a constant diffusion coefficient,

$$C = \frac{C_1 \ln(r_2/r) + C_2 \ln(r/r_1)}{\ln(r_2/r_1)}. \tag{56}$$

Here C refers to Mg concentration (100 Mg/(Ca + Mg)) in calcite, r is radial distance from the center of the cylinder, subscript 1 refers to the calcite–dolomite interface, and subscript 2 to a point where the concentration gradient decreases to zero. The curves in Figure 5.27 follow this equation.

Dolomite rods and plates contain more Fe and less Sr than the surrounding *calcite*. The factors, which are 4.7 and 0.50 respectively, may be compared with equilibrium distribution coefficients of 3.4 and 0.40 for high-grade marble, as discussed in Sec. 3.8.4. Thus, in agreement with the results of Long and Agrell (1965) and Puustinen (1974), Fe and Sr, in trace concentrations, have also participated in the exsolution process.

An attempt was made to recover the original (pre-exsolution) composition of calcite in calcite–dolomite marble by separating and analyzing calcite crystals together with the embedded dolomite. The large variation that was found (2.4 to 8.6 mole % $MgCO_3$) may be explained by the migration of some *dolomite* to crystal boundaries, e.g., to the relatively large and exterior crystals of dolomite shown in Figure 5.25. An estimate of the Mg content of the calcite phase in calcite–dolomite marble was also obtained (by use of X-rays) and was found to range from 7.6 mol % to trace amounts; assuming saturation at the peak of metamorphism, these results imply much variation in reaction progress within the study area.

From the shape of the dolomite microcrystals (Figure 5.26) it is clear that although the crystallographic orientation of the calcite–dolomite interface varies greatly, basal (0001) and rhombohedral $(10\bar{1}1)$ orientations are rather common, especially where plates and small rhombohedra have crystallized. The cell dimensions of magnesian calcite and dolomite are sufficiently different that the (0001) and $(10\bar{1}1)$ interfaces are likely to be semi-coherent (Sec. 4.4.1), as proposed by Grover and Kubanek (1983), rather than coherent. An examination of these interfaces will serve to illustrate the procedure for calculating the density of edge dislocations on semi-coherent interfaces.

First, the a and c cell dimensions of magnesian calcite and dolomite were estimated for combinations of temperature and pressure extending from 700 °C, 7 kbar to 100 °C, 1 kbar (a possible cooling path), based on information on thermal expansion, compressibility, and (for calcite) the equilibrium Mg content. The spacing of cations along a in calcite (d_a^{Cal}) is then compared with the spacing in dolomite (d_a^{Dol}) by defining a misfit parameter,

$$\delta_a = (d_a^{\mathrm{Cal}} - d_a^{\mathrm{Dol}})/d_a^{\mathrm{Dol}}. \tag{57}$$

The spacing of dislocations along a is therefore

$$D_a = a_{\mathrm{calcite}}/\delta_a, \tag{58}$$

where a_{calcite} is the a dimension of the rhombohedral unit cell of calcite. Values of dislocation density along different lines can now be used to calculate the dislocation density on different planes, and the results for (0001) and $(10\bar{1}1)$ interfaces are presented in Figure 5.28.

Following Porter and Easterling (1981), The interfacial energy per unit area (σ) is viewed as consisting of a structural part (σ_{st}) which may be equated to coherency strain, and a chemical part (σ_{ch}) which is determined by the enthalpy and entropy of the interface phase:

$$\sigma = \sigma_{\mathrm{st}} + \sigma_{\mathrm{ch}}. \tag{59}$$

Assuming that σ_{st} is proportional to dislocation density, we have, from Figure 5.28,

$$\sigma_{\mathrm{st}}(0001) < \sigma_{\mathrm{st}}(10\bar{1}1). \tag{60}$$

Now in view of excellent cleavage on $(10\bar{1}1)$, evidently

$$\sigma_{\mathrm{ch}}(0001) > \sigma_{\mathrm{ch}}(10\bar{1}1). \tag{61}$$

It is possible, therefore, that,

$$\sigma(0001) \sim \sigma(10\bar{1}1). \tag{62}$$

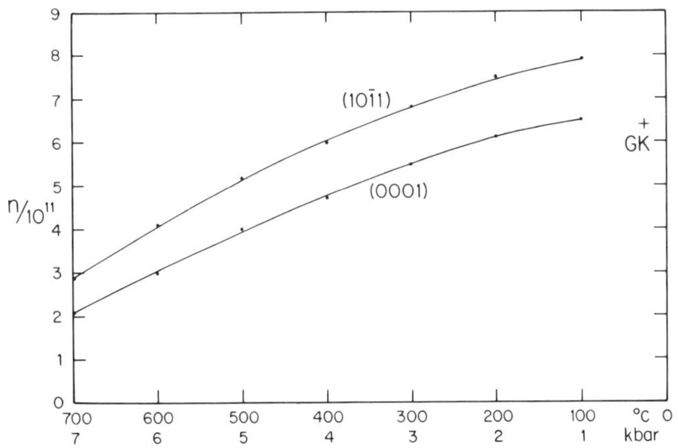

Figure 5.28 Calculated number of dislocations/cm^2 (symbol n) on (10$\bar{1}$1) and (0001) semi-coherent interfaces between coaxial dolomite and calcite, as a function of the indicated combinations of temperature and pressure (data from Kretz 1988). Data point GK is for (0001) at 25 °C, 1 bar, from Grover and Kubanek (1983)

and that both {0001} and {10$\bar{1}$1} are present in the equilibrium shape (i.e., the shape with least interfacial energy) of a particle of dolomite embedded in calcite. Thus a truncated rhombohedron (Figure 5.26f) or a plate with rhombohedral terminations (Figure 5.26d) may approximate the equilibrium shape. But many of the microcrystals, particularly the rods, are evidently bounded by non-coherent interfaces, and one must conclude that the influence of interfacial energy in determining the shape of exsolved particles in carbonates is not as strong as in some other minerals.

Estimates of cell volumes obtained for the above calculations can be used to determine the volume change for the exsolution reaction. The reaction may be written as follows:

$$\mathrm{Ca}_{1-x_1}\mathrm{Mg}_{x_1}\mathrm{CO}_3 \rightarrow 2\left[X_1 - X_2\left(\frac{1-2X_1}{1-2X_2}\right)\right]\mathrm{Ca}_{0.5}\mathrm{Mg}_{0.5}\mathrm{CO}_3 + \left(\frac{1-2X_1}{1-2X_2}\right)\mathrm{Ca}_{1-X_2}\mathrm{Mg}_{X_2}\mathrm{CO}_3,$$

where X_1 and X_2 are the initial and final mole fractions of $MgCO_3$ in calcite. We find that the volume change resulting from exsolution between 600 °C, 6 kbar and 500 °C, 5 kbar is −0.04%, i.e., a contraction by the factor 0.0004. Although the volume change is extremely small, there is a rather large (anomalous) expansion of a (calcite) upon cooling. Consequently, upon exsolution in a calcite crystal of a specified volume and shape, a minimum of distortion of the crystal occurs when the exsolved dolomite is arranged as plates normal to crystallographic c, as found.

In the following interpretation of the above observations, the exsolution process is viewed as occurring in three stages. During stage I, which extended from peak metamorphic temperatures (650 to 700 °C) to very approximately 550 °C, the Mg content of calcite decreased from 10 to 6 mole % $MgCO_3$ as 8 volume % of dolomite was released (Figure 5.29). From the low Mg content of many samples of combined calcite and inclusions, the dolomite released during stage I is inferred to have crystallized on primary dolomite at calcite–dolomite boundaries. The migration of Mg^{2+} to these boundaries, in exchange for Ca^{2+}, resulting in the crystallization of dolomite, does not require a nucleation event, and the only barrier which might prevent the reaction from keeping pace with the drop in temperature is diffusion in calcite, and possibly the rearrangement of atoms to convert the calcite structure to the dolomite structure, as the boundary migrated into calcite. It is possible that where small grains of dolomite occur on calcite–calcite boundaries, all of this dolomite was originally dissolved in calcite; if so, a nucleation of dolomite on the boundary did occur.

During stage II (Figure 5.29), which extended to approximately 480 °C, nuclei of dolomite formed *within* crystals of calcite, followed by growth, to produce plates and rods. Imperfections must have existed in the calcite crystals, and these, together with twin planes and foreign particles, could have provided favorable nucleation sites. But, given a random distribution of Mg atoms in a calcite structure, some small clusters, approximating a dolomite configuration, must exist, and homogeneous nucleation is also a possibility. A statistical test, similar to that used in Sec. 5.2, but confined to two dimensions, confirmed that the rods illustrated in Figure 5.26b (excluding a narrow strip along the bottom of the image) and plates in another sample, are distributed at random or nearly so. Thus a high density of potential nucleation sites within calcite may be inferred, the actual sites having been chosen through random atomic fluctuations.

Where concentration gradients were created during stage I, nucleation of plates and rods may have occurred selectively where Mg was high (in grain centers) and this would explain the occurrence of rods radiating outward from crystal centers.

Why the growth of dolomite nuclei produced plates in one crystal and rods in another (Figure 5.25) or both in the same crystal (Figure 5.26d) is not understood. Beginning with a truncated rhombohedron (Figure 5.30a), coaxial with the enclosing calcite, a plate is formed when the less-coherent interfaces, $(10\bar{1}1)$, advance more rapidly than the more-coherent interfaces, (0001), as in Figure 5.30b; the stated relation of growth rate to coherency has been observed in some metals (Porter and Easterling 1981). Or, if the specific interfacial energy ($\sigma_{st} + \sigma_{ch}$, combined) is less on (0001) than on $(10\bar{1}1)$, then the formation of plates can be understood as an attempt to achieve a minimum of interfacial energy. Finally, strain energy resulting from volume change (as noted above) may have determined that plates must form (normal to *c* of calcite) rather than equidimensional rhombohedra.

The non-segmented rods must be bounded by non-coherent interfaces, and σ_{st} evidently did not play a role in determining the shape of these crystals. With regard

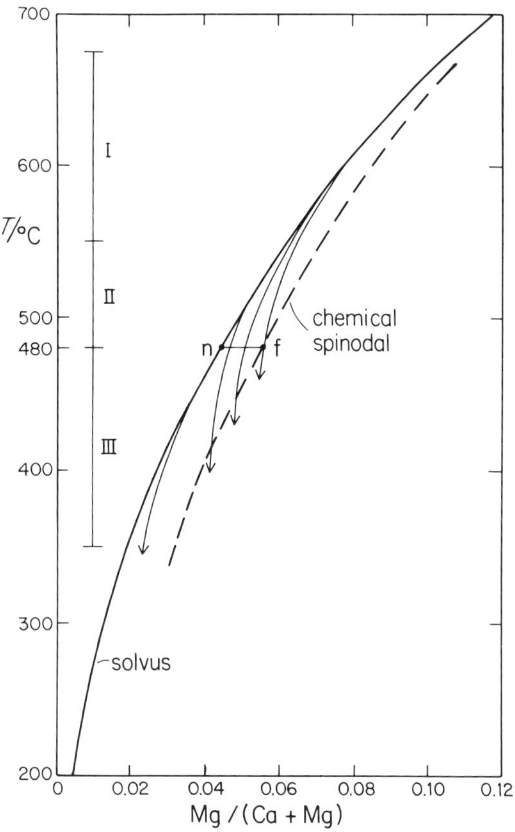

Figure 5.29 Proposed stages (I to III), and cooling paths, resulting in carbonate exsolution within the solvus (from Ch. 3, Figure 3.28) and within the spinodal region; position of chemical spinodal is schematic; Mg/(Ca + Mg) is mole fraction $MgCO_3$ in calcite

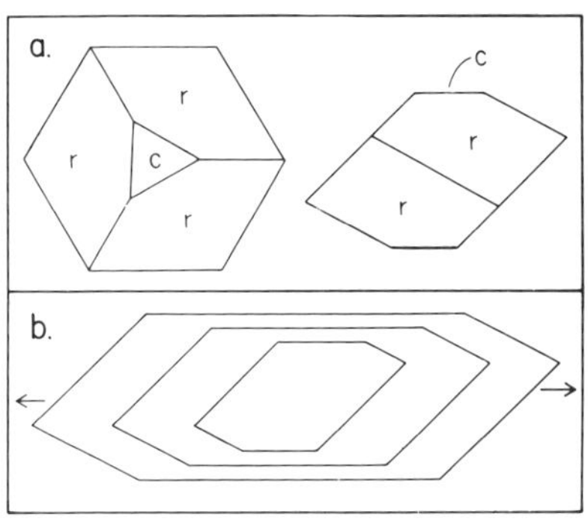

Figure 5.30 (a) Idealized truncated rhombohedron (plan and elevation); (b) growth of a truncated rhombohedron to produce a plate. From Kretz (1988)

to segmented rods (Figure 5.26c), which may be viewed as linked rhombohedra, one possibility is that, following the crystallization of one rhombohedron, a second began to grow at one of the six lateral vertices (Figure 5.30a) and then grew to a size similar to that of the first, and this was followed by a third rhombohedron, and so on. Meanwhile, some precipitation occurred along the rods to produce a smooth surface. It is also possible that the rods grew as more regular cylinders, and that segmentation occurred during or following growth, as an adjustment toward a reduction in interfacial energy.

Stage II extended to approximately 480 °C, which is the temperature estimate obtained from the composition of *calcite* near plates and rods (Figure 5.27); this is shown as point n in Figure 5.29. At this temperature, which marks the beginning of stage III, volumes of supersaturated calcite away from the plates and rods (points f, Figure 5.29) entered the spinodal region, where virtually no energy barriers existed to exsolution (Sec. 4.4.2), and a massive precipitation of very small particles occurred (Figure 5.26e). With a continued decrease in temperature, various small volumes of calcite, which at 480 °C were progressively less supersaturated, entered the spinode (Figure 5.29), resulting in further precipitation of fine dolomite. But spinodal precipitation did not occur everywhere (e.g. Figure 5.26a, b), and volumes of calcite, which during stages I and II were more strongly purged of Mg, would not have entered the spinodal region (Figure 5.29). Stage III may have continued to about 350 °C, where atomic mobility declined and the reaction was 'left behind', as temperature continued to drop. But, because the present Mg content of calcite, as determined by use of X-rays, is not uniformly low, and in places is surprisingly high, the point on the composition or temperature scale at which the exsolution reaction ceased to operate evidently varied from place to place in the study area; this also remains unexplained.

Behavior similar to that proposed above is found in the Fe–C system where, on controlled cooling, the precipitation of ferrite, exsolved from austenite, occurs first on grain boundaries, and finally as microcrystals within grains of austenite (Shewmon 1969).

In conclusion, an astonishing diversity exists in the shape, size, and distribution of microcrystals of dolomite in calcite from high-grade marble. The single uniformity is a tendency for the crystallographic axes of the microcrystals to lie parallel to those of the host calcite, which together with the presence of stranded diffusion gradients about plates and rods would appear to confirm that the microcrystals were produced by an exsolution process. All of the study samples are from a single metamorphic terrane, and although very small differences in cooling rate are not impossible, the principal cause of the diversity in microstructure must be sought in subtle initial differences in the chemical composition and structure (including imperfections and strain) of the calcite crystals. Thus numerous questions related to exsolution kinetics in carbonate crystals remain for further study.

5.5 ROLE OF INTERFACIAL ENERGY IN CHANGING THE SHAPE AND SIZE OF CRYSTALS, AND A NOTE ON PLASTIC DEFORMATION AND RECRYSTALLIZATION

In Chapter 1 (Sec. 1.4), the term 'granular microstructure' was introduced to refer to the shape, size, orientation, and distribution of crystals in samples of metamorphic rock, and in earlier sections of the present chapter, some microstructures were described and interpreted as being the direct products of chemical reaction. We now consider modifications of earlier microstructures that were produced, as above, by chemical reaction, or by sedimentary or igneous processes, with emphasis on modifications that are evidently driven by interfacial energy. Early statements of the importance of interfacial energy in terrestrial processes were expressed by Becke (1913), Buerger (1947), Verhoogen (1948), Ramberg (1952a), Rosenfeld (1955), DeVore (1959), Griggs et al. (1960), and Voll (1960).

Plastic deformation and the release of strain energy are important microstructure-forming processes in metamorphic terranes. By analogy with metals, plastic deformation may be followed by recrystallization (driven by strain energy), followed by grain growth (driven by interfacial energy). Although we are presently concerned mainly with the stage of grain growth, resulting in changes in crystal shape and size, it is appropriate to begin our study with a brief survey of the preceding processes — deformation and recrystallization; this large and important subject was reviewed by Nicolas and Poirier (1976), and by several authors in a book edited by H.-R. Wenk (Wenk 1985).

5.5.1 Plastic deformation and recrystallization

While carrying out field studies of metamorphic terranes, one is frequently struck by the overwhelming evidence that appears particularly on the scale of 10 cm to 10 m, for rock bodies having experienced solid-state (plastic) deformation, often of a very large magnitude. The evidence occurs principally as deformed planar features, including bedding, dikes, and foliation planes. The term 'plastic' (or 'ductile') deformation implies permanent deformation (or strain) without fracturing, and analogies with cold-worked metal and glacier ice immediately come to mind.

An example of plastic deformation is presented in Figure 5.31a. It is reasonable to suppose that the illustrated quartz–feldspar dike was initially planar, and was subsequently folded, and in places, fractured and disjointed, as deformation in the enclosing amphibolite was entirely plastic in nature. Figure 5.31b presents an example of localized strain in granite, and an impression of the magnitude of strain can be had from the quartz–feldspar dike, which was drawn into the shear zone; an enhancement and reorientation of crystal foliation within the shear zone is apparent.

The study of deformation (both natural and experimental) in silicate and carbonate crystals and rocks has derived much benefit from parallel studies of

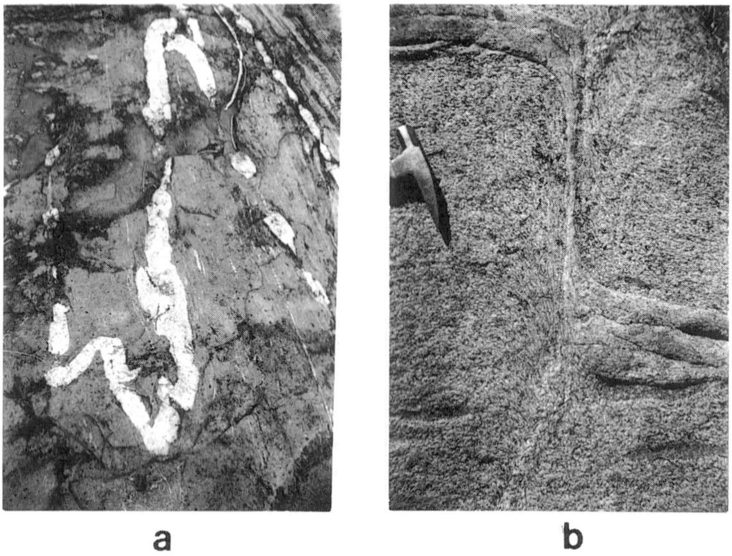

Figure 5.31 (a) Plastic deformation in amphibolite, revealed by a folded quartz–feldspar dike. (b) Plastic strain in granite, concentrated in a shear zone; magnitude of strain is revealed by the quartz–feldspar dike. Location, Northern Indian Lake, Manitoba (Kretz 1967)

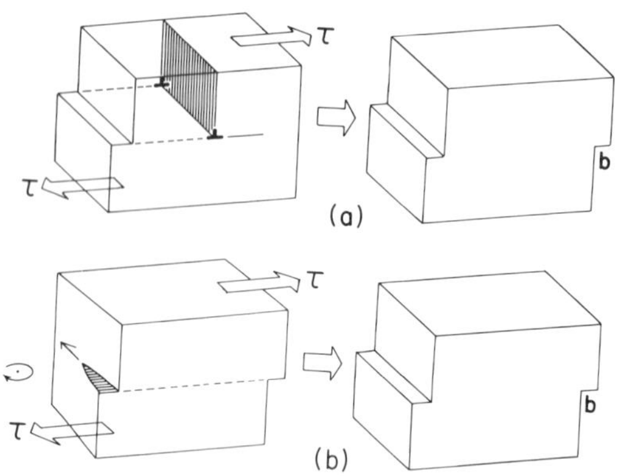

Figure 5.32 Slip, by the movement of an edge dislocation (a), and a screw dislocation (b), under the influence of a shear stress, τ. Small arrow is the direction of motion; **b** is the Burgers vector. From Hayden, Moffatt and Wulff (1965). Reproduced by permission of John Wiley & Sons, Inc.

metals. Plastic deformation in metals occurs by the migration of dislocations and by the migration of atoms and vacancies, and the several mechanisms that are known to occur in metals and other crystals are reviewed by Hayden, Moffatt and Wulff (1965), Hobbs, Means and Williams (1976), Nicolas and Poirier (1976), and Ranalli (1987). In the process known as 'slip' (or 'translational gliding') displacement occurs on certain crystallographic planes and in certain directions. One of the slip systems in quartz, for example, is (0001) [1000], which means 'slip on a (0001) plane in the direction of one of the *a* axes'. Well-known slip systems for various minerals are compiled by Spry (1969) and Hobbs, Means and Williams (1976). Slip occurs by the migration of edge or screw dislocations (Figure 5.32) or a combination of both. Shewman (1969) has noted that slip is like moving a large rug across the floor by forming a ripple, and moving the ripple. An example of a crystal evidently deformed by slip is illustrated in Figure 5.33.

At relatively high temperatures (amphibolite and granulite facies) and low rates of strain, plastic deformation can conceivably occur by the diffusion of atoms, a process known in metallurgy as 'diffusional creep'. Atoms move from points of high compressive stress to points of lower stress, as vacancies migrate in the opposite direction (Figure 5.34); the result is a change in shape, without a change

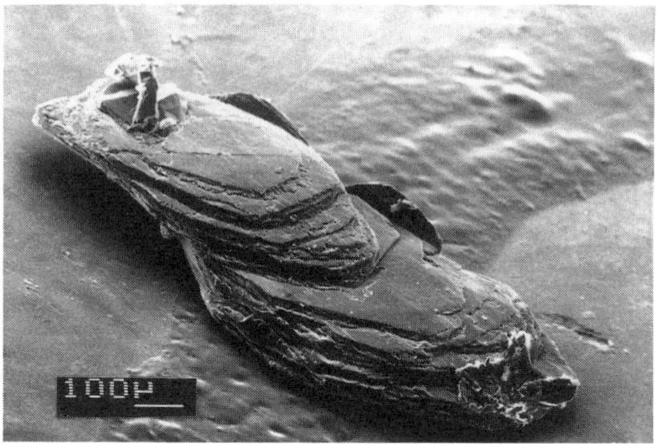

Figure 5.33 A possible example of slip; deformed graphite crystal separated from Grenville marble

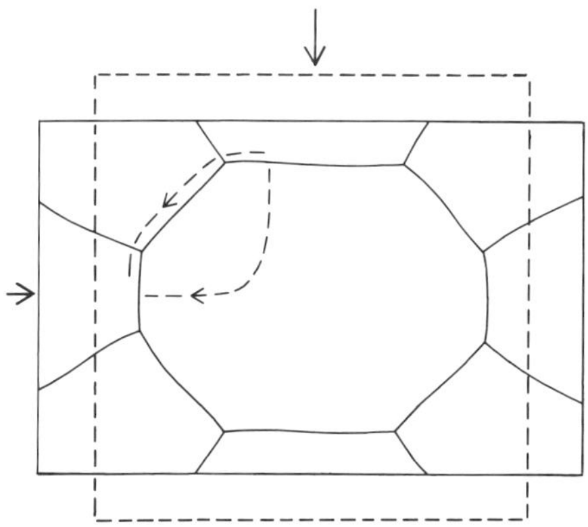

Figure 5.34 Illustration of diffusional creep; arrows represent migration of atoms, through the central crystal, and along grain boundaries; vacancies migrate in the opposite direction

in crystallographic orientation. Diffusion may occur through a crystal (Nabarro–Herring creep) or through an adjoining grain boundary (Coble creep), leading to rates of strain that are proportional to the coefficients of lattice diffusion and boundary diffusion respectively. A possible example of diffusional creep (in garnet gneiss) is illustrated in Figure 5.35.

The plastic deformation of metals at low temperatures (cold working) results in a large increase in dislocation density, for example, from 10^7 to 10^{11} dislocation lines/cm^2 (Hayden, Moffatt and Wulff 1965). During subsequent heating, these dislocations migrate to form clusters, thus reducing the elastic strain energy; this is referred to as 'recovery'. The clusters may form walls, separating slightly misoriented domains (polygonization) as is commonly observed in quartz. A further reduction in strain energy occurs as new crystals with a low dislocation density 'nucleate' and grow at the expense of the strained crystals. This change is referred to as 'recrystallization' or, if it occurs while plastic deformation is in progress, as 'syntectonic recrystallization'. The beginning of growth of the new crystals must differ from classical nucleation, as defined in Sec. 4.4.1, and is not fully understood. Enlargement of the crystals occurs by migration of the boundary between these and the host, the driving force being strain energy. In general, the orientation of the new crystals is found to differ by only a few degrees from that of the host.

An example of plastic deformation, recovery, and recrystallization is provided by Olsen and Kohlstedt (1985) who studied the Nordre Strømfjord zone in west Greenland, with emphasis on the behavior of plagioclase. The shear zone is 7 to 18 km wide, >150 km long, and cuts Precambrian gneiss and amphibolite; the lateral displacement is estimated at 120 km. Deformation in the zone occurred during metamorphism, at 700 to 800 °C and 6 to 8 kbar.

In a hornblende gneiss unit that was chosen for detailed study, plagioclase (25 to 48 mole % anorthite) occurs as crystals 0.2 to 2.0 mm in diameter. The crystals are

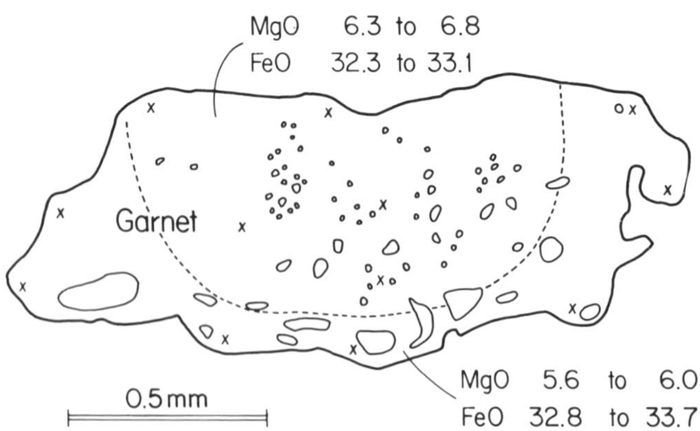

Figure 5.35 One of many 'flattened' garnet crystals in garnet–biotite–quartz–feldspar gneiss of the Otter Lake terrane, a possible illustration of deformation by diffusional creep

twinned (albite and pericline twins) and show irregular extinction. Many smaller crystals, 0.01 to 0.15 mm in diameter, are found as clusters in the margins of the larger ones, and some of these are also twinned. The large crystals are oriented with (010) nearly parallel to the shear zone, and the orientation of the smaller crystals differs slightly from that of the adjoining larger crystals. The anorthite content of the small crystals exceeds that of the larger ones by about 7 mole %. The results are interpreted as follows: (1) plastic deformation of the large plagioclase crystals occurred principally by slip and twinning, (2) the presence of low-angle boundaries (irregular extinction) is the result of recovery, (3) small plagioclase crystals formed by recrystallization, which was syntectonic, (4) the cause of the change in composition during recrystallization is not known. These interpretations, which are based largely on the granular microstructure, were supported by a TEM study of the fine microstructure.

Mylonites, in which the crystal size is uniformly or predominantly very small, evidently represent a more advanced stage of recrystallization, compared with that described above, in which all or most of the original, plastically strained, crystals have disappeared. In common gneiss and amphibolite of highly deformed metamorphic terranes, a bimodal size distribution for plagioclase and other minerals, as described above, is usually not present. Thus, if recrystallization occurred, to produce a large number of small, strain-free crystals, modifications of the microstructure occurred subsequently, possibly by grain growth (coarsening), driven by interfacial energy.

5.5.2 Interfacial tension and interfacial energy defined

The tendency for a small drop of water or mercury to form a sphere can be understood by supposing that forces exist within the water–air or mercury–air interfaces. The interface itself can be viewed as a geometric curved plane, or as a layer of finite width (Figure 5.36a). Consider a line of length L lying within the boundary plane; then the force acting normal to this line (per unit length measured along L) is referred to as the 'surface' or 'interfacial tension' (dyne/cm), and is denoted σ (Figure 5.36b). Thus, an increase in the area of the interface by a small amount, $dA = L\,dx$, involves an expenditure of work,

$$w = \sigma L \, dx = \sigma dA \,. \tag{63}$$

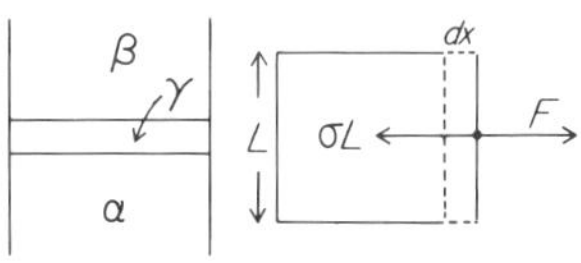

Figure 5.36 (a) Illustration of a system with phases α and β, separated by a boundary phase, γ; (b) illustration of surface tension (σ), normal to a line of length L lying in the interface (F is the force that opposes the tension); at equilibrium, $F = \sigma L$

The term σ dA has units of energy (dyne- cm; erg) and it is convenient to assign to a particular interface a quantity of energy (per unit area), referred to as the 'specific excess interfacial free energy', or simply 'interfacial energy' (ergs/cm^2; J/cm^2).

Following Aveyard and Haydon (1973), interfacial energy may be expressed in terms of internal, Helmholtz, or Gibbs energy. For an entire system (Figure 5.36a), composed of phases α and β and a boundary phase, γ, a change in internal energy (U) is expressed by the equation (from Sec. 2.1),

$$\mathrm{d}U = T\,\mathrm{d}S - P\,\mathrm{d}V. \tag{64}$$

Suppose that a change occurs in the volume of α and β, and also in the area of the interface. Then the work done by the system is

$$w = -P\,\mathrm{d}V + \sigma\,\mathrm{d}A\,, \tag{65}$$

and, for the entire system,

$$\mathrm{d}U = T\,\mathrm{d}S - P\,\mathrm{d}V + \sigma\,\mathrm{d}A\,. \tag{66}$$

Thus, one definition for interfacial tension is

$$\sigma = \left(\frac{\partial U}{\partial A}\right)_{S,V} \tag{67}$$

The internal energy of the system as a whole is

$$U = U^{\alpha} + U^{\beta} + U^{\gamma}, \tag{68}$$

and, for the boundary phase alone,

$$\mathrm{d}U^{\gamma} = T\,\mathrm{d}S^{\gamma} - P\,\mathrm{d}V^{\gamma} + \sigma\,\mathrm{d}A\,. \tag{69}$$

The Helmholtz energy of a system is, by definition (Sec. 2.1),

$$F = U - TS, \tag{70}$$

and, for a boundary phase,

$$F^{\gamma} = U^{\gamma} - TS^{\gamma}. \tag{71}$$

Thus,

$$\mathrm{d}F^{\gamma} = \mathrm{d}U^{\gamma} - T\,\mathrm{d}S^{\gamma} - S\,\mathrm{d}T^{\gamma}. \tag{72}$$

By combining (72) with (69),

$$\mathrm{d}F^{\gamma} = -S\,\mathrm{d}T^{\gamma} - P\,\mathrm{d}V^{\gamma} + \sigma\,\mathrm{d}A\,, \tag{73}$$

or, for the entire system,

$$\mathrm{d}F = -S\,\mathrm{d}T - P\,\mathrm{d}V + \sigma\,\mathrm{d}A\,. \tag{74}$$

This equation provides a second definition of interfacial tension,

$$\sigma = \left(\frac{\partial F}{\partial A}\right)_{T,V}. \tag{75}$$

Recall (Sec. 2.1), that the Gibbs energy of a system is, by definition,

$$G = U - TS + PV. \tag{76}$$

Thus, for the boundary phase,

$$\mathrm{d}G^{\gamma} = \mathrm{d}U^{\gamma} - T\,\mathrm{d}S^{\gamma} - S^{\gamma}\,\mathrm{d}T + P\,\mathrm{d}V^{\gamma} + V^{\gamma}\,\mathrm{d}P. \tag{77}$$

By combining (77) with (69),

$$\mathrm{d}G^{\gamma} = -S^{\gamma}\mathrm{d}T + V^{\gamma}\mathrm{d}P + \sigma\mathrm{d}A, \tag{78}$$

and, for the system as a whole,

$$\mathrm{d}G = -S\,\mathrm{d}T + V\,\mathrm{d}P + \sigma\,\mathrm{d}A. \tag{79}$$

This equation provides a third definition of interfacial tension,

$$\sigma = \left(\frac{\partial G}{\partial A}\right)_{T,P}. \tag{80}$$

For a single-component system, for example, α = liquid water, β = water vapor, interfacial tension and specific interfacial energy are numerically equal. Thus, with σ = 500 dyne/cm, F^{γ} = 500 erg/cm^2. In a system of two or more components, atoms of a second component are commonly concentrated (adsorbed) in the boundary phase, and the equality does not hold. Then, by adding to equation (73) the term

$$\sum_i \mu_i\,\mathrm{d}n_i$$

which expresses a change in F^{γ} resulting from the addition to γ of a number (dn) of components (i), each of chemical potential μ_i, we obtain

$$\mathrm{d}F^{\gamma} = -S^{\gamma}\,\mathrm{d}T - P\,\mathrm{d}V^{\gamma} + \sigma\,\mathrm{d}A + \sum_i \mu_i\,\mathrm{d}n_i^{\gamma}. \tag{81}$$

Integrating at constant T, P, μ_i, σ,

$$F^{\gamma} = -PV^{\gamma} + \sigma A + \sum_i \mu_i n_i^{\gamma}. \tag{82}$$

The volume of the boundary phase (V^{γ}) is very small; therefore by removing the first term on the right, and dividing by A,

$$F^{\gamma}/A = \sigma + \sum_i \mu_i \Gamma_i^{\gamma}, \tag{83}$$

where Γ_i^γ is the surface excess concentration of component i. In metals, grain-boundary adsorption is common, for example, concentrations of Ni, Cr, and Mn on Fe–Fe boundaries are greater than those in adjacent crystals by a factor that ranges from 10 to 1000 (Porter and Easterling 1981), and interfacial tension is therefore numerically less than specific interfacial energy. For a detailed treatment of the thermodynamics of interfaces, which emphasizes the work of Gibbs (1878), the reader is referred to Cahn (1979).

There is a striking similarity in the microstructure of a mosaic of soap bubbles (where cell shape is certainly governed by interfacial tension), annealed metals, and some high-grade metamorphic rocks (Figure 5.37a, b, c). A tendency obviously exists in all of these materials for the formation of triple junctions with interface angles close to 120°. Different possibilities for interfacial angles in crystal mosaics were examined by Smith (1952) and Herring (1953), and are illustrated in Figure 5.38.

Figure 5.38a represents three crystals of the same phase (denoted a) meeting along a line, normal to the page. Assuming that the interfacial tension in each of

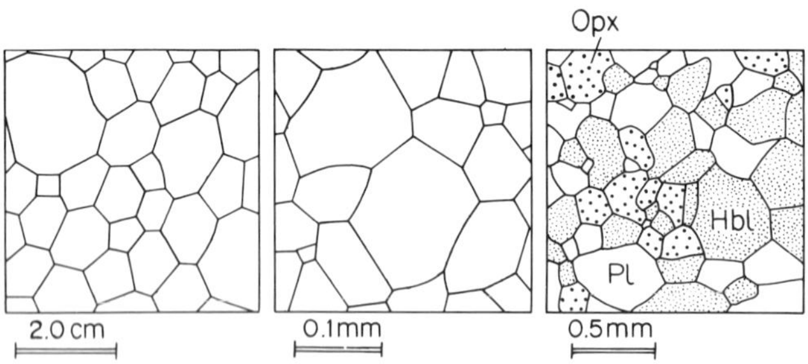

Figure 5.37 Illustration of mosaic microstructures; (a) soap bubbles between parallel sheets of glass, (b) annealed nickel (cold worked, followed by stages of recrystallization and grain growth); from Moffatt, Pearsall and Wulff (1964), reproduced by permission of John Wiley & Sons, Inc.; (c) granulite (meta-gabbro) composed of orthopyroxene (Opx), hornblende (Hbl), and plagioclase (Pl), from Kretz, Jones and Hartree (1989). Reproduced by permission of Springer-Verlag

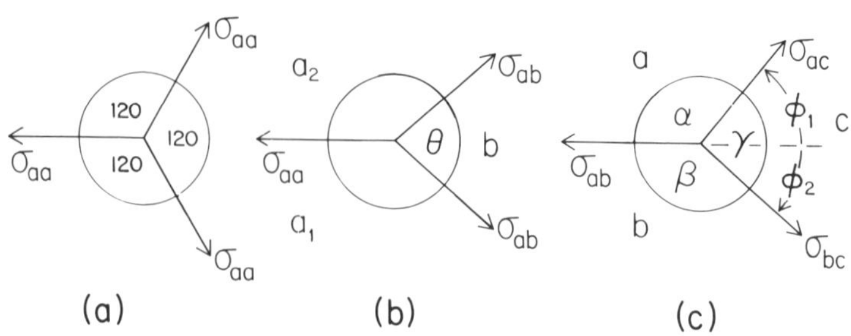

Figure 5.38 Balance of interfacial tensions in (a) three crystals of the same kind, (b) two crystals of a and one of b (θ is the dihedral angle), and (c) three different crystals

the three interfaces is independent of the crystallographic orientation of the crystals, a static balance of the three forces (σ_{aa}) obviously requires that all interfacial angles be 120°.

Figure 5.38b represents the meeting of two crystals of phase a, and one of phase b. Assuming again that σ is orientation-independent, a static balance is obtained when

$$\sigma_{aa} = 2\sigma_{ab}\cos(\theta/2). \tag{84}$$

The angle, θ, is referred to as the 'dihedral angle'.

Figure 5.38c represents the meeting of three crystals (a, b, c) each of a different phase. Now, by use of the law of sines,

$$\frac{\sigma_{ab}}{\sin\gamma} = \frac{\sigma_{ac}}{\sin\beta} = \frac{\sigma_{bc}}{\sin\alpha}, \tag{85}$$

or

$$\sigma_{ab} - \sigma_{ac}\cos\phi_1 - \sigma_{bc}\cos\phi_2 = 0. \tag{86}$$

In general, σ is expected to depend on the crystallographic orientation of the two adjoining crystals, this variation being defined by

$$\frac{\partial\sigma_{ac}}{\partial\phi_1} \text{ and } \frac{\partial\sigma_{bc}}{\partial\phi_2}.$$

An analysis of this problem leads to the Herring (1951) equation (also derived by Swalin (1962)),

$$\sigma_{ab} - \sigma_{ac}\cos\phi_1 - \sigma_{bc}\cos\phi_2 + \left[\frac{\partial\sigma_{ac}}{\partial\phi_1}\sin\phi_1 + \frac{\partial\sigma_{bc}}{\partial\phi_2}\sin\phi_2\right] = 0. \tag{87}$$

In the absence of an orientation effect, the expression in brackets disappears, and equations (87) and (86) become identical.

The recognition that crystal–vapor interfaces may be said to possess an amount of excess energy, per unit area, has led to the proposal that the total interfacial energy of a particular crystal (e.g., MgO) of a given volume in a vapor or vacuum, will be reduced to a minimum when the crystal is bounded by planes of a certain orientation. This proposal was expressed by Gibbs (1878), and was developed by Wulff (1901) in the form of a theorem. In the case of MgO, the equilibrium shape is not a sphere, as for mercury, representing a minimum of surface area, but a cube, {100}.

The Wulff theorem can be expressed as follows. Construct a large number of lines extending from the center of a crystal, and let the length of each line be proportional to the interfacial energy of the surface plane which lies normal to that line. The result is a surface (the γ-plot) that for mercury is a sphere, but for crystals contains cusps representing normals to planes of low energy. The polyhedron that is just enclosed by the γ-plot represents the equilibrium shape. The concept applied to a crystal of biotite embedded in calcite (in place of a vapor) is illustrated in Figure 5.39. The equilibrium shape for some crystals can be determined by heating

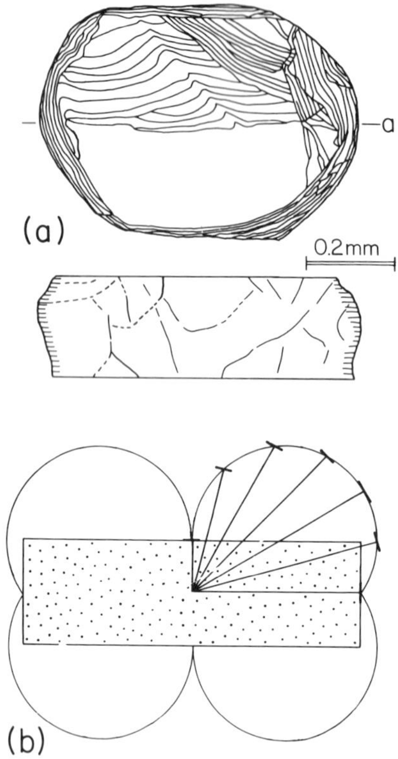

Figure 5.39 (a) Plan and elevation of a biotite (phlogopite) crystal in marble (growth steps are shown); from Kretz (1968a) reproduced by permission of the Mineralogical Society, London; (b) representation of the construction of a γ-plot for biotite; each line radiating from the center is proportional to the energy of the surface (per unit area) normal to that line; the enclosed body represents the equilibrium shape

a small particle, ground to the shape of a sphere, and observing the faces that slowly develop. Herring (1953) has pointed out that because of energy barriers, only very small crystals can aquire an equilibrium shape, given that this shape was not produced during crystal growth.

5.5.3 Crystal shape in medium- to high-grade metamorphic terranes

It is instructive to compare the microstructure of some samples of granulite and metagabbro (e.g. Figure 5.37c) with that of gabbro, particularly with regard to the shape of plagioclase crystals. In typical gabbro, plagioclase crystals are blocky, resulting from crystallization in a liquid, while the microstructure of many granulites is smooth (as Picasso's *Woman with a book* is to Modigliani's *Bohémienne a l'enfant*). During the change from gabbro to granulite, plagioclase that was not consumed by reaction must pass through the stages of recrystallization and grain growth, resulting in grain-boundary curvature and angles that are possibly determined by interfacial tension.

Table 5.4 is a listing of boundary-angle measurements from samples of granulite. These are true angles, obtained by mounting thin sections on a universal stage, and adjusting the stage until the line of meeting of three crystals is vertical. Variations, expressed as standard deviation, are also reported.

A mosaic microstructure, similar to that of Figure 5.37c could conceivably arise by the random nucleation and growth of crystals during primary metamorphic crystallization or recrystallization, in complete defiance of the small interfacial tensions acting on the boundaries. A two-dimensional model of this behavior (Figure 5.40) does indeed produce triple junctions, with many angles close to 120°, but the variation in the angle is 30° (one standard deviation from 120°), much greater than that found in the natural mosaics (Table 5.4). Therefore the evidence for interfacial tension having an influence in the development of natural microstructures is not the 120° mean angle, but the small variation in this angle, expressed by a standard deviation of only about 10°.

Table 5.4 also lists representative dihedral angles, which in most cases are <120°, implying (equation (84) of Sec. 5.5.2) a smaller tension on 'unlike' than on 'like' boundaries. This is in agreement with data from metals, where, for example, $(\sigma_{\mathrm{CuZn}}/\sigma_{\mathrm{CuCu}}) = 0.78$ at 700 °C (Smith 1948). We may note in passing that where the dihedral angle approaches zero, the second phase will tend to spread along boundaries of the first phase. An example is θ (galena versus sphalerite–sphalerite) $\rightarrow 0$ at 900 °C, described by Stanton and Gorman (1968).

Where three 'like' crystals are in contact and the interfacial tensions are not balanced, a balance can be achieved by local crystal-boundary migration, i.e., by across-interface atomic movement, and this may result in boundary curvature. Thus the observed variation in boundary angles, for example, $120° \pm 2s = 120° \pm 20°$

Table 5.4 Some measured angles at triple junctions in granulite

Minerals	Mean angle	Standard deviation	Source	σ_{ab}/σ_{aa}
Scp	120	8	1	
Pl	120	10	2	
Cpx	120	12	2	
Ol	120	7	2	
	θ			
Cpx v. Scp–Scp	128	~13	1	1.14
Scp v. Cpx–Cpx	109	~9	1	0.86
Cpx v. Pl–Pl	109	16	2	0.86
Pl v. Cpx–Cpx	99	19	2	0.77

Sources: 1. Ca-pyroxene (Cpx)–scapolite (Scp) granulite, Quebec (Kretz 1966a)
2. Mafic and ultramafic granulite, with Ca pyroxene (Cpx), olivine (Ol), and plagioclase (Pl), Scotland (Vernon 1970).

$\sigma_{ab}/\sigma_{aa} = 1/(2\cos(\theta/2))$; equation (84), Sec. 5.5.2.

See Figure 5.38b for definition of dihedral angle (θ).

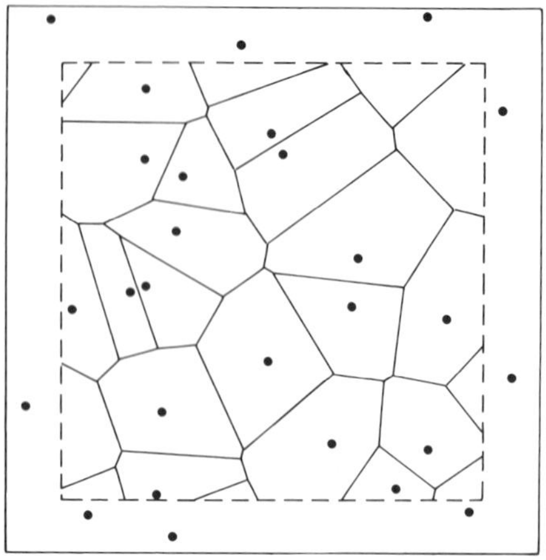

Figure 5.40 Two-dimensional model of a phase, 'nucleated' instantaneously and at random, and permitted to grow at a constant rate, to completion; note triple junctions, and variation in boundary angles. This is the Meijering cell model (Gilbert 1962), also known as a Voronoi polygon (Crain 1976)

for plagioclase may be attributed to departure from equilibrium with regard to the static balance of interfacial tensions. Vernon (1970) has noted that with reference to dihedral angles, a balance of tensions is particularly difficult where small grains of the second phase occur on boundaries between large crystals of the first phase. Another possible cause for variation in grain-boundary and dihedral angles is the orientation effect, as expressed by equation (87) in Sec. 5.5.2.

The above results raise the question of the three-dimensional shape of crystals or cells in an efficient filling of space. The problem was solved by Thomson (1887) who determined that space can be occupied completely by packing 'crystals' of uniform size, each with the shape of a truncated octahedron, referred to as the 'α-tetrakaidecahedron'. This body is bounded by six square and eight hexagonal 'faces', and is evidently not found anywhere but in certain biological tissues of unusual regularity (Smith 1954). Williams (1968) observed that in brass, five-sided 'faces' occur frequently, and he proposed the β-tetrakaidecahedron, which is bounded by two squares, four hexagons, and eight pentagons. An SEM image of a hornblende crystal from a hornblende-rich amphibolite is presented in Figure 5.41; the crystal is bounded by planar to slightly curved, four-to six-sided surfaces, with straight to slightly curved edges.

Grain-boundary angles at junctions of three hornblende crystals in high-grade amphibole can depart greatly from 120°. In the example presented in Figure 5.42, the departure can be related to the orientation of the crystals. Thus, boundary ac lies parallel to (110) in crystal a, and boundary bc bisects the angle between (110) of crystal b and (110) of crystal c. There is some uncertainty here as to whether interfacial tension has indeed played a role in determining the disposition of the

Figure 5.41 SEM image of a crystal of hornblende, separated (intact) from hornblende-rich amphibolite

crystal boundaries, but if it has, we see in this example, a demonstration of the orientation effect, as expressed by equation (87).

At this point in our study, it is more convenient to examine interfaces in terms of interfacial energy. For a large number a possible orientations of an interface between crystal a and c in Figure 5.42, that orientation which lies parallel to (110) of either one or the other of the two crystals, presumably represents the lowest specific interfacial energy. A further and more familiar example is found where two biotite crystals are in contact, and the boundary plane almost invariably lies parallel to (001) of one of the crystals. In the example illustrated in Figure 5.43,

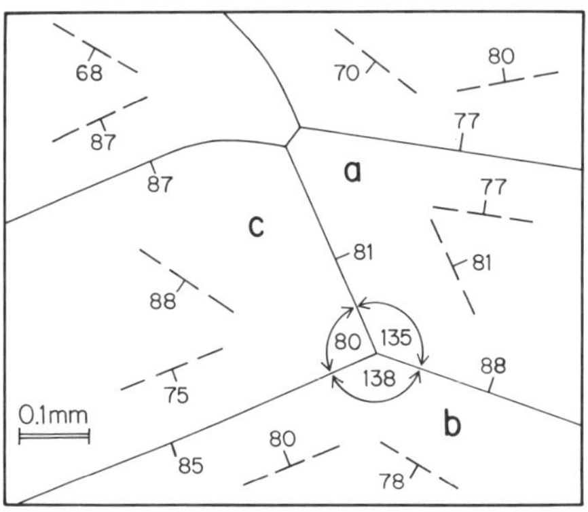

Figure 5.42 Distribution of crystal boundaries in an aggregate of hornblende crystals in high-grade amphibolite; broken lines give the 'strike and dip' of (110) cleavage planes. From Kretz (1966a). Reproduced by permission of Oxford University Press

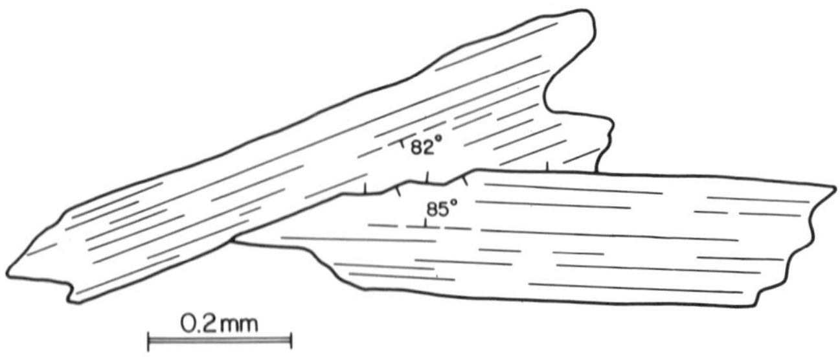

Figure 5.43 Boundary between two biotite crystals in high-grade gneiss; the boundary lies alternately parallel to (001) of one and then the other crystal. From Kretz (1966a). Reproduced by permission of Oxford University Press

the boundary is corrugated, being alternately parallel to (001) of one crystal, and (001) of the other crystal. Compared with a smooth boundary plane, the corrugated boundary has a larger area but smaller values of both specific and total interfacial energy.

Another possible demonstration of interfacial energy is found in some inclusions of quartz and feldspar in crystals of garnet and other ferromagnesian minerals. Immediately after a grain of quartz has been engulfed by a growing garnet crystal, the shape of the grain is expected to be somewhat irregular. But we occasionally see, in high-grade gneisses (Figure 5.44), that the grains are bounded by planes that are 'faces' of garnet, not of quartz, i.e., the garnet has imposed its crystal form on the quartz inclusions. These and other boundary adjustments conceivably take place very slowly, as bodies of rock reside for millions of years at mid-to-deep crustal levels.

The question of whether some natural crystals represent an equilibrium shape can be examined in relation to crystals of mica in marble and schist. Consider a crystal of biotite or muscovite embedded in a matrix of calcite or quartz, and assume that (001) and six prismatic faces, designated (*hk*0), are present at equilibrium,

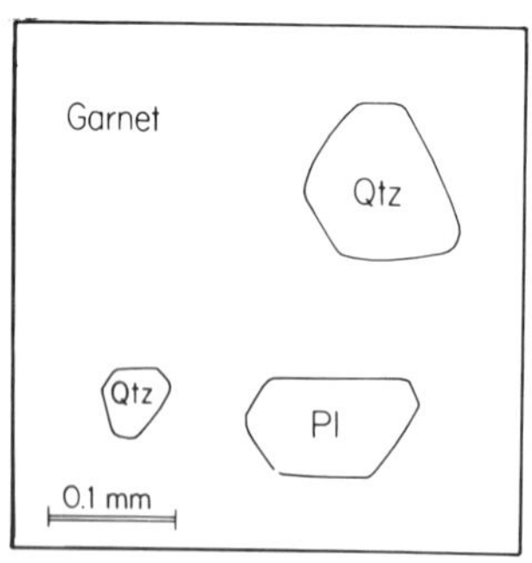

Figure 5.44 Inclusions of quartz (Qtz) and plagioclase (Pl) in a garnet crystal in high-grade gneiss. The 'faces' on the inclusions are those of garnet. From Kretz (1966a). Reproduced by permission of Oxford University Press

i.e., consider the crystal shown in Figure 5.39a of Sec. 5.5.2 as an approximation to the equilibrium shape. Then the total mica–calcite or mica–quartz interfacial energy is

$$F^{\gamma} = f_{001}A_{001} + f_{hk0}A_{hk0}, \tag{88}$$

where f_{001} and f_{hk0} stand for the specific interfacial energy of the (001)-matrix and $(hk0)$-matrix respectively (neglecting a possible orientation effect from the matrix crystals), and A is area. Thus, for a crystal of diameter a and height c,

$$F^{\gamma} = f_{001}(3a^2/\tan 60°) + f_{hk0}(6ac/\tan 60°). \tag{89}$$

Differentiating this equation, we find that F^{γ} is a minimum when

$$\frac{a}{c} = \frac{f_{hk0}}{f_{001}}. \tag{90}$$

Measurements of true a and c dimensions for 615 biotite crystals separated from a sample of high-grade marble (Kretz 1966a) gave $a = 0.10$ to 2.4 mm, and $(a/c)_{\text{mean}} = 2.8$, with 75% of the aspect ratios falling between 2 and 4. The results are not entirely consistent with the concept of equilibrium, which requires (a/c) to remain constant, regardless of crystal size, and the shape of the biotite crystals was possibly influenced by both interfacial energy and growth kinetics.

Rivers and Fyson (1977) reported on the shape of muscovite crystals in schist, extending from the chloritoid–staurolite zone to the sillimanite–K-feldspar zone, within the Fernleigh Belt of the Grenville Province (Canadian Shield). Muscovite crystals are alined, and from field evidence it is clear that the terrane was highly deformed.

The ratio of the apparent length of muscovite crystals (L) as seen in thin sections, to the true width (W) was measured for 100 crystals in each of 47 samples, and the mean aspect ratio, $(L/W)_{\text{mean}}$, is plotted against distance (increasing grade) in Figure 5.45. The remarkable decrease in L/W with increasing grade (i.e., the crystals become more equidimensional) is accompanied by a large increase in mean size, W_{mean}, from 0.02 to 0.32 mm.

The change in aspect ratio, approaching $(L/W) = 4$ with increasing grade, as well as a smaller variation in this ratio in samples of highest grade, is viewed by the authors as an approach to equilibrium, represented by a disc or hexagonal prism, with an aspect ratio (modified to allow for L being an apparent dimension) of approximately 5.0.

In the biotite and muscovite studies, described above, it is difficult to determine with confidence if aspect ratios were governed entirely by interfacial energy, by a combination of interfacial energy and growth kinetics, or entirely by growth kinetics. Thus, in a study similar to the muscovite study, described above, but with biotite, Jones and Galwey (1972) also reported a decrease in (L/W) with increasing metamorphic grade. At Glen Clova, Scotland, for example, the ratio extends from 4.5 to 2.8. The authors interpreted these results in terms of rates of crystal growth, influenced by temperature.

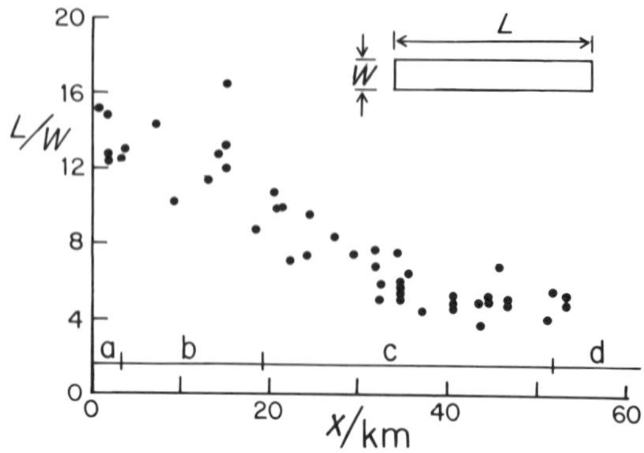

Figure 5.45 Decrease in the mean length (L) to width (W) ratio (mean for each sample) of muscovite crystals with increase in metamorphic grade (x); a, b, c, d denote the chloritoid–staurolite, staurolite–kyanite, sillimanite, and sillimanite–K-feldspar mineral zones. From Rivers and Fyson (1977)

In Sec. 4.4.1 we noted that in metals the energy associated with non-coherent boundaries is 500 to 1500 erg/cm^2. Few data are available for silicates, but Oglesby, Gutshall and Phillips (1976) have compiled surface energies for some minerals. For example, the surface–vacuum energies for phlogopite (001) and muscovite (001) are 6060 and 5125 erg/cm^2 respectively. By analogy with metals, crystal–crystal energies will be less, possibly by a factor of 1/3. It is important to emphasize again that these numbers are very small. For example, 500 erg/cm^2($= 5 \times 10^{-4}$ J/cm^2) is less than the energy released in the conversion of 60 g of SiO_2 from cristobalite to quartz (at STP), by a factor of 1×10^{-9}. We have attempted to show above that interfacial energy, although small in magnitude, *does* appear to be responsible, under favorable conditions, for some aspects of the shape of crystals in the middle and deep Crust.

5.5.4 The coarsening of crystal size

The tendency for single-phase aggregates of crystals to coarsen with time is clearly demonstrated in glacier ice and in metals, and has no doubt been recognized since Antiquity. The increase in crystal size of calcite when limestone changes to marble was described by G. Arduino in 1779 (Chapter 1, Sec. 1.2), and since 1886 and possibly earlier (Rinne and Boecke 1908), several attempts have been made to simulate the process experimentally; a study by Tullis and Yund (1982) is representative of recent work. Grain coarsening, which is also known as 'Ostwald ripening', is generally attributed to interfacial tension, as expressed by the Gibbs–Thomson equation. A form of this equation (the Thomson–Freundlich equation) was referred to in Sec. 5.4.2, above, in relation to the coarsening of exsolution lamellae in cryptoperthite, but in its more general form, the equation applies also to coarsening in single-phase aggregates. Because of the importance of this equation, the derivation, according to Porter and Easterling (1981) is here reproduced.

Consider a sphere of a liquid phase, β, near a much larger bulk mass of β, both in contact with a vapor phase, α. Let r and A stand for the radius and area of the sphere, n for the number of moles of β in the sphere, G for the Gibbs energy of the sphere, and V^β for the molar volume of β. The interfacial tension (σ) at the surface of the sphere is defined (Sec. 5.5.2, equation (80)) as

$$\sigma = \frac{\mathrm{d}G}{\mathrm{d}A} \tag{91}$$

Suppose that the molar Gibbs energy of β in the bulk is less than in the sphere, by an amount ΔG. Then the transfer of dn moles of β from the bulk to the sphere will lead to an increase in the total Gibbs energy of the sphere, resulting from the increase in mass, equal to

$$\mathrm{d}G = \Delta G\,\mathrm{d}n. \tag{92}$$

By combination of (91) and (92),

$$\Delta G = \frac{\sigma\,\mathrm{d}A}{\mathrm{d}n}, \tag{93}$$

or

$$\Delta G = \sigma\frac{\mathrm{d}A/\mathrm{d}r}{\mathrm{d}n/\mathrm{d}r}. \tag{94}$$

Initially, the volume of the sphere was

$$nV^\beta = (4/3)\pi r^3. \tag{95}$$

By rearrangement,

$$n = (1/V^\beta)(4/3)\pi r^3, \tag{96}$$

and, by differentiation of (96),

$$\frac{\mathrm{d}n}{\mathrm{d}r} = (1/V^\beta)4\pi r^2. \tag{97}$$

The initial area of the sphere was

$$A = 4\pi r^2, \tag{98}$$

and, by differentiation,

$$\frac{\mathrm{d}A}{\mathrm{d}r} = 8\pi r. \tag{99}$$

By substitution of (98) and (99) in (94), we obtain

$$\Delta G = (2V^\beta\sigma)/r, \tag{100}$$

or, for the spontaneous transfer of matter from the sphere to the bulk,

$$\Delta G = -(2V^\beta\sigma)/r. \tag{101}$$

This equation (the Gibbs–Thomson equation) relates the excess Gibbs energy that a small spherical body possesses, as a consequence of interfacial tension.

Burke and Rosolowski (1973) have emphasized that equation (101) also applies when the small mass of β is a polyhedral crystal. Thus, for a cube of edge length $2r$, the equation remains unchanged, assuming negligible adsorption on the crystal surface.

In a mosaic of cells, as in Figure 5.37a, the Gibbs energy of small cells, bounded by concave interfaces, will, by equation (101), possess a higher Gibbs energy than larger cells, and boundaries will tend to migrate, to eliminate small cells. In equation (101), ΔG can be equated to the work done, i.e., force (F) times distance ($\mathrm{d}x$), where F is the force that pulls a boundary toward a cell of higher energy. The number of moles of atoms entering the growing cell is $\mathrm{d}x(1/V)$, where V is molar volume, and the energy released is $\Delta G(\mathrm{d}x/V)$. Thus,

$$\Delta G(\mathrm{d}x/V) = F\,\mathrm{d}x, \tag{102}$$

$$F = \Delta G/V, \tag{103}$$

i.e., the force on the boundary is the Gibbs energy difference per unit volume of matter.

The migration of grain boundaries and subsequent grain coarsening can be actually observed in some experiments with organic crystals, and an example of an experiment of this kind is presented in Figure 5.46. Initially (Figure 5.46a), crystals labelled 6 and 10, which have only five vertical sides, could be selected as ones likely to disappear, and they do so; but for others, with six or seven sides, it is difficult to predict their behavior. Crystal 9, for example, begins with

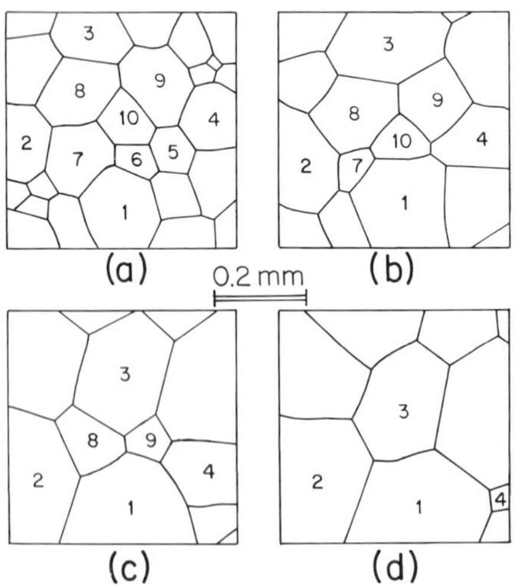

Figure 5.46 Stages, (a) to (d), in the coarsening of an aggregate of crystals of octachloropropane, between parallel plates of glass. From McCrone (1949)

eight sides, but these are reduced to five (Figure 5.46b and c) before the crystal disappears.

An approximate expression for the rate of increase of mean crystal diameter ($\overline{d}$) can be obtained as follows (Shewmon 1969). The rate of advance of a grain boundary (U) is assumed to be proportional to ΔG (equation (101)), and inversely proportional to the grain-boundary width (δ):

$$U = -M(\Delta G/\delta). \tag{104}$$

The mobility (M) is proportional to the coefficient for diffusion of atoms across the boundary, and is therefore a function of temperature. Consider a sample of unit volume, made up of a range of cells or crystals (as in Figure 5.37a and b), of mean diameter $\overline{d}$, extending to $2\overline{d}$. The radius of curvature (r) of a moving boundary depends on crystal size, and is approximately

$$r = \alpha\overline{d}, \tag{105}$$

where α is a constant. Then by combining equations (101), (104), and (105),

$$U \propto \frac{\mathrm{d}\overline{d}}{\mathrm{d}t} = \frac{M2V^{\beta}\sigma}{\delta\alpha\overline{d}}. \tag{106}$$

Thus, the rate of coarsening decreases as mean size increases and as temperature (and M) increase, in approximate agreement with observations in metals. Integrating (106) at constant temperature,

$$\overline{d}^2 - \overline{d}_0^2 = kt, \tag{107}$$

where k is a constant. For small initial size ($\overline{d}_0$),

$$\overline{d} \propto k^{1/2}t^{1/2}, \tag{108}$$

i.e., the rate of increase of size decreases with time.

Data on the mean size of quartz crystals in quartzite within the Ballachulish contact aureole (Scotland), as a function of distance from the granitic pluton, were obtained by Buntebarth and Voll (1991), and results are presented in Figures 5.47 and 5.48. Mean crystal diameter as measured in sections, and therefore somewhat smaller than true diameter, increases with increasing temperature, in qualitative agreement with equation (106). Assuming different temperature paths at different points from the contact, similar to those plotted in Figure 1.28 of Sec. 1.10, then points progressively nearer the pluton experienced not only higher temperatures, but also longer durations of time above a threshold temperature, at which growth is initiated. Both of these effects combined should, in view of equation (108), produce a size–distance relation that is less parabolic and possibly nearly linear, as observed (Figure 5.48). In examining Figure 5.47, the early mosaic microstructure becomes progressively more irregular (with crystal boundaries that are not planar or gently curved), and the authors have suggested that boundary migration in these

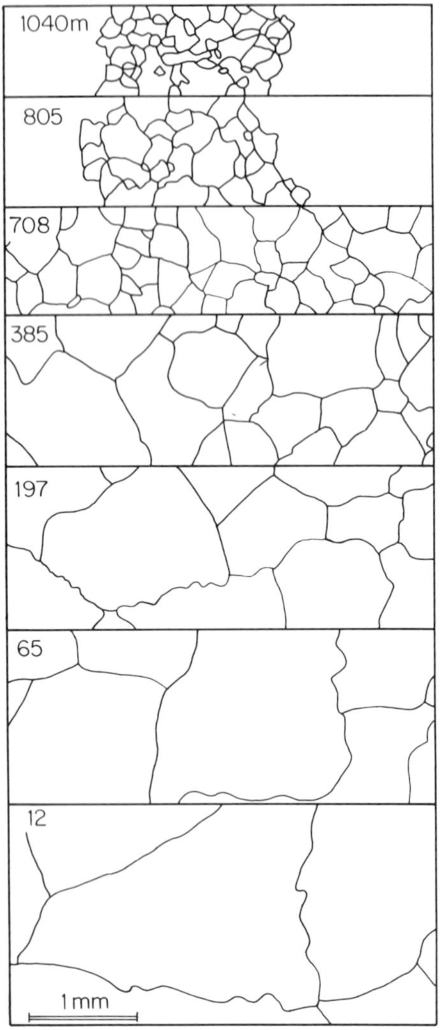

Figure 5.47 Coarsening of. quartzite in the Ballachulish contact aureole in Scotland; numbers are distance in meters from the contact; the scale in each frame is the same. Tracings of photographs presented by Buntebarth and Voll (1991), reproduced by permission of Springer-Verlag

samples occurred in response to both interfacial and strain energy. The numerous natural and experimental results on grain coarsening that are now available have been reviewed by Joesten (1991b).

The Gibbs–Thomson equation may be applied to crystal enlargement in a variety of metamorphic situations; three of these are here mentioned.

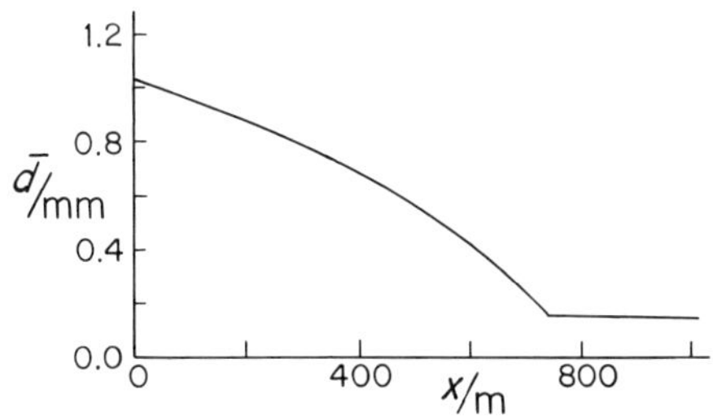

Figure 5.48 Mean diameter ($\bar{d}$) of quartz grains in samples of quartzite, related to distance (x) from the contact of the Balluchulish granitic pluton in Scotland. From Buntebarth and Voll (1991). Reproduced by permission of Springer-Verlag

(1) Where one phase, β, occurs as near-equidimensional exsolved particles within a host phase, the geometry is identical to that assumed above in the derivation of the Gibbs–Thomson equation. An approximate theory exists, as reviewed by Raghaven and Cohen (1975) and Joesten (1991b), to predict that the growth rate of those crystals of a population that are smaller than the mean will be negative, while the growth rate of those crystals that are larger than the mean will be positive. If, at any time, $\bar{r}$ is the radius of the crystal of average size, then $\bar{r}$ increases (at constant temperature) as (time)$^{1/3}$. The $\bar{r} \propto t^{1/3}$ relation has been found in the coarsening of Cu precipitated from Fe–Cu solid solution (Raghaven and Cohen 1975), but only at high temperatures (730 to 830 °C) and small particle size (0.02 to 0.20 μm). Evidently, coarsening does occur in cryptoperthite (Sec. 5.4.2), but in a study of dolomite particles in calcite (Sec. 5.4.3) evidence for coarsening was not detected.

(2) Recall the Rivers and Fyson data on muscovite, described above (Sec. 5.5.3), and the reported increase in mean crystal width from 0.02 mm to 0.32 mm, with increasing metamorphic grade. The authors have attributed this change in size to interfacial energy. With muscovite occupying nearly half of the total rock volume, many crystals were obviously in contact from an early stage, and this, together with a rise in temperature and contemporaneous deformation, evidently created conditions especially favorable for the diffusion needed to bring about crystal enlargement, driven by interfacial energy.

(3) Cashman and Ferry (1988) and Miyazaki (1991) have proposed that coarsening also occurs with respect to garnet crystals embedded in a quartz–feldspar matrix. In the garnet microstructure described in Sec. 5.2, small, near-euhedral crystals occur, just touching much larger crystals, and evidence for negative growth of the small crystals was not detected. Also, according to the kinetic theory, referred to briefly above, during coarsening, the smallest crystals of a population become smaller and the largest ones only grow larger, while certain crystals of intermediate size first become larger and then smaller, as the mean size shifts toward a larger size. With all of this variation, it seems unlikely that the Mn and Fe profiles

(dC/dr) would finally be the *same* in all crystals, regardless of size, as found by chemical analysis (Figure 5.7 of Sec. 5.2). This is not to deny that, under favorable conditions, coarsening of garnet, staurolite, kyanite, and other minerals can occur.

In summary, coarsening driven by interfacial energy, is a widespread process. It operates conspicuously in monomineralic rocks (e.g., marble), but also, under favorable conditions, in rocks composed of two or more minerals. The kinetics of coarsening is complex, especially if the process takes place during deformation and while temperature increases, and this may explain why very different relations between size and distance have been found in metamorphic aureoles and terranes.

5.6 THE SPATIAL DISTRIBUTION OF CRYSTALS, WITH SPECIAL REFERENCE TO GNEISS AND GRANULITE

Everywhere within the solid Earth, excluding occurrences of monomineralic rock, the crystals of two or more minerals are intermingled to form many different patterns of crystal distribution. Some of these patterns, for example, the zonal arrangement of crystals described in Sec. 5.3, provide information on mineral reactants, while others (e.g. Fig. 1.9b of Sec. 1.4) are complex and not easily understood. Crystal concentrations or clusters, where found, also represent certain elemental concentrations, and it is necessary, but not always possible, to distinguish those that have their origin in the precursor sedimentary or igneous rocks from the various clusters that have formed by metamorphic processes. We shall first examine aspects of the distribution of crystals in small samples of nearly homogeneous rock ($\sim$1 to 10 cm^3 in volume), and in the next section (5.7), proceed to an examination of certain mineral layers and veins that are generally viewed as products of metamorphic differentiation.

From field and microscopic observations, it is clear that although centimeter-scale variations in mineral proportions are not uncommon, it is frequently possible to delineate an area within a thin-section of metamorphic rock that is approximately homogeneous. A condition of near-homogeneity — perfect homogeneity is an ideal concept — can be confirmed by obtaining and comparing modal or chemical analyses of different portions, taking analytical error into consideration. Obviously, the chemical analysis must be non-destructive, because we need the sample for further study.

Very many possibilities exist for the distribution of crystals in nearly homogeneous study volumes. To bring the problem into focus, consider the microstructure of a granulite, illustrated in Figure 5.49. The following questions arise. (1) Are crystals of the principal minerals, Ca pyroxene and scapolite, arranged at random? (2) Are the smaller crystals of sphene distributed at random? (3) Do preferred mineral associations exist, for example, between sphene and Ca pyroxene? Answers to questions of this kind can possibly be understood in terms of nucleation sites, or the role of interfacial energy during rock-forming processes.

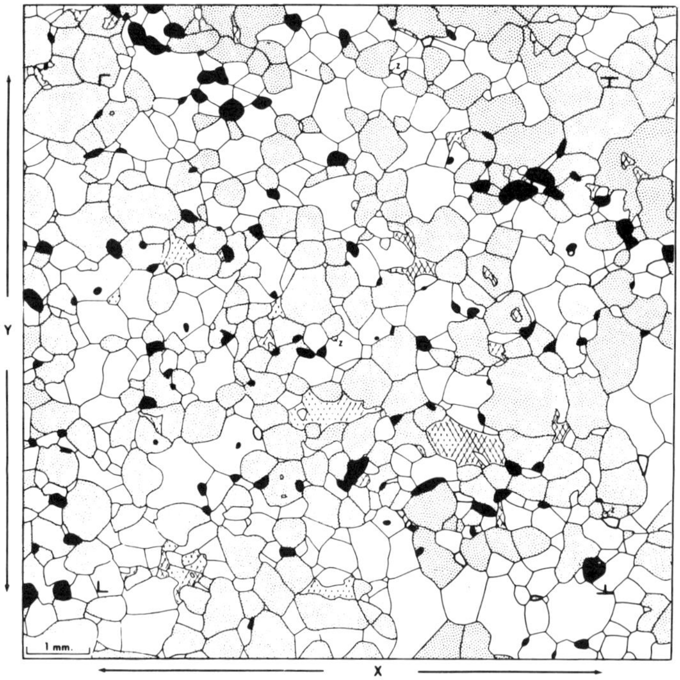

Figure 5.49 Microstructure of a sample of scapolite (clear)-Ca-pyroxene (light stippled)-Ca-amphibole (cross-hatch)-sphene (black) granulite, with minor apatite (dark stippled) and zircon (z). The distribution of sphene can be viewed as a distribution of points on a plane, and the distribution of Ca pyroxene and scapolite can be viewed in relation to the frequency of 'like' and 'unlike' contacts. From Kretz (1969), with permission of Elsevier Science Publishers

The concept of spatial randomness is illustrated in Figure 5.50, where x and y coordinates to 'growth' centers were taken from a table of random numbers. Notice that a certain amount of clustering is an inherent property of a group of random points on a plane, and by extension, in a volume. (A group of random points can also be obtained by holding a sheet of cardboard in the rain.) Random numbers were then used to label each point in Figure 5.50 as clear or stippled. Each point was then permitted to 'grow' at a constant rate until all of the background space was filled. Notice that a certain amount of clustering of clear and stippled 'crystals' is an inherent property of a random mosaic.

Consider first the distribution of crystals of a mineral that forms less than a few percent of the total rock volume. Then each crystal can be viewed as a point, and a definition of randomness can be formulated, as above, in terms of random coordinates. In Sec. 5.2, the three-dimensional distribution of garnet crystals in a sample of schist was described and a test for randomness was applied, but normally information is available only on sections through a volume, and therefore tests are needed that apply to points on a plane. If the points (sections through crystals)

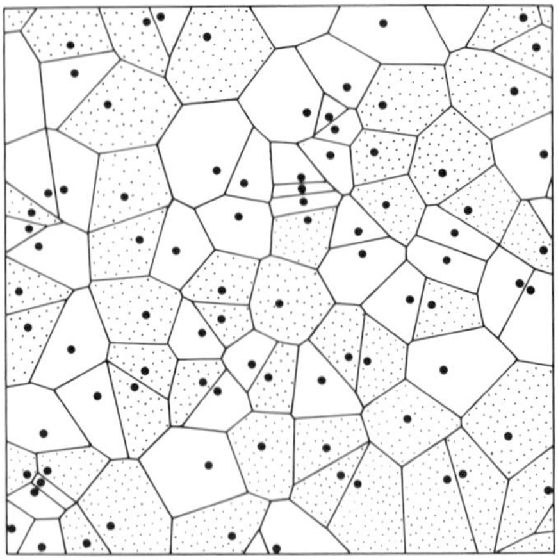

Figure 5.50 A mosaic produced by the 'growth' of clear and stippled 'crystals' from randomly distributed points; the labelling of points, as clear or stippled, was also randomly determined; note that some clustering of 'crystals' is a property of this random distribution. From Kretz (1969), with permission of Elsevier Science Publishers

are not random, then the distribution is either clustered or regular (more widely spaced); an analogy with trees in a forest and a neighboring orchard, comes to mind.

Suppose, for example, that we wish to test the null hypothesis that the crystals of sphene in Figure 5.49 are distributed at random. The results of two tests are presented in Table 5.5; one is based on measurements from each crystal to its nearest neighbor, and the other on measurements from randomly located points to the nearest crystal (Clark and Evans 1954, Moore 1954, Ripley 1981). In each case, information must be available on crystals that lie slightly beyond the area chosen

Table 5.5 Two methods for measuring the distribution of points on a plane, applied to the distribution of sphene in granulite (Figure 5.49)

Area	XY (Figure 5.49)
Size of area (A)	64.0 mm^2
Number of sphene crystals (N)	85
Density (N/A)	1.328 mm^{-2}

Nearest-neighbor method		Random-point method	
Sum of distances (Σr)	39.36 mm	Number of measurements (n)	100
Mean distance ($\overline{r}_0 = \Sigma r/N$)	0.463 mm	Mean square distance	
Mean distance expected		$\overline{w} = (\Sigma r^2)/n$	0.2495 mm^2
($\overline{r}_E = 1/2(N/A)^{1/2}$)	0.434 mm	$2n\pi(N/A)\overline{w}$	208
($\overline{r}_0/\overline{r}_E$)	1.067	Degrees of freedom (d.f.) = $2n$	200
Standard variate		$\chi^2(0.95)$	234
($c = N(\overline{r}_0 - \overline{r}_E)/0.261A^{1/2}$)	1.18	$\chi^2(0.05)$	168
Distribution ($c < 1.96$)	Random	Distribution (234 > 208 > 168)	Random
		α index ($\alpha = \pi(N/A)\overline{w}$)	1.04

for study. Both tests lead to the conclusion that at the 0.95 level of confidence, the null hypothesis is acceptable.

Results for some other samples and minerals are listed in Table 5.6. Randomness in the distribution of garnet appears to be common, and can be understood in terms of nucleation theory, as discussed in Sec. 4.4. Clustering occurs in some samples of gneiss, including one for which information was obtained in three dimensions by serial sectioning. Garnet clusters also occur in schist, as described, for example, by Kamineni (1978), who used microprobe analyses to confirm the location of boundaries. Several regular distributions were described by Carlson (1989) who adopted and developed an interpretation of Rogers and Bogy (1958), namely that the presence of diffusion haloes about a growing crystal decreases the probability that a new crystal of the same kind will nucleate in the vicinity.

Returning to Figure 5.50, we shall proceed to the more difficult problem of the distribution of crystals of major minerals, represented by clear (W) and stippled (G) cells. Now, departure from randomness could be visualized in terms of the frequency of contacts between 'like' and 'unlike' cells. Thus, departure toward clustering would necessarily increase W–W and G–G contacts, while departure toward a more regular (checkerboard) distribution would increase the W–G contacts. In a simpler mosaic of hexagonal cells of uniform size, one may show (and the reader is invited to confirm) that for a random distribution, the probabilities of W–W, G–G, and W–G contacts are

$$
\begin{aligned}
p_{\mathrm{W-W}} &= a_{\mathrm{W}}^2, \\
p_{\mathrm{G-G}} &= a_{\mathrm{G}}^2, \\
p_{\mathrm{W-G}} &= 2a_{\mathrm{W}}a_{\mathrm{G}},
\end{aligned}
\tag{109}
$$

where a is area fraction. This model was applied to granitic rocks by Rogers and Bogy (1958), who observed that, in relation to equations (109), K-feldspar–K-feldspar contacts are less frequent than expected. The general problem was analyzed by Lafeber (1963).

A slightly different approach was initiated by Derek Flinn in 1957 (Flinn 1969), who recorded the sequence of crystals as they appear along a line through a rock volume (across a thin-section), the line being broken or bent to cover a given area. This method, when applied to the mosaic of Figure 5.50, produces the following *transition-tally matrix* of 75 transitions from 'crystal' i to 'crystal' j, along a zig-zag traverse across the area;

		to (j) G	W	Σ = sum	
from (i)	G	22	18	40	(110)
	W	19	16	35	
	Σ = sum	41	34	75	

Thus the observed frequency of G-to-G transitions is $22/40 = 0.550$, and this and other frequencies, similarly calculated, are expressed as a *transition-probability*

Table 5.6 Distribution of crystals, random-point method*

Sample***		Crystal	A/mm^2	N	$(N/A)/mm^{-2}$	n	$\overline{w}/mm^2$	$2n\pi(N/A)\overline{w}$**	Distribution	α
296-61-1	garnet–biotite schist	garnet	900	57	0.633	100	4.93	196.3	Random	0.98
225-56	garnet–biotite–sillimanite gneiss	garnet	458	38	0.083	50	3.28	85.4	Random	0.85
314-55	garnet–biotite–K-feldspar gneiss	garnet	2180	43	0.020	100	14.53	180.0	Random	0.90
195-56	garnet–biotite–sillimanite gneiss	garnet	238	68	0.286	100	3.37	605.6	Clustered	3.0
296-61-3	plagioclase–cummingtonite schist	cummingtonite	8.3	506	61.2	100	0.00524	201.5	Random	1.01
296-61-2	cordierite–biotite schist	cordierite nodules	900	32	0.0356	50	5.14	57.4	Regular	0.57
32-57	biotite–Ca-pyroxene granulite (Figure 5.51)	biotite	16.5	420	25.5	100	0.0108	172.8	Random	0.86
G11-66	pyroxene marble	Ca pyroxene	412	72	0.175	100	1.952	214.3	Random	1.07

*See Table 5.5 for definition of symbols.

**Compare, $\chi^2(0.95, n = 100) = 234$, $\chi^2(0.05, n = 100) = 168$ (d.f. $= 2n$),
$\chi^2(0.95, n = 50) = 124$, $\chi^2(0.05, n = 50) = 78$ (d.f. $= 2n$).

***Samples of schist are from the Slave Province (near Yellowknife), the remainder from the Grenville Province (Otter Lake terrane), Quebec.

matrix, P,

$$P = \begin{pmatrix} 0.550 & 0.450 \\ 0.543 & 0.457 \end{pmatrix}. \tag{111}$$

Now, in a random sequence, as defined by a zero-order Markov chain, the probability of G-to-G and W-to-G transitions depends only on the overall frequency of G, e.g., $p\,(\text{W to G}) = p\,(\text{G}) = 41/75 = 0.557$, where p is probability. We thus have, for comparison with (111), a *fixed-probability vector*, Q,

$$Q = (0.557 \quad 0.453). \tag{112}$$

Comparing (111) with (112), we see that the observed and expected transition frequencies are in agreement.

Thus results obtained by the method proposed by Flinn (1969) — here referred to as the *transition frequency method* — may be analyzed in terms of Markov chain theory, and this was done in particular by Vistelius (1966, 1972) in his study of granite microstructures. Students wishing to pursue this approach are encouraged to study the work of Vistelius (1992) and Whitten, Dacey and Thompson (1975).

We shall now apply the transition frequency method to a sample of fine-grained granulite (Figure 5.51) that consists of plagioclase (0.601 ± 0.010), biotite (0.193 ± 0.007), Ca pyroxene (0.173 ± 0.006), and minor hornblende, ilmenite, pyrrhotite, and apatite (0.03). The numbers in brackets are volume fractions based on point-count analysis (2025 points); these do not enter the calculation, but are referred to below.

Symbols Pl, Bt, and Cpx will be used for the major minerals, and O for the remainder, grouped together. Transition traverses were extended normal to and

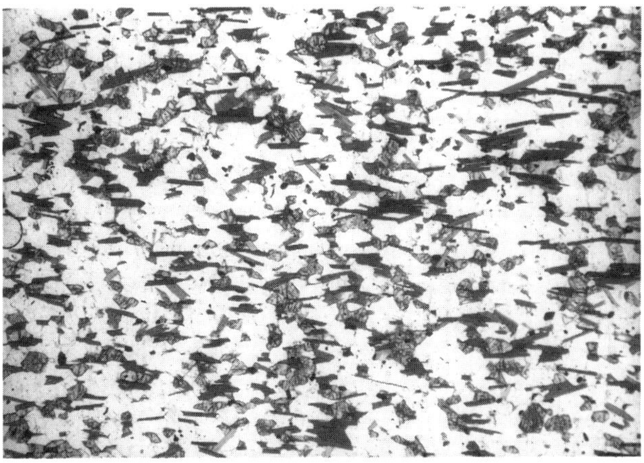

Figure 5.51 Sample of plagioclase (white)–Ca-pyroxene (gray)–biotite (dark) granulite (32–57) from the Grenville Province, selected for study of the frequency of crystal contacts; area is 5×6 mm

parallel to the biotite foliation, with similar results; the combined data are as follows:

		to				
		Pl	Bt	Cpx	O	Σ
from	Pl	196	193	109	63	561
	Bt	187	74	49	13	323
	Cpx	119	41	25	12	197
	O	57	17	14	9	97
	Σ	559	325	197	97	1178

(113)

$$P = \begin{pmatrix} 0.349 & 0.344 & 0.194 & 0.112 \\ 0.579 & 0.229 & 0.152 & 0.040 \\ 0.604 & 0.208 & 0.127 & 0.061 \\ 0.588 & 0.175 & 0.144 & 0.093 \end{pmatrix}, \tag{114}$$

$$Q = (0.474 \quad 0.276 \quad 0.167 \quad 0.082). \tag{115}$$

Note that the fraction of points falling on plagioclase, biotite, and Ca pyroxene during point-count analysis (above) is 0.60, 0.19, and 0.17, while fractions in the transition traverse, from (115), are 0.47, 0.28, and 0.17. Differences in these pairs of numbers reflect differences in the mean size of plagioclase and biotite crystals. In a transition traverse, relatively small crystals are encountered more frequently than relatively large crystals of equal modal volume.

By comparison of (114) with (115), we see that the frequencies of 'like' transitions (Pl–Pl, Bt–Bt, Cpx–Cpx) are fewer than expected, and the frequencies of 'unlike' transitions (involving Pl, Bt, Cpx) are more numerous than expected. Although it is obvious that the hypothesis of randomness must be rejected, this decision could be confirmed (Agterberg 1974, p. 432) by calculating the variable

$$2\sum_{i=1}^{m}\sum_{j=1}^{m} t_{ij} \ln(p_{ij}/p_i), \tag{116}$$

where t_{ij} is the i-to-j entry in the transition-tally matrix (113) of m rows and m columns, p_{ij} is the i-to-j probability in the transition-probability matrix (114), and p_j is the probability of j in the fixed-probability vector (115). This variable, for a random sequence of crystals is distributed as chi-squared (χ^2) with $(m-1)^2$ degrees of freedom.

The finding that crystals of plagioclase and Ca pyroxene in granulite are more frequently in contact than is expected was first described by Flinn (1969) in his study of several samples from Siberia. His results and those presented above in (113) to (115) can be characterized by use of a *transition ratio*, t_L, which compares the observed 'like' transitions (e.g., Pl–Pl) with those prescribed by the random model, or, by use of t_U, which compares the observed 'unlike' transitions (e.g., Pl–Cpx and Cpx–Pl, combined) with those prescribed by the random model. For example, from equations (114) and (115),

$$t_L \text{ (Pl–Pl)} = 0.349/0.474 = 0.74,$$

$$t_U \text{ (Pl–Cpx)} = [(0.194/0.167) + (0.604/0.474)]/2 = 1.22.$$

The property of a microstructure that results in $t_L < 1.0$ and $t_U > 1.0$ is here referred to as the '$t_U > t_L$ property'. Values of t_L and t_U for the Siberian granulites and for the sample described above (Figure 5.51) are listed in Table 5.7. Numerous additional data on Adirondack granulites, which contain four or five major minerals, were obtained by Jen (1975), who found that complex patterns are present, with t_L and t_U showing much variation for different minerals and mineral pairs.

Rogers and Bogy (1958), from measurements of contact frequencies rather than transitions, detected the $t_U > t_L$ property in granitic rocks of presumed igneous crystallization, and Mahan and Rogers (1968) found it to be present also in gneissic (metamorphic) granite, where it is less pronounced. It is of interest therefore to extend the comparison to mafic rocks and compare amphibolite with gabbro. The transition frequency method was applied to two samples from the Grenville swarm of gabbro dikes, and two of high-grade amphibolite from the Otter Lake terrane (Grenville Province) and the results are listed in Table 5.7. The $t_U > t_L$ property is strongly present in *both* rock types. It is possible, therefore, that the spatial distribution of atoms that was created during crystallization of a melt was preserved to the extent that the original $t_U > t_L$ condition was not altered greatly during metamorphism and deformation.

Table 5.7 Frequency of crystal transitions relative to the frequency expected for a random distribution of crystals

				t_L		t_U
Sample	*N*	A	B	A–A	B–B	A–B
Random mosaic, Figure 5.50	75	white	gray	1.01	0.99	0.98
*Siberian granulite, mean of 35 data sets.	>3500	plagioclase	Ca pyroxene	0.78	0.73	1.18
32-57, fine-grained Grenville granulite (Figure 5.51)	1178	plagioclase	Ca pyroxene	0.74	0.76	1.22
80-55, from gabbro dike	686	plagioclase	Ca pyroxene	0.85	0.55	1.31
42-55, from gabbro dike	878	plagioclase	Ca pyroxene	0.88	0.35	1.33
133-56 Grenville amphibolite	613	plagioclase	hornblende	0.66	0.80	1.24
122-57 Grenville amphibolite	415	plagioclase	hornblende	0.87	0.77	1.19
*Shetland granitic gneisses, mean of five	>5000	plagioclase	biotite	0.72	0.91	1.12
1594 Grenville hornblende gneiss (Figure 5.52)	1232	plagioclase	hornblende	1.10	1.02	0.94
268-61 Yellowknife meta-sedimentary schist	500	plagioclase	biotite	0.62	0.61	1.34

N: number of transitions recorded
t: transitions observed/transitions expected
*From Flinn (1969)

In Sec. 1.6, gneiss was defined as a rock in which some clustering of crystals occurs, to define a foliation or lineation. But it is important to note that the clusters may or may not result in a high frequency of contacts between crystals of the same kind. For example, crystals of a mineral may be localized in alternating planes (one-grain layering) with only few points of contact. Indeed, Mahan and Rogers (1968) and Flinn (1969) found the $t_U > t_L$ property to be present in the granitic gneisses they examined. The results of Flinn (1969) from Shetland gneiss are included in Table 5.7.

It is of interest, therefore, to apply the transition method to a sample wherein a dramatic clustering of crystals occurs (Figure 5.52), to produce a conspicuous gneissic microstructure. This sample, which is also from the Otter Lake terrane, is particularly suitable for study because, apart from a trace of apatite, which is disregarded, the only minerals present are plagioclase (Pl), hornblende (Hbl), and minor magnetite (Mt), and with the exception of Mt–Mt all crystal contacts are very clearly visible. Transitions were recorded on traverses extending across and parallel to the gneissic foliation, and both produced similar transition-probability matrices. The combined results are as follows:

		to			
		Pl	Hbl	Mt	Σ
from	Pl	219	260	16	495
	Hbl	265	399	27	691
	Mt	9	35	2	46
	Σ	493	694	45	1232

(117)

Figure 5.52 Portion of a sample of plagioclase (white)–hornblende (gray)–magnetite (black) gneiss (1594) from the Grenville Province, selected for a study of the frequency of crystal contacts; note conspicuous gneissic clustering of hornblende and plagioclase crystals; area is 5 × 6 mm

$$P = \begin{pmatrix} 0.442 & 0.526 & 0.032 \\ 0.384 & 0.577 & 0.039 \\ 0.196 & 0.761 & 0.043 \end{pmatrix}, \tag{118}$$

$$Q = \begin{pmatrix} 0.400 & 0.563 & 0.037 \end{pmatrix}, \tag{119}$$

$$\begin{aligned} t_L \ (\mathrm{Pl{-}Pl}) &= 1.10, \\ t_L \ (\mathrm{Hbl{-}Hbl}) &= 1.02, \\ t_U \ (\mathrm{Pl{-}Hbl}) &= 0.94, \\ t_U \ (\mathrm{Pl{-}Mt}) &= 0.67, \\ t_U \ (\mathrm{Hbl{-}Mt}) &= 1.20. \end{aligned} \tag{120}$$

We find that the t ratios involving Pl and Hbl are all very close to 1.0, as expected for a random distribution. Yet the crystals of hornblende and plagioclase that are relatively large are certainly clustered. The explanation lies in the presence of small crystals of Pl in Hbl clusters, and small crystals of Hbl in Pl clusters (Figure 5.52), giving a disproportionately high frequency of Hbl-to-Pl and Pl-to-Hbl transitions. The data (equations (120)) also reveal a preferred association of magnetite with hornblende ($t_U = 1.20$) rather than with plagioclase ($t_U = 0.67$).

From the above studies it is clear that the transition method by itself is not a satisfactory test for randomness and departure from randomness. Nevertheless, the t_U ratios that are obtained by this method could be used as a rough measure of progressive crystal clustering, extending from approximately 1.32 (no clustering) to 1.0 (conspicuous clustering). Obviously, modifications of the transition method, or entirely new methods, are needed to characterize fully the spatial distribution of crystals in small volumes of rock. Spectral analysis, as described by Agterberg (1974), which examines the distribution of minerals rather than crystals, could be applied more frequently, to search for subtle layering and other departures from randomness. Whatever new methods are developed they must of course be applied three-dimensionally, i.e., on several sections through the chosen volume of rock; the study of Whitten, Dacey and Thompson (1975) could be viewed as a model in this regard.

Under favorable conditions, measurements of the kind described above, and many others, can now be obtained rapidly with the aid of a computer. The procedures, which fall under the heading of 'image analysis', have been detailed by Fabbri (1984), who has applied them to some metamorphic microstructures. Numerous publications on this subject have appeared, including a collection of papers edited by Petruk (1989) and a book by Russ (1990). Another promising approach is the computer simulation of microstructure described by Harvey and Ferguson (1978) and Sempels (1978).

In view of the very many distribution patterns that exist in small volumes of metamorphic rock, only a few of which were illustrated in this section, it is not possible to list here all the possible causes for that diversity. The following lists some of the interpretations that can be considered.

1. We have seen above that the crystal and therefore the atomic distribution in amphibolite could be *inherited* to some extent from that which existed in gabbro prior to metamorphism. Figure 5.53 illustrates more clearly the change of what was originally a large Ca pyroxene crystal in gabbro, to a cluster of hornblende crystals, with relics of Ca pyroxene. During the plastic deformation of meta-gabbro, as illustrated by Ramsay and Graham (1970) and Myers (1978), the mafic 'clots' may be drawn out to form a gneissic foliation or lineation. Similarly, as noted above, the regular distribution that is created during the formation of granite may be preserved as granite is changed to gneiss. Additional possibilities of this kind were considered by Vernon (1974).

2. The distribution of favorable or unfavorable *nucleation sites* is almost certainly important in determining that under some conditions clusters of crystals are formed, while elsewhere the crystals are randomly or regular distributed, as noted above with reference to garnet. The cause of conspicuous clusters of other minerals (sometimes radiating), including tremolite, cordierite, and epidote, apparently lies in the nucleation process, but the details are not understood.

3. The *mechanical segregation* of more ductile crystals (e.g., mica, quartz) from less ductile crystals (e.g., plagioclase) during heterogeneous deformation was proposed by Schmidt (1932). From more recent information on metals, and the absence of segregation where it might be expected (e.g., on limbs of folds), Shelley (1974) has found this interpretation to be questionable. Plastic deformation may, however, modify a particular distribution, for example, by causing clustering of garnet, provided the crystals remain passive and rigid.

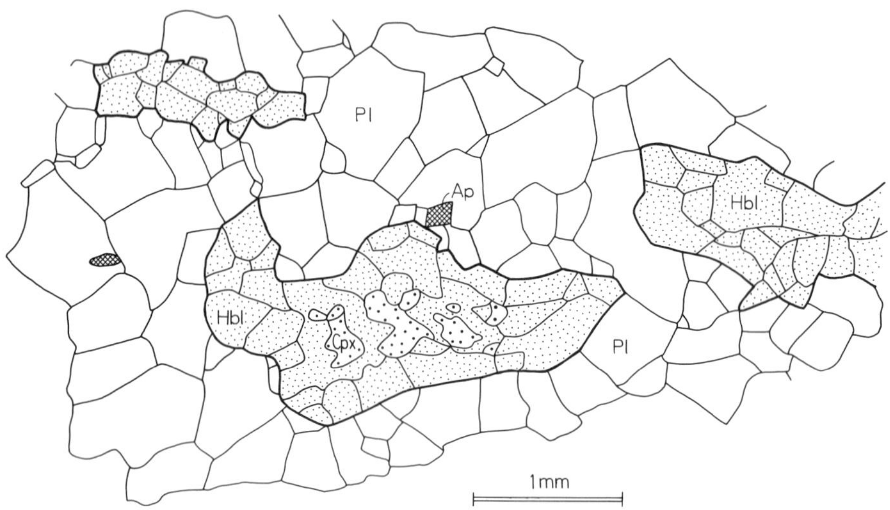

Figure 5.53 Conspicuous clustering (gneissic microstructure) of plagioclase (Pl) and hornblende (Hbl) crystals in deformed meta-gabbro; hornblende clusters evidently formed from single crystals of Ca pyroxene (Cpx), now occurring as relics. From Kretz, Jones and Hartree (1989)

4. The possible role of *interfacial energy* in determining the distribution of crystals was evidently first proposed by Seng (1936), and was considered by Laffitte (1957) and especially by DeVore (1956). It was adopted by Flinn (1969) to explain the large number of unlike contacts in gneiss and granulite, using as supporting evidence the finding that in metals the specific interfacial energy on boundaries between two unlike crystals is less than on the boundaries between two like crystals. This interpretation assumes that the $t_U > t_L$ property as measured, was formed during metamorphic annealing and is not inherited. Ramberg (1952a) has questioned whether the very small amounts of energy involved are adequate to cause, for example, a crystal of plagioclase to move between two crystals of Ca pyroxene, as proposed by Flinn (1969); a balance of interfacial tensions, as discussed in Sec. 5.5.2, would appear to hinder the process from occurring readily. Interfacial energy may, however, influence the selection of nucleation sites during recrystallization, and cause new crystals of plagioclase to nucleate preferably on Ca-pyroxene–Ca-pyroxene boundaries.

Coarsening is driven by interfacial energy, as discussed in Sec. 5.5.4, and a possible interpretation of the microstructure of hornblende gneiss that was investigated above (Figure 5.52) is that coarsening of hornblende and plagioclase occurred non-uniformly, with hornblende coarsening at sites now represented by hornblende clusters, and similarly for plagioclase. This explains the spatial association of coarse hornblende with fine plagioclase and vice versa. Thus plastic deformation possibly plays a role in determining the sites of preferred positive and negative crystal growth during coarsening.

5. DeVore (1969) has suggested and demonstrated that in an assembly of crystals of two minerals of different elastic properties, a minimum of *elastic strain energy* is obtained when the crystals are arranged in monomineral layers. This proposal is worthy of additional study, particularly with regard to gneisses.

5.7 METAMORPHIC DIFFERENTIATION

The term 'metamorphic differentiation' was introduced by Stillwell (1918) to refer to the 'segregation' of hornblende and plagioclase in certain bodies of amphibolite that he examined in Antarctica. The term should be used only if it is reasonably certain that the mineral concentrations under examination have formed during metamorphism (Figure 5.54), and do not represent heterogeneities that existed prior to metamorphism. Stillwell (1918) viewed metamorphic differentiation as occurring by solid-state diffusion, but the term is here used regardless of the mechanism of transport. There is, in fact, little agreement regarding mechanisms; aqueous dissolution and precipitation, melting and melt segregation, and diffusion in a fluid or solid crystal-boundary phase have all been considered as possibilities (Yardley 1978, Ashworth 1985, Williams 1990). Following are some examples of mineral layers and veins that have been interpreted as products of metamorphic differentiation.

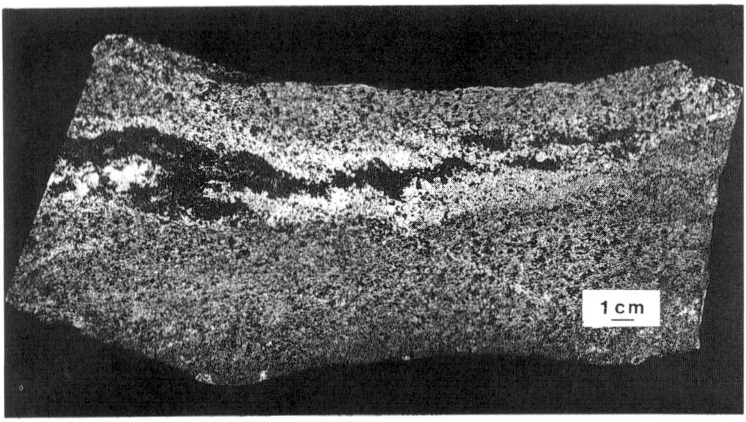

Figure 5.54 Clusters of hornblende (dark) and plagioclase (light) in amphibolite; an example of metamorphic differentiation. Sample is 15 × 35 cm

5.7.1 Metamorphic layering in meta-sediments of the greenschist facies

It is frequently difficult to determine if the layering one sees in metamorphic terranes has formed during metamorphism and deformation, or if it existed, in some form, prior to metamorphism. But in low-grade rocks, where bedding is clearly recognizable, and where mineral layers cross the bedding planes, a secondary origin for the layers is certain. Examples of such layering, oriented nearly parallel to the axial planes of folded meta-graywacke and meta-shale, occur at Bermugui, Australia (Figure 5.55), where they were studied by Williams (1972, 1990). The grade is lower greenschist facies.

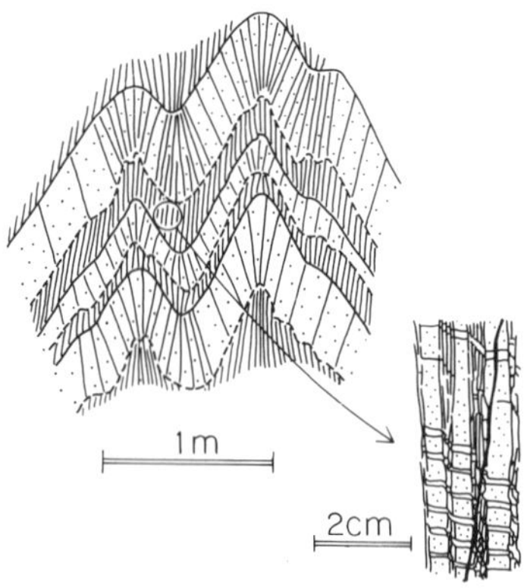

Figure 5.55 Metamorphic layers alternately rich in quartz and sheet silicates (inset) in interlayered and folded meta-shale and meta-graywacke at Bermugui, Australia. From Williams (1972). Reprinted by permission of *American Journal of Science*

The layers range in width from a few micrometers to 2 cm (increasing with increasing grainsize of the meta-sediment) and are alternately rich in quartz and in sheet silicates. Where present in meta-shale, crystal foliation differs in the two kinds of layers, being parallel to bedding in the quartz-rich layers and nearly parallel to axial planes in the quartz-poor layers (Figure 5.55).

Williams (1990) proposed that the layering was produced either by a transfer of SiO_2 from quartz-poor to quartz-rich layers, or by the selective removal of SiO_2 from the site of layer formation, resulting in a decrease in rock volume.

Further examples of differentiation in low-grade terranes are described by Ramberg (1961), Talbot and Hobbs (1968), Soula and Debat (1976), and Marlow and Etheridge (1977).

5.7.2 Quartz–plagioclase veins in amphibolite

Common amphibolite of the amphibolite facies is frequently found to enclose quartz–plagioclase veins that almost certainly have formed during metamorphism. In an example from Australia (Figure 5.56) the amphibolite consists of plagioclase, hornblende, and quartz, while the veins consist of the same minerals, but with higher proportions of quartz and plagioclase.

Observe first (Figure 5.56) the presence of hornblende-rich zones adjacent to the veins, presenting evidence that the vein minerals or portions thereof were extracted from the amphibolite, i.e., the veins and the zones are the result of a local rearrangement of atoms and crystals. A volume-percent profile extending across the vein that is visible in the upper right-hand corner of Figure 5.56 is presented in Figure 5.57. A local balance of material appears to exist approximately, but because of vein

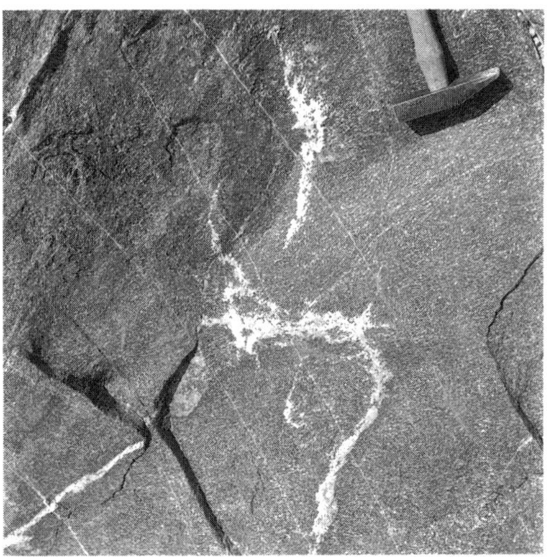

Figure 5.56 Irregular quartz–plagioclase veins in amphibolite at Einasleigh, Australia; note hornblende-rich margins. From Kretz (1966b), reproduced by permission of Blackwell Scientific Publications, Inc.

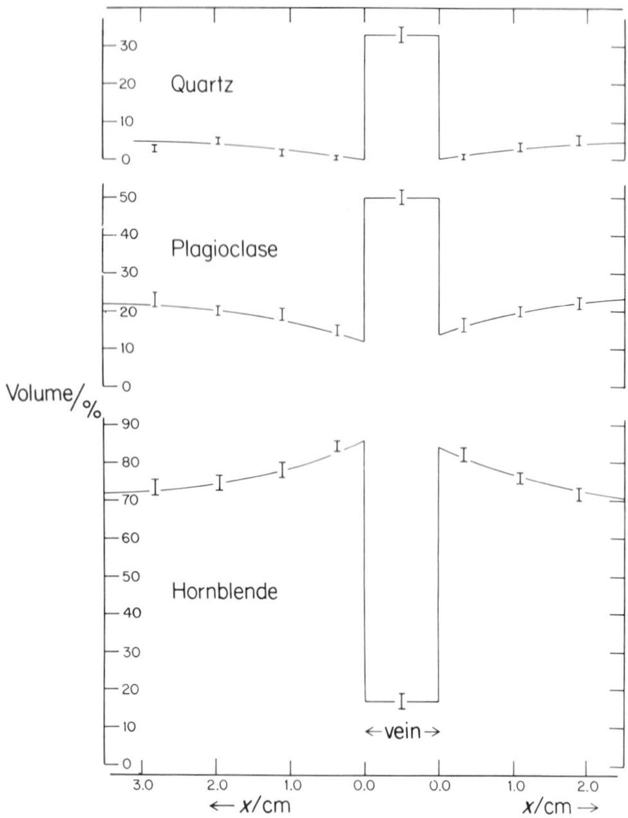

Figure 5.57 Variation in volume % of minerals (C) on a traverse across the vein shown in the upper right corner of Figure 5.56; error bars are 0.95 confidence limits; equations for the error-function curves are given in text

irregularities, in two and three dimensions, perfect balance cannot be expected on a single-line traverse; this was emphasized by Chakraborty (1977).

Additional support for a local source of the vein-forming crystals is found in new chemical analyses of hornblende and plagioclase, as listed in Table 5.8. Taking analytical precision into consideration, hornblende crystals in amphibolite, hornblende zone, and vein are all compositionally indistinguishable. Minor millimeter-scale variation was found in plagioclase composition, and although further results are needed to determine if the mean composition is the same throughout, the most sodic analyses (Table 5.8) are certainly similar.

The local transfer of atoms needed for vein formation possibly occurred by crystal-boundary diffusion. If so, then concentration gradients existed on crystal boundaries in the hornblende-rich zones–gradients of Mg, Fe, Ti, and other elements away from the vein, and of Na, Ca, Al, Si, and other elements toward the vein. Mineral fractions within the hornblende zones may be mimicking these gradients,

Table 5.8 Analyses of minerals in amphibolite (A), mafic margin (M) and vein (V) (Figure 5.57)

	Hornblende			Plagioclase		
	A	M	V	A	M	V
SiO_2/wt%	43.4	43.7	43.6	56.9	57.6	57.5
Al_2O_3	11.2	11.1	10.7	26.1	26.3	27.0
TiO_2	1.7	1.8	1.7			
FeO*	20.6	20.8	20.3			
MgO	8.1	8.3	8.2			
MnO	0.31	0.30	0.28			
CaO	11.3	11.2	11.4	8.7	8.7	9.2
Na_2O	1.3	1.5	1.4	6.6	6.4	6.3
K_2O	0.68	0.66	0.68	0.1	0.1	0.1

*total iron as FeO

as they existed, and Figure 5.57 shows that the mineral percentage (C) versus distance (x) profiles can in fact be expressed by a diffusion equation of the kind presented in Sec. 4.3.2. Referring to this figure, the equations are
Quartz:

$$C = 5\,\mathrm{erf}[x/2] \qquad \text{both sides}$$

Plagioclase:

$$C = 12 + 10[\mathrm{erf}(x/2)] \quad \text{left side}$$

$$C = 14 + 10[\mathrm{erf}(x/2)] \quad \text{right side}$$

Hornblende:

$$C = 86 - 14[\mathrm{erf}(x/2)] \quad \text{left side}$$

$$C = 84 - 14[\mathrm{erf}(x/2)] \quad \text{right side}$$

Some possibilities concerning the driving force for vein formation will be considered below.

Other examples of metamorphic differentiation in amphibolite were described by Reitan (1959) and by Bowes and Park (1966).

5.7.3 Veins in aluminous schist in a terrane of increasing metamorphic grade

For an example of mineral veins in aluminous schist, and of the correlation of vein minerals with metamorphic grade, we turn to a study by Vidale (1974), who examined the distribution of four principal kinds of veins in a terrane that ranges from the chlorite zone to the sillimanite–K-feldspar zone. The study area covers 1700 km^2 of dominantly lower Paleozoic meta-shale in and about Dutchess County, New York. Most of the veins range in width from a few millimeters to 10 cm, and consist of different combinations of calcite, quartz, plagioclase ($X_{\mathrm{an}} = 0.20$ to

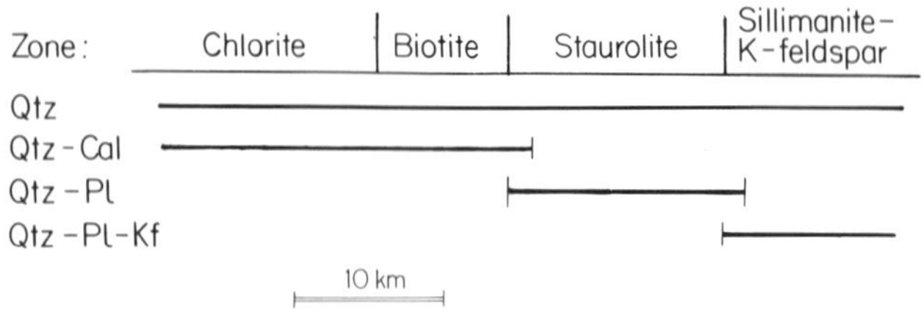

Figure 5.58 Mineral content of metamorphic veins in relation to metamorphic grade, Dutchess County, NY, USA; the composition of plagioclase is $X_{an} = 0.20$ to 0.50. Data from Vidale (1974)

0.50), and K-feldspar. In the chlorite zone, crystals of vein quartz and albite (which is locally present) commonly extend fully across the veins, and in the higher-grade zones, vein quartz and feldspar form a mosaic microstructure.

Quartz–calcite veins are virtually confined to the chlorite and biotite zones, quartz–plagioclase veins to the staurolite zone, and quartz–plagioclase–K-feldspar veins to the sillimanite–K-feldspar zone, with minor overlap, as shown in Figure 5.58. Quartz veins occur throughout. Careful field and microscopic observations have confirmed that with the exception of calcite, the minerals found in a given vein also occur in the enclosing schist. Moreover, the composition of plagioclase in a vein and in the enclosing rock is nearly identical.

Based on her observations, Vidale (1974) has proposed that the vein-forming elements in most veins were extracted from the immediately surrounding schist, at or near the time of peak metamorphic temperatures. Vein 'opening' and the crystallization of vein minerals occurred simultaneously.

Additional examples of veins in aluminous schist are provided by Read (1933), Chapman (1950), and McLellan (1989).

5.7.4 Quartz–feldspar veins in granitic gneiss (veined gneiss)

Examples of layers and veins in rock of granitic composition are provided by the Otter veined-gneiss Complex within the Grenville Province of the Canadian Shield. The complex consists mainly of quartz–plagioclase–K-feldspar–biotite–gneiss (with local garnet and rare sillimanite) and smaller proportions of other rock types, including a set of intensely folded and disjointed dikes of garnet–amphibolite. Where the gneiss is relatively free of veins, the chemical composition is close to that of average granite. But nearly everywhere, quartz–feldspar veins and layers are present, mainly parallel to the biotite foliation, and some of these are also folded. Two representative samples are illustrated in Figure 5.59.

In sample A (Figure 5.59a) two kinds of litho-units can be recognized. The even-numbered units are relatively high in biotite and low in K-feldspar (these are referred to as 'gneiss'), and the odd-numbered units are relatively low in biotite and high in K-feldspar (referred to as 'veins'). The variation in mineral proportions (measured by point-count analysis on the etched and strained surface of sample A)

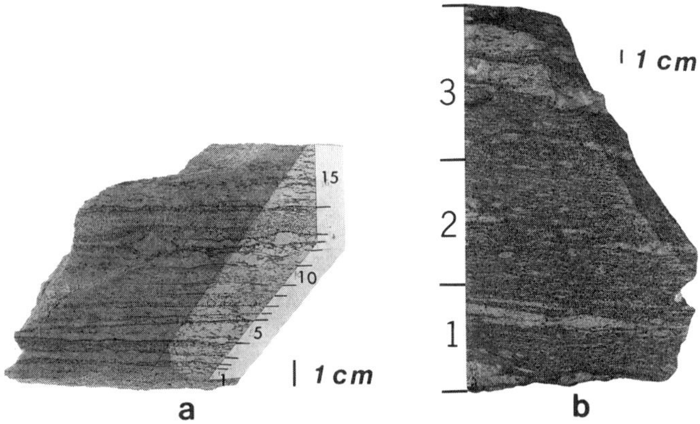

Figure 5.59 Representative samples of veined gneiss from the Otter Complex, Grenville Province. From Kretz (1994)

is illustrated in Figures 5.60 and 5.61. The inverse relation of volume-percent K-feldspar (C_{Kf}) and biotite (C_{Bt}) is shown in Figure 5.62, and can be expressed approximately by the hyperbolic equation

$$C_{Kf} \cdot C_{Bt} = 130.$$

The variation in vein width (from 2 mm to 1 cm) that is frequently observed is illustrated in sample B (Figure 5.59b). The microstructure of gneiss in this sample is illustrated in Figure 1.9a of Chapter 1, and that of a portion of a vein and the adjacent gneiss in Figure 5.63. Garnet is present in the gneiss and in the veins.

To test the hypothesis that veins in the Otter Complex were locally derived, chemical analyses were obtained of portions 1 and 3 of sample B (Figure 5.59b), including the several veins that are present in these portions, and of portion 2, which is nearly free of veins, and the results are presented in Table 5.9. Taking analytical error into consideration, portions 1 and 3 are indistinguishable from 2 with regard to eleven elements; the only demonstrated differences are in Ti, Al, Fe, Y, and Zr, and these differences are very small. Had the veins been introduced from afar, the observed compositional similarities would not be expected.

Additional support for segregation is found in the chemical composition of minerals that are common to veins and gneiss. Microprobe analyses of garnet and plagioclase crystals encountered on a traverse that crosses two veins in zone 1 of sample B are presented in Figure 5.64. With regard to Mn and Mg in garnet, a slight decrease in concentration does occur along the traverse, but this variation cannot be related to the presence of the two veins. With regard to Ca in garnet, and Ca and Na in plagioclase, no vein–gneiss differences can be detected.

Results from the two samples described above and six others are summarized as follows.

(1) In a highly deformed, metamorphic complex that consists mainly of veined

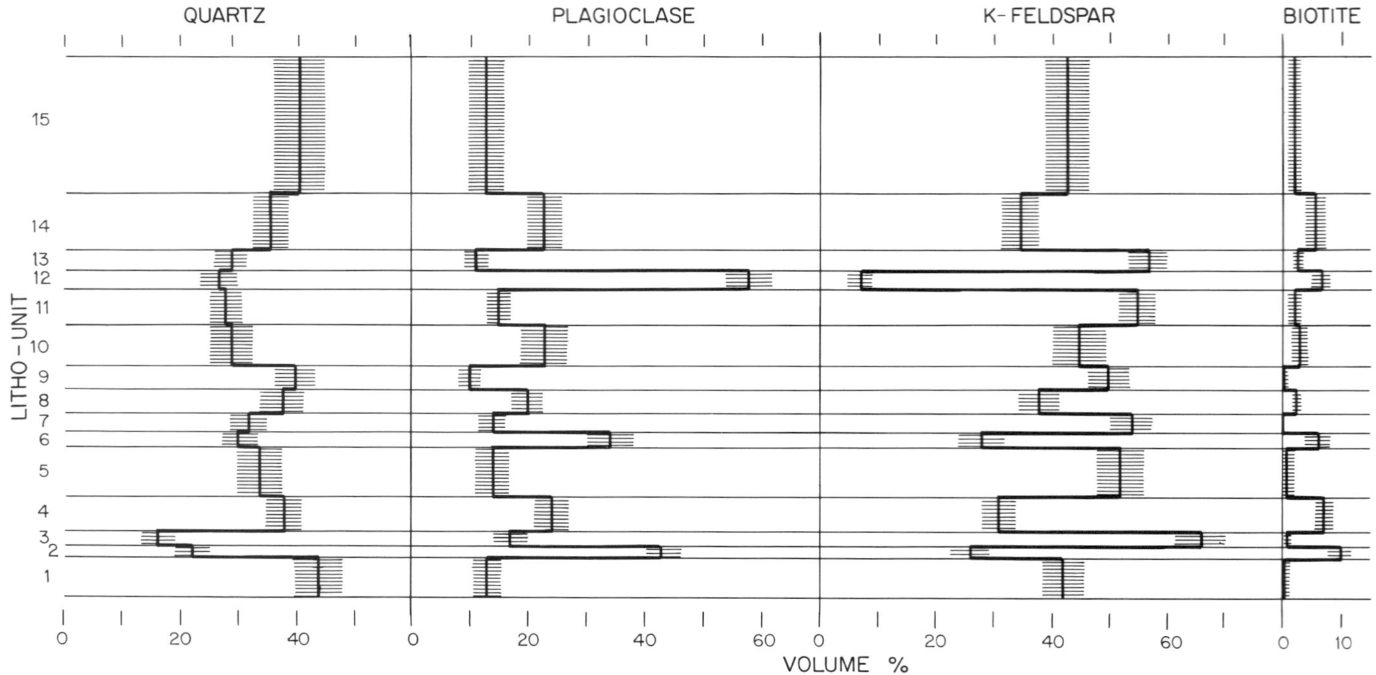

Figure 5.60 Volume % of minerals in litho-units 1 to 15 of sample A (Figure 5.59a); 0.95 confidence limits are indicated

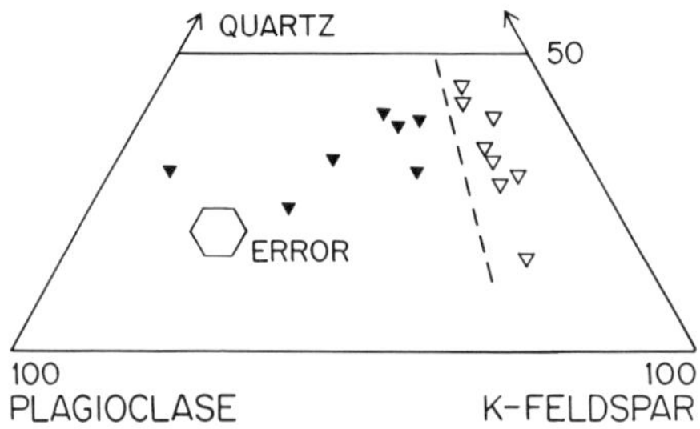

Figure 5.61 Quartz : plagioclase : K-feldspar volume proportions in the 15 litho-units of sample A (Figure 5.59a); solid symbols, gneiss; open symbols, veins. From Kretz (1994)

Table 5.9 Gneiss with many veins compared with adjacent gneiss with few veins, sample B, Figure 5.59b

	Few veins 2	Many 1 and 3	*c*
SiO_2/wt%	68.5	69.2	±0.6
TiO_2	0.49	0.37*	0.02
Al_2O_3	13.7	14.2**	0.20
FeO	4.32	3.53*	0.04
MgO	0.63	0.54	0.08
MnO	0.07	0.06	0.01
CaO	1.41	1.35	0.03
Na_2O	2.79	3.10	0.20
K_2O	4.74	4.94	0.10
P/ppm	740	610	±100
Rb	150	150	12
Sr	100	110	12
Y	150	120*	6
Zr	500	380*	10
Ba	1260	1230	20

FeO total iron as FeO
c 95% confidence limits to analyses
**significantly larger than number to the left
*significantly smaller than number to the left

gneiss of granitic composition, the mineral assemblage of the veins mimics that of the enclosing gneiss. Thus in addition to quartz, plagioclase, K-feldspar, and biotite (which are common to both), wherever veins contain notable garnet, magnetite, or sillimanite, the same is true of the enclosing gneiss, and in general the minor minerals, apatite, allanite, and zircon, are also common to both.

(2) Quartz content of the veins lies between 15 and 45 volume % and locally is higher, while the volume ratio K-feldspar/(plagioclase + K-feldspar) varies greatly (Figure 5.65) and in general also mimics that of the enclosing gneiss. Thus where

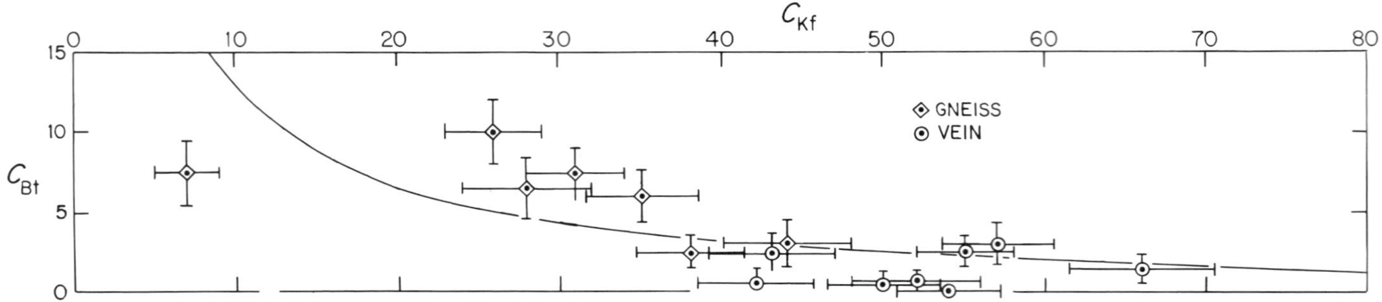

Figure 5.62 Inverse relation of volume % biotite (C_{Bt}) and K-feldspar (C_{Kf}) in litho-units of sample A (Figure 5.59a); the curve is for the equation, $C_{Bt} \cdot C_{Kf} = 130$. From Kretz (1994)

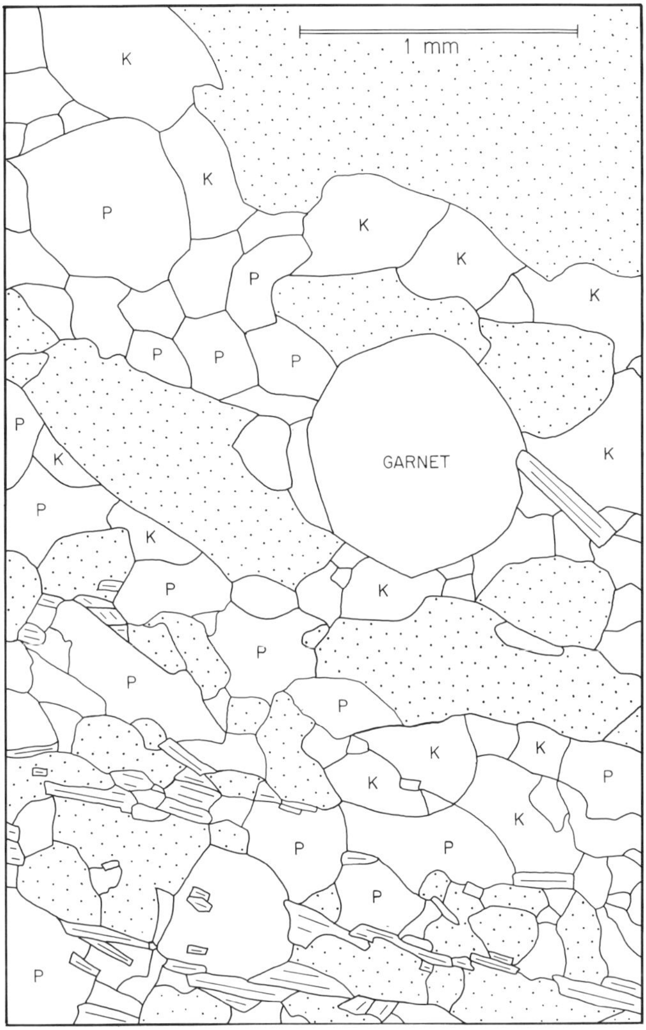

Figure 5.63 Microstructure of portion of a vein and adjacent gneiss (at bottom) in sample B (Figure 5.59b). Minerals are biotite (line pattern), quartz (dot pattern), feldspar (clear: plagioclase (P), K-feldspar (K))

this ratio in the veins is exceptionally high, exceptionally low, or highly variable, the same applies to the neighboring gneiss. The veins invariably contain more K-feldspar and less biotite than adjacent gneiss, and the inverse K-feldspar–biotite relation for the various litho-units as a whole is approximately hyperbolic rather than linear.

(3) For each of several elements, the bulk concentration in portions of gneiss that contain many veins and in adjoining portions that are nearly free of veins, is (within analytical error) identical.

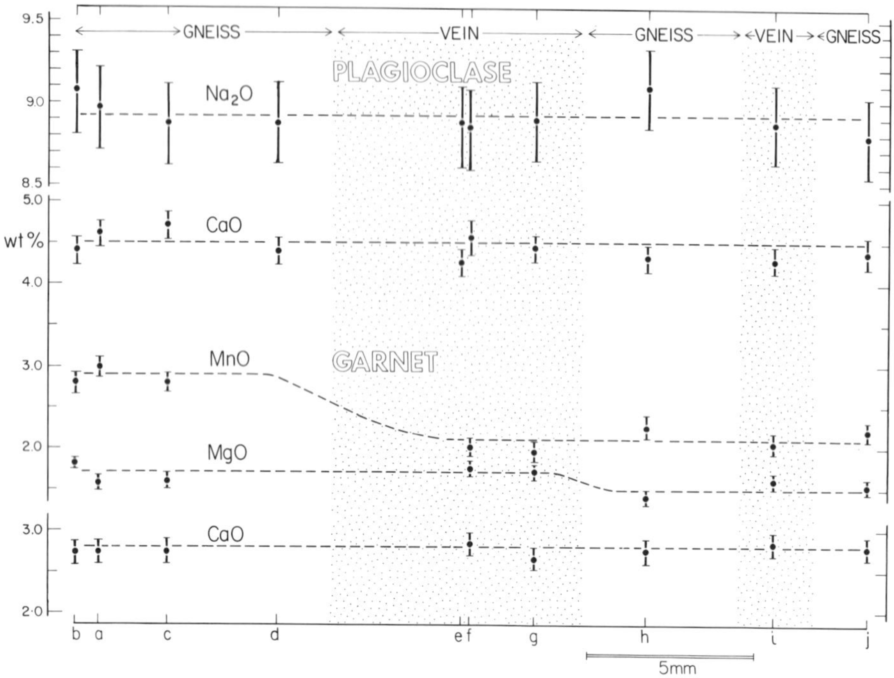

Figure 5.64 Variation in the composition of plagioclase and garnet crystals on a traverse (b to j) across two veins and adjacent gneiss in sample B (zone 1, Figure 5.59b); error bars are 0.95 confidence limits, analytical precision. From Kretz (1994)

(4) The composition of plagioclase in veins and in adjacent gneiss at two localities is indistinguishable. For garnet, in one sample (Figure 5.64), a vein–gneiss difference in composition could not be detected. In another sample, Mn and Ca are identical, but vein garnet contains slightly less Fe and more Mg than gneiss garnet, and a similar style of compositional zoning (for Ca and Fe) is present in both.

The above relations support the proposal that quartz–feldspar veins of the study area were locally derived.

Experimental studies, as reviewed by Johannes (1985) have demonstrated that for rocks of granitic composition, partial melting is possible at the conditions of high-grade metamorphism, provided an abundant supply of H_2O molecules is maintained at the site of melting. The possibility that elements, when transported from gneiss to veins existed in a molten state, will therefore be considered.

The nature of phase relations in the system SiO_2–$NaAlSi_3O_8$–$KAlSi_3O_8$–H_2O, i.e., the presence of a cotectic and a melt minimum (Figure 5.65), lead to the expectation that partial melting of K-feldspar gneiss, followed by partial or complete segregation of the melt, will result in a vein composition that falls near the minimum melt composition. This expectation is not modified significantly by the presence

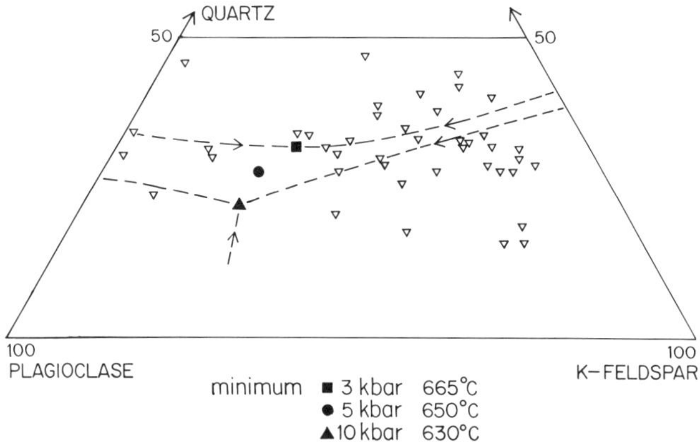

Figure 5.65 Experimental results (from Johannes 1985) on cotectic (line) and minimum melt (point) positions in the system SiO_2-$NaAlSi_3O_8$-$KAlSi_3O_8$ (H_2O saturated), and superimposed quartz: Na plagioclase: K-feldspar volume proportions for *veins* in samples of veined gneiss from the Otter Complex, including results from sample A (Figure 5.61). From Kretz (1994)

of biotite and about 20% of anorthite component in plagioclase (Johannes 1985), nor by a reduction in the activity of H_2O in melt from 1.0 to 0.5 (Ebadi and Johannes 1991). In Figure 5.65 we see, however, that vein composition, in terms of quartz:plagioclase:K-feldspar proportions, varies greatly, and shows no tendency to lie near the cotectic or melt-minimum locations. Moreover, given that near-equilibrium partial melting of plagioclase should leave the crystals enriched in anorthite and melt in albite (Bowen 1928), the finding that plagioclase crystals in veins and adjacent gneiss have the same composition (as found here and elsewhere by Harris (1974) and others) is not in agreement with the melt-segregation hypothesis.

Another possibility for crystal segregation is the *pressure-transfer* hypothesis of Ramberg (1952a, 1956), which states that mass transfer in vein formation occurs by diffusion, resulting from gradients of pressure that arise during regional deformation. The model was applied and extended by Reitan (1959), Gresens (1966), DeVore (1969). Vidale (1974), Stephansson (1974), Robin (1979), Ortoleva, Merino and Strickholm (1982), and others.

Following Ramberg (1956), two processes are considered regarding the transfer of matter during vein formation. In the *exchange process* a vein nucleates and then grows by replacing gneiss, i.e., most of the atoms needed for growth are taken from the site where growth occurs, with, however, the addition of some atoms (e.g., K) from the surrounding gneiss, in exchange for others (e.g., Mg and Fe). In the *dilation process,* vein nucleation occurs at certain places, and as the vein grows and the walls move apart, all of the atoms needed for growth are obtained in a one-way migration from the surrounding gneiss; sites within the gneiss, from which atoms are removed, collapse by plastic flow.

With regard to the exchange process, some vein-forming reactions are now proposed. In relation to sample A, the problem is to define vein–gneiss cation exchange processes, to produce the various litho-units within a volume of rock that initially was nearly homogeneous. Negligible (non-dilational) changes in volume are assured by supposing that a migration of oxygen has not occurred.

An example of exchange is provided by the Orville (1963) equation,

$$\underset{\text{K-feldspar}}{KAlSi_3O_8} + [Na^+] \underset{\text{in vein}}{\overset{\text{in gneiss}}{\rightleftarrows}} \underset{\text{albite}}{NaAlSi_3O_8} + [K^+]$$

which effectively transfers K-feldspar from gneiss to vein, in exchange for albite, this occurring by Na $\rightleftarrows$ K exchange across the vein–gneiss interface. The 'transfer' of K-feldspar of composition, Na/(K + Na) = 0.1, from gneiss to vein, and of plagioclase, of composition Na/(Na + Ca) = 0.8, from vein to gneiss is described by an analogous equation,

$$\underset{\text{K-feldspar}}{(KSi)_{0.9}(NaSi)_{0.1}AlSi_2O_8} + [(CaAl)_{0.2} + (NaSi)_{0.8}] \underset{\text{in vein}}{\overset{\text{in gneiss}}{\rightleftarrows}}$$

$$\underset{\text{plagioclase}}{(CaAl)_{0.2}(NaSi)_{0.8}AlSi_2O_8} + [(NaSi)_{0.1} + (KSi)_{0.9}] \qquad (121)$$

The reaction produces complementary changes in the K-feldspar:plagioclase ratio, as shown by arrows (1) in Figure 5.66. Assuming that cations move independently, the expressions in square brackets can be rewritten as

$$[Ca_{0.2} + Al_{0.2} + Na_{0.7}]^{1.7+} \rightleftarrows [K_{0.9} + Si_{0.2}]^{1.7+}$$

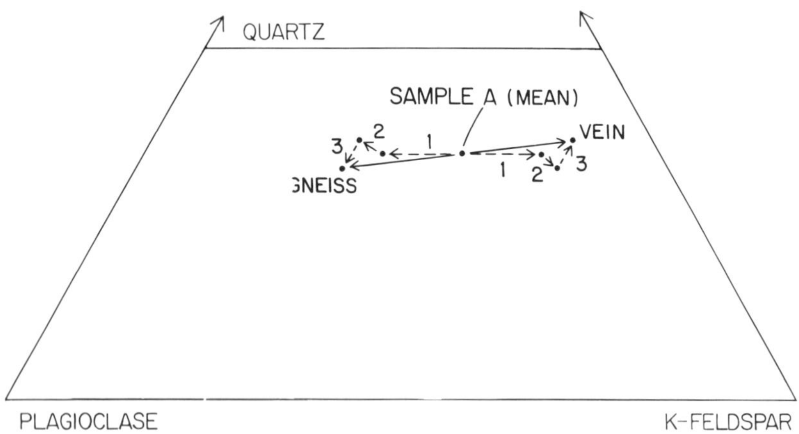

Figure 5.66 Example of differentiation produced by model ion-transfer reactions (121), (122), and (123), beginning with a composition represented by the mean of the 15 litho-units in Sample A (Figure 5.61). From Kretz (1994)

The observed decline in biotite content during vein formation can be described by the Fisher (1970) equation, which was also used by Olsen (1977). The equation, written to locally conserve oxygen, is

$$\underset{\text{K-feldspar (163 cm}^3\text{)}}{1.5KAlSi_3O_8} + [3Mg^{2+} + 2H^+] \underset{\text{in vein}}{\overset{\text{in gneiss}}{\rightleftarrows}}$$

$$\underset{\text{biotite (150 cm}^3\text{)}}{KMg_3AlSi_3O_{10}(OH)_2} + 0.5[K^+ + Al^3 + 3Si^{4+}] \qquad (122)$$

where simplified formulas are written, and Mg stands for both Mg and Fe. The process implied by this equation effectively displaces a volume of biotite from a vein to the adjacent gneiss, in exchange for an equal volume of K-feldspar. Thus reaction (122), operating alone, would produce a trend that in Figure 5.62 would appear as a line of slope -1. The slope that was actually observed can be explained by reactions (121) and (122) operating together. In relation to Figure 5.66, reaction (122) is described by a change in K-feldspar volume, relative to that of plagioclase and quartz, i.e., by arrows labelled 2 in this figure.

A third and more erratic vein–gneiss difference in sample A and others is in the quartz content. One possibility for an increase in quartz at sites of vein formation is that the required oxygen was taken from plagioclase while Si was obtained in part from associated plagioclase and in part from the adjacent gneiss:

$$\underset{\text{quartz}}{4SiO_2} + [(CaAl)_{0.2} + (NaSi)_{0.8} + Al] \underset{\text{in vein}}{\overset{\text{in gneiss}}{\rightleftarrows}}$$

$$\underset{\text{plagioclase}}{(CaAl)_{0.2}(NaSi)_{0.8}AlSi_2O_8} + [2Si] \qquad (123)$$

This results in a change in the quartz:plagioclase volume ratio in vein and gneiss, as shown by arrows labelled 3 in Figure 5.66. Thus equations (121), (122), and (123) provide an interpretation for heterogeneity in sample A (Figure 5.59a) and other samples, in terms of metamorphic differentiation.

With regard to garnet, where the volume fraction of this mineral in a vein is similar to that in the adjacent gneiss, the crystals were possibly not involved in the vein-forming process. Where volume fractions are small, a reaction similar to (122) could be proposed to 'transfer' garnet to gneiss, or, alternatively, the vein could have formed, in part, by dilation, which caused the crystals simply to move apart, as quartz and feldspar were added. In the dilation process, all atoms, including oxygen, must move to the site of vein formation.

We inquire now if diffusion coefficients and fluxes are adequate to produce, by the exchange process, a vein 2 cm wide in a reasonable period of geological time.

The experimental results of Christoffersen, Yund and Tullis (1983) on Na–K exchange diffusion in (Na, K) feldspar (Sec. 4.3.7) form the basis for an estimate of the rate of crystal-boundary diffusion in gneiss at 600 °C. For a crystal of composition K/(Na + K) = 0.4, and for diffusion ⊥ (001), a diffusion coefficient ($\tilde{D}$) of 1×10^{-13} cm^2 s^{-1} was measured at 900 °C. This number, extrapolated to 600 °C is approximately 1×10^{-17} cm^2 s^{-1}. In metal systems, the coefficient for grain-boundary diffusion is commonly found to exceed that for volume diffusion by a factor of 10^5 (Sec. 4.3.5), and by applying this factor, D' (K–Na exchange, feldspar boundary, 600 °C) $\sim 1 \times 10^{-12}$ cm^2 s^{-1}. The factor of 10^5 also appears when diffusion of Ni and Cr on NiO–NiO boundaries at 600 °C is compared with diffusion in NiO crystals (Atkinson and Taylor 1986). Here D' (Ni, self-diffusion, NiO–NiO boundary, 600 °C) = 1×10^{-12} cm^2 s^{-1}.

A second estimate is obtained by assuming that the crystal boundaries possess a structure (Sec. 1.5) and that the structure is somewhat similar to that of hydrated zeolite. Experiments of Barrer, Bartholomew and Rees (1963) on the self-diffusion of K^+ in K-chabazite at 55 °C produced a diffusion coefficient of 19.4×10^{-12} cm^2 s^{-1}, which by an Arrhenius extrapolation to 600 °C gives approximately 2×10^{-8} cm^2 s^{-1}.

The above estimates of $\sim 1 \times 10^{-12}$ and $\sim 1 \times 10^{-8}$ cm^2 s^{-1} are now assumed to define limits for various cation-exchange diffusion coefficients within gneiss during vein formation. These numbers translate to a mean displacement, $\bar{x}$, where $\bar{x} = (Dt)^{1/2}$, of 3 cm and 3 m respectively for a time (t) of 0.3 My, and to 30 cm and 30 m respectively for a time of 30 My (Figure 4.12, Sec. 4.3).

But more important than mean displacement is the question of the flux of ions, which in view of narrow boundary widths, may be small. Consider the flux (J) of K^+ ions across 1 cm^2 of vein–gneiss interface. In a necessary simplification, the gneiss is viewed as consisting of cubic 1 mm crystals, and the crystal-boundary width is taken to be a uniform 20 Å, which is the NiO–NiO grain boundary width (7 Å; Atkinson and Taylor 1986) multiplied by an arbitrary factor of 3. The tortuosity coefficient in this simple model is 1.0. The cross-sectional area of the crystal-boundary phase per cm^2 of vein–gneiss interface is therefore 4×10^{-6} cm^2. Now by use of the first equation of diffusion, $J = -D\, dC/dx$, and by assuming a steady-state concentration gradient (dC/dx) of -0.01 g cm^{-2}, a diffusion coefficient (D) of 10^{-9} cm^2 s^{-1} results in a flux of 40 mg of potassium in a time of 30 My. For a vein that is 2 cm wide, vein formation requires an addition of approximately 1.5 wt% K_2O or approximately 40 mg of K per cm^3 of vein. These estimates and calculations indicate therefore that diffusional rates and fluxes are adequate to accomplish the transfer of matter that is needed for the formation of small-to-medium veins by metamorphic differentiation.

The formation of dilation veins requires that oxygen as well as cations be transported to the site of vein crystallization. In oxides, the diffusion of oxygen is considerably slower than that of cations (Oishi and Kingery 1960) but Nagy and Giletti (1986; see Sec. 4.3.7) have measured D' (oxygen, K-feldspar–albite boundary, 600 °C) $\sim 1 \times 10^{-9}$ cm^2 s^{-1} (assuming a boundary width of 10 Å), which is similar to the above estimates for cation diffusion. In this and other studies of oxygen diffusion, some uncertainty exists regarding the role of H and whether the diffusing species is O, OH, or H_2O. In any case, the experimental data do not rule out the possibility that diffusion was the transport mechanism in the formation dilation, as well as replacement veins.

5.7.5 Theory of stress-induced metamorphic differentiation

In the above examples of mineral veins (Secs. 5.7.1 to 5.7.4) we considered the sources and migrations of vein-forming atoms, but did not identify possible causes of the migrations. We shall now examine a theory of differentiation that was developed by Robin (1979), and is based of the pressure-transfer proposal of Ramberg (1952a), on the Coble (1963) concept of diffusional creep, and on the knowledge that during deformation, some rocks are more competent than others.

Consider a volume of layered quartz–muscovite rock, with quartz concentrated in some layers (Q) and muscovite in others (M), and suppose that the maximum principal stress (σ_1) is directed normal to the layers, as in Figure 5.67. The maximum stress (σ_1) must be the same in all layers, but if the competency of the layers is not equal, the differential stress ($\sigma_1 - \sigma_3$) will be *greater* in the more competent layers. This was confirmed by Strömgärd (1973) and is demonstrated in the frequently observed breaking of competent layers, to form boudinage structures. Assuming now that the Q layers are more competent than the M layers, it follows that during deformation, the mean stress, $\overline{\sigma} = (\sigma_1 + \sigma_2 + \sigma_3)/3$ is greater in the M than in the Q layers, and that pressure gradients will exist from M to Q layers — pressure being identified, simply and approximately with mean stress ($\overline{\sigma}$).

Recall (Sec. 2.1) the change in Gibbs energy (G) of a crystal with changes in temperature (T) and pressure (P),

$$dG = -S\,dT + V\,dP.$$

At constant temperature,

$$\left(\frac{\partial G}{\partial P}\right)_T = V,$$

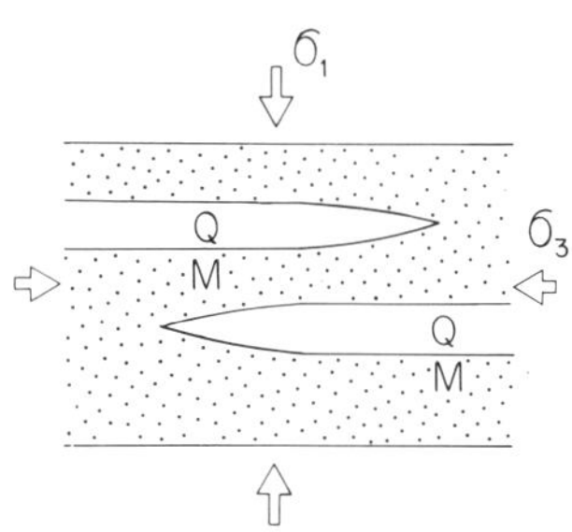

Figure 5.67 Disposition of less competent, muscovite-rich layers (M) and more competent, quartz-rich layers (Q) in relation to the principal compressive stress (σ_1) in the stress-induced metamorphic differentiation model of Robin (1979). Because of competency contrast, $\sigma_3(M) > \sigma_3(Q)$. Quartz migrates from M to Q, in the direction of decreasing mean pressure. From Kretz (1994)

where V is the molar volume of the crystal. Thus the Gibbs energy of crystals of quartz and muscovite in the M layers is greater than corresponding crystals in the Q layers, and a tendency will exist for quartz and muscovite to migrate from M to Q, provided that atomic migration is possible. Although different mechanisms of plastic deformation may occur within both layers, suppose that diffusional creep on crystal boundaries (Coble creep) is one of these mechanisms. Diffusional creep in quartz, for example, is driven by Gibbs energy gradients from portions of crystals affected by σ_1 (where dissipation occurs) to portions affected by σ_2 or σ_3 (where growth occurs). Similarly, but on a larger scale, diffusion may occur from σ_1 points on quartz in M layers to σ_3 points on quartz in adjoining Q layers. This is the essence of the pressure-transfer mechanism of metamorphic differentiation, proposed by Ramberg (1952a).

Although energy gradients are necessary for atomic migration to occur, rates of migration are determined, instead, by energies of activation associated with the migration process. Robin (1979) has argued that because quartz commonly occurs in pressure shadows and in boudin fillings, the energy of activation associated with the diffusion of ions resulting in the transfer of quartz is lower than that associated with the transfer of muscovite. Thus although both quartz and muscovite would 'like' to move from M to Q layers, quartz 'wins', resulting in an increase in the volume fractions of quartz in Q layers and of muscovite in M layers.

Robin considered that the segregation process could be *initiated* by small concentrations of muscovite in an original, near-homogeneous rock. That the process, once started, does not generally proceed to completion is explained by supposing that, with increasing muscovite content, the competency of the M layers begins to *increase*, and a condition of stability could be created as M–Q competency contrast decreases to zero.

The above model can be modified and applied to various examples of metamorphic differentiation, including those described above. For example, in veined gneiss (Sec. 5.7.4), Q layers are quartz–feldspar veins and M layers are biotite–quartz–feldspar gneiss. All of these minerals except quartz are solid solutions, and we must now speak of chemical potential gradients, e.g., that of the albite component in plagioclase. Thus, from Sec. 2.1, with regard to the albite component (ab) in a crystal of plagioclase, the chemical potential (μ) is

$$\mu_{\mathrm{ab}} = G^{\mathrm{Ab}} + RT \ln X_{\mathrm{ab}} + RT \ln \gamma_{\mathrm{ab}},$$

Where X is mole fraction and γ is activity coefficient. Differentiating at constant temperature,

$$\frac{\partial \mu_{\mathrm{ab}}}{\partial P} = \frac{\partial G^{\mathrm{Ab}}}{\partial P} + RT \frac{\partial \ln X_{\mathrm{ab}}}{\partial P} + RT \frac{\partial \ln \gamma_{\mathrm{ab}}}{\partial P}.$$

The second term on the right is zero and the third term is very small, and therefore,

$$\frac{\partial \mu_{\mathrm{ab}}}{\partial P} \simeq \frac{\partial G^{\mathrm{Ab}}}{\partial P} = V^{\mathrm{Ab}}$$

Thus chemical potential gradients, $d\mu/dP$, are always positive and are equal to the molar volume of the component in its pure form. This means that for plagioclase, where $V^{Ab} \sim V^{An}$, the chemical potential gradient for the albite and anorthite components will be similar, but whether or not a change in plagioclase composition will occur as crystals are transplanted from sites of high to lower pressure, will depend on the activation energy for the different diffusing species.

In the above quartz–muscovite segregation, as conceived by Robin, diffusion of SiO_2 (possibly as ions) occurs from M to Q layers, with no diffusion occurring in the opposite direction, and it is not clear what happens at sites in M from which SiO_2 is removed. Q–M interfaces must advance into M, and presumably a physical or chemical collapse occurs throughout the M layers, causing muscovite to take up the spaces vacated by SiO_2.

The ion-exchange equations, (121), (122), (123) in Sec. 5.7.4, describe a transfer of atoms from gneiss (M) to veins (Q), *accompanied* by a transfer of other atoms in the reverse direction. Thus, if we wish to attribute this two-way diffusion to a pressure gradient, up-pressure diffusion must occur for certain ions. This is not impossible, provided that crystal-boundary concentrations of those ions that diffuse down-pressure, create up-pressure gradients in concentration or chemical potential for the other ions.

A further and more advanced theoretical treatment of the problem of metamorphic differentiation is presented by Ortoleva, Merino and Strickholm (1982) and Dewers and Ortoleva (1990).

5.8 METASOMATIC CRYSTALLIZATION

It is not uncommon to find, in metamorphic terranes, bodies of rock that have obviously experienced a change in chemical composition. A simple example is a layer of nearly pure diopside, between dolomite marble on one side and quartzite on the other. It is reasonable to suppose that the diopside skarn has formed by the reaction

$$\underset{\text{marble}}{CaMgC_2O_6} + \underset{\text{quartzite}}{2SiO_2} \rightarrow \underset{\text{skarn}}{CaMgSi_2O_6} + 2CO_2,$$

in a change that is entirely analogous to the reaction $MgO + Al_2O_3 \rightarrow MgAl_2O_4$ of Sec. 5.3.1. In examining zones between rock units, as well as those between crystals, it is frequently difficult to determine whether the reaction zone (e.g., skarn) has replaced the rock on one side (marble) or on the other (quartzite), but it is clear that for any sample taken from the zone, a change in chemical composition has occurred. The change is referred to as a 'metasomatic process'.

In the above example, the source of the skarn atoms can be identified with confidence (Si from quartzite, Ca and Mg from marble) but in many occurrences of metasomatic rock, all that can be said is that certain atoms were introduced, without knowing where they came from. An example is a zone of tourmaline-rich schist, bordered on one side by normal schist, and on the other, by a narrow quartz vein. Obviously, boron was transported, possibly as a vapor, along a fracture, now occupied by the quartz vein, and penetrated the adjacent schist, resulting in the metasomatic crystallization of tourmaline.

Theoretical studies of metasomatism have been carried out by Korzhinskii (1936b, 1970), Ramberg (1952a), Thompson (1959), Fletcher and Hofmann (1974), Weare, Stephens and Eugster (1976), Brady (1977), Frantz and Mao (1979), Volkova and Sheplev (1980), Helgeson (1987), and others, and some important experimental results were recently obtained by Zharikov and Zaraiski (1991).* Natural occurrences of metasomatism were described, for example, by Shaw et al. (1963), Schrijver, (1973), Joesten (1974), Thompson (1975), Ono (1977), Sanford (1982), Sack (1982), Rice (1983), Glassley (1983), Hogarth and Lapointe (1984), Frisch and Helgeson (1984), Shearer et al. (1984), Andersen (1989), and Lentz (1991, 1992). The evidence for metasomatism in the Mantle is reviewed in a volume edited by Menzies and Hawkesworth (1987).

Some examples of small-scale metasomatism will now be examined. One of the advantages of working on a small scale is that the pre-metasomatic conditions can be reconstructed with reasonable confidence, although difficulties in defining metasomatic processes exist, regardless of scale.

5.8.1 Crystallization of zoned skarn

Skarn, also referred to as 'calc–silicate rock', is composed principally of various Ca- or (Mg, Fe)-bearing minerals, including Ca pyroxene, Ca garnet, wollastonite, biotite, epidote, and scapolite, and is one of the most common products of metasomatic crystallization. It forms layers or irregularly shaped bodies that range in size from a few millimeters to several meters or more. Commonly it occurs at contacts between marble on one side and a silicate rock (granite, schist, amphibolite, quartzite) on the other, but it also occurs as layers and veins in silicate rocks and as inclusions in marble. A zonal arrangement of the minerals may be present.

Occurrences of zoned skarn in the Sierra Nevada of California were studied by Kerrick (1977), and a sample from a layer, 2 cm wide, that has crystallized at a contact between marble and amphibolite is illustrated in Figure 5.68a. The skarn consists principally of Ca garnet, Ca pyroxene, epidote, Ca amphibole, and plagioclase, and from the distribution of these minerals, three zones can be recognized (units 2, 3, 4), as shown in Figure 5.68a. The composition of garnet is $Ca_3(Al, Fe^{3+})_2Si_3O_{12}$, with mole fraction grossular ~0.60, varying irregularly across zone 2. The mole fraction of diopside in Ca pyroxene is estimated at ~0.80. Presumably the epidote in zones 3 and 4 contains minor Fe^{3+}, and the Ca amphibole in zone 4 is actinolite.

The variation in rock composition that occurs on moving from unit 1 (marble) to unit 5 (amphibolite) can be expressed approximately in terms of a variation in $CaO:Al_2O_3:MgO$ proportions, as is illustrated in Figure 5.68b. This figure emphasizes that the zone boundaries in the natural skarn represent discontinuities in rock composition; for example, a zone defined by the assemblage garnet (Grs)–Ca pyroxene (Di)–epidote (Ep) is absent.

* reference missing in reference list

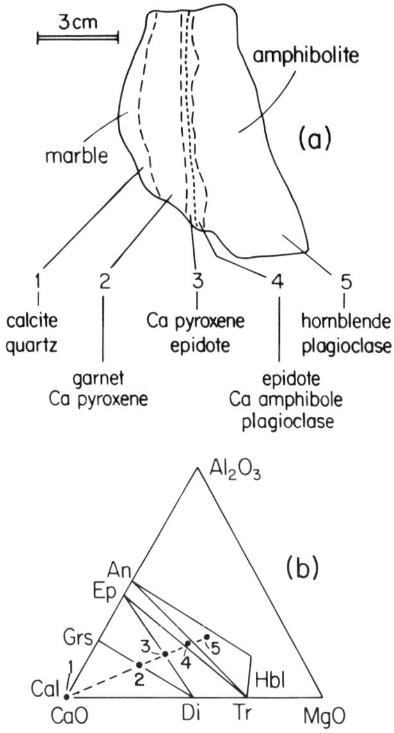

Figure 5.68 (a) Zoned skarn (zones 2, 3, 4) between marble (1) and amphibolite (5) from the Sierra Nevada (California); (b) mineral association of (1) to (5) in a simplified system (free of Fe and Na). From Kerrick (1977). Reproduced by permission of Oxford University Press

The formation of metasomatic zoning, with each zone consisting of only one or a few minerals was considered by Thompson (1959) in relation to the formation of mineral zones between periclase and quartz. We shall now examine this simple system and compare it with that described above. The approach here differs slightly from that of Thompson (1959) but leads to the same conclusion.

Consider two monomineralic 'rocks' in contact, one composed of periclase (MgO) and the other of quartz (SiO_2). Provided that each of quartz, periclase, enstatite ($MgSiO_3$), and forsterite (Mg_2SiO_4) is intrinsically stable at the arbitrarily chosen temperature and pressure, we note, by applying the phase rule, that no more than two of the four phases (and only in certain combinations) are permitted to exist at equilibrium. The spatial system, delineated to include crystals of periclase and quartz is obviously far from equilibrium, and assuming that a reaction of quartz with periclase is initiated, whether enstatite or forsterite is produced will be determined by reaction kinetics. Suppose, however, that crystals of both minerals are produced, by way of a metastable reaction, such as

$$\begin{array}{ccccccc} 4 \text{ periclase} & + & 3 \text{ quartz} & \rightarrow & 1 \text{ forsterite} & + & 2 \text{ enstatite}, \\ 45 \text{ cm}^3 & & 68 & & 44 & & 63 \end{array} \qquad (124)$$

to form a narrow zone of forsterite and enstatite crystals between periclase and quartz,

periclase	enstatite and forsterite	quartz

as in Figure 5.69b. Assume now, that the forsterite crystals in the right margin of the central zone begin to react with adjacent quartz,

$$\begin{array}{ccccc} 1 \text{ forsterite} & + & 1 \text{ quartz} & \rightarrow & 2 \text{ enstatite}, \\ 44 \text{ cm}^3 & & 23 & & 63 \end{array} \qquad (125)$$

while the enstatite crystals in the left margin react with adjacent periclase,

$$\begin{array}{ccccc} 2 \text{ enstatite} & + & 2 \text{ periclase} & \rightarrow & 2 \text{ forsterite}, \\ 63 \text{ cm}^3 & & 22 & & 87 \end{array} \qquad (126)$$

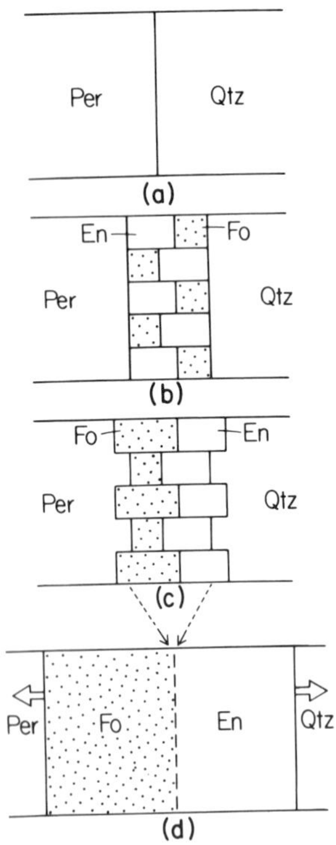

Figure 5.69 Illustration of the development, (a) to (d), of zones of forsterite (Fo) and enstatite (En) between periclase (Per) and quartz (Qtz)

with enstatite crystals growing into quartz and forsterite into periclase, as in Figure 5.69c, to produce an arrangement

periclase | forsterite | enstatite | quartz

that avoids direct contact of incompatible phases. Reaction must now continue by way of two half-reactions, one occurring at the enstatite–quartz interface,

$$\text{quartz} + [\text{a}] \rightarrow \text{enstatite} + [\text{b}], \tag{127}$$

and the other at the forsterite–periclase interface,

$$\text{periclase} + [\text{b}] \rightarrow \text{forsterite} + [\text{a}], \tag{128}$$

and in a broadening of the enstatite and forsterite zones (Figure 5.69d). As noted in Sec. 5.3.1 with regard to the reaction $MgO + Al_2O_3 \rightarrow MgAl_2O_4$, the half-reactions cannot be defined intuitively, because different possibilities exist for the migration of cations and anions through the product zones. That is why the mobile atoms, (a and b in (127) and (128)), have not been specified.

Equilibrium in the entire four-phase spatial system (Figure 5.69d) can only be achieved with the disappearance of two phases, but, as noted by Thompson (1959), a 'local equilibrium' can be assumed to exist at boundaries between adjacent zones. The question of whether this concept may be applied as the rate of the reaction expressed by (127) and (128) decreases to zero, will be left with the reader for consideration.

A similar model can be constructed for the zoned skarn that was described above (Figure 5.68), and reaction equations analogous to (124) to (128) can be proposed. In doing so, we shall simplify the system by removing Fe and Na, and by disregarding Ca pyroxene; then the minerals of primary interest are anorthite, calcite, quartz, grossular, and Fe-free epidote. The initial reaction to occur at the amphibolite–marble interface was possibly

$$\begin{aligned} &6\ \text{anorthite} + 7\ \text{calcite} + 3\ \text{quartz} + H_2O \\ &\rightarrow 3\ \text{grossular} + 2\ \text{epidote} + 7CO_2, \end{aligned} \tag{129}$$

which is analogous to (124). The two reactions that followed, comparable to (125) and (126), are

$$\begin{aligned} &1\ \text{grossular} + 5\ \text{anorthite} + 2H_2O \\ &\rightarrow 4\ \text{epidote} + 1\ \text{quartz}, \end{aligned} \tag{130}$$

resulting in the initiation of zone 3, and

$$\begin{aligned} &2\ \text{epidote} + 5\ \text{calcite} + 3\ \text{quartz} \\ &\rightarrow 3\ \text{grossular} + 2H_2O + 5CO_2, \end{aligned} \tag{131}$$

resulting in the initiation of zone 2 (Figure 5.68a). Growth of the two zones (the garnet zone advancing into marble, and the epidote zone into amphibolite) would then occur by half-reactions analogous to (127) and (128), these occurring by unspecified cation and anion diffusion to and from the two advancing interfaces.

Reactions (129) to (131) differ from (124) to (126) in the participation of some components (H_2O, CO_2) that migrated to and from the site of skarn formation. From data on phase equilibrium, as compiled and calculated by Kerrick (1977), reactions (129) to (131) could occur at ~450 °C and (necessarily) at $CO_2/(H_2O + CO_2) < 0.04$, or somewhat larger than 0.04 for the natural (Fe-bearing) crystals. The reactions could, in fact, be initiated and driven by an introduction of H_2O to the amphibolite–marble boundary. Kerrick has suggested that H_2O fluid inclusions, which are present in the skarn, are a sample of this introduced water. The entire reaction must have been more complex than is implied above, because it resulted, also, in the crystallization of Ca pyroxene and the alteration of some hornblende to actinolite.

Kerrick (1977) proposed that for contact metasomatic zones of the kind described above, the dominant mechanism of atomic transport is exchange diffusion along crystal boundaries. With regard to other bodies of skarn in the Sierra Nevada, notably those that are bounded on one side by granitic rock that is assumed to have existed, during skarn formation, as a silicate melt, a mechanism of transport of ions carried in solution in a permeating fluid or vapor phase is favored.

5.8.2 Metasomatic crystallization of muscovite in granite and tourmaline in schist at contacts with pegmatite

Striking examples of metasomatism on a small scale are found in the southern Slave Province of the Canadian Shield (Figure 1.17), where bodies of muscovite–biotite granite invaded a thick succession of deformed Archean meta-shale and meta-graywacke. Countless pegmatite dikes occur within the granite bodies and the adjacent cordierite-zone schist, for example, within and to the east of the Sparrow pluton, 40 km northeast of Yellowknife. Where pegmatite dikes cut granite, the adjacent granite is enriched in muscovite (Figure 5.70a), and where dikes cut schist, the adjacent rock is locally enriched in tourmaline (Figure 5.70b). The problem of identifying the elements that were added and removed, and defining the metasomatic mineral reactions at these localities will now be considered.

In any metasomatic process, changes in both density and volume may occur, and in attempting to identify, by use of chemical analysis, those elements that were introduced or removed, these changes must not be ignored. In a hypothetical case of extreme metasomatism, where migration of every element occurred, resulting in a large increase or decrease in volume, a solution to the problem would be possible only if inert markers were present, to delineate a volume of rock as it existed prior to metasomatism. In the absence of markers, a solution to the problem must be based on the assumption that the volume change was negligible, or that certain elements (to be identified) were virtually immobile. In addition, a sample of unaltered rock must be available; where metasomatic zones are small and occur in homogeneous rock this is not a serious matter, but some uncertainty in composition prior to metasomatism cannot be avoided.

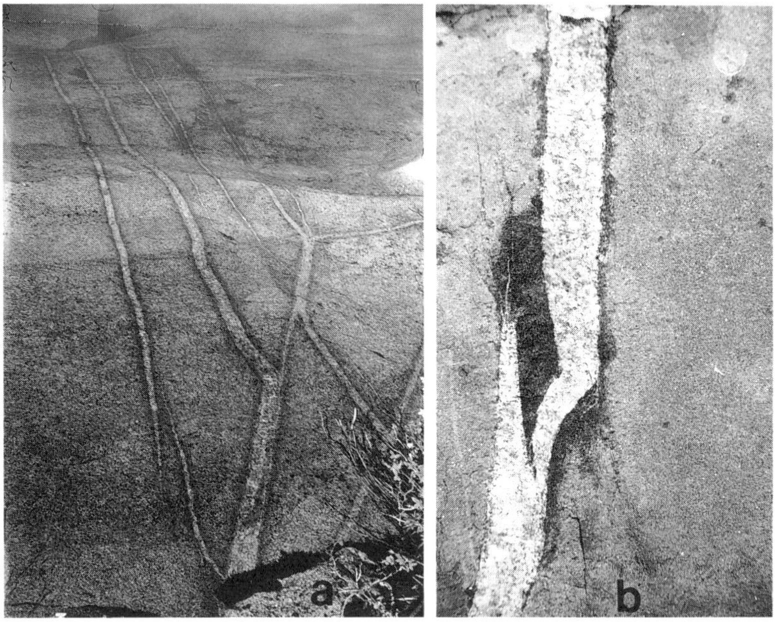

Figure 5.70 (a) Branching pegmatite dikes in granite (largest dikes are ~10 cm wide) showing dark (muscovite-rich) alteration zones; (b) pegmatite dike in quartz–plagioclase–biotite schist (dike is 5 cm wide) showing irregular development of tourmaline-rich zone (dark); location: near Yellowknife, Slave Province, Canadian Shield. From Kretz (1968b)

The problem of calculating additions and removals, taking changes of volume and density into consideration was treated by Gresens (1967) as follows. Let

M^{α} = mass of a body of rock of composition α, prior to metasomatism, here chosen as a mass of 1.0 g, bounded by an imaginary inert marker,

M^{β} = mass (g) of the same body of rock, i.e., bounded by the same inert marker, after metasomatism, and of composition β,

x_{A}^{α} = mass fraction (wt%/100) of element A in α, i.e., in the rock chosen as the unaltered equivalent of β,

x_{A}^{β} = mass fraction of element A in altered rock, β,

$R_{\mathrm{A}}^{\alpha/\beta}$ = concentration ratio, $x_{\mathrm{A}}^{\alpha}/x_{\mathrm{A}}^{\beta}$,

d^{α} = density (g cm^{-3}) of α at STP,

d^{β} = density (g cm^{-3}) of β at STP,

d^{α} and d^{β} are measured at STP and taken as estimates of density at the temperature and pressure of metasomatism,

$D^{\alpha/\beta}$ = density ratio, d^{α}/d^{β},

$V^{\beta/\alpha}$ = fraction volume change,

m_A = mass (g) of element A added to (+) or removed from (-) the original 1 g mass of α.

Now, for each element, A,

$$x_A^\alpha M^\alpha + m_A = x_A^\beta M^\beta. \qquad (132)$$

The volume change is

$$V^{\beta/\alpha} = \frac{(1/d^\beta)M^\beta}{(1/d^\alpha)M^\alpha} = D^{\alpha/\beta}(M^\beta/M^\alpha), \qquad (133)$$

and by rearranging terms,

$$M^\beta = (V^{\beta/\alpha}/D^{\alpha/\beta})M^\alpha. \qquad (134)$$

By substitution of (134) in (132), recalling that M^α = 1.0 g,

$$m_A = [(V^{\beta/\alpha}/D^{\alpha/\beta})x_A^\beta] - [x_A^\alpha]. \qquad (135)$$

For a given metasomatic zone and a particular element, A, x_A^β is obtained by chemical analysis, and an estimate of x_A^α is obtained by analysis of the unaltered rock, immediately beyond the zone. Similarly, by weighing fragments of α and β in air and in water, values of d^β and d^α, and hence $D^{\alpha/\beta}$, are obtained. This leaves $V^{\beta/\alpha}$ as an unknown. The procedure adopted here, regarding this term, is to assume (initially) that $V^{\beta/\alpha} = 1.0$ (no volume change), and then to determine if this assumption is consistent with a certain group of elements (e.g., Al, Ti, Si) remaining stationary. This is done by noting that when m_A in equation (135) is zero,

$$V^{\beta/\alpha} = R^{\alpha/\beta}D^{\alpha/\beta}. \qquad (136)$$

Thus, if $R^{\alpha/\beta}D^{\alpha/\beta}$ is found to be the same number for three or more elements, say 1.05, the initial assumption of a zero or very small volume change is strengthened, but remains, of course, unproven.

A large sample (labelled M1) consisting of a pegmatite dike, 4 cm wide, similar to those shown in Figure 5.70a, the adjoining muscovite-rich zones, and the adjacent unaltered granite was examined with regard to minerals present and variation in mineral proportions, and the results for one of the two adjacent zones are presented in Figure 5.71. In moving from right to left, from apparently unaltered granite to pegmatite contact, in a distance of 18 cm, the most conspicuous changes are an increase in muscovite and a decrease in K-feldspar abundance. An increase in apatite and tourmaline is also apparent near the pegmatite. The increase

---

Figure 5.71 Changes in volume % of minerals in granite, in relation to distance (x) from a pegmatite dike (sample M1). From Kretz, Hartree and Jones (1989). Reproduced by permission of Springer-Verlag

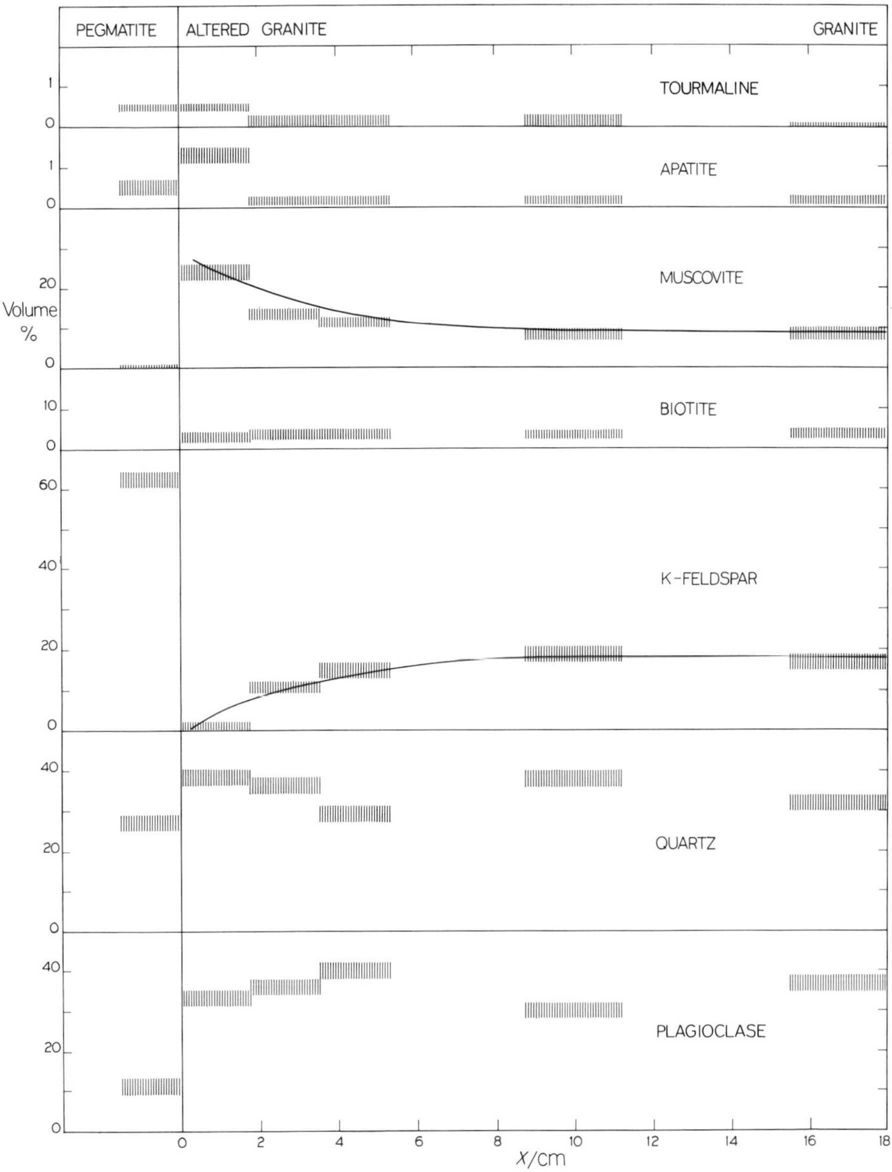
PEGMATITE
ALTERED GRANITE
GRANITE
TOURMALINE
APATITE
MUSCOVITE
BIOTITE
K-FELDSPAR
QUARTZ
PLAGIOCLASE
Volume %
X/cm
0
2
4
6
8
10
12
14
16
18
1
0
20
10
60
40

in muscovite is gradational, and the profile resembles the diffusion profiles of Figure 4.11 (p. 282); the volume concentration (C_{Ms}) versus distance (x) relation can indeed be expressed by use of the same kind of error-function equation,

$$C_{Ms} = 10 + 15[1 - \text{erf}(0.20x)], \tag{137}$$

where C_{Ms} is volume % muscovite and x is distance in centimeters from the pegmatite contact.

Chemical analyses of unaltered granite (x = 6 to 13 cm from the pegmatite contact, referred to as zone 3), partly altered granite (x = 2 to 6 cm, zone 2) and most altered granite (x = 0 to 2 cm, zone 1) are presented in Table 5.10, where we see that, assuming only minor changes in density and volume, little or no addition or removal occurred for most elements; the exceptions are K (removed), P (added), H (added), and B (added).

An estimate of the number of atoms added or removed from 1 g of granite during metasomatism is illustrated in Table 5.11. First, measurements of density

Table 5.10 Chemical analyses of pegmatite, unaltered granite (3), partly altered granite (2) and most-altered granite (1) is sample M1, Figure 5.71

	Pegmatite	1	2	3
SiO_2/wt%	69.3	74.1	73.6	73.4
TiO_2	0.00	0.13	0.13	0.14
Al_2O_3	16.0	14.6	14.5	14.6
Fe_2O_3		0.11	0.12	0.15
FeO	0.12	0.96	1.00	0.96
MnO	0.02	0.04	0.04	0.03
MgO	0.00	0.24	0.21	0.23
CaO	0.41	0.64	0.63	0.64
Na_2O	2.00	3.56	3.81	3.66
K_2O	10.5	3.03	3.78	4.52
P_2O_5	0.85	0.51	0.35	0.30
H_2O		1.18	0.80	0.70
sum		99.1	99.0	99.3
Li/ppm	80	370	300	320
Be	80	7	8	7
B	150	60	15	17
Zn	34	98	93	67
Rb	830	390	330	340
Sr	70	30	70	60
Zr	<10	70	70	70
Ba	260	190	270	320
$d/(g/cm^3)$		2.695	2.665	2.653
x/cm		0–2	2–6	6–13

d density
x distance from pegmatite contact
Zone 1: most-altered granite
2: partly altered granite
3: unaltered granite

Table 5.11 Calculated addition (+) and removal (−) of elements in the formation of muscovite zones adjacent granite (sample M1)

	$R^{\alpha/\beta}$	$R^{\alpha/\beta}D^{\alpha/\beta}$	$\pm 3s$	$(m_A)_{est} = 1.00x^{\beta} - x^{\alpha}$	$(m_A)_{est}$ μmol/g
Si	0.991	0.976	0.04		
Ti	1.08	1.06	0.06		
Al	1.00	0.985	0.03		
Fe^{3+}	1.36	1.34	0.4		
Fe^{2+}	1.00	0.985	0.03		
Mn	0.77	0.76	0.3		
Mg	0.959	0.945	0.08		
Ca	1.00	0.985	0.03		
Na	1.03	1.02	0.10		
K	1.49	1.47	0.05	−0.0123	−320
P	0.59	0.58	0.4	+0.0007	+23
H	0.591	0.582		+0.00054	+540
Li	0.86	0.85	0.4		
Be	1.00	0.99	0.4		
B	0.283	0.279	0.4	+0.000043	+4
Zn	0.68	0.67	0.4		
Rb	0.92	0.91	1.5		
Sr	2.0	1.97	0.4	−0.00003	−1/2
Zr	1.0	0.99	0.4		
Ba	1.68	1.65	0.4	−0.00013	−1

α = zone 3, β = zone 1 (Table 5.10)

$D^{\alpha/\beta} = 2.653/2.695 = 0.985$

Chosen $(R^{\alpha/\beta}D^{\alpha/\beta}) = 0.985$(Al, Fe^{2+}, Ca) $= V_{est}^{\beta/\alpha}$

then $V_{est}^{\beta/\alpha}/D^{\alpha/\beta} = 0.985/0.985 = 1.0$

for zones 3 to 1 (2.653, 2.665, 2.695) indicate a small increase in density, with $D^{\alpha/\beta}(\alpha = \text{zone } 3, \beta = \text{zone } 1) = 0.985$. Then, for each of the 20 elements, $R^{\alpha/\beta}$ is calculated (column 2) and also $(R^{\alpha/\beta}D^{\alpha/\beta})$ (column 3). From equation (136), $(R^{\alpha/\beta}D^{\alpha/\beta})$ for those elements that remained stationary could provide an estimate of $V^{\beta/\alpha}$. We find that for each of Al, Fe^{2+}, and Ca, $(R^{\alpha/\beta}D^{\alpha/\beta}) = 0.985$, and assuming that these elements remained stationary, then $(V^{\beta/\alpha}/D^{\alpha/\beta})$ in equation (135) is equal to $0.985/0.985 = 1.00$. This means that a small increase in density was compensated by a small decrease in volume; hence, from equation (135), additions and removals of elements (in grams) are obtained, simply, as

$$m_A = x_A^{\beta} - x_A^{\alpha}.$$

It is important to note that when analytical precision is considered (column 4 in Table 5.11), many of the $(R^{\alpha/\beta}D^{\alpha/\beta})$ numbers in column 3 do not differ significantly from 0.985. Rather broad limits ($\pm 3 \times$ standard deviation, analytical error) were chosen, to allow for the possibility that the composition of the granite at the site of metasomatism was, prior to metasomatism, not exactly as that in unaltered granite in zone 3. Thus, addition or removal can be demonstrated for six elements only, and the changes for these elements, expressed as micromoles per 1 g of original rock, are listed in the last column of Table 5.11. The progressive changes in potassium, phosphorus, and hydrogen are shown in Figure 5.72, and these profiles can also be approximated by error-function equations.

In another and very similar sample (M2), minor Ti, Fe, and Ca were evidently introduced to the muscovite zones, in addition to the changes described above, and some Na was removed. Combining this information with chemical analyses of muscovite, K-feldspar, and plagioclase, it is possible to express the metasomatic change in terms of a mineral reaction, as follows:

$$\begin{aligned} &660 \text{ K-feldspar} + 24 \text{ biotite} + 190 \text{ plagioclase} \\ &\quad + (770\text{H} + 36\text{P} + 3\text{Ti} + 13\text{Fe} + 13\text{Ca}) \\ &\rightarrow 400 \text{ muscovite} + 1100 \text{ quartz} + 11 \text{ apatite} \\ &\qquad + (240\text{Na} + 260\text{K}), \end{aligned} \tag{138}$$

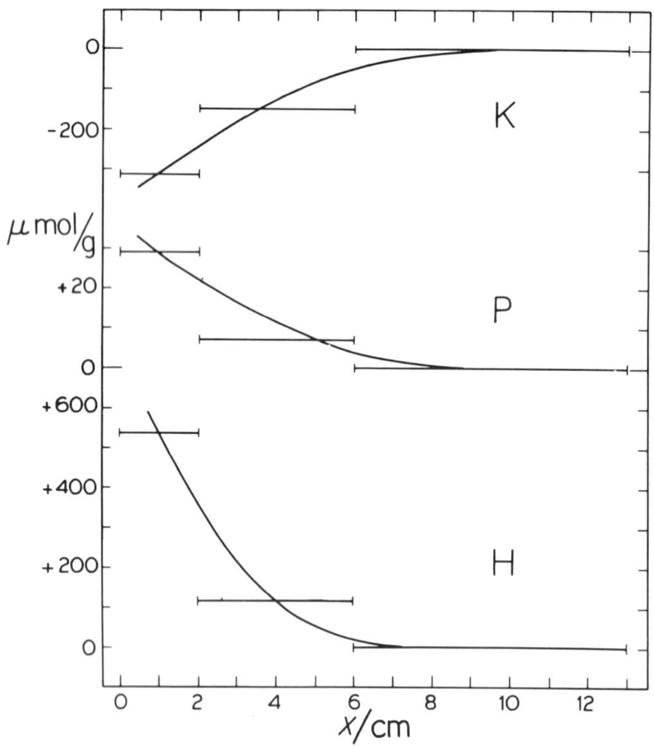

Figure 5.72 Calculated addition (+) and removal (−) of hydrogen (H), phosphorus (P), and potassium (K) in altered granite (sample M1, Figure 5.71), expressed as micromoles per gram of unaltered granite. From Kretz, Hartree and Jones (1989). Reproduced by permission of Springer-Verlag

Table 5.12 Balance sheet for reaction (7) in M2. Units are μmols/gram of unaltered granite (α)

	Unaltered granite (α)				Altered granite (β)				Sum	
	660 Kf	24 Bt	190 Pl	(+)	400 Ms	1100 Qtz	11 Ap	(−)	α	β
Si	1990	72	560		1230	1100			2620	2320
Al	670	24	202		1100				900	1100
Ti		4		3	11				7	11
Fe		50		13	38				63	38
Mg		18			32				18	32
Ca	1		7	13	1		55		21	56
Na	25	2	179		28			240	210	270
K	603	22	1		350			260	630	610
P				36			34		36	34
H		48		770	800				820	810

α = unaltered granite, β = metasomatically altered granite
(+), (−): estimated addition and removal, using the procedure illustrated in Table 5.11
Kf K-feldspar, Bt biotite, Pl plagioclase, Ms muscovite, Qtz quartz, Ap apatite.

where numbers are μmol relative to 1.0 g of original granite. Table 5.12 is a balance sheet for this metasomatic reaction. Equation (138) may be compared with the Hemley and Jones (1964) equation,

$$\underset{\text{K-feldspar}}{1.5\text{KAlSi}_3\text{O}_8} + \text{H}^+ \rightarrow \underset{\text{muscovite}}{0.5\text{KAl}_3\text{Si}_3\text{O}_{10}(\text{OH})_2} + \underset{\text{quartz}}{3\text{SiO}_2} + \text{K}^+, \tag{139}$$

which was studied experimentally at 500 °C and 2.4 kbar, i.e., under conditions similar to those estimated for the natural reaction. The essence of both reactions is the crystallization of muscovite from K-feldspar, resulting from an increase in the activity of H^+ relative to K^+, i.e., a decrease in the ratio $a(K^+)/a(H^+)$. In the natural reaction, additional changes include the breakdown of minor biotite to provide Mg, Fe, and Ti for muscovite and tourmaline, and the less-certain breakdown of minor plagioclase to provide Ca for apatite.

Turning now to the tourmaline-rich zones adjacent to pegmatite dikes in schist (Figure 5.70b), they are more variable in width than the muscovite zones, and are also, locally, associated with quartz veins. In general, the transition from schist (plagioclase, quartz, biotite) to tourmaline rock (tourmaline, quartz, apatite) is gradational, as in Figure 5.73, from right to left. In another sample (T2) volume percent of minerals was determined as a function of distance, and the results are presented in Figure 5.74. At this location the metasomatic reaction obviously consumed biotite, plagioclase, and boron, and produced tourmaline and quartz, and within a narrow zone, 0 to 6 mm from the pegmatite contact, the reaction progressed nearly to completion. As in Figure 5.71, the relation of volume percent of product (in this case, tourmaline) to distance (x', in mm) can be expressed by

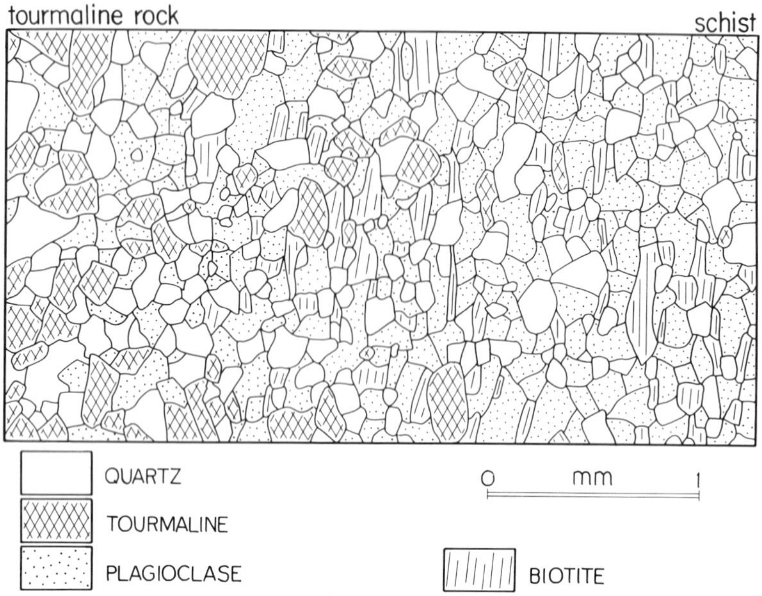

Figure 5.73 Microstructure to illustrate the progressive and pervasive metasomatic crystallization of tourmaline and quartz at the expense of biotite and plagioclase (right margin to left margin); pegmatite dike lies immediately to the left. From Kretz, Hartree and Jones (1989). Reproduced by permission of Springer-Verlag

use of an error-function equation,

$$C_{\mathrm{Trm}} = 50[1 - \mathrm{erf}(0.157x')]. \tag{140}$$

Here distance (x') is set at zero where $x = 7.5$ mm (Figure 5.74).

In another, larger, sample (T3) taken from the edge of a dike, 76 cm wide, chemical analyses of unaltered schist (7 to 9 cm from the pegmatite contact, zone 3) and most altered schist (0 to 2 cm from the contact, zone 1) produced the results of Table 5.13, where mean mineral proportions are also indicated. It is obvious that even if rather large changes in volume are assumed, Ca, P, Zn, and B were added during metasomatism, and K and Rb were removed. Applying now the procedure described above (Table 5.11), based on the Gresens equation, the densities of zones 3 and 1 in sample T3 were found to be 2.761 and 2.884 g/cm^3 respectively, giving $D^{\alpha/\beta}(\alpha = \text{zone } 3, \beta = \text{zone } 1) = 0.9574$. Assuming (initially) that $V^{\beta/\alpha} = 1.0$ (no volume change) then $(R^{\alpha/\beta}D^{\alpha/\beta})$ could provide an identification of elements that were added or removed. Many of these numbers depart greatly from 1.0 and to regard any of the corresponding elements as stationary would imply a very large change in volume. But for Si, Al, and Mg, $(R^{\alpha/\beta}D^{\alpha/\beta})$ is very close to 1.02, and by taking $(R^{\alpha/\beta}D^{\alpha/\beta}) = 1.02 = V^{\beta/\alpha}$ (implying a slight increase in volume), then

$$V^{\beta/\alpha}/D^{\alpha/\beta} = 1.02/0.9574 = 1.065.$$

Table 5.13 Data on unaltered schist (3) and tourmaline schist (1) in T3, and estimate of addition (+) and removal (−) of elements $(m_A)_{est}$

	Pegmatite	Zone 1	Zone 3		$(m_A)_{est}$ μmol/g
SiO_2/wt%	71.8	60.1	63.5	Si	
TiO_2		0.56	0.61	Ti	
Al_2O_3	16.1	14.2	15.1	Al	
FeO	0.21	5.72	5.71	Fe	+75
MnO		0.15	0.08	Mn	+11
MgO		2.77	3.10	Mg	
CaO	0.66	4.09	2.77	Ca	+280
Na_2O	6.0	0.57	2.89	Na	−740
K_2O	3.1	1.03	3.05	K	−410
P_2O_5	1.0	3.82	0.33	P	+530
				B	+1100
Zn/ppm	30	440	130	Zn	+3
Rb	470	330	890	Rb	−7
Sr	90	140	370	Sr	−3
Zr	10	130	140	Zr	
Ba	70	230	570	Ba	−2
Quartz/volume %		43	24		
Plagioclase		—	43		
Biotite		12	32		
Tourmaline		40	—		
Apatite		5	1		
d/g cm^{-3}		2.884	2.761		
x/cm		0 to 2	7 to 9		

d: density
x: distance from pegmatite contact
Total iron expressed as FeO
Boron addition estimated from tourmaline abundance

On this argument and by use of equation (135), m_A is

$$m_A = 1.065x_A^{\beta} - x_A^{\alpha},$$

and the addition or removal for all elements can be calculated. The results, in terms of micromoles relative to 1 g of unaltered schist appear in the last column of Table 5.13. We find that if Si and Al were indeed stationary, or nearly so, then atoms of Fe, Mn, Ca, P, B, and Zn were added and atoms of Na, K, Rb, Sr, and Ba were removed.

In another sample (T4) where the tourmaline zone is only 6 mm wide and where mineral volume proportions and variations are similar to those described above, chemical analyses of tourmaline, biotite, and plagioclase were used, together with an assumption of negligible change in rock volume, to construct the reaction equation

$$\begin{aligned} &700 \text{ biotite} + 960 \text{ plagioclase} + (1500\text{B} + 600\text{H} + 100\text{P}) \\ &\rightarrow 490 \text{ tourmaline} + 900 \text{ quartz} + 33 \text{ apatite} + (400\text{Na} + 640\text{K}), \qquad (141) \end{aligned}$$

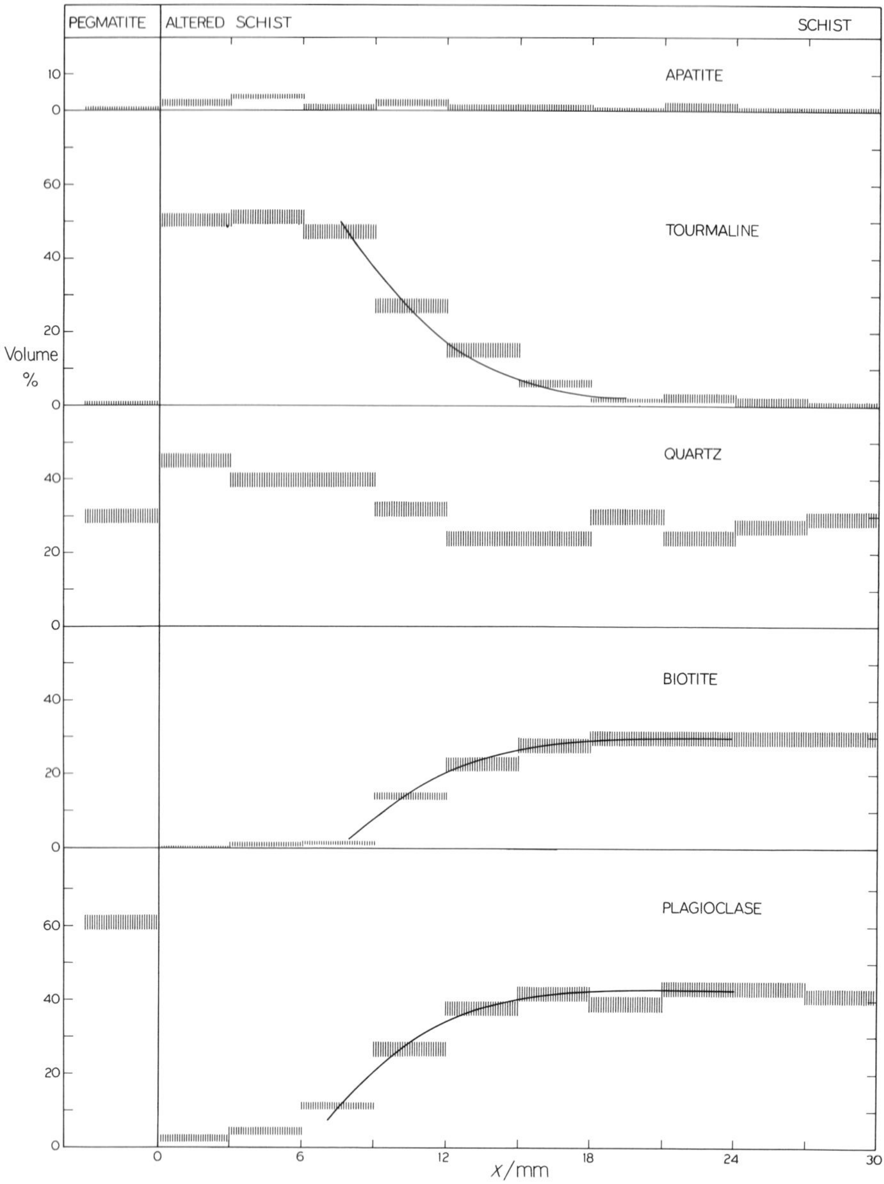
PEGMATITE
ALTERED SCHIST
SCHIST
APATITE
TOURMALINE
QUARTZ
BIOTITE
PLAGIOCLASE
Volume %
X/mm
0
6
12
18
24
30
10
20
40
60

where numbers are μmol relative to 1 g of unaltered rock. The mineral formulas are as follows:
Biotite:

$$(K_{0.92}Na_{0.02}Ca_{0.01})(Mg_{1.01}Fe_{1.32}Mn_{0.02}Ti_{0.12}Al_{0.60})Al_{1.19}Si_{2.81}O_{22}(OH_{0.9}F_{0.10})_2,$$
$$Mg/(Mg + Fe) = 0.43$$

Plagioclase:
core: $Na_{0.74}Ca_{0.27}K_{0.004}Al_{1.32}Si_{2.69}O_8$, $Na/(Na + Ca) = 0.73$
rim: $Na_{0.83}Ca_{0.21}K_{0.005}Al_{1.19}Si_{2.80}O_8$, $Na/(Na + Ca) = 0.80$

Tourmaline:
core: $Na_{0.56}Ca_{0.17}Mg_{1.72}Fe_{0.90}Ti_{0.11}Al_{6.52}\ Si_{6.00}O_{18}[(OH)_{0.96}F_{0.04}]_4$
$$Mg/(Mg + Fe) = 0.65$$
$$Na/(Na + Ca) = 0.77$$
rim: $Na_{0.65}Ca_{0.11}Mg_{1.59}Fe_{1.03}Ti_{0.10}Al_{6.55}\ Si_{6.0}O_{18}[(OH)_{0.96}F_{0.04}]_4$
$$Mg/(Mg + Fe) = 0.61$$
$$Na/(Na + Ca) = 0.86$$

Equation (141) is balanced approximately for most major elements and for P, B, F, H. The principal atomic relocations are as follows; from biotite: Mg, Fe, Ti, Al, Si, F to tourmaline, with the release and escape of K; from plagioclase: Ca to apatite, Al to tourmaline, Si to additional quartz, and Na, partly to tourmaline, the remainder removed. Notice that the Mg:Fe proportions in biotite and tourmaline are not equal, implying a minor gain of Mg or loss of Fe.

Although several elements were involved in the metasomatic processes that were described in the present section, the principal ones are hydrogen, boron, phosphorus, potassium, and sodium. Where metasomatism was most intense, some 600 μmol of H were introduced to 1 g of granite, as 300 μmol of K were removed, and some 1100 μmol of B and 400 μmol of P were introduced to 1 g of schist as 1200 μmol of K and Na were removed. These migrations were apparently related to pegmatite emplacement, but the uniform width of the muscovite zones, irrespective of dike width, and the erratic distribution of the tourmaline zones raise the possibility that metasomatism occurred immediately prior to pegmatite emplacement. There is no doubt, however, that the mobile elements penetrated the granite or schist from the plane that marks the interface between metasomatic rock and pegmatite.

Although the mechanism of transport of hydrogen and boron, and the ionic or molecular forms that these atoms assumed during transport are not known, the nature of the concentration–distance profiles (Figures 5.71, 5.72, 5.74) and the finding that the profiles can be described by use of diffusion equations, are viewed as evidence that transport occurred by diffusion. If this is true, then we have in the

---

Figure 5.74 Changes in volume % of minerals in schist, in relation to distance from a pegmatite dike (sample T2). From Kretz, Hartree and Jones (1989). Reproduced by permission of Springer-Verlag

metasomatic zones, examples of diffusion accompanied by chemical reaction and partial immobilization, as discussed in Sec. 4.3.6.

The following sequence of events is proposed, referring, for example, to the crystallization of tourmaline in schist, as in equation (141), which in simplified form is

$$\text{biotite} + \text{plagioclase} + \text{B} \rightarrow \text{tourmaline} + \text{quartz} + \text{K}.$$

(1) The presence of boron at the plane that is now the pegmatite contact, increased rapidly to a set value, causing boron to penetrate the schist on crystal boundaries, producing within the crystal-boundary phase immediately at the contact plane, a fixed boron concentration, C'.

(2) The further penetration of schist by boron then occurred by the intrinsic crystal-boundary diffusion of boron, as neutral molecular species. Reaction to produce tourmaline has not yet started, and the concentration (C) versus distance (x) profile at time (t) was of the form

$$C = C'\left[1 - \operatorname{erf}\left(\frac{x}{2(D't)^{1/2}}\right)\right], \tag{142}$$

(Sec. 4.3.2) where D' is the crystal-boundary diffusion coefficient.

(3) As the concentration of boron in the boundary phase at any point, x, increased to a low threshold number, C_0, the tourmaline-forming reaction was initiated, resulting in the partial immobilization of boron. Reaction was initiated earlier or later, depending on distance from the source (x).

(4) At any point along x, and within a cross-sectional segment of width Δx, the rate of reaction, i.e., the rate of removal of boron was approximately proportional to the concentration of boron in the boundary phase (C) within that segment, i.e.,

$$s = RC, \tag{143}$$

where s is the concentration of immobilized boron (present in tourmaline) in the segment, and R is a constant. Then from Sec. 4.3.6, the appropriate diffusion equation is

$$\frac{\partial C}{\partial t} = \left(\frac{D'}{R+1}\right)\frac{\partial^2 C}{\partial x^2}, \tag{144}$$

where D' is the diffusion coefficient in the absence of immobilization, and $(D'/(R+1))$ is the 'effective' diffusion coefficient, for that fraction of boron that is not immobilized by tourmaline crystallization. For the conditions stated above, i.e., $C = C'$ for $x = 0$, at $t \geq 0$, the solution to (144) is the same as that for the second equation of diffusion (given by equation (142)) but with D' replaced by $D'/(R+1)$. By substitution in (143) we have

$$s = RC'\left[1 - \operatorname{erf}\left(\frac{x}{2\{D'/(R+1)\}^{1/2}t^{1/2}}\right)\right]. \tag{145}$$

Because R is constant, this equation is analogous to (142) and, when plotted, produces similar concentration profiles.

(5) After a period of time, C' decreased to zero, leaving a 'stranded' profile in terms of immobilized boron (s) or tourmaline abundance, expressible by an error-function equation, in agreement with observation.

We have examined, above, a few examples of metasomatism on a small scale. From a survey of recent studies it is clear that both small- and large-scale metasomatic processes are complex and not fully understood. As a result, different interpretations are commonly stated regarding the mechanism of atomic transport. Several theoretical models have been constructed, and a need now exists for experimental investigations, particularly on chemical reaction accompanying mass transfer. Detailed studies of natural occurrences of metasomatic crystallization should also be encouraged.

Appendices

A1. QUANTITIES AND UNITS

Quantity	Unit name	Unit symbol	Note
length	meter	m	
	centimeter	cm	
	ångström	Å	1 Å = 10^{-8} cm
mass	kilogram	kg	
	gram	g	
time	second	s	
	year	y	1 y = 3.16×10^7 s
temperature	kelvin	K	*
	degree celsius	°C	*
amount	mole	mol	**
	gram-formula weight	gfw	**
force	newton	N	
pressure	pascal	Pa	1 Pa = 1 N/m^2
	megapascal	MPa	1 MPa = 10^6 Pa
	bar	bar	1 bar = 10^5 Pa
	kilobar	kbar	1 kbar = 100 MPa
energy***	joule	J	1 J = 1 N · m
	erg	erg	1 erg = 10^{-7} J
	calorie	cal	1 cal = 4.186 J
entropy***	joule/kelvin	J/K	
volume***	cubic centimeter	cm^3	
	joule/bar	J/bar	1 J/bar = 10 cm^3

*Temperature (symbol T) as used in thermodynamic expressions refers to the absolute scale (K) where the unit of temperature (1 kelvin) is defined as 1/273.16 of the triple point of water. When reporting experimental and natural data, symbol T is often used for temperature on the Celsius scale (°C). The units of temperature on both scales are equal. The conversion is, x °C = x + 273.15 K, e.g., 500 °C = 773 K.

**1 mole of a substance contains the same number of atoms (or molecules or formula units) as in 12 g of pure carbon 12, e.g., 1 mole of H_2O = 18.016 g H_2O (Appendix A5). In general, the atoms in minerals (e.g., Mg_2SiO_4) do not occur as molecules, and the term gram-formula weight (gfw), which is analogous to the mole, is sometimes used. For example, 1 gfw of forsterite, expressed as Mg_2SiO_4 = 140.70 g of forsterite, and 1 gfw of forsterite, expressed as $MgSi_{0.5}O_2$ = 70.35 g of forsterite.

***Symbols G (Gibbs energy), H (enthalpy), S (entropy), and V (volume) in this book refer to one mole or one gram-formula weight of a phase, with some (obvious) exceptions.

A2. SI PREFIXES (APPLIED, FOR EXAMPLE, TO LENGTH)

1	gigameter	Gm	$= 10^9$ m
1	megameter	Mm	$= 10^6$ m
1	kilometer	km	$= 10^3$ m
1	meter	m	$= 10^0$ m
1	decimeter	dm	$= 10^{-1}$ m
1	centimeter	cm	$= 10^{-2}$ m
1	millimeter	mm	$= 10^{-3}$ m
1	micrometer	μm	$= 10^{-6}$ m
1	nanometer	nm	$= 10^{-9}$ m

A3. CONSTANTS

N_A	Avogadro number	6.022×10^{23} mol^{-1}
R	Gas constant	8.314 J $\cdot$ mol$^{-1} \cdot$ K^{-1}
		1.99 cal $\cdot$ mol$^{-1} \cdot$ K^{-1}
k	Boltzmann constant	1.3807×10^{-23} J $\cdot$ K^{-1}
h	Planck constant	6.626×10^{-34} J $\cdot$ s

A4. QUANTITY CALCULUS (McGLASHAN 1973, GUGGENHEIM 1977)

Note the following equation:

symbol for physical quantity = number × symbol for unit.

Example:

$$r = 5 \text{ mm } (r \text{ stands for radius of a sphere}),$$
$$r = 5 \times \text{mm},$$
$$r/\text{mm} = 5.$$

Example:

$$T = 900 \text{ K},$$
$$T = 900 \times \text{K},$$
$$T/\text{K} = 900.$$

Example:

$$T = 900 \text{ K},$$
$$T = 900 \times \text{K},$$
$$\frac{1}{T} = \frac{1}{900} \times \frac{1}{\text{K}},$$
$$\frac{1000}{T} = \frac{1000}{900} \times \frac{1}{\text{K}},$$
$$1000 \text{ K}/T = 1.11.$$

This notation is useful in figures and tables.

A5. ATOMIC AND MOLECULAR WEIGHTS

Aluminum	Al	26.982	Al_2O_3	101.96
Barium	Ba	137.33	BaO	153.33
Beryllium	Be	9.01	BeO	25.01
Boron	B	10.81	B_2O_3	69.62
Calcium	Ca	40.08	CaO	56.08
Carbon	C	12.01	CO_2	44.01
Cesium	Cs	132.91		
Chlorine	Cl	35.45		
Chromium	Cr	52.00		
Fluorine	F	19.00		
Gallium	Ga	69.72		
Germanium	Ge	72.59		
Hafnium	Hf	178.49		
Helium	He	4.003		
Hydrogen	H	1.008	H_2O	18.016
Indium	In	114.82		
Iodine	I	126.90		
Iron	Fe	55.847	FeO	71.85
			Fe_2O_3	159.70
Krypton	Kr	83.80		
Lithium	Li	6.94	Li_2O	29.88
Magnesium	Mg	24.305	MgO	40.31
Manganese	Mn	54.94	MnO	70.94
			Mn_2O_3	157.88
Neon	Ne	20.18		
Nickel	Ni	58.70		
Niobium	Nb	92.91		
Nitrogen	N	14.01		

Oxygen	O	16.00		
Phosphorus	P	30.97		
Potassium	K	39.098	K_2O	94.20
Rubidium	Rb	85.468	Rb_2O	186.94
Scandium	Sc	44.96		
Silicon	Si	28.086	SiO_2	60.09
Sodium	Na	22.99	Na_2O	61.98
Strontium	Sr	87.62	SrO	103.62
Sulfur	S	32.06		
Thorium	Th	232.04		
Tin	Sn	118.69		
Titanium	Ti	47.90	TiO_2	79.90
Tungsten	W	183.85		
Uranium	U	238.03		
Vanadium	V	50.94		
Yttrium	Y	88.91		
Zinc	Zn	65.38		
Zirconium	Zr	91.22	ZrO_2	123.22

A6. IONIC RADIUS (WHITTAKER AND MUNTUS 1970)

	Coordination	Radius/Å
Al^{3+}	4	0.47
	6	0.61
B^{3+}	4	0.20
Ba^{2+}	8	1.50
Be^{2+}	4	0.35
Ca^{2+}	6	1.08
	8	1.20
Cr^{3+}	6	0.70
Cs^{+}	12	1.96
F^{-}	4	1.23
Fe^{2+}	6	0.69 LS
	6	0.86 HS
Fe^{3+}	6	0.63 LS
	6	0.73 HS
K^{+}	8	1.59
	12	1.68
Li^{+}	6	0.82
Mg^{2+}	6	0.80

	Coordination	Radius/Å
Mn^{2+}	6	0.75 LS
	6	0.91 HS
Na^{+}	6	1.10
	8	1.24
Ni^{2+}	6	0.77
O^{2-}	4	1.30
	6	1.32
	8	1.34
Rb^{+}	8	1.68
	12	1.81
Sc^{3+}	6	0.83
Si^{4+}	4	0.34
Sr^{2+}	8	1.33
Ti^{3+}	6	0.75
Ti^{4+}	6	0.69
V^{3+}	6	0.72
Y^{3+}	6	0.98
Zn^{2+}	6	0.83
Zr^{4+}	6	0.80

LS low spin, HS high spin

A7. EARTH DATA (KRAUSKOPF 1979)

Mean radius	6371 km
Radius of Core	3471 km
Mass	5.98×10^{27} g
Volume	1.083×10^{27} cm^3
Density	
Mean	5.52 g/cm^3
Mean for Crust	2.8 g/cm^3
Mean for Mantle	4.5 g/cm^3
Mean for Core	10.7 g/cm^3
Age	4.5×10^{9} y

A8. GREEK ALPHABET

Greek letter	Greek name	English equivalent
A α	Alpha	a
B β	Beta	b
Γ γ	Gamma	g
Δ δ	Delta	d
E ϵ	Epsilon	ĕ
Z ζ	Zeta	z
H η	Eta	ē
Θ θ	Theta	th
I ι	Iota	i
K κ	Kappa	k
Λ λ	Lambda	l
M μ	Mu	m
N ν	Nu	n
Ξ ξ	Xi	x
O o	Omicron	ŏ
Π π	Pi	p
P ρ	Rho	r
Σ σ	Sigma	s
T τ	Tau	t
Υ υ	Upsilon	u
Φ ϕ	Phi	ph
X χ	Chi	ch
Ψ ψ	Psi	ps
Ω ω	Omega	ō

A9. PROPAGATION OF ERROR

Ideally, every number, for example, the weight % of Mg or MgO in a crystal or rock, should be accompanied by two statements, one expressing the accuracy (systematic error) and the other expressing the precision (random error). Accuracy can be evaluated by co-analyzing one or more standards or by analyzing some samples by two or more different methods (Youden 1959). Precision can be evaluated by calculating the standard deviation estimate (s) from a large number (n) of determinations (x) of a single sample,

$$s^2 = \frac{\Sigma(x - \overline{x})^2}{n - 1}, \tag{1}$$

where $\overline{x}$ is the mean value of x. The standard deviation can also be estimated when duplicate determinations (x_1, x_2) are available for many (N) similar samples,

$$s^2 = \frac{\Sigma(x_1 - x_2)^2}{2N}. \tag{2}$$

When s is well known from past experience, then the confidence limits ($\pm c$) to a single determination can be written:

$$x \pm 1.96s \qquad (0.95 \text{ confidence}).$$

But when two or more (n) repeated measurements are obtained, the confidence limits ($\pm c$) to the mean of these determinations ($\overline{x}$) are smaller by a factor of ($1/n^{1/2}$),

$$\overline{x} \pm (1.96s)/n^{1/2} \qquad (0.95 \text{ confidence}),$$

(Day and Underwood 1974). When confidence limits are calculated in this way, it becomes apparent that, in general, not more than three figures (e.g., 2.39, 23.9) should be used in reporting chemical determinations.

Commonly, the numbers obtained by chemical analysis (e.g., x, y) are combined to form certain sums, products, or quotients, here referred to as 'concentration variables' (u). Propagation-of-error calculations show how uncertainty in each of x and y (expressed as s_x and s_y) are propagated to u. In general, if u is a function of x and y,

$$u = f\,(x, y), \tag{3}$$

then the uncertainty in u, expressed as s_u is obtained from the equation

$$s_u^2 = \left(\frac{\partial u}{\partial x}\right)^2 s_x^2 + \left(\frac{\partial u}{\partial y}\right)^2 s_y^2, \tag{4}$$

provided that s_x and s_y are uncorrelated. Solutions to equation (4) give the following results (Kretz 1985):

$u = kx$ (k = constant):

$$s_u = ks_x, \tag{5}$$

$u = x + y$:

$$s_u = \left[s_x^2 + s_y^2\right]^{1/2}, \tag{6}$$

$u = x/y$:

$$s_u = u\left[(s_x^2/x^2) + s_y^2/y^2)\right]^{1/2}, \tag{7}$$

$u = x/(x + y)$:

$$s_u = \frac{\left[(y^2 s_x^2 + x^2 s_y^2)\right]^{1/2}}{(x + y)^2}, \tag{8}$$

$u = x/(x + y + z)$:

$$s_u = \frac{\left[(y + z)^2 s_x^2 + x^2 (s_y^2 + s_z)^2\right]^{1/2}}{(x + y + z)^2}. \tag{9}$$

Detailed discussions of the treatment of numerical data are presented by Agterberg (1974) and Le Maitre (1982).

References

Note: Adv. Phys. Geochem. refers to the series 'Advances in Physical Geochemistry' published by Springer Verlag, and Rev. Miner. refers to 'Reviews in Mineralogy' from the Mineralogical Society of America.

Adams, F.D. 1938. *The birth and development of geological sciences.* Dover (1954), New York.

Adams, L.H. 1953. A note on the stability of jadeite. Am. J. Sci. 251, 299-308.

Agrell, S.O., Bown, M.G. and McKie, D. 1965. Deerite, howieite, and zussmanite, three new minerals from the Franciscan of Laytonville District, Mendocino Co., California. Am. Miner. 50, 278.

Agrell, S.O., Long, J.V.P. and Ogilvie, R.E. 1963. Nickel content of kamacite near the interface with taenite in iron meteorites. Nature 198, 749-750.

Agterberg, F.P. 1974. *Geomathematics.* Elsevier, Amsterdam.

Agullo-Lopez, F., Catlow, C.R.A. and Townsend, P.D. 1988. *Point defects in materials.* Academic Press, London.

Albee, A.L., 1965. Phase equilibria in three assemblages of kyanite-zone pelitic schists, Lincoln Mountain Quadrangle, Central Vermont. J. Petrology 6, 246-301.

Albee, A.L., 1968. Metamorphic zones in Northern Vermont. In *Studies of Appalachian geology*, ed. by E-An Zen, W.S. White, J.B. Hadley, J.B. Thompson Jr. Interscience, New York.

Albee, A.L. and Chodos, A.A. 1969. Minor element content of coexisting Al_2SiO_5 polymorphs. Am. J. Sci. 267, 310-316.

Albee, A.L., Gancarz, A.J. and Chodos, AA. 1973. Metamorphism of Apollo 16 and 17 and Luna 20 metaclastic rocks at about 3.95 AE. Proc. 4th Lunar Sci. Conf., Geochim. Cosmochim. Acta, Suppl. Four, 1, 569-595.

Althaus, E., Karotke, E., Nitsch, K.H. and Winkler, H.G.F. 1970. An experimental reexamination of the upper stability of muscovite plus quartz. Neues Jb. Miner. Mh. 7, 325-336.

Amelinckx, S. 1952. La croissance hélicoidale de cristaux de biotite. Comp. Rend. Acad. Sci. Paris 234, 971-973.

Amelinckx, S., Delavignette, P. and Heerschap, M. 1965. Dislocations and stacking faults in graphite. Chem. Phys. Carbon, 1, 1-71, Marcel Dekker, New York.

Andersen, D.J. and Lindsley, D.H. 1988. Internally consistent solution models for Fe-Mg-Mn-Ti oxides: Fe-Ti oxides. Am. Miner. 73, 714-726.

Andersen, T. 1989. Carbonatite-related contact metasomatism in the Fen complex, Norway: effects and petrogenetic implications. Miner. Mag. 53, 395-414.

Anderson, P.J. and Horlock, R.F. 1962. Thermal decomposition of magnesium hydroxide Trans. Faraday Soc. 58, 1993-2004.

Angino, E.E. and Billings, G.K. 1972. *Atomic absorption spectrometry in geology.* Elsevier, Amsterdam.

Annersten, H. 1968. A mineral chemical study of a metamorphosed iron formation in Northern Sweden. Lithos 1, 374-397.

Annersten, H. and Ekström, T. 1971. Distribution of major and minor elements in coexisting minerals from a metamorphosed iron formation. Lithos 4, 185-204.

Anovitz, L.M. and Essene, E.J. 1987. Phase equilibria in the system $CaCO_3$-$MgCO_3$-$FeCO_3$. J. Petrology 28, 389-414.

Anovitz, L.M., Essene, E.J. and Dunham, W.R. 1988. Order-disorder experiments in orthopyroxenes: implications for the orthopyroxene geospeedometer. Am. Miner. 73, 1060-1073.

Appleyard, E.C. 1969. Nepheline gneisses of the Wolfe Belt, Lyndoch Township, Ontario. II. Textures and mineral paragenesis. Can J. Earth Sci., 6, 689-717.

Aranovich, L. Ya. and Podlesskii, K.K. 1983. The cordierite-garnet-sillimanite-quartz equilibrium: experiments and applications. Adv. Phys. Geochem. 3, 173-198.

Aranovich, L. Ya., Lavrent'eva, I.V. and Kosyakova, N.A., 1988. Calibration of the biotite-garnet and biotite-orthopyroxene geothermometers corrected for the variable Al level in biotite. Geochemistry, 50-59.

Armbruster, T. and Oberhänsli, R. 1988. Crystal chemistry of double-ring silicates: structural, chemical, and optical variation in osumilites. Am. Miner. 73, 585-594.

Ashworth, J.R. 1985. *Migmatites*. Blackie, Glasgow.

Atkins, F.B. 1969. Pyroxenes of the Bushveld Intrusion, South Africa. J. Petrology 10, 222-249.

Atkinson, A. and Taylor, R.I. 1986. Impurity diffusion in NiO grain boundaries. J. Phys. Chem. Solids 47, 315-323.

Atlas, L. 1952. The polymorphism of $MgSiO_3$ and solid-state equilibria in the system $MgSiO_3$-$CaMgSi_2O_6$. J. Geol. 60, 125-147.

Attoh, K. and Vander Meulen, M.J. 1984. Metamorphic temperatures in the Michigamme formation compared with the thermal effect of an intrusion, northern Michigan. J. Geol. 91, 417-432.

Aveyard, R. and Haydon, D.A. 1973. *An introduction to the principles of surface chemistry*. Cambridge University Press.

Avrami, M. 1939. Kinetics of phase change. I. General theory. J. Chem. Phys. 7, 1103-1112.

Avrami, M. 1940. Kinetics of phase change II. Transformation-time relations for random distribution of nuclei. J. Chem. Phys. 8, 212-224.

Avrami, M. 1941. Kinetics of phase change. III. Granulation, phase change, and microstructure. J. Chem. Phys. 9, 177-184.

Azāroff, L.V. 1960. *Introduction to solids*. McGraw-Hill, New York.

Back, M.H. and Laidler, K.J. 1967. *Selected readings in chemical kinetics*. Pergamon Press, Oxford.

Baedecker, P.A. and McKown, D.M. 1987. Instrumental neutron activation analysis of geological samples. In *Methods for Geochemical Analysis*, ed. by P.A. Baedecker. U.S. Geol. Surv. Bull. 1770, H1-H14.

Baltatzis, E. 1970. Distribution of Fe and Mg between garnet and biotite in Scottish Barrovian metamorphic zones. Miner. Mag. 43, 155-157.

Barker, A.J. 1990. *Introduction to metamorphic textures and microstructures*. Blackie, Glasgow.

Barrer, R.M. and Fender, B.E.F. 1961. The diffusion and sorption of water in zeolites-II. Intrinsic and self-diffusion. J. Phys. Chem. Solids 21, 12-24.

Barrer, R.M. and Rees, L.V.C. 1964. Self- and exchange diffusion coefficients. J. Phys. Chem. Solids 25, 1035-1038.

Barrer, R.M. and Sammon, D.C. 1955. Exchange equilibria in crystals of chabazite. J. Chem. Soc., 2838-2849.

Barrer, R.M., Bartholomew, R.F. and Rees, L.V.C. 1963. Ion exchange in porous crystals. Part I. Self- and exchange-diffusion of ions in chabazites. J. Phys. Chem. Solids 24, 51-62.

Barrow, G. 1893. On an intrusion of muscovite-biotite gneiss in the south-eastern Highlands of Scotland, and its accompanying metamorphism. Quart. J. Geol. Soc. 49, 330-358.

Barth, T.F.W. 1934. Polymorphic phenomena and crystal structure. Am. J. Sci. 27, 273-286.

Barth, T.F.W. 1951. The feldspar geologic thermometers. N. Jb. Miner. Abh. 82, 143–154.
Barth, T.F.W. 1956. Studies in gneiss and granite. Skrift. Norske Vidensk.-Akad. Oslo, 1. Mat.-Nat. Kl., no. 1.
Barth, T.F.W. 1961. The feldspar lattices as solvents of foreign ions. Inst. 'Lucas Mallada', Cursillos y Conferencias, Fasc. 8, Madrid, pp. 3–8.
Barth, T.F.W. 1962. The feldspar geologic thermometers. Norsk Geol. Tidsskr. 42, (Supplement), 330–339.
Barth, T.F.W. 1969. *Feldspars*. J. Wiley & Sons, New York.
Bartholomé, P. 1961. Coexisting pyroxenes in igneous and metamorphic rocks. Geol. Mag. 98, 346–348.
Basu, A.R. and MacGregor, I.D. 1975. Chromite spinels from ultramafic xenoliths. Geochim. Cosmochim. Acta 39, 937–945.
Becke, F. 1913. Über Mineralbestand und Struktur der kristallinischen Schiefer. Denkschr. Akad. Wiss., Wien, 75, 1–53.
Becker, R. and Döring, W. 1935. Kinetische behandlung der keimbildung in übersättigten dämpfen. Ann. Phys. 24, 719–752.
Bell, P.M. and Nord, G. 1974. Microscopic and electron diffraction study of fibrolitic sillimanite. Carnegie Inst. Washington, Year Book 73, 444–448.
Bell, T.H. 1985. Deformation partitioning and porphyroblast rotation in metamorphic rocks: a radical reinterpretation. J. Metamorphic Geol. 3, 109–118.
Belonoshko, A.B. 1989. The thermodynamics of aqueous carbon dioxide fluid within thin pores. Geochim. Cosmochim. Acta 53, 2581–2590.
Bent, H.A. 1965. *The second law*. Oxford University Press, New York.
Beran, A., Rossman, G.R. and Grew, E.S. 1989. The hydrous component of sillimanite. Am. Miner. 74, 812–817.
Berger, A.S., Seidman, D.N. and Balluffi, R.W. 1973. A quantitative study of vacancy defects in quenched platinum by field ion microscopy and electrical resistivity-I. Experimental results. Acta Met. 21, 123–136.
Berman, R.G. 1988. Internally-consistent thermodynamic data for minerals in the system Na_2O-K_2O-CaO-MgO-FeO-Fe_2O_3-Al_2O_3-SiO_2-TiO_2-H_2O-CO_2. J. Petrology 29, 445–522.
Berman, R.G. 1990. Mixing properties of Ca-Mg-Fe-Mn garnets. Am. Miner. 75, 328–344.
Berman, R.G. and Brown, T.H. 1985. Heat capacity of minerals in the system Na_2O-K_2O-CaO-MgO-FeO-Fe_2O_3-Al_2O_3-SiO_2-TiO_2-H_2O-CO_2; representation, estimation, and high temperature extrapolation. Contr. Miner. Petrol. 89, 168–183.
Berman, R.G. and Koziol, A.M. 1991. Ternary excess properties of grossular-pyrope-almandine garnet and their influence in geothermobarometry. Am. Miner. 76, 1223–1231.
Bernotat, W.H. and Morteani, G. 1982. The microcline/sanidine transformation isograd in metamorphic regions: Western Tauern Window and Merano-Mules-Anterselva complex (Eastern Alps). Am. Miner. 67, 43–53.
Beswick, A.E. 1973. An experimental study of alkali metal distributions in feldspars and micas. Geochim. Cosmochim. Acta 37, 183–208.
Bethe, H.A. 1935. Statistical theory of superlattices. Proc. Roy. Soc. (London), 150, 552–575.
Binns, R.A. 1962. Metamorphic pyroxenes from the Broken Hill district, New South Wales. Miner. Mag. 33, 320–338.
Birch, F. 1966. Compressibility; elastic constants. In *Handbook of physical constants*, ed. by S.P. Clark Jr. Geol. Soc. Amer. Mem. 97.
Birch, F. and LeComte, P. 1960. Temperature-pressure plane for albite composition. Am. J. Sci. 258, 209–217.
Blander, M. 1972. Thermodynamic properties of orthopyroxenes and clinopyroxenes based on the ideal two-site model. Geochim. Cosmochim. Acta 36, 787–799.

Boettcher, A.L. and Wyllie, P.J. 1968. The calcite-aragonite transition measured in the system $CaO-CO_2-H_2O$. J. Geol. 76, 314-330.

Bohlen, S.R. and Essene, E.J. 1977. Feldspar and oxide thermometry of granulites in the Adirondack Highlands. Contr. Miner. Petrol. 62, 153-169.

Bohlen, S.R., Montana, A., Kerrick, D.M., Huggins, C. and Manning, C.E. 1989. The kyanite = sillimanite transition revisited. EOS 70, 1392.

Bons, A.J., Drury, M.R., Schryvers, D. and Zwart, H.J. 1990. The nature of grain boundaries in slates. Phys. Chem. Miner. 17, 402-408.

Borg, R.J. and Dienes, G.J. 1988. *An introduction to solid state diffusion*. Academic Press, Boston.

Bowen, N.L. 1928. *The evolution of the igneous rocks*. Princeton University Press, Princeton, NJ.

Bowen, N.L. 1940. Progressive metamorphism of siliceous limestone and dolomite. J. Geol. 48, 225-274.

Bowen, N.L. and Tuttle, O.F. 1950. The system $NaAlSi_3O_8-KAlSi_3O_8-H_2O$. J. Geol. 58, 489-511.

Bowes, D.R. and Park, R.G. 1966. Metamorphic segregation banding in the Loch Kerry basite sheet from the Lewisian of Gairloch, Ross-shire, Scotland. J. Petrology 7, 306-330.

Boyd, F.R. and Schairer, J.F. 1964. The system $MgSiO_3-CaMgSi_2O_6$. J. Petrology 5, 275-309.

Brace, W.F., Silver, E., Hadley, K. and Goetze, C. 1972. Cracks and pores: a closer look. Science 178, 162-164.

Brady, J.B. 1977. Metasomatic zones in metamorphic rocks. Geochim. Cosmochim. Acta 41, 113-125.

Brady, J.B. 1983. Intergranular diffusion in metamorphic rocks. Am. J. Sci. 283-A, 181-200.

Brady, J.B. 1987. Coarsening of fine-scale exsolution lamellae. Am. Miner. 72, 697-706.

Bragg, Sir L. and Claringbull, G.F. 1965. *Crystal structure of minerals*. G. Bell and Sons, London.

Bragg, W.L. and Williams, E.J. 1934. The effect of thermal agitation on atomic arrangement in alloys. Proc. Roy. Soc. (London) 145, 699-730.

Bray, J.M. 1942. Spectroscopic distribution of minor elements in igneous rocks from Jamestown, Colorado. Geol. Soc. Am. Bull. 53, 765-814.

Brown, E.H. 1971. Phase relations of biotite and stilpnomelane in the greenschist facies. Contr. Miner. Petrol. 31, 275-299.

Brown, G.C. and Fyfe, W.S. 1971. Kyanite andalusite equilibrium. Contr. Miner. Petrol. 33, 227-231.

Brown, G.M. 1957. Pyroxenes from the early and middle stages of fractionation of the Skaergaard Intrusion, East Greenland. Miner. Mag. 31, 511-543

Brown, M.E., Dollimore, D. and Galwey, A.K. 1980. Reactions in the solid state. In *Comprehensive chemical kinetics*, ed. by C.H. Bamford and C.F.H. Tipper. Elsevier, Amsterdam.

Brown, W.L. and Parsons, I. 1989. Alkali feldspars: ordering rates, phase transformations and behaviour diagram for igneous rocks. Miner. Mag. 53, 25-42.

Brown, W.L., Becker, S.M. and Parsons, I. 1983. Cryptoperthites and cooling rate in a layered syenite pluton: a chemical and TEM study. Contr. Miner. Petrol. 82, 13-25.

Brunton, C.J. and Helgeson, H.C. 1983. Thermodynamic consequences of differences between fluid and geostatic pressure in hydrothermal systems. Am. J. Sci. 283-A, 540-588.

Buchwald, V.F. 1975. *Handbook of iron meteorites, vol. 1*. University of California Press, Berkeley.

Buckley, H.E. 1951. *Crystal growth*. J. Wiley & Sons, New York.

Buddington, A.F. and Lindsley, D.H. 1964. Iron-titanium oxide minerals and synthetic equivalents. J. Petrology 5, 310-357.

Buening, D.K. and Buseck, P.R. 1973. Fe-Mg lattice diffusion in olivine. J. Geophys. Res. 78, 6852-6862.

Buerger, M.J. 1947. The relative importance of the several faces of a crystal. Am. Miner. 32, 593-606.

Bunch, T.E. and Olsen, E. 1974. Restudy of pyroxene-pyroxene equilibration temperatures for ordinary chondrite meteorites. Contr. Miner. Petrol. 43, 83-90.

Bunch, T.E. and Olsen, E. 1975. Distribution and significance of chromium in meteorites. Geochim. Cosmochim. Acta 39, 911-927.

Bundy, F.P., Strong, H.M. and Wentorf, R.H. Jr. 1973. Methods and mechanisms of synthetic diamond growth. Phys. Chem. Carbon 10, 213-263.

Buntebarth, G. and Voll, G. 1991. Quartz grain coarsening by collective crystallization in contact quartzites. In *Equilibrium and kinetics in contact metamorphism*, ed. by G. Voll, J. Töpel, D.R.M. Pattison and F. Seifert. Springer-Verlag, Berlin.

Burkart-Baumann, I. 1972. Vanadium: element and geochemistry. In *The encyclopedia of geochemistry and environmental sciences*, ed. by R.W. Fairbridge. Van Nostrand Reinhold Co., New York.

Burke, J. 1965. *The kinetics of phase transformations in metals*. Pergamon, Oxford.

Burke, J.E. and Rosolowski, J.H. 1973. Sintering. In *Treatise on solid state chemistry, Volume 4, Reactivity of solids*, ed. by N.B. Hannay. Plenum Press, New York.

Burke, J.G. 1966. *Origins of the science of crystals*. University of California Press, Berkeley.

Burnham, C.W. 1973. Order-disorder relationships in some rock-forming silicate minerals. Ann. Rev. Earth Planet. Sci. 1, 313-338.

Burnham, C.W., Holloway, J.R. and Davis, N.F. 1969. Thermodynamic properties of water to 1000 °C and 10 000 bars. Geol. Soc. Am. Special Paper 132.

Burns, R.G. 1970. *Mineralogical applications of crystal field theory*. Cambridge University Press.

Burns, R.G. 1973. The partitioning of trace transition elements in crystal structures: a provocative review with applications to mantle geochemistry. Geochim. Cosmochim. Acta 37, 2395-2403.

Burton, W.K., Cabrera, N. and Frank, F.C. 1951. The growth of crystals and the equilibrium structure of their surfaces. Phil. Trans. Roy. Soc. (London) 243, 299-358.

Buseck, P.R. and Bo-Jun, H. 1985. Conversion of carbonaceous material to graphite during metamorphism. Geochim. Cosmochim. Acta 49, 2003-2016.

Buseck, P.R. and Iijima, S. 1974. High resolution electron microscopy of silicates. Am. Miner. 59, 1-21.

Butler, P. 1969. Mineral compositions and equilibria in the metamorphosed iron formation of the Gagnon region, Quebec, Canada. J. Petrology 10, 56-101.

Cahn, J.W. 1961. On spinodal decomposition. Acta Metal. 9, 795-801.

Cahn, J.W. 1979. Thermodynamics of solid and fluid surfaces. In *Interfacial segregation*, ed. by W.C. Johnson and J.M. Blakeley. Am. Soc. for Metals.

Cahn, J.W. 1982. Transitions and phase equilibria among grain boundary structures. J. de Phys. 43, C6-199-C6-214.

Cameron, E.N. 1975. Postcumulus and subsolidus equilibration of chromite and coexisting silicates in the Eastern Bushveld Complex. Geochim. Cosmochim. Acta 39, 1021-1033.

Carlson, W.D. 1989. The significance of intergranular diffusion to the mechanisms and kinetics of porphyroblast crystallization. Contr. Miner. Petrol. 103, 1-24.

Carlson, W.D. and Rosenfeld, J.L. 1981. Optical determination of topotactic aragonite-calcite growth kinetics: metamorphic implications. J. Geol. 89, 615-638.

Carmichael, D.M. 1969. On the mechanism of prograde metamorphic reactions in quartz-bearing pelitic rocks. Contr. Miner. Petrol. 20, 244-267.

Carmichael, D.M. 1970. Intersecting isograds in the Whetstone Lake Area, Ontario. J. Petrology 11, 147-181.

Carpenter, A.B. 1967. Mineralogy and petrology of system $CaO-MgO-CO_2-H_2O$ at Crestmore, California. Am. Miner. 52, 1341-1363.

Carpenter, M.A. 1985. Order-disorder transformations in mineral solid solutions. Rev. Miner. 14, 187-217.

Cashman, K.V. and Ferry, J.M. 1988. Crystal size distribution (CSD) in rocks and the kinetics and dynamics of crystallization. Contr. Miner. Petrol. 99, 401-415.

Chaix-Pluchery, O., Bouillot, J., Freund, F. and Niepce, J.C. 1984. Precursor states in the phase transformation accompanying solid-state reaction. In *Phase transformations in solids*, ed. by T. Tsakalakos. North-Holland, New York, pp. 69-74.

Chakraborty, K.R. 1977. A new method of material balance evaluation in metamorphic differentiated systems. Contr. Miner. Petrol. 65, 101-110.

Chandrasekhar, S., 1943. Stochastic problems in physics and astronomy. Rev. Mod. Phys. 15, 1-89.

Chao, E.C.T. 1968. Pressure and temperature histories of impact metamorphosed rocks—based on petrographic observations. In *Shock metamorphism of natural materials*, ed. by B.M. French and N.M. Short. Mono Book Corp., Baltimore.

Chao, E.C.T., Fahey, J.J. and Littler, J. 1962. Stishovite, SiO_2, a very high pressure new mineral from Meteor Crater, Arizona. J. Geophys. Res. 67, 419-421.

Chao, E.C.T., Shoemaker, E.M. and Madsen, B.M. 1960. First natural occurrence of coesite. Science 132, 220-222.

Chapman, C.A. 1950. Quartz veins formed by metamorphic differentiation of aluminous schists. Am. Miner. 35, 693-710.

Chatillon-Colinet, C., Newton, R.C., Perkins, D. and Kleppa, O.J. 1983. Thermochemistry of (Fe^{2+}, Mg) SiO_3 orthopyroxene. Geochim. Cosmochim. Acta 47, 1597-1603.

Chatterjee, N., 1987. Evaluation of the thermochemical data on Fe-Mg olivine, orthopyroxene, spinel, and Ca-Fe-Mg-Al garnet. Geochim. Cosmochim. Acta 51, 2515-2525.

Chatterjee, N.D. 1972. The upper stability limit of the assemblage paragonite + quartz and its natural occurrences. Contr. Miner. Petrol. 4, 288-303.

Chatterjee, N.D. and Froese, E. 1975. A thermodynamic study of the pseudobinary join muscovite-paragonite in the system $KAlSi_3O_8-NaAlSi_3O_8-Al_2O_3-SiO_2-H_2O$. Am. Miner. 60, 985-993.

Chatterjee, N.D. and Johannes, W. 1974. Thermal stability and standard thermodynamic properties of synthetic 2 M_1-muscovite, $KAl_2[AlSi_3O_{10}(OH)_2]$. Contr. Miner. Petrol. 48, 89-114.

Chelishchev, N.F. 1967. Laboratory studies in the distribution of rare alkalies between potassium minerals and aqueous solutions at high temperatures and pressures. Dok. Acad. Sci. USSR, Earth Sci. Sec. 175, 205-207.

Cheney, J.T. and Guidotti, C.V. 1979. Muscovite-plagioclase equilibria in sillimanite + quartz bearing metapelites, Puzzle Mountain area, Northwest Maine. Am. J. Sci. 279, 411-434.

Chernosky, J.V. Jr., Day, H.W. and Caruso, L.J. 1985. Equilibria in the system $MgO-SiO_2-H_2O$: experimental determination of the stability of Mg-anthophyllite. Am. Miner. 70, 223-236.

Chernov, A.A. 1989. Formation of crystals in solutions. Contemporary Phys. 30, 251-276.

Cherry, M.E. and Trembath, L.T. 1979. Order-disorder paths in alkali feldspars. Am. Miner. 64, 66-70.

Chinner, G.A. 1960. Pelitic gneisses with varying ferrous/ferric ratios from Glen Clova, Angus, Scotland. J. Petrology 1, 178-217.

Chinner, G.A. 1961. The origin of sillimanite in Glen Clova, Angus. J. Petrology 2, 312-323.

Chinner, G.A. and Dixon, J.E. 1973. Some high-pressure parageneses of the Allalin gabbro, Valais, Switzerland. J. Petrology 14, 185-202

Chinner, G.A., Smith, J.V. and Knowles, C.R. 1969. Transition-metal contents of Al_2SiO_5 polymorphs. Am. J. Sci. 267A, 96-113.

Chipman, D. and Warren, B.E. 1950. X-ray measurement of long range order in β-brass. J. App. Phys. 21, 696-697.

Chopin, C. 1984. Coesite and pure pyrope in high-grade blueschists of the Western Alps: a first record and some consequences. Contr. Miner. Petrol. 86, 107-118.

Chopin, C., Klaska, R., Medenbach, O. and Dron, D. 1986. Ellenbergerite, a new high-pressure Mg-Al-(Ti,Zr)-silicate with a novel structure based on face-sharing octahedra. Contr. Miner. Petrol. 92, 316-321.

Christian, J.W. 1975. *The theory of transformations in metals and alloys.* Pergamon Press, Oxford.

Christie, O.H.J. 1968. Spinodal precipitation in silicates, 1. Introductory application to exsolution in feldspars. Lithos 1, 187-192.

Christoffersen, R., Yund, R.A. and Tullis, J. 1983. Inter-diffusion of K and Na in alkali feldspars: diffusion couple experiments. Am. Miner. 68, 1126-1133.

Clark, P.J. and Evans, F.C. 1954. Distance to nearest neighbor as a measure of the spatial relationships in populations. Ecology 35, 445-453.

Clark, S.R. Jr. and Ringwood, A.E. 1964. Density distribution and constitution of the mantle. Rev. Geophys. 2, 35-88.

Clemm, P.J. and Fisher, J.C. 1955. The influence of grain boundaries in the nucleation of secondary phases. Acta Metal. 3, 70-73.

Coble, R.L. 1963. A model for boundary-diffusion controlled creep in polycrystalline materials. J. App. Phys. 34, 1679-1682.

Cogulu, E. 1967. Etude pétrographique de la région de Mihaliccik (Turquie). Schweizer. Miner. Petrograph. Mitt. 47, 683-824.

Coleman, R.G. and Lee, D.E. 1962. Metamorphic aragonite in the glaucophane schists of Cazadero, California. Am. J. Sci. 260, 577-595.

Collerson, K.D. 1976. Composition and structural state of alkali feldspars from high-grade metamorphic rocks, Central Australia. Am. Miner. 61, 201-211.

Compagnoni, R., Lombardo, B. and Prato, R. 1974. Andalousite et sillimanite aux contacts du granite central de l'Argentera (Alps Maritimes). Rendiconti della Società Italiana di Mineralogia e Petrologia, 30, 32-54.

Condie, K.C. 1967. Oxygen, carbon dioxide, and sulfur fugacities during diagenesis and low-grade metamorphism of late Precambrian subgraywackes from Northern Utah. Am. Miner. 52, 1153-1160.

Condie, K.C. 1989. *Plate tectonics and crustal evolution.* Pergamon Press, Oxford.

Coomāraswāmy, A.K. 1902. The crystalline limestones of Ceylon. Quart. J. Geol. Soc. (London) 58, 399-424.

Crain, I.K. 1976. Statistical analysis of geotectonics. In *Random processes in geology*, ed. by D.F. Merriam. Springer-Verlag, New York.

Crank, J. 1975. *The mathematics of diffusion.* Clarendon Press, Oxford.

Crawford, M.L. and Hollister, L.S. 1986. Metamorphic fluids: evidence from fluid inclusions. Adv. Phys. Geochem. 5, 1-35.

Cressey, G., Schmid, R. and Wood, B.J., 1978. Thermodynamic properties of almandine-grossular garnet solid solutions. Contr. Miner. Petrol. 67, 397-404.

Dachille, F. and Roy, R. 1964. Effectiveness of shearing stresses in accelerating solid phase reactions at low temperatures and high pressures. J. Geol. 72, 243-246.

Dahl, O., 1969. Irregular distribution of iron and magnesium among coexisting biotite and garnet. Lithos 2, 311-322.

Dallmeyer, R.D., 1974. Metamorphic history of the northeastern Reading Prong, New York and Northern New Jersey. J. Petrology 15, 325-359.

Dal Negro, A., De Pieri, R. and Quareni, S. 1978. The crystal structures of nine K feldspars from the Adamello Massif (Northern Italy). Acta Crystal. B34, 2699-2707.

Danielsson, A. 1950. Das calcit-wollastonitgleichgewicht. Geochim. Cosmochim. Acta 1, 55-69.

Darken, L.S. 1948. Diffusion, mobility and their interrelation through free energy in binary metallic systems. Trans. Am. Inst. Min. Met. Engrs. 175, 184-201.

Darken, L.S. and Gurry, R.W. 1953. *Physical chemistry of metals*. McGraw-Hill, New York.

Davidson, L.R. 1968. Variation in ferrous iron-magnesium distribution coefficient of metamorphic pyroxenes from Quairading, Western Australia. Contr. Miner. Petrol. 19, 239-259.

Davidson, P.M. and Lindsley, D.H. 1985. Thermodynamic analysis of quadrilateral pyroxenes. Part II: Model calibration from experiments and applications to geothermometry. Contr. Miner. Petrol. 91, 390-404.

Davis, B.T.C. and Boyd, F.R. 1966. The join $Mg_2Si_2O_6$-$CaMgSi_2O_6$ at 30 kilobars pressure and its application to pyroxenes from kimberlites. J. Geophys. Res. 71, 3567-3576.

Day, H.W. 1972. Geometrical analysis of phase equilibria in ternary systems of six phases. Am. J. Sci. 272, 711-734.

Day, H.W. 1973. The high temperature stability of muscovite plus quartz. Am. Miner. 58, 255-262.

Day, H.W., Chernosky, J.V. and Kumin, H.J. 1985. Equilibria in the system MgO-SiO_2-H_2O: a thermodynamic analysis. Am. Miner. 70, 237-248

Day, R.A. and Underwood, A.L. 1974. *Quantitative analysis*. Prentice-Hall, Englewood Cliffs, NJ.

Debe, M.K. and King, D.A. 1979. The clean thermally induced $W\{001\}(1\times1) \rightarrow (\sqrt{2}\times\sqrt{2})$ R 45° surface structure transition and its crystallography. Surf. Sci. 81, 193-237.

Deer, W.A., Howie, R.A. and Zussman, J. 1963. *Rock-forming minerals*. Vols. I to V, Longmans, London.

Deer, W.A., Howie, R.A. and Zussman, J. 1978. *Rock-forming minerals*. Vol. 2A, Single-chain silicates. J. Wiley & Sons, New York.

Delor, C.P. and Leyreloup, A.F. 1986. Chromium-rich kyanite in an eclogite from the Rouergue area, French Massif Central. Miner. Mag. 50, 535-537.

Denbigh, K.G. 1951. *The thermodynamics of the steady state*. Methuen & Co., London.

Denbigh, K. 1961. *The principles of chemical equilibrium*. Cambridge University Press.

Dent, L.S. and Smith J.V. 1958. Crystal structure of chabazite, a molecular sieve. Nature 181, 1794.

DeVore, G.W. 1955. The role of adsorption in the fractionation and distribution of elements. J. Geol. 63, 159-190.

DeVore, G.W. 1956. Surface chemistry as a chemical control on mineral association. J. Geol. 64, 31-55.

DeVore, G.W. 1959. Role of minimum interfacial free energy in determining the macroscopic features of mineral assemblages. J. Geol. 67, 211-227.

DeVore, G.W. 1969. Preferred mineral distributions of polymineralic rocks related to non-hydrostatic stresses as expressions of mechanical equilibrium. J. Geol. 77, 26-38.

Dewers, T. and Ortoleva, P. 1990. Geochemical self-organization. III: A mechano-chemical model of metamorphic differentiation. Am. J. Sci. 290, 473-521.

Dobretsov, N.L. 1966. Significance of the partial pressure of H_2O and CO_2 in metamorphic systems. Dok. Acad. Sci. USSR, Earth Sci. Sec. 166, 154-157.

Dorogokupets, P.I., Karpov, I.K. and Lashkevich, V.V. 1988. Thermodynamics of silica polymorphs. Int. Geol. Rev. 30, 1278-1287.

Dougan, T.W. 1974. Cordierite gneisses and associated lithologies of the Guri area, Northwest Guayana Shield, Venezuela. Contr. Miner. Petrol. 46, 169-188.

Dowty, E. 1980. Crystal-chemical factors affecting the mobility of ions in minerals. Am. Miner. 65, 174-182.

Drost-Hansen, W. 1971. Structure of water near solid interfaces. In *Chemistry and physics of interfaces II*. Am. Chem. Soc., Washington.

Dudek, K. and Kienast, J.R. 1989. Deerite from Ile de Groix, Brittany, France. Miner. Mag. 53, 603-612.

Duffy, D.M. 1986. Grain boundaries in ionic crystals. J. Phys. C: Solid State Phys. 19, 4393-4412.

Dupuy, C. 1972. Coefficients de partage du strontium entre phases leucocrates des ignimbrites de Toscane (Italie). Bull. Soc. fr. Minéral. Cristallogr. 95, 322-329.

Ebadi, A. and Johannes, W. 1991. Beginning of melting and composition of first melts in the system Qz-Ab-Or-H_2O-CO_2. Contr. Miner. Petrol. 106, 286-295.

Edmunds, W.M. and Atherton, M.P. 1971. Polymetamorphic evolution of garnet in the Fanad Aureole, Donegal, Eire. Lithos, 4, 147-161.

Eggleton, R.A. and Banfield, J.F. 1985. The alteration of granitic biotite to chlorite. Am. Miner. 70, 902-910.

Eggleton, R.A. and Buseck, P.R. 1980. The orthoclase-microcline inversion: a high-resolution transmission electron microscope study and strain analysis. Contr. Miner. Petrol. 74, 123-133.

Elkins, L.T. and Grove, T.L. 1990. Ternary feldspar experiments and thermodynamic models. Am. Miner. 75, 544-559.

Ellis, D.J. and Green, D.H. 1979. An experimental study of the effect of Ca upon garnet-clinopyroxene Fe-Mg exchange equilibrium. Contr. Miner. Petrol. 71, 13-22.

Endo, Y. and Sunagawa, I. 1968. Surface structures of garnet from Wada-toge Pass, Nagona Prefecture (in Japanese). J. Miner. Soc. Japan 9, 69-80.

Engelhardt, W.v. and Stöffler, D. 1968. Stages of shock metamorphism in crystalline rocks of the Ries Basin, Germany. In *Shock metamorphism of natural materials*, ed. by B.M. French and N.M. Short. Mono Book Corp., Baltimore.

Engi, M. 1983. Equilibria involving Al-Cr spinel: Mg-Fe exchange with olivine. Experiments, thermodynamic analysis, and consequences for geothermometry. Am. J. Sci. 283A, 29-71.

Engi, M. 1992. Thermodynamic data for minerals: a critical assessment. In *The stability of minerals*, ed. by G.D. Price and Nancy L. Ross. Chapman & Hall, London.

Engi, M. and Evans, B. 1980. A re-evaluation of the olivine-spinel geothermometer: discussion. Contr. Miner. Petrol. 73, 201-203.

England, P.C. and Thompson, A.B. 1984. Pressure-temperature-time paths of regional metamorphism, Part I: Heat transfer during the evolution of regions of thickened continental crust. J. Petrology 25, 894-928.

Ernst, W.G. 1971. Do mineral parageneses reflect unusually high-pressure conditions of Franciscan metamorphism? Am. J. Sci. 270, 81-108.

Ernst, W.G. 1976. *Petrologic phase equilibria*. W.H. Freeman, San Francisco.

Ernst, W.G., Seki, Y., Onuki, H. and Gilbert, M.C. 1970. Comparative study of lowgrade metamorphism in the California Coast Range and the Outer Metamorphic Belt of Japan. Geol. Soc. Am. Mem. 124.

Esbensen, H. 1978. Coronites from the Fongen gabbro complex, Trondheim Region, Norway: role of water in the olivine-plagioclase reaction. N. Jb. Miner. Abh. 132, 113-135.

Eskola, P. 1915. On the relation between the chemical and mineralogical composition in the metamorphic rocks of the Orijärvi region. Bull. Comm. Géol. Finlande, no. 44.

Eskola, P. 1922. On contact phenomena between gneiss and limestone in western Massachusetts. J. Geol. 30, 265-294.

Essene, E.J. 1983. Solid solutions and solvi among metamorphic carbonates with applications to geologic thermobarometry. Rev. Miner. 11, 77-96.

Etheridge, M.A., Wall, V.J. and Vernon, R.H. 1983. The role of the fluid phase during regional metamorphism and deformation. J. Metamorphic Geol. 1, 205-226.

Eugster, H.P. 1955. The cesium-potassium equilibrium in the system sanidine-water. Carnegie Inst. Washington, Year Book 1954-1955; 112-113.

Eugster, H.P. 1959. Reduction and oxidation in metamorphism. In *Researches in geochemistry*, ed. by P.H. Abelson. J. Wiley & Sons, New York.

Eugster, H.P. 1972. Reduction and oxidation in metamorphism (II). 24th Inter. Geol. Congress, Sec. 10, 3-11.

Eugster, H.P. and Skippen, G.B. 1967. Igneous and metamorphic reactions involving gas equilibria. In *Researches in Geochemistry*, ed. by P.H. Abelson. J. Wiley & Sons, New York.

Eugster, H.P. and Wones, D.R. 1962. Stability relations of the ferruginous biotite, annite. J. Petrology 3, 82-125.

Evans, B.W. 1965. Application of a reaction rate method to the breakdown equilibria of muscovite and muscovite plus quartz. Am. J. Sci. 263, 647-667.

Evans, B.W. 1977. Metamorphism of Alpine peridotite and serpentinite. Ann. Rev. Earth Planet Sci. 5, 397-447.

Evans, B.W. and Frost, B.R. 1975. Chrome-spinel in progressive metamorphism — a preliminary analysis. Geochim. Cosmochim. Acta 39, 959-972.

Evans, B.E. and Guidotti, C.V. 1966. The sillimanite-potash feldspar isograd in Western Maine, USA. Contr. Miner. Petrol. 12, 25-62.

Evans, B.E. and Wright, T.L. 1972. Composition of liquidus chromite from the 1959 (Kilauea Iki) and 1965 (Makaopuhi) eruptions of Kilauea volcano, Hawaii. Am. Miner. 57, 217-230.

Fabbri, A.G. 1984. *Image processing of geological data*. Van Nostrand Reinhold, New York.

Fabriès, J. 1979. Spinel-olivine geothermometry in peridotites from ultramafic complexes. Contr. Miner. Petrol. 69, 329-336.

Fast, J.D., 1962. *Entropy*. McGraw-Hill, New York.

Fei, Y., Saxena, S.K. and Eriksson, G. 1986. Some binary and ternary silicate solution models. Contr. Miner. Petrol. 94, 221-229.

Ferry, J.M. 1986. Reaction progress: a monitor of fluid-rock interaction during metamorphic and hydrothermal events. Adv. Phys. Geochem. 5, 60-88.

Ferry, J.M. 1991. Dehydration and decarbonation reactions as a record of fluid infiltration. Rev. Miner. 26, 351-393.

Ferry, J.M. and Spear, F.S. 1978. Experimental calibration of the partitioning of Fe and Mg between biotite and garnet. Contr. Miner. Petrol. 66, 113-117.

Finlay, C.A. and Kerr, A. 1987. Evidence for differences in growth rate among garnets in pelitic schists from northern Sutherland, Scotland. Miner. Mag. 51, 569-576.

Fischmeister, H.F. 1985. Structure and properties of high angle grain boundaries. J. de Phys. 46, C4-3-C4-23.

Fisher, G.W. 1970. The application of ionic equilibria to metamorphic differentiation: An example. Contr. Miner. Petrol. 29, 91-103.

Fisher, G.W. 1973. Nonequilibrium thermodynamics as a model for diffusion-controlled metamorphic processes. Am. J. Sci. 273, 897-924.

Fisher, G.W. and Medaris, L.G. Jr. 1969. Cell dimensions and X-ray determinative curve for synthetic Mg-Fe olivines. Am. Miner. 54, 741-753.

Fisher, J.C. 1951. Calculation of diffusion penetration curves for surface and grain boundary diffusion. J. App. Phys. 22, 74-77.

Fleet, M.E. 1974. 1974. Mg-Fe^{2+} site occupancies in coexisting pyroxenes. Contr. Miner. Petrol. 47, 207-214.

Fleet, M.E. 1977. The crystal structure of deerite. Am. Miner. 62, 990-998.

Fleet, M.E. and Ribbe, P.H. 1963. An electron-microscope investigation of a moonstone. Phil. Mag. 8, 1179-1187.

Fleet, S.G. and Megaw, H.D. 1962. Crystal structure of yoderite. Acta Crystal. 15, 721-728

Fletcher, R.C. and Hofmann, A.W. 1974. Simple models of diffusion and combined diffusion-infiltration metasomatism. In *Geochemical transport and kinetics*, ed. by

A.W. Hofmann, B.J. Giletti, H.S. Yoder, Jr. and R.A. Yund. Carnegie Institution of Washington.

Flinn, D. 1969. Grain contacts in crystalline rocks. Lithos 3, 361–370.

Flowers, G.C. and Helgeson, H.C., 1983. Equilibrium and mass transfer during progressive metamorphism of siliceous dolomites. Am. J. Sci. 283, 230–286.

Fonarev, V.I. and Graphchikov, A.A. 1982. Experimental study of Fe–Mg- and Ca-distribution between coexisting ortho- and clinopyroxenes at P = 294 MPa, T = 750 and 800 °C. Contr. Miner. Petrol. 79, 311–318.

Fonarev, V.I. and Graphchikov, A.A. 1991. Two-pyroxene thermometry: a critical evaluation. In *Progress in metamorphic and magmatic petrology*, ed. by L.L. Perchuk. Cambridge University Press, Cambridge.

Fowler, A.D. 1990. Self-organized mineral textures of igneous rocks: the fractal approach. Earth Science Rev. 29, 47–55.

Fowler, C.M.R., Nisbet, E.G. and Pandit, B.I. 1988. The transfer of heat by conduction in the Earth's crust: physical constraints. In *Short course on heat, metamorphism, and tectonics*, ed. by E.G. Nisbet and C.M.R. Fowler. Mineral. Assoc. Canada.

Frank, F.C. 1949. The influence of dislocations on crystal grwowth. Disc. Faraday Soc., No 5, 48–54.

Frank, F.C. 1955. Chemistry of crystal dislocations. In *Chemistry of the solid state*, ed. by W.E. Garner. Butterworths, London.

Frantz, J.D. and Mao, H.K. 1979. Bimetasomatism resulting from integranular diffusion: II: Prediction of multimineralic zone sequences. Am. J. Sci. 279, 302–323.

Fraser, D.G. 1990. Applications of the high-resolution scanning proton microprobe in the Earth sciences: an overview. Chem. Geol. 83, 27–37.

Freer, R. 1980. Self diffusion and impurity diffusion in oxides. J. Mater. Sci. 15, 803–824.

Freer, R. 1981. Diffusion in silicate minerals and glasses: a data digest and guide to the literature. Contr. Miner. Petrol. 76, 440–454.

French, B.M. 1966. Some geological implications of equilibria between graphite and a C–H–O gas phase at high temperatures and pressures. Rev. Geophys. 4, 223–253.

French, B.M. and Eugster, H.P. 1965. Experimental control of oxygen fugacities by graphite–gas equilibriums. J. Geophys. Res. 70, 1529–1539.

Frenkel, J. 1946. *Kinetic theory of liquids*. Oxford University Press, Oxford.

Freund, F. 1987. Hydrogen and carbon in solid solution in oxides and silicates. Phys. Chem. Miner. 15, 1–18.

Frey, M. 1970. The step from diagenesis to metamorphism in pelitic rocks during Alpine orogenesis. Sedimentology 15, 261–270.

Frey, M., Bucher, K., Frank, E. and Mullis, J. 1980. Alpine metamorphism along the geotraverse Basel–Chiasso — a review. Eclogae geol. Helv. 73, 527–546.

Frisch, C.J. and Helgeson, H.C. 1984. Metasomatic phase relations in dolomites of the Adamello Alps. Am. J. Sci. 284, 121–185.

Froese, E. 1973. The assemblage quartz–K feldspar–biotite–garnet–sillimanite as an indicator of P_{H_2O}–T conditions. Can. J. Earth Sci. 10, 1575–1579.

Froese, E. and Gordon, T.M. 1974. Activity coefficients of coexisting pyroxenes. Am. Miner. 59, 204–205.

Froese, E. and Winkler, H.G.F. 1966. The system $CaCO_3$–$SrCO_3$ at high pressures and 500 °C to 700 °C. Can. Miner. 8, 551–566.

Frost, B.R. 1985. On the stability of sulfides, oxides, and native metals in serpentinite. J. Petrology 26, 31–63.

Frost, B.R. and Lindsley, D.H. 1992. Equilibria among Fe–Ti oxides, pyroxenes, olivine, and quartz: Part II. Applications. Am. Miner. 77, 1004–1020.

Fuhrman, M.L. and Lindsley, D.H. 1988. Ternary-feldspar modeling and thermometry. Am. Miner. 73, 201–215.

Fyfe, W.S., Price, N.J. and Thompson, A.B. 1978. *Fluids in the Earth's Crust*. Elsevier, Amsterdam.

Fyfe, W.S., Turner, F.J. and Verhoogen, J. 1958. *Metamorphic reactions and metamorphic facies*. Geol. Soc. Am. Memoir 73.

Galwey, A.K. and Jones, K.A. 1963. An attempt to determine the mechanism of a natural mineral-forming reaction from examination of the products. J. Chem. Soc., 5681-5686.

Ganguly, J. 1982. Mg-Fe order-disorder in ferromagnesian silicates II. Thermodynamics, kinetics, and geological applications. Adv. Phys. Geochem. 2, 58-99.

Ganguly, J. and Kennedy, G.C. 1974. The energetics of natural garnet solid solution. I. Mixing of the aluminosilicate end members. Contr. Miner. Petrol. 48, 137-148.

Ganguly, J. and Saxena, S.K., 1987. *Mixtures and mineral reactions*. Springer-Verlag, Berlin.

Ganguly, J., Cheng, W. and O'Neil, H.St.C. 1993. Synthesis, volume, and structural changes of garnets in the pyrope-grossular join: implications for stability and mixing properties. Am. Miner. 78, 583-593.

Geiger, C.A., Newton, R.C. and Kleppa, O.J., 1987. Enthalpy of mixing of synthetic almandine-grossular and almandine-pyrope garnets from high-temperature solution calorimetry. Geochim. Cosmochim. Acta 51, 1755-1763.

Geikie, A. 1905. *The founders of geology*. Dover (1962), New York.

Ghent, E.D. 1976. Plagioclase-garnet-Al_2SiO_5-quartz: a potential geobarometer-geothermometer. Am. Miner. 61, 710-714.

Ghent, E.D., Stout, M.Z. and Parrish, R.R. 1988. Determination of metamorphic pressure-temperature-time (P-T-t) paths. In *Short course on heat, metamorphism, and tectonics*, ed. by E.G. Nisbet and C.M.R. Fowler. Mineralogical Association of Canada.

Ghiorso, M.S. 1984. Activity/composition relations in the ternary feldspars. Contr. Miner. Petrol. 87, 282-296.

Ghose, S. 1962. The nature of Mg^{2+}-Fe^{2+} distribution in some ferromagnesian silicate minerals. Am. Miner. 47, 388-394.

Ghose, S. 1965. Mg^{2+}-Fe^{2+} order in an orthopyroxene, $Mg_{0.93}Fe_{1.07}Si_2O_6$. Z. Kristal. 122, 81-99.

Ghose, S. 1982. Mg-Fe order-disorder in ferromagnesian silicates. I. Crystal chemistry. Adv. Phys. Geochem. 2, 4-57.

Ghose, S. and Schindler, P. 1969. Determination of the distribution of trace amounts of Mn^{2+} in diopsides by electron paramagnetic resonance. Miner. Soc. Am. Spec. Paper 2, 51-58.

Gibbs, J.W. 1878. On the equilibrium of heterogeneous substances. Trans. Connecticut Acad. III, 1875-1876 (pp. 108-248), 1877-1878 (pp. 343-524).

Gilbert, E.N. 1962. Random subdivisions of space into crystals. Ann. Math. Stat. 33, 958-972.

Gilmer, G.H. 1980. Computer models of crystal growth. Science 208, 355-363.

Glassley, W.E. 1983. Deep crustal carbonates as CO_2 fluid sources: evidence from metasomatic reaction zones. Contr. Miner. Petrol. 84, 15-24.

Gleiter, H. and Chalmers, B. 1972. High-angle grain boundaries. In *Progress in materials science* 16. Pergamon Press, Oxford.

Goffé, B. and Chopin, C. 1986. High-pressure metamorphism in the Western Alps: zoneography of metapelites, chronology and consequences. Schweizer. Miner. Petrograph. Mitt. 66, 41-52.

Goldschmidt, V.M. 1911. Die Kontaktmetamorphose im Kristianiagebiet. Oslo Vidensk. Skr., I, Math.-Naturv. Kl., No. 11, 405 pp.

Goldschmidt, V.M. 1954. *Geochemistry*, Oxford University Press.

Goldsmith, J.R. 1960. Exsolution of dolomite from calcite. J. Geol. 68, 103-109.

Goldsmith, J.R. 1980. Melting and breakdown reactions of anorthite at high pressures and temperatures. Am. Miner. 65, 272-284.

Goldsmith, J.R. and Graf, D.L. 1958. Structural and compositional variations in some natural dolomites. J. Geol. 66, 678–693.

Goldsmith, J.R. and Heard, H.C. 1961. Subsolidus phase relations in the system $CaCO_3$–$MgCO_3$. J. Geol. 69, 45–74.

Goldsmith, J.R. and Newton, R.C. 1969. *P–T–X* relations in the system $CaCO_3$–$MgCO_3$ at high temperatures and pressures. Am. J. Sci. 267A, 160–190.

Goldsmith, J.R., Graf, D.L., Witters, J. and Northrop, D.A. 1962. Studies in the system $CaCO_3$–$MgCO_3$–$FeCO_3$. J. Geol. 70, 659–688.

Gordon, T.M. 1971. Some observations on the formation of wollastonite from calcite and quartz. Can. J. Earth Sci. 8, 844–851.

Gordon, T.M. and Greenwood, H.J. 1970. The reaction dolomite + quartz + water = talc + calcite + carbon dioxide. Am. J. Sci. 268, 225–242.

Grambling, J.A. and Williams, M.L. 1985. The effect of Fe^{3+} and Mn^{3+} on aluminum silicate phase relation in north-central New Mexico, U.S.A. J. Petrology 26, 324–354.

Grant, S.M. 1988. Diffusion models for corona formation in metagabbros from the Western Grenville Province, Canada. Contr. Miner. Petrol. 98, 49–63.

Gray, N.H. 1969. Kinetics of contact metamorphic processes. Earth Planet. Sci. Letters 11, 205–210.

Green, D.H. 1964. The petrogenesis of the high-temperature peridotite intrusion in the Lizard area, Cornwall. J. Petrology 5, 134–188.

Green, N.L. and Usdansky, S.I. 1986. Ternary-feldspar mixing relations and thermobarometry. Am. Miner. 71, 1100–1108.

Greenwood, H.J. 1962. Metamorphic reactions involving two volatile components. Carnegie Inst. Washington, Year Book 61, 82–85.

Greenwood, H.J. 1963. The synthesis and stability of anthophyllite. J. Petrology 4, 317–351.

Greenwood, H.J. 1967a. Mineral equilibria in the system MgO–SiO_2–H_2O–CO_2. In *Researches in geochemistry*, ed. by P.H. Abelson. J. Wiley & Sons, New York.

Greenwood, H.J. 1967b. Wollastonite: stability in H_2O–CO_2 mixtures and occurrence in a contact metamorphic aureole near Salmo, BC, Canada. Am. Miner. 52, 1669–1680.

Greenwood, H.J. 1972. Al^{iv}–Si^{iv} disorder in sillimanite and its effect on phase relations of aluminum silicate minerals. Geol. Soc. Am. Mem. 132, 553–571.

Greenwood, H.J. 1973. Thermodynamic properties of gaseous mixtures of H_2O and CO_2 between 450° and 800°C and 0 to 500 bars. Am. J. Sci. 73, 561–571.

Greenwood, H.J. 1975. Buffering of pore fluids by metamorphic reactions. Am. J. Sci. 272, 573–593.

Greenwood, H.J. 1976. Metamorphism at moderate temperatures and pressures. In *The evolution of the crystalline rocks*, ed. by D.K. Bailey and R. Macdonald. Academic Press, London.

Gresens, R.L. 1966. The effect of structurally produced pressure gradients on diffusion in rocks. J. Geol. 74, 307–321.

Gresens, R.L. 1967. Composition–volume relations in metasomatism. Chem. Geol. 2, 47–65.

Grew, E.S. 1974. Carbonaceous material in some metamorphic rocks of New England and other areas. J. Geol. 82, 50–73.

Griffin, D.T., Gosney, T.C. and Phillips, Wm.R. 1982. The chemical formula of natural staurolite. Am. Miner. 67, 292–297.

Griffin, L.J. 1950. Observation of unimolecular growth steps on crystal surfaces. Phil. Mag. 41, 196–199.

Griffin, W.L. and Heier, K.S. 1973. Petrological implications of some corona structures. Lithos 6, 315–335.

Griggs, D.T., Paterson, M.S., Heard, H.C. and Turner, F.J. 1960. Annealing recrystallization in calcite crystals and aggregates. Geol. Soc. Am. Mem. 79, 21–37.

Grove, T.L., Baker, M.B. and Kinzler, R.J. 1984. Coupled CaAl-NaSi diffusion in plagioclase feldspar: experiments and applications to cooling rate speedometry. Geochim. Cosmochim. Acta 48, 2113-2121.

Grover, J. 1977. Chemical mixing in multicomponent solutions: an introduction to the use of Margules and other thermodynamic excess functions to represent non-ideal behaviour. In *Thermodynamics in Geology*, ed. by D.G. Fraser. D. Reidel, Dordrecht.

Grover, J. and Kubanek, F. 1983. The formation of ordered dolomite from high-magnesian calcite at 250 °C to 350 °C and 1500 bars: epitactic growth with optimal phase orientation, and implications for carbonate diagenesis. Am. J. Sci. 283A, 514-539.

Grover, J. and Orville, P. 1969. The partitioning of cations between coexisting single- and multi-site phases with applications to the assemblages: orthopyroxene-clinopyroxene and orthopyroxene-olivine. Geochim. Cosmochim. Acta 33, 205-226.

Guggenheim, E.A. 1937. Theoretical basis of Raoult's law. Trans. Faraday Soc. 33, 151-159.

Guggenheim, E.A. 1977. *Thermodynamics*. North-Holland Pub. Co., Amsterdam.

Guidotti, C.V. 1984. Micas in metamorphic rocks. Rev. Miner. 13, 357-467.

Guitard, G. 1966. Le métamorphisme et les faciès minéralogiques des marbres dérivant des dolomies siliceuses dans les massifs du Canigou et de la Carança (Pyrénées-Orientales). Comp. Rend. Acad. Sci. Paris, 262, Ser. D., 245-247.

Gummer, W.K. and Burr, S.V. 1946. Nephelinized paragneisses in the Bancroft area, Ontario. J. Geol. 54, 137-168.

Gurevich, V.M. and Khlyustov, V.G. 1979. Calorimeter for determining of low-temperature heat capacity of minerals. Quartz heat capacity under the temperature 9-300 K. Geochemistry 1979, 829-839.

Hafner, S. and Laves, F. 1957. Ordnung/Unordnung und Ultrarotabsorption. II Variation der Lage und Intensität einiger Absorptionen von Feldspäten. Zur Struktur von Orthoklas und Adular. Z. Kristal. 109, 204-225.

Hall, A. 1967. The distribution of some major and trace elements in feldspars from the Rosses and Ardara granite complexes, Donegal, Ireland. Geochim. Cosmochim. Acta 31, 835-847.

Halliwell, H.F. and Nyburg, S.C. 1963. Negative degrees of freedom and the phase rule. Nature 200, 670-671.

Ham, F.S. 1958. Theory of diffusion-limited precipitation. J. Phys. Chem. Solids 6, 335-351.

Hannay, N.B. 1976. *Solid-state chemistry*. Prentice-Hall, Englewood Cliffs, NJ.

Hariya, Y. and Kennedy, G.C. 1968. Equilibrium study of anorthite under high pressure and high temperature. Am. J. Sci. 266, 193-203.

Harker, A. 1932. *Metamorphism*. Methuen, London.

Harker, R.I. and Tuttle, O.F. 1956. Experimental data on the $P_{CO_2}-T$ curve for the reaction calcite + quartz = wollastonite + CO_2. Am. J. Sci. 254, 239-256.

Harris, N.B.W. 1974. Some migmatite types and their origins, from the Barousse massif, Central Pyrenees. Geol. Mag. 111, 319-328.

Hart, E.W. 1972. Grain boundary phase transformations. In *The nature and behavior of grain boundaries*, ed. by H. Hu. Plenum, New York.

Hartman, P. (ed.) 1973. *Cystal growth: an introduction*. North-Holland Pub., Amsterdam.

Harvey, P.K. and Ferguson, C.C. 1978. A computer simulation approach to textural interpretation in crystalline rocks. In *Recent advances in geomathematics*, ed. by D.F. Merriam. Pergamon Press, Oxford.

Haselton, H.T. and Newton, R.C., 1980. Thermodynamic properties of pyrope-grossular garnets and their stabilities at high temperatures and high pressures. J. Geophys. Res. 85, 6973-6982.

Haselton, H.T., Hovis, G.L., Hemingway, B.S. and Robie, R.A. 1983. Calorimetric investigation of the excess entropy of mixing in analbite-sanidine solid solutions: lack of evidence for Na, K short-range order and implications for two-feldspar thermometry. Am. Miner. 68, 398-413.

Hay, R.S. and Evans, B. 1988. Intergranular distribution of pore fluid and the nature of high-angle grain boundaries in limestone and marble. J. Geophys. Res. 93, 8959–8974.

Hayden, H.W., Moffatt, W.G. and Wulff, J. 1965. *The structure and properties of materials. Volume III, Mechanical behavior*. J. Wiley & Sons, New York.

Hays, J.F. 1967. Lime–alumina–silica. Carnegie Inst. Washington Year Book 65, 234–239.

Hazen, R.M. 1976. Effect of temperature and pressure on the crystal structure of forsterite. Am. Miner. 61, 1280–1293.

Hazen, R.M., 1985. Comparative crystal chemistry and the polyhedral approach. Rev. Miner. 14, 317–346.

Heier, K.S. 1957. Phase relations in potash feldspar in metamorphism. J. Geol. 65, 468–479.

Heier, K.S. 1962. Trace elements in feldspar–a review. Norsk Geol. Tidsskr. 42 (Supplement), 415–454.

Helgeson, H.C. 1987. Application of moderation theorems to metasomatic processes. In *Chemical transport in metasomatic processes*, ed. by H.C. Helgeson. D. Reidel, Dordrecht.

Helgeson, H.C., Delany, J.M., Nesbitt, H.W. and Bird, D.K. 1978. Summary and critique of thermodynamic properties of rock-forming minerals. Am. J. Sci. 278A, 1–229.

Hemingway, B.S., 1987. Quartz heat capacity from 340 to 1000 K and revised values for the thermodynamic properties. Am. Miner. 72, 273–279.

Hemingway, B.S., Robie, R.A., Evans, H.T. Jr. and Kerrick, D.M. 1991. Heat capacities and entropies of sillimanite, fibrolite, andalusite, kyanite, and quartz and the Al_2SiO_5 phase diagram. Am. Miner. 76, 1597–1613.

Hemley, J.J. and Jones, W.R. 1964. Chemical aspects of hydrothermal alteration with emphasis on hydrogen metasomatism. Econ. Geol. 59, 538–569.

Hemley, J.J., Montoya, J.W., Shaw, D.R. and Luce, R.W. 1977. Mineral equilibria in the $MgO–SiO_2–H_2O$ system: II. Talc–antigorite–forsterite–anthophyllite–enstatite stability relations and some geologic implications in the system. Am. J. Sci. 277, 353–383.

Henderson, J.B. 1985. Geology of the Yellowknife–Hearne Lake area, District of Mackenzie: a segment across an Archean basin. Geol. Surv. Canada Mem. 414.

Hensen, B.J. and Green, D.H. 1973. Experimental study of the stability of cordierite and garnet in pelitic compositions at high pressures and temperatures. Contr. Miner. Petrol. 38, 151–166.

Hermeling, J. and Schmalzried, H. 1984. Tracerdiffusion of Fe-cations in olivine $(Fe_xMg_{1-x})_2SiO_4$ (III). Phys. Chem. Miner. 11, 161–166.

Herring, C. 1951. Some theorems on the free energies of crystal surfaces. Phys. Rev. 82, 87–93.

Herring, C. 1953. The use of classical macroscopic concepts in surface-energy problems. In *Structure and properties of solid surfaces*, ed. by R. Gomer and C.S. Smith. University of Chicago Press.

Hietanen, A. 1956. Kyanite, andalusite, and sillimanite in the schist in Boehls Butte Quadrangle, Idaho. Am. Miner. 41, 1–27.

Hinton, R.W. 1990. Ion microprobe trace-element analysis of silicates: measurement of multi-element glasses. Chem. Geol. 83, 11–25.

Hirth, J.P. and Lothe, J. 1982. *Theory of dislocations*. J. Wiley & Sons, New York.

Hlabse, J. and Kleppa, O.J. 1968. The thermochemistry of jadeite. Am. Miner. 53, 1281–1292.

Hoare, M.R. and Pal, P. 1972. Statistics and stability of small assemblies of atoms. J. Crystal Growth 17, 77–96.

Hobbs, B.E., Means, W.D. and Williams, P.F. 1976. *An outline of structural geology*. J. Wiley & Sons, New York.

Hodges, K.V. and Spear, F.S. 1982. Geothermometry, geobarometry and the Al_2SiO_5 triple point at Mt. Moosilauke, New Hampshire. Am. Miner. 67, 1118–1134.

Hoersch, A.L. 1981. Progressive metamorphism of the chert-bearing Durness limestone in the Beinn an Dubhaich aureole, Isle of Skye, Scotland: a reexamination. Am. Miner. 66, 491-506.

Hoffer, E. 1976. The reaction sillimanite + biotite + quartz = cordierite + K-feldspar + H_2O, and partial melting in the system K_2O-FeO-MgO-Al_2O_3-H_2O. Contr. Miner. Petrol. 55, 127-130.

Hoffman, R.E. and Turnbull, D. 1951. Lattice and grain boundary self-diffusion in silver. J. App. Phys. 22, 634-639.

Hoffmann, C. 1970. Die Glaukophangesteine, ihre stofflichen Äquivalente und Umwandlungs-Produkte in Nordcalabrien (Süditalien). Contr. Miner. Petrol. 27, 283-320.

Hogarth, D.D. and Lapointe, P. 1984. Amphibole and pyroxene development in fenite from Cantley, Quebec. Can. Miner. 22, 281-295.

Hoinkes, G., 1986. Effect of grossular-content in garnet on the partitioning of Fe and Mg between garnet and biotite. Contr. Miner. Petrol. 92, 393-399.

Holdaway, M.J. 1971. Stability of andalusite and the aluminium silicate phase diagram. Am. J. Sci. 271, 97-131.

Holdaway, M.J. and Lee, S.M. 1977. Fe-Mg cordierite stability in high-grade pelitic rocks based on experimental, theoretical, and natural observations. Contr. Miner. Petrol. 63, 175-198.

Holdaway, M.J. and Mukhopadhyay, B. 1993. A reevaluation of the stability relations of andalusite: thermochemical data and phase diagram for the aluminum silicates. Am. Miner. 78, 298-315.

Holdaway, M.J., Mukhopadhyay, B., Dyar, M.D., Dutrow, B.L., Rumble, D. and Grambling, J.A. 1991. A new perspective on staurolite crystal chemistry: use of stoichiometric and chemical end-members for a mole fraction model. Am. Miner. 76, 1910-1919.

Holland, T.J.B. 1980. The reaction albite = jadeite + quartz determined experimentally in the range 600-1200 °C. Am. Miner. 65, 129-134.

Holland, T.J.B. and Powell, R. 1990. An enlarged and updated internally consistent thermodynamic dataset with uncertainties and correlations. J. Metamorphic Geol. 8, 89-124.

Holland, T.J.B., Navrotsky, A. and Newton, R.C. 1979. Thermodynamic parameters of $CaMgSi_2O_6$-$Mg_2Si_2O_6$ pyroxenes based on regular solution and cooperative disordering models. Contr. Miner. Petrol. 69, 337-344.

Hollister, L.S. 1966. Garnet zoning: an interpretation based on the Rayleigh fractionation model. Science, 154, 1647-1651.

Holloway, J.R. and Wood, B.J., 1988. *Simulating the Earth.* Unwin Hyman, London.

Hornbogen, E. 1973. Der Beginn der Phasenbildung in realen Festkörpern. Fortschritte Miner. 50, 270-311.

Hoschek, G. 1969. The stability of staurolite and chloritoid and their significance in metamorphism of pelitic rocks. Contr. Miner. Petrol. 22, 208-232.

Hounslow, A.W. and Moore, J.M. 1967. Chemical petrology of Grenville schists near Fernleigh, Ontario. J. Petrology 8, 1-28.

Howells, S. and O'Hara, M.J. 1975. Palaeogeotherms and the diopside-enstatite solvus. Nature 254, 406-408.

Howie, R.A. 1955. The geochemistry of the charnockite series of Madras, India. Roy. Soc. Edin. Trans. 62, 725-768.

Hsu, L.C. 1968. Selected phase relations in the system Al-Mn-Fe-Si-O-H: a model for garnet equilibria. J. Petrology 9, 40-83.

Huebner, J.S. 1980. Pyroxene phase equilibria at low pressure. Rev. Miner. 7, 213-288.

Hull, D. and Bacon, D.J. 1984. *Dislocations in crystals.* Pergamon Press, Oxford.

Hutcheon, I. and Moore, J.M. 1973. The tremolite isograd near Marble Lake, Ontario. Can. J. Earth Sci. 10, 936-947.

Iiyama, J.T. 1966. Contribution à l'étude des équilibres sub-solidus du système ternaire orthos-albite-anorthite à l'aide des réactions d'echange d'ions Na-K au contact d'une solution hydrothermale. Bull. Soc. Fr. Mineral. Cristal. 89, 442-454.

Iiyama, J.T. 1968. Étude expérimentale de la distribution d'éléments en traces entre deux feldspaths. Feldspath potassique et plagioclase coexistants. I. Distribution de Rb, Cs, Sr et Ba à 600 °C. Bull. Soc. fr. Minéral. Cristal. 91, 130-140.

Iiyama, J.T. and Volfinger, M. 1976. A model for trace-element distribution in silicate structures. Miner. Mag. 40, 555-564.

Iiyama, J.T., Wyart, J. and Sabatier, G. 1963. Équilibre des feldspaths alcalins et des plagioclases a 500, 600, 700, et 800 °C sous une pression d'eau de 1000 bars. Comp. Rend. Acad. Sci. Paris 256, 5016-5020.

Indares, A. and Martignole, J. 1985. Biotite-garnet geothermometry in the granulite facies: the influence of Ti and Al in biotite. Am. Miner. 70, 272-278.

Irvine, T.N. 1965. Chromian spinel as a petrogenetic indicator. Part I. Theory. Can. J. Earth Sci. 2, 648-672.

Irving, A.J. 1978. A review of experimental studies of crystal/liquid trace element partitioning. Geochim. Cosmochim. Acta 42, 743-770.

Jacobson, R.L. and Usdowski, H.E. 1976. Partitioning of strontium between calcite, dolomite and liquids. Contr. Miner. Petrol. 59, 171-185.

Jaeger, J.C. 1957. The temperature in the neighborhood of a cooling intrusive sheet. Am. J. Sci. 255, 306-318.

Jaeger, J.C. 1959. Temperature outside a cooling intrusive sheet. Am. J. Sci. 257, 44-54.

Jaffe, H.W., Robinson, P. and Tracy, R.J. 1978. Orthoferrosilite and other iron-rich pyroxenes in microperthite gneiss of the Mount Marcy area, Adirondack Mountains. Am. Miner. 63, 1116-1136.

James, H.L. 1955. Zones of regional metamorphism in the Precambrian of northern Michigan. Geol. Soc. Am. Bull. 60, 1455-1488.

Jamieson, H.E. and Roeder, P.L. 1984. The distribution of Mg and Fe^{2+} between olivine and spinel at 1300 °C. Am. Miner. 69, 283-291.

Jamieson, J.C. 1953. Phase equilibrium in the system calcite-aragonite. J. Chem. Phys. 21, 1385-1390.

Jamieson, R.A. 1988. Textures, sequences of events, and assemblages in metamorphic rocks. In *Short course in heat, metamorphism, and tectonics*, ed. by E.G. Nisbet and C.M.R. Fowler. Mineral. Assoc. Canada.

Jamieson, R.A. and Beaumont, C. 1988. Deformation and $P-T-t$ paths in convergent orogens. In *Short course in heat, metamorphism, and tectonics*, ed. by E.G. Nisbet and C.M.R. Fowler. Mineral. Assoc. Canada.

Jaoul, O., Houlier, B. and Abel, F. 1983. Study of ^{18}O diffusion in magnesium orthosilicate by nuclear microanalysis. J. Geophys. Res. 88, 613-624.

Jaoul, O., Poumellec, M., Froidevaux, C. and Havette, A. 1981. Silicon diffusion in forsterite: a new constraint for understanding mantle deformation. In *Anelasticity in the Earth*, ed. by F.D. Stacey, M.S. Paterson and A. Nicoals, Am. Geophys. Union Geodyn. Ser., 4, 95-100. AGU, Washington, DC.

Jen, L.S. 1975. *Spatial distribution of crystals and phase equilibria in charnockitic granulites from the Adirondack Mountains, New York.* Ph.D. Thesis, University of Ottawa, Ottawa, Canada.

Jen, L.S. and Kretz, R. 1981. Mineral chemistry of some mafic granulites from the Adirondack region. Can. Miner. 19, 479-491.

Jiang, J., Clayton, R.N. and Newton, R.C. 1988. Fluids in granulite facies metamorphism: a comparative oxygen isotope study on the south India and Adirondack high-grade terrains. J. Geol. 96, 517-533.

Joesten, R. 1974. Local equilibrium and metasomatic growth of zoned calc-silicate nodules from a contact aureole, Christmas Mountains, Big Bend region, Texas. Am. J. Sci. 274, 876-901.

Joesten, R. 1977. Evolution of mineral assemblage zoning in diffusion metasomatism. Geochim. Cosmochim. Acta 41, 649-670.

Joesten, R. 1986. The role of magmatic reaction, diffusion and annealing in the evolution of coronitic microstructure in troctolitic gabbro from Risör, Norway. Miner. Mag. 50, 441-467.

Joesten, R. 1991a. Grain-boundary diffusion kinetics in silicate and oxide minerals. Adv. Phys. Geochem. 8, 345-395.

Joesten, R.L. 1991b. Kinetics of coarsening and diffusion-controlled mineral growth. Rev. Miner. 26, 507-582.

Johannes, W. 1979. Ternary feldspars: kinetics and possible equilibria at 800 °C. Contr. Miner. Petrol. 68, 221-230.

Johannes, W. 1985. The significance of experimental studies for the formation of migmatites. In *Migmatites*, ed. by J.R. Ashworth. Blackie, Glasgow.

Johannes, W. and Puhan, D. 1971. The calcite-aragonite transition, reinvestigated. Contr. Miner. Petrol. 31, 28-38.

Johannes, W. and Schreyer, W. 1981. Experimental introduction of CO_2 and H_2O into Mg-cordierite. Am. J. Sci. 281, 299-317.

Johannes, W., Bell, P.M., Mao, H.K., Boettcher, A.L., Chipman, D.W., Hays, J.F., Newton, R.C. and Seifert, F. 1971. An interlaboratory comparison of piston-cylinder pressure calibration using the albite breakdown reaction. Contr. Miner. Petrol. 32, 24-38.

Johansson, S.A.E. 1989. PIXE: a novel technique for elemental analysis. Endeavour 13, 48-53.

Johnson, C.D. and Carlson, W.D. 1990. The origin of olivine-plagioclase coronas in metagabbros from the Adirondack Mountains, New York. J. Metamorphic Geol. 8, 697-717.

Johnson, M.R.W. 1963. Some time relations of movement and metamorphism in the Scottish Highlands. Geol. en Mijnb. 42, 121-142.

Johnson, R.G. and King, B.-S.L. 1987. Energy-dispersive X-ray fluorescence spectrometry. In *Methods for geochemical analysis*, ed. by P.A. Baedecker. US Geol. Surv. Bull. 1770, F1-F5.

Johnson, W.A. and Mehl, R.F. 1939. Reaction kinetics in processes of nucleation and growth. Trans. Am. Inst. Min. Metal. Engrs. 135, 416-458.

Johnston, J.H. and Knedler, K.E. 1979. A Mössbauer spectroscopic study of the cooling history of hypersthenes from selected members of the Taupo Pumice formation, New Zealand. Miner. Mag. 43, 279-285.

Jones, K.A. and Galwey, A.K. 1966. Size distribution, composition, and growth kinetics of garnet crystals in some metamorphic rocks from the west of Ireland. Quart. J. Geol. Soc. (London) 122, 29-44.

Jones, K.A. and Galwey, A.K. 1972. Variation in the length/thickness ratio of biotite crystals in metamorphic rocks. J. Geol. 80, 112-120.

Joplin, G.A. 1935. The exogenous contact zone at Ben Bullen, New South Wales. Geol. Mag. 72, 385-400.

Jost, W. 1960. *Diffusion in solids, liquids, gases.* Academic Press, New York.

Kamineni, D.C. 1978. A study of some garnet aggregates from Yellowknife, Canada. N. Jb. Miner. Mh., 455-462.

Käse, H.-R. and Metz, P. 1980. Experimental investigation of the metamorphism of siliceous dolomites IV. Equilibrium data for the reaction: 1 diopside + 3 dolomite = 2 forsterite + 4 calcite + $2CO_2$. Contr. Miner. Petrol. 73, 151-159.

Kaur, I. and Gust, W. 1989. *Fundamentals of grain and interphase boundary diffusion.* Ziegler Press, Stuttgart.

Kelley, K.K. 1943. Specific heat at low temperatures of magnesium orthosilicate and magnesium metasilicate. J. Amer. Chem. Soc. 65, 339-341.

Kerrick, D.M. 1968. Experiments on the upper stability limit of pyrophyllite at 1.8 kb and 3.9 kb water pressure. Am. J. Sci. 266, 204-214.

Kerrick, D.M. 1972. Experimental determination of the muscovite + quartz stability with $P_{H_2O} < P_{total}$. Am. J. Sci. 272, 946-958.

Kerrick, D.M. 1974. Review of metamorphic mixed-volatile (H_2O-CO_2) equilibria. Am. Miner. 59, 729-762.

Kerrick, D.M. 1977. The genesis of zoned skarns in the Sierra Nevada, California. J. Petrology 18, 144-181.

Kerrick, D.M. 1987. Fibrolite in contact aureoles of Donegal, Ireland. Am. Miner. 72, 240-254.

Kerrick, D.M. 1990. The Al_2SiO_5 polymorphs. Rev. Miner. 22, 1-406.

Kerrick, D.M. (ed.) 1991. *Contact metamorphism*. Rev. Miner. 26.

Kerrick, D.M. and Darken, L.S. 1975. Statistical thermodynamic models for ideal oxide and silicate solid solutions, with applications to plagioclase. Geochim. Cosmochim. Acta 38, 1431-1442.

Kerrick, D.M. and Jacobs, G.K. 1981. A modified Redlich-Kwong equation for H_2O, CO_2 and H_2O-CO_2 mixtures at elevated pressures and temperatures. Am. J. Sci. 21, 735-767.

Kerrick, D.M. and Speer, J.A. 1988. The role of minor element solid solution on the andalusite-sillimanite equilibrium in metapelites and peraluminous granitoids. Am. J. Sci. 288, 152-192.

Kerrick, D.M. and Woodsworth, G.J. 1989. Aluminum silicates in the Mount Raleigh pendant, British Columbia. J. Metamorphic Geol. 7, 547-563.

Kerrick, D.M., Lasaga, A.C. and Raeburn, S.P. 1991. Kinetics of heterogeneous reactions. Rev. Miner. 26, 583-671.

Kieffer, S.W. 1985. Heat capacity and entropy: systematic relations to lattice vibrations. Rev. Miner. 14, 65-126.

Kielland, J. 1935. Thermodynamics of base-exchange equilibria of some different kinds of clays. J. Soc. Chem. Ind. (London) 54, 232T-234T.

Kingery, W.D. 1959. Introduction. In *Kinetics of high temperature processes*, ed. by W.D. Kingery. The Technology Press (MIT), and J. Wiley & Sons, New York.

Kirkpatrick, R.J. 1975. Crystal growth from the melt: a review. Am. Miner. 60, 798-814.

Kirkpatrick, R.J. 1981. Kinetics of crystallization in igneous rocks. Rev. Miner. 8, 321-398.

Kitayama, K. and Katsura, T. 1968. Activity measurements in orthosilicate and metasilicate solutions-I. Mg_2SiO_4-Fe_2SiO_4 and $MgSiO_3$-$FeSiO_3$ at 1204 °C. Bull. Chem. Soc. Japan 41, 1146-1151.

Kitsul, V.I. 1964. Regularity in association of minerals in metamorphic cabonate rocks. Akad. Nauk SSSR, Yakut Fil. Sibirsk Inst. Geol. 1964, 5-17.

Klein, C. Jr. 1966. Mineralogy and petrology of the metamorphosed Wabush iron formation, southwestern Labrador. J. Petrology, 7, 246-305.

Knight, C.A. 1971. Breeding of crystal nuclei by classical nucleation: theory and some observations and experiments. J. Crystal Growth 11, 201-217.

Korzhinskii, D.S. 1936a. Paragenetic analysis of quartz-bearing low-calcium schists of Archaean complex of Southern Pribaikalje. Zapiski Mineral Obsch., Ser. 2, 65, 247-280.

Korzhinskii, D.S. 1936b. Mobility and inertness of components in metasomatism. Izv. Akad. Nauk SSSR, ser. geol. 1, 36-60.

Korzhinskii, D.S. 1959. *Physiochemical basis of the analysis of the paragenesis of minerals.* Consultants Bureau, New York.

Korzhinskii, D.S. 1970. *Theory of metasomatic zoning*. Trans. by J. Agrell. Clarendon Press, Oxford.

Koziol, A.M. 1990. Activity-composition relationships of binary Ca-Fe and Ca-Mn garnets determined by reversed, displaced equilibrium experiments. Am. Miner. 75, 319-327.

Koziol, A.M. and Bohlen, S.R. 1992. Solution properties of almandine-pyrope garnet as determined by phase equilibrium experiments. Am. Miner. 77, 765-773.

Koziol, A.M. and Newton, R.C. 1988. Redetermination of the anorthite breakdown reaction and improvement of the plagioclase-garnet-Al_2SiO_5-quartz geobarometer. Am. Miner. 73, 216-223.

Koziol, A.M. and Newton, R.C. 1989. Grossular activity-composition relationships in ternary garnets determined by reversed, displacement equilibrium experiments. Contr. Miner. Petrol. 103, 423-433.

Kracek, F.C. and Neuvonen, K.J. 1952. Thermochemistry of plagioclase and alkali feldspars. Am. J. Sci. Bowen volume, 293-318.

Kranck, S.H. 1961. A study of phase equilibria in a metamorphic iron formation. J. Petrology 2, 137-184.

Krauskopf, K.B. 1979. *Introduction to geochemistry*. McGraw-Hill, New York.

Kretz, R. 1959. Chemical study of garnet, biotite, and hornblende from gneisses of Southwestern Quebec, with emphasis on the distribution of elements in coexisting minerals. J. Geol. 67, 371-402.

Kretz, R. 1961. Some applications of thermodynamics to coexisting minerals of variable composition. J. Geol. 69, 361-387.

Kretz, R. 1964. Analysis of equilibrium in garnet-biotite-sillimanite gneisses from Quebec. J. Petrology 5, 1-20.

Kretz, R. 1966a. Interpretation of the shape of mineral grains in metamorphic rocks. J. Petrology 7, 68-94.

Kretz, R. 1966b. Metamorphic differentiation at Einasleigh, Northern Queensland. J. Geol. Soc. Australia 13, 561-582.

Kretz, R. 1967. Granite and pegmatite studies at Northern Indian Lake, Manitoba. Geol. Surv. Canada, Bull. 148.

Kretz, R. 1968a. Growth of phlogopite crystals in marble from Quebec. Int. Miner. Assoc. 85-93.

Kretz, R. 1968b. Study of pegmatite bodies and enclosing rocks, Yellowknife-Beaulieu region, District of Mackenzie. Geol. Surv. Canada, Bull. 159.

Kretz, R. 1969. On the spatial distribution of crystals in rocks. Lithos 2, 39-65.

Kretz, R. 1974. Some models for the rate of crystallization of garnet in metamorphic rocks. Lithos 7, 123-131.

Kretz, R. 1978. Distribution of Mg, Fe^{2+} and Mn in some calcic pyroxene-hornblende-biotite-garnet gneisses and amphibolites from the Grenville Province. J. Geol. 86, 599-619.

Kretz, R. 1980. Occurrence, mineral chemistry, and metamorphism of Precambrian carbonate rocks in a portion of the Grenville province. J. Petrology 21, 573-620.

Kretz, R. 1981. Site-occupancy interpretation of the distribution of Mg and Fe between orthopyroxene and clinopyroxene in metamorphic rocks. Can. Miner. 19, 493-500.

Kretz, R. 1982a. Transfer and exchange equilibria in a portion of the pyroxene quadrilateral, as deduced from natural and experimental data. Geochim. Cosmochim. Acta 46, 411-421.

Kretz, R. 1982b. A model for the distribution of trace elements between calcite and dolomite. Geochim. Cosmochim. Acta 46, 1979-1981.

Kretz, R. 1985. Calculation and illustration of uncertainty in geochemical analyses. J. Geol. Educ. 33, 40-44.

Kretz, R. 1988. SEM study of dolomite microcrystals in Grenville marble. Am. Miner. 73, 619-631.

Kretz, R. 1990. Biotite and garnet compositional variation and mineral equilibria in Grenville gneisses of the Otter Lake area, Quebec. J. Metamorphic Geol. 8, 493-506.

Kretz, R. 1993. A garnet population in Yellowknife schist, Canada. J. Metamorphic Geol. 11, 101-120.

Kretz, R. 1994. Petrology of veined gneisses of the Otter complex, Southern Grenville Province. Can. J. Earth Sci. (in press).

Kretz, R., Hartree, R. and Jones, P. 1989. Metasomatic crystallization of muscovite in granite and tourmaline in schist related to pegmatite emplacement near Yellowknife, Canada. Contr. Miner. Petrol. 102, 191-204.

Kretz, R., Jones, P. and Hartree, R. 1989. Grenville metagabbro complexes of the Otter Lake area, Quebec. Can. J. Earth Sci. 26, 215-230.

Kretz, R., Loop, J. and Hartree, R. 1989. Petrology and Li-Be-B geochemistry of muscovite-biotite granite and associated pegmatite near Yellowknife, Canada. Contr. Miner. Petrol. 102, 174-190.

Kridelbaugh, S. 1973. The kinetics of the reaction: calcite + quartz → wollastonite + carbon dioxide at elevated temperatures and pressures. Am. J. Sci. 273, 757-777.

Kröger, F.A, 1985. Point defects in solids: physics, chemistry, and thermodynamics. In *Point defects in minerals*, ed. by R.N. Schock. Geophys. Monograph 31, 1-7, Am. Geophys. Union.

Kroll, H. and Müller, W.F. 1980. X-ray and electron-optical investigation of synthetic high-temperature plagioclases. Phys. Chem. Miner. 5, 255-277.

Kurata, H. and Banno, S. 1974. Low-grade progressive metamorphism of pelitic schists of the Sazare area, Sanbagawa Metamorphic Terrane in central Sikoku, Japan. J. Petrology 15, 361-382.

Lacroix, M. 1889. Contribution à l'étude des gneiss à pyroxène et des roches wernérite. Bull. Soc. Franc. Miner. XII, 83-359.

Lafeber, D. 1963. On the spatial distribution of fabric elements in rock and soil fabrics. Proc. Austr. N.Z. Conf. Soil Mech. Found. Eng., 4th, Adelaide 11, 185-199.

Laffitte, P. 1957. *Introduction à l'étude des roches métamorphiques et des gîtes métallifères.* Masson et Cie, Paris.

Laidler, K.J. 1969. *Theory of chemical reaction rates.* McGraw-Hill, New York.

Laidler, K.J. 1987. *Chemical kinetics.* Harper and Row, New York.

Lamb, W.M. and Valley, J.W. 1988. Granulite facies amphibole and biotite equilibria, and calculated peak-metamorphic water activities. Contr. Miner. Petrol. 100, 349-360.

Lamb, W.M., Valley, J.W. and Brown, P.E. 1987. Post-metamorphic CO_2-rich fluid inclusions in granulites. Contr. Miner. Petrol. 96, 485-495.

Langer, J.S. 1980. Instabilities and pattern formation in crystal growth. Rev. Mod. Phys. 52, 1-28.

Langer, J.S. 1989. Dendrites, viscous fingers, and the theory of pattern formation. Science 243, 1150-1156.

Lasaga, A.C. 1981. The atomistic basis of kinetics: defects in minerals. Rev. Miner. 8, 261-319.

Laves, F. 1952. Phase relations of the alkali feldspar. J. Geol. 60, 436-450, 549-574.

Laves, F. and Baskin, Y. 1956. On the formation of the rhombohedral graphite modification. Z. Kristal. 107, 337-356.

Le Claire, A.D. 1963. The analysis of grain boundary diffusion measurements. Brit. J. Appl. Phys. 14, 351-356.

Le Claire, A.D. 1976. Diffusion. In *Treatise on solid state chemistry* vol. 4, *Reactivity of solids*, ed. by N.B. Hannay. Plenum Press, New York.

Leelanandam, C. 1967. Chemical study of pyroxenes from the charnockitic rocks of Kondapalli (Andhra Pradesh) India, with emphasis on the distribution of elements in coexisting minerals. Miner. Mag. 36, 153-170.

Le Maitre, R.W. 1982. *Numerical petrology.* Elsevier, Amsterdam.

Lentz, D.R. 1991. Radioelement distribution in U, Th, Mo, and rare-earth-element pegmatites, skarns, and veins in a portion of the Grenville Province, Ontario and Quebec. Can. J. Earth Sci. 28, 1-12.

Lentz, D.R. 1992. *Petrogenesis of U-, Th-, Mo- and REE-bearing pegmatites, skarns, and veins in the Central Metasedimentary Belt of the Grenville Province, Ontario and Quebec.* Unpub. Ph.D. Thesis, The University of Ottawa, Ottawa, Canada.

Lerman, A. 1965. Paleoecological problems of Mg and Sr in biogenic calcites in light of recent thermodynamic data. Geochim. Cosmochim. Acta 29, 977-1002.

Lin, L.S., Goldstein, J.I. and Williams, D.B. 1977. Analytical electron microscopy study of the plessite structure in the Carleton iron meteorite. Geochim. Cosmochim. Acta 41, 1861-1874.

Lindsley, D.H. 1983. Pyroxene thermometry. Am. Miner. 68, 477-493.

Lindsley, D.H. and Andersen, D.J. 1983. A two-pyroxene thermometer. Proc. 13th Lunar Planet. Sci. Conf. Part 2, J. Geophys. Res. V. 88 Sup., A887-A906.

Lindsley, D.H. and Dixon, S.A. 1976. Diopside-enstatite equilibria at 850 °C to 1400 °C, 5 to 35 kb. Am. J. Sci. 276, 1285-1301.

Lindsley, D.H. and Frost, B.R. 1992. Equilibria among Fe-Ti oxides, pyroxenes, olivine, and quartz: Part I. Theory. Am. Miner. 77, 987-1003.

Lindsley, D.H., Grover, J.E. and Davidson, P.M. 1981. The thermodynamics of the $Mg_2Si_2O_6$-$CaMgSi_2O_6$ join: a review and an improved model. Adv. Phys. Geochem. 1, 149-175.

Lipson, H. and Stokes, A.R. 1942. The structure of graphite. Proc. Roy. Soc. (London) 181, 101-185.

Liu, L.-G. and Bassett, W.A. 1986. *Elements, oxides, and silicates. High-pressure phases with implications for the Earth's interior.* Oxford University Press, New York.

Liu, M. and Yund, R.A. 1992. NaSi-CaAl interdiffusion in plagioclase. Am. Miner. 77, 275-283.

Liu, M. and Yund, R.A. 1993. Transformation kinetics of polycrystalline aragonite to calcite: new experimental data, modelling and implications. Contr. Miner. Petrol. 114, 465-478.

Lloyd, R.V., Lumsden, D.N. and Gregg, J.M. 1985. Relationship between paleotemperatures of metamorphic dolomites and ESR determined Mn (II) partitioning ratios. Geochim. Cosmochim. Acta 49, 2565-2568.

Loewinson-Lessing, F.Y. 1954. *A historical survey of petrology.* Oliver & Boyd, Edinburgh.

Long, J.V.P. and Agrell, S.O. 1965. The cathodo-luminescence of minerals in thin sections. Miner. Mag. 34, 318-326.

Lonker, S.W. 1981. The P-T-X relations of the cordierite-garnet-sillimanite-quartz equilibrium. Am. J. Sci., 281, 1056-1090.

Lonker, S.W. 1988. An occurrence of grandidierite, kornerupine, and tourmaline in southeastern Ontario, Canada. Contr. Miner. Petrol. 98, 502-516.

Lorimer, G.W. and Champness, P.E. 1973. The origin of the phase distribution in two perthitic alkali-feldspars. Phil. Mag. 28, 1391-1403.

Louisnathan, S.J. and Smith, J.V. 1968. Cell dimensions of olivine. Miner. Mag. 37, 1123-1134.

Lovering, T.S. 1935. Theory of heat conduction applied to geological problems. Geol. Soc. Am. Bull. 46, 69-94.

Lumsden, J. 1952. *Thermodynamics of alloys.* The Institute of Metals, London.

Luth, W.C. 1974. Analysis of experimental data on alkali feldspars: unit cell parameters and solvi. In *The feldspars*, ed. by W.S. MacKenzie and J. Zussman. Manchester University Press, pp. 249-296.

Luth, W.C. and Fenn, P.M. 1973. Calculation of binary solvi with special reference to the sanidine-high albite solvus. Am. Miner. 58, 1009-1015.

Luth, W.C. and Tuttle, O.F. 1966. The alkali feldspar solvus in the system Na_2O-K_2O-Al_2O_3-SiO_2-H_2O. Am. Miner. 51, 1359-1373.

Lyons, J.B. and Morse, S.A. 1970. Mg/Fe partitioning in garnet and biotite from some granitic, peltic, and calcic rocks. Am. Miner. 55, 231-244.

Mahan, S.M. and Rogers, J.J.W. 1968. A study of grain contacts in some high-grade metamorphic rocks. Am. Miner. 53, 323-327.

Manning, J.R. 1968. *Diffusion kinetics for atoms in crystals.* Van Nostrand, Princeton, NJ.

Marlow, P.C. and Etheridge, M.A. 1977. Development of layered crenulation cleavage in mica schists of the Kammantoo Group near Macclesfield, South Australia. Geol. Soc. Am. Bull. 88, 873-882.

Martin, B. and Fyfe, W.S. 1970. Some experimental and theoretical observations on the kinetics of hydration reactions with particular references to serpentinization. Chem. Geol. 6, 185-202.

Martin, J.W. and Doherty, R.D. 1976. *Stability of microstructure in metallic systems.* Cambridge University Press.

Maruyama, S., Liou, J.G. and Suzuki, K. 1982. The peristerite gap in low-grade metamorphic rocks. Contr. Miner. Petrol. 81, 268-276.

Mason, R. 1978. *Petrology of the metamorphic rocks.* George Allen & Unwin, London.

Mason, R.A. 1982. Trace element distributions between the perthite phases of alkali feldspar from pegmatites. Miner. Mag. 45, 101-106.

Mather, J.D. 1970. The biotite isograd and the lower greenschist facies in the Dalradian rocks of Scotland. J. Petrology 11, 253-275.

Matsui, Y. and Banno, S. 1965. Intracrystalline exchange equilibrium in silicate solid solutions. Proc. Japan. Acad. 41, 461-466.

Matsui, Y. and Banno, S. 1970. Partition of divalent transition metals between coexisting ferromagnesian minerals. Chem. Geol. 5, 259-265.

McCallister, R.H., Finger, L.W. and Ohashi, Y. 1976. Intracrystalline Fe^{2+}-Mg equilibria in three natural Ca-rich clinopyroxenes. Am. Miner. 61, 671-676.

McConnell, J.D.C. 1965. Electron optical study of effects associated with partial inversion in a silicate phase. Phil. Mag. 11, 1289-1301.

McCrone, W.C. 1949. Boundary migration and grain growth. Disc. Faraday Soc. 5, 158-166.

McGlashan, M.L. 1973. Internationally recommended names and symbols for physicochemical quantities and units. Ann. Rev. Phys. Chem. 24, 51-76.

McIntire, W.L. 1963. Trace element partition coefficients—a review of theory and applications to geology. Geochim. Cosmochim. Acta 27, 1209-1264.

McKie, D. 1959. Yoderite, a new hydrous magnesium iron alumino-silicate from Mautia Hill, Tanganyika. Miner. Mag. 32, 282-307.

McLellan, E.L. 1989. Sequential formation of subsolidus and anatectic migmatites in response to thermal evolution, eastern Scotland. J. Geol. 97, 165-182.

Medaris, L.G. Jr. 1975. Coexisting spinel and silicates in alpine peridotites of the granulite facies. Geochim. Cosmochim. Acta 39, 947-958.

Menzies, M.A. and Hawkesworth, C.J. (eds.). 1987. *Mantle metasomatism.* Academic Press, New York.

Merkel, G.A. and Blencoe, J.G. 1982. Thermodynamic procedures for treating the monoclinic/triclinic inversion as a high-order phase transition in equations of state for binary analbite-sanidine feldspars. Adv. Phys. Geochem. 2, 243-284.

Merkle, K.L. 1991. High-resolution electron microscopy of interfaces in fcc materials. Ultramicroscopy 37, 130-152.

Merkle, K.L. and Smith, D.J. 1987. Atomic structure of symmetric tilt grain boundaries in NiO. Phys. Rev. Letters 59, 2887-2890.

Merlet, C. and Bodinier, J.-L. 1990. Electron microprobe determination of minor and trace transition elements in silicate minerals: a method and its application to mineral zoning in the peridotite nodule PHN 1611. Chem. Geol. 83, 55-69.

Merz, C., Thelin, P. and Persoz, F.-P. 1989. Influences respectives du métamorphisme et de la déformation sur l'état des feldspaths potassiques du granite de Medel (Massif du Gothard, Alpes Centrales Suisses). Schweizer. Miner. Petrograph. Mitt. 69, 359-375.

Metz, P. 1970. Experimentelle Untersuchung der Metamorphose von kieselig dolomitischen Sedimenten, II. Die Bildungsbedingungen des Diopsids. Contr. Miner. Petrol. 28, 221-250.

Metz, P. 1976. Experimental investigation of metamorphism of siliceous dolomites III. Equilibrium data for the reaction 1 tremolite + 11 dolomite = 8 forsterite + 13 calcite + 9 CO_2 + H_2O for total pressures of 3000 and 5000 bars. Contr. Miner. Petrol. 58, 137-148.

Metz, P. and Puhan, D. 1970. Experimentelle Untersuchung der Metamorphose von kieselig dolomitischen Sedimenten I. Die Gleichgewichts daten der Reaction 3 Dolomit + 4 Quarz + $1H_2O$ = 1 Talk + 3 Calcit + $3CO_2$ für die Gesamtgasdrucke von 1000, 3000, und 5000 Bar. Contr. Miner. Petrol. 26, 302-314.

Metz, P.W. and Winkler, H.G.F. 1963. Experimentelle Gesteinsmetamorphose. VII. Die Bildung von Talk aus kieseligem Dolomit. Geochim. Cosmochim. Acta 27, 431-457.

Miller, G.H., Rossman, G.R. and Harlow, G.E. 1987. The natural occurrence of hydroxide in olivine. Phys. Chem. Miner. 14, 461-472.

Misener, D.J. 1974. Cationic diffusion in olivine to 1400 °C and 35 kbar. In *Geochemical transport and kinetics*, ed. by A.W. Hofmann, B.J. Giletti, H.S. Yoder, Jr. and R.A. Yund. Carnegie Inst. Washington, Pub. 634.

Miyashiro, A. 1953. Calcium-poor garnet in relation to metamorphism. Geochim. Cosmochim. Acta 4, 179-208.

Miyashiro, A. 1961. Evolution of metamorphic belts. J. Petrology 2, 277-311.

Miyashiro, A. 1964. Oxidation and reduction in the Earth's crust with special reference to the role of graphite. Geochim. Cosmochim. Acta 28, 717-729.

Miyashiro, A. 1973. *Metamorphism and metamorphic belts*. George Allen & Unwin, London.

Miyazaki, K. 1991. Ostwald ripening of garnet in high P/T metamorphic rocks. Contr. Miner. Petrol. 108, 118-128.

Moffatt, W.G., Pearsall, G.W. and Wulff, J. 1964. *The structure and properties of materials* vol. 1. *Structure*. J. Wiley & Sons, New York.

Molin, G.M., Saxena, S.K. and Brizi, E. 1991. Iron-magnesium order-disorder in an orthopyroxene crystal from the Johnstown meteorite. Earth Planet. Sci. Letters 105, 260-265.

Montaser, A. and Golightly, D.W. 1987. *Inductively coupled plasma in analytical atomic spectrometry*. VCH Publishers, New York.

Moore, J.N. and Kerrick, D.M. 1976. Equilibria in siliceous dolomites of the Alta Aureole, Utah. Am. J. Sci. 276, 502-524.

Moore, P.B., Sen Gupta, P.K. and Schlemper, E.O. 1989. Kornerupine: chemical crystallography, comparative crystallography, and its cation relation to olivine and to Ni_2In intermetallic. Am. Miner. 74, 642-655.

Moore, P.G. 1954. Spacing in plant populations. Ecology 35, 222-227.

Mori, T. and Green, D.H. 1975. Pyroxenes in the system $Mg_2Si_2O_6$-$CaMgSi_2O_6$ at high pressure. Earth Planet. Sci. Letters 26, 277-286.

Morioka, M. 1981. Cation diffusion in olivine-II. Ni-Mg, Mn-Mg, Mg and Ca. Geochim. Cosmochim. Acta 45, 1573-1580.

Morioka, M. and Nagasawa, H. 1991. Ionic diffusion in olivine. Adv. Phys. Geochem. 8, 176-197.

Moxham, R.L. 1965. Distribution of minor elements in coexisting hornblendes and biotites. Can. Miner. 8, 204-240.

Mueller, R.F. 1960. Compositional characteristics and equilibrium relations in mineral assemblages of a metamorphosed iron formation. Am. J. Sci. 258, 449-497.

Mueller, R.F. 1961a. Analysis of relations among Mg, Fe, and Mn in certain metamorphic minerals. Geochim. Cosmochim. Acta 25, 267-296.

Mueller, R.F. 1961b. Oxidation in high temperature petrogenesis. Am. J. Sci. 259, 460-480.

Mueller, R.F. 1962. Energetics of certain silicate solutions. Geochim. Cosmochim. Acta 26, 581-598.

Mueller, R.F. 1963. Interaction of chemistry and mechanics in magmatism. J. Geol. 71, 759-772.

Mueller, R.F. 1964. Theory of immiscibility in mineral systems. Miner. Mag. 33, 1015-1023.

Mueller, R.F. 1967a. Model for order-disorder kinetics in certain quasi-binary crystals of continuously variable composition. J. Phys. Chem. Solids 28, 2239-2243.

Mueller, R.F. 1967b. Mobility of the elements in metamorphism. J. Geol. 75, 565-582.

Mueller, R.F., 1972. Stability of biotite: a discussion. Am. Miner. 57, 300-316.

Myers, J. and Eugster, H.P. 1983. The system Fe-Si-O; oxygen buffer calibrations to 1500 K. Contr. Miner. Petrol. 82, 75-90.

Myers, J.S. 1978. Formation of banded gneisses by deformation of igneous rocks. Precambrian Res. 6, 43-64.

Nabarro, F.R.N. 1940. The influence of elastic strain on the shape of particles segregating in an alloy. Proc. Phys. Soc. 52, 90-93.

Nafziger, R.H. and Muan, A. 1967. Equilibrium phase composition and thermodynamic properties of olivines and pyroxenes in the system $MgO-FeO-SiO_2$. Am. Miner. 52, 1364-1385.

Naggar, M.H. and Atherton, M.P. 1970. The composition and metamorphic history of some aluminium silicate-bearing rocks from the aureoles of the Donegal granites. J. Petrology 11, 549-589.

Nagle, J.F. 1975. Critical phenomena and phase transitions. In *Treatise on solid state chemistry*, vol. 5, *Change of state*, ed. by N.B. Hannay. Plenum Press, New York.

Nagy, K.L. and Giletti, B.J. 1986. Grain boundary diffusion of oxygen in a macroperthite feldspar. Geochim. Cosmochim. Acta 50, 1151-1158.

Narayan, C. and Goldstein, J.I. 1985. A major revision of iron meteorite cooling rates-an experimental study of the growth of the Widmanstätten pattern. Geochim. Cosmochim. Acta 49, 397-410.

Navrotsky, A. 1971. The intracrystalline cation distribution and the thermodynamics of solid solution formation in the system $FeSiO_3-MgSiO_3$. Am. Miner. 56, 201-211.

Navrotsky, A., 1977. Progress and new directions in high temperature calorimetry. Phys. Chem. Miner. 2, 89-104.

Navrotsky, A. 1987. Models of crystalline solutions. Rev. Miner. 17, 35-69.

Navrotsky, A. and Kleppa, O.J. 1967. The thermodynamics of cation distributions in simple spinels. J. Inorg. Nuc. Chem. 29, 2701-2714.

Navrotsky, A., Newton, R.C. and Kleppa, O.J. 1973. Sillimanite-disordering enthalpy by calorimetry. Geochim. Cosmochim. Acta 37, 2497-2508.

Neddermeyer, H. 1990. STM studies of nucleation and the initial stages of film growth. Critical Reviews in Solid State and Material Sciences 16, 309-335.

Nehru, C.E. and Wyllie, P.J. 1974. Electron microprobe measurement of pyroxenes coexisting with H_2O-undersaturated liquid in the join $CaMgSi_2O_6-Mg_2Si_2O_6-H_2O$ at 30 kilobars with applications to geothermometry. Contr. Miner. Petrol. 48, 221-228.

Nesbitt, B.E. and Essene, E.J. 1982. Metamorphic thermometry and barometry in a portion of the southern Blue Ridge province. Am. J. Sci., 282, 701-729.

Newton, R.C. 1966. Kyanite-sillimanite equilibrium at 750 °C. Science 151, 1222-1225.

Newton, R.C. 1985. Geobarometry of high-grade metamorphic rocks. Am. J. Sci. 285-A, 1-28.

Newton, R.C. 1986. Fluids of granulite facies metamorphism. Adv. Phys. Geochem. 5, 36-59.

Newton, R.C. and Fyfe, W.S. 1976. High-pressure metamorphism. In *The evolution of the crystalline rocks*, ed. by D.K. Bayley and R. Macdonald. Academic Press, London.

Newton, R.C. and Haselton, H.T. 1981. Thermodynamics of the garnet-plagioclase-Al_2SiO_5-quartz geobarometer. Adv. Phys. Geochem. 1, 131-147.

Newton, R.C. and Smith, J.V. 1967. Investigations concerning the breakdown of albite at depth in the Earth. J. Geol. 75, 268-286.

Newton, R.C. and Wood, B.J. 1979. Thermodynamics of water in cordierite and some petrologic consequences of cordierite as a hydrous phase. Contr. Miner. Petrol. 68, 391-405.

Newton, R.C. and Wood, B.J., 1980. Volume behavior of silicate solid solutions. Am. Miner. 65, 733-745.

Newton, R.C., Charlu, T.V. and Kleppa, O.J. 1977. Thermochemistry of high pressure garnets and clinopyroxenes in the system $CaO-MgO-Al_2O_3-SiO_2$. Geochim. Cosmochim. Acta 41, 369-377.

Newton, R.C., Charlu, T.V. and Kleppa, O.J. 1980. Thermochemistry of high structural state plagioclases. Geochim. Cosmochim. Acta 44, 933-941.

Newton, R.C., Goldsmith, J.R. and Smith, J.V. 1969. Aragonite crystallization from strained calcite at reduced pressures and its bearing on aragonite in low-grade metamorphism. Contr. Miner. Petrol. 22, 335-348.

Nickel, E.H. 1955. The distribution of major and minor elements among some co-existing ferromagnesian silicates. Am. Miner. 39, 486-493.

Nickel, E.H. and Nichols, M.C. 1991. *Mineral reference manual.* Van Nostrand Reinhold, New York.

Nicolas, A. and Poirier, J.P. 1976. *Crystalline plasticity and solid state flow in metamorphic rocks.* J. Wiley & Sons, London.

Niggli, P. 1954. *Rocks and mineral deposits.* Trans. by R.L. Parker. Freeman, San Francisco.

Nishiyama, T. 1983. Steady diffusion model for olivine-plagioclase corona growth. Geochim. Cosmochim. Acta 47, 283-294.

Nix, F.C. and Shockley, W. 1938. Order-disorder transformations in alloys. Rev. Mod. Phys. 10, 1-71.

Nixon, P.H. and Boyd, F.R. 1973. Petrogenesis of the granular and sheared ultrabasic nodule suite in kimberlites. In *Lesotho kimberlites*, ed. by P.H. Nixon. Lesotho National Development Corp.

Nockolds, S.R. and Mitchell, R.L. 1948. The geochemistry of some Caledonian plutonic rocks: a study in the relationship between the major and trace elements in igneous rocks and their minerals. Trans. Royal Soc. Edinburgh 61, 533-575.

Oglesby, M.L., Gutshall, P.L. and Phillips, J.M. 1976. Cleavage surface energy of selenite. Am. Miner. 61, 295-298.

Ohashi, Y. and Burnham, C.W. 1972. Electrostatic and repulsive energies of the M1 and M2 cation sites in pyroxenes. J. Geophys. Res. 77, 5761-5765.

Oishi, Y. and Kingery, W.D. 1960. Self-diffusion of oxygen in single crystal and polycrystalline aluminum oxide. J. Chem. Phys. 33, 480-486.

Okrusch, M. 1969. Die Gneishornfelse um Steinach in der Oberpfalz. Contr. Miner. Petrol. 22, 32-72.

Okrusch, M. and Evans, B.E. 1970. Minor element relationships in coexisting andalusite and sillimanite. Lithos 3, 261-268.

Olsen, S.N. 1977. Origin of the Baltimore gneiss migmatites at Piney Creek, Maryland. Geol. Soc. Am. Bull. 88, 1089-1101.

Olsen, T.S. and Kohlstedt, D.L. 1985 Natural deformation and recrystallization of some intermediate plagioclase feldspars. Tectonophys. 111, 107-131.

O'Nions, R.K. and Smith, D.G.W. 1973. Bonding in silicates: an assessment of bonding in orthopyroxene. Geochim. Cosmochim. Acta 37, 249-257.

Ono, A. 1975. Spatial distribution, size and chemistry of garnet in a hornfels from the Takato area, Ryoke metamorphic belt, central Japan. J. Japanese Assoc. Miner. Petrol. Econ. Geol. 70, 169-173.

Ono, A. 1977. Chemical reactions at the boundary between gneiss and amphibolite in the Ryoke metamorphic terrain at Takato, Japan. J. Geol. Soc. Japan 83, 33-40

Onsager, L. 1931. Reciprocal relations in irreversible processes. Phys. Rev. 37, 405-426; 38, 2265-2279.

Orr, R.L. 1953. High temperature heat contents of magnesium orthosilicate and ferrous orthosilicate. J. Am. Chem. Soc. 75, 528-529.

Ortoleva, P., Merino, E. and Strickholm, P. 1982. Kinetics of metamorphic layering in anisotropically stressed rocks. Am. J. Sci. 282, 617-643.

Orville, P.M. 1963. Alkali ion exchange between vapour and feldspar phases. Am. J. Sci. 261, 201-237.

Orville, P.M. 1972. Plagioclase cation exchange equilibria with aqueous chloride solution: results at 700 °C and 2000 bars in the presence of quartz. Am. J. Sci. 272, 234-272.

Osborne, F.F. 1930. The nepheline-gneiss complex near Egan Chute, Dungannon Township, and its bearing on the origin of nepheline syenite. Am. J. Sci. 20, 33-60.

Owen, D.C. and McConnell, J.D.C. 1971. Spinodal behaviour in an alkali feldspar. Nature 230, 118-119.

Ozawa, K. 1983. Evaluation of olivine-spinel geothermometry as an indicator of thermal history of peridolites. Contr. Miner. Petrol. 82, 52-65.

Paktunç, A.D. 1983. *Petrology and geochemistry of some ultramafic rocks of the Thompson Nickel Belt and the Cuthbert Lake Dikes of the Pikwitonei Region, Northern Manitoba.* Ph.D. Thesis, The University of Ottawa, Ottawa, Canada.

Paktunç, A.D. 1984. Metamorphism of the ultramafic rocks of the Thompson Nickel Belt, Northern Manitoba. Can. Miner. 22, 77-91.

Papike, J.J., Ross, M. and Clark, J.R. 1969. Crystal-chemical characterization of clinoamphiboles based on five new structure refinements. Miner. Soc. Am. Special Paper 2, 117-136.

Parsonage, N.G. and Staveley, L.A.K. 1978. *Disorder in crystals.* Clarendon Press, Oxford.

Parsons, I. and Brown, W.L. 1991. Mechanisms and kinetics of exsolution-structural control of diffusion and phase behavior in alkali feldspars. Adv. Phys. Geochem. 8, 304-344.

Patrick, B.E. and Day, H.W. 1989. Controls on the first appearance of jadeitic pyroxene, northern Diablo Range, California. J. Metamorphic Geol. 7, 629-639.

Pattison, D.R.M. 1992. Stability of andalusite and sillimanite and the Al_2SiO_5 triple point: constraints from the Ballachulish aureole, Scotland. J. Geol. 100, 423-446.

Pauling, L. 1935. The structure and entropy of ice and of other crystals with some randomness of atomic arrangement. J. Am. Chem. Soc. 57, 2680-2684.

Pearson, D.G., Davies, G.R., Nixon, P.H. and Milledge, H.J. 1989. Graphitized diamonds from a peridotite massif in Morocco and implications for anomalous diamond occurrences. Nature 338, 60-62.

Pearson, G.R. and Shaw, D.M. 1960. Trace elements in kyanite, sillimanite and andalusite. Am. Miner. 45, 808-817.

Perchuk, L.L. 1989. Consistency in some Fe-Mg geothermometers based on Nernst's law: revision. Geochem. Int. 26, no 12, 1-10.

Perchuk, L.L. and Lavrent'eva, I.V. 1983. Experimental investigation of exchange equilibria in the system cordierite-garnet-biotite. Adv. Phys. Geochem. 3, 199-239.

Perchuk, L.L and Ryabchikov, I.D. 1968. Mineral equilibria in the system nepheline-alkali feldspar-plagioclase and their petrologic significance. J. Petrology 9, 123-167.

Perchuk, L.L. and Surikov, V.V. 1971. Thermodynamic problems of distribution of cations in superlattices of minerals. Int. Geol. Rev. 13, 1657-1670.

Perkins, D., Essene, E.J. and Marcotty, L.A. 1982. Thermometry and barometry of some amphibolite-granulite facies rocks from the Otter Lake area, southern Quebec. Can. J. Earth Sci. 19, 1759-1774.

Petersen, E.U., Essene, E.J., Peacor, D.R. and Marcotty, L.A. 1989. The occurrence of högbomite in high-grade metamorphic rocks. Contr. Miner. Petrol. 101, 350-360.

Petruk, W. (ed.) 1989. *Short course on image analysis applied to mineral and Earth sciences.* Mineralogical Association of Canada.

Philpotts, A.R. 1990. *Principles of igneous and metamorphic petrology.* Prentice-Hall, Englewood Cliffs, NJ.

Phinney, W.C. 1963. Phase equilibria in the metamorphic rocks of St. Paul Island and Cape North, Nova Scotia. J. Petrology, 4, 90-130.

Pitcher, W.S. 1965. The aluminium silicate polymorphs. In *Controls of metamorphism*, ed. by W.S. Pitcher and G.W. Flinn. Oliver and Boyd, Edinburgh.

Pitcher, W.S. and Read, H.H. 1963. Contact metamorphism in relation to manner of emplacement of the granites of Donegal, Ireland. J. Geol. 71, 261-296.

Porter, D.A. and Easterling, K.E. 1981. *Phase transformations in metals and alloys*. Van Nostrand Reinhold, New York.

Potts, P.J., Tindle, A.G. and Issacs, M.C. 1983. On the precision of electron microprobe data: a new test for the homogeneity of mineral standards. Am. Miner. 68, 1237-1242.

Povarennykh, A.S. 1972. *Crystal chemical classification of minerals*. Plenum Press, New York.

Powell, R., Condliffe, D.M. and Condliffe, E. 1984. Calcite-dolomite geothermometry in the system $CaCO_3$-$MgCO_3$-$FeCO_3$; an experimental study. J. Metamorphic Geol. 2, 33-41.

Pownceby, M.I., Wall, V.I. and O'Neill, H.St.C. 1987. Fe-Mn partitioning between garnet and ilmenite; experimental calibration and applications. Contr. Miner. Petrol. 97, 116-126.

Prigogine, I. and Defay, R. 1954. *Chemical thermodynamics* (translated and revised by D.H. Everett). Longmans, Green & Co., London.

Prissok, F. and Lehmann, G. 1986. An EPR study of Mn^{2+} and Fe^{3+} in dolomites. Phys. Chem. Miner. 13, 331-336.

Puhan, D. 1976. Metamorphic temperature determined by means of the dolomite-calcite solvus geothermometer-examples from the central Damara orogen (South West Africa). Contr. Miner. Petrol. 58, 23-28.

Puhan, D. 1984. Microtexture of dolomite within exsolved magnesian calcite — examples from the Damara orogen (Namibia). Contr. Miner. Petrol. 87, 98-99.

Puhan, D. and Hoffer, E. 1973. Phase relations of talc and tremolite in metamorphic calcite-dolomite sediments in the southern portion of the Damara belt (South West Africa). Contr. Miner. Petrol. 40, 207-214.

Putnis, A. 1992. *Introduction to mineral sciences*. Cambridge University Press.

Putnis, A. and McConnell, J.D.C. 1980. *Principles of mineral behaviour*. Blackwell Scientific Publications, Oxford.

Puustinen, K. 1974. Dolomite exsolution textures in calcite from the Siilinjärvi carbonatite complex, Finland. Bull. Geol. Soc. Finland 46, 151-159.

Raase, P. and Morteani, G. 1976. The potassic feldspar in metamorphic rocks from the western Hohe Tauern area, Eastern Alps. Geol. Rund. 65, 422-436.

Raghaven, V. and Cohen, M. 1975. Solid-state phase transformations. In *Treatise on solid state chemistry*, vol. 5, ed. by N.B. Hannay. Plenum Press, New York.

Ramberg, H. 1948. Titanic iron ores formed by dissociation of silicates in granulite facies. Econ. Geol. 43, 553-570.

Ramberg, H. 1952a. *The origin of metamorphic and metasomatic rocks*. University of Chicago Press.

Ramberg, H. 1952b. Chemical bonds and the distribution of cations in silicates. J. Geol. 60, 331-355.

Ramberg, H. 1956. Pegmatites in West Greenland. Geol. Soc. Am. Bull. 67, 185-214.

Ramberg, H. 1961. A study of veins in Caledonian rocks around Trondheim Fjord, Norway. Norsk Geol. Tidsskr. 41, 1-43.

Ramberg, H. and DeVore, G. 1951. The distribution of Fe^{++} and Mg^{++} in coexisting olivines and pyroxenes. J. Geol. 59, 193-210.

Ramsay, C.R. 1973. The origin of biotite in Archean meta-sediments near Yellowknife, NWT, Canada. Contr. Miner. Petrol. 42, 43-54.

Ramsay, J.G. and Graham, R.H. 1970. Strain variation in shear belts. Can. J. Earth Sci. 7, 786-813.

Ranalli, G. 1987. *Rheology of the Earth*. Allen & Unwin, Boston.

Ratner, A.P. 1933. On the theory of the distribution of elecrolytes between a solid crystalline and a liquid phase. J. Chem. Phys. 1, 789-794.

Read, H.H. 1933. On quartz-kyanite rocks in Unst, Shetland Islands, and their bearing on metamorphic differentiation. Miner. Mag. 23, 317-328.

Read, H.H. 1957. *The granite controversy*. Interscience, New York.

Reed, S.J.B. 1990. Recent developments in geochemical microanalysis. Chem. Geol. 83, 1-9.

Reeder, R.J. 1983. Crystal chemistry of the rhombohedral carbonates. Rev. Miner. 11, 1-47.

Reeder, R.J. and Wenk, H.-R. 1983. Structure refinements of some thermally disordered dolomites. Am. Miner. 68, 769-776.

Reichert, J. 1966. Verteilung anorganischer Fremdionen bei der Kristallization von alkalichloriden. Contr. Miner. Petrol. 13, 134-160.

Reitan, P. 1959. Pegmatite veins and the surrounding rocks III. Structural control of small pegmatites in amphibolite, Rytterholmen, Kragerøfjord, Norway. Norsk Geol. Tidsskr. 39, 175-195.

Ribbe, P.H. 1984. Average structures of alkali and plagioclase feldspars. In *Feldspars and feldspathoids*, ed. by W.L. Brown. D. Reidel, Dordrecht.

Rice, J.M. 1983. Metamorphism of rodingites: Part I. Phase relations in a portion of the system $CaO-MgO-Al_2O_3-SiO_2-CO_2-H_2O$. Am. J. Sci. 283-A, 121-150.

Rich, R.A. 1979. Fluid inclusion evidence for Silurian evaporites in southeastern Vermont: Summary Geol. Soc. Am. Bull. pt. 1. 90, 901-902.

Richardson, S.W. 1968. Staurolite stability in part of the system Fe-Al-Si-O-H. J. Petrology, 9, 467-488.

Ringwood, A.E. 1975. *Composition and petrology of the Earth's mantle*. McGraw-Hill, New York.

Rinne, F. and Boecke, H.E. 1908. Über Thermometamorphose und Sammel krystallisation. Tschermak Min. Pet. Mitteilungen 27, 393-398.

Ripley, B.D. 1981. *Spatial statistics*. J. Wiley & Sons, New York.

Rivers, T. and Fyson, W.K. 1977. Shape, size, and orientation of muscovite crystals in a schist of variable metamorphic grade. Can. J. Earth Sci. 14, 185-195.

Rivers, T. and Mengel, F.C. 1988. Contrasting assemblages and petrogenetic evolution of corona and noncorona gabbros in the Grenville Province of Western Labrador. Can. J. Earth Sci., 25, 1629-1648.

Robertson, F.C., Birch, F. and MacDonald, G.J.F. 1957. Experimental determination of jadeite stability relations to 25,000 bars. Am. J. Sci. 255, 115-137.

Robie, R.A. and Hemingway, B.S. 1984. Entropies of kyanite, andalusite, and sillimanite: additional constraints on the pressure and temperature of the Al_2SiO_5 triple point. Am. Miner. 69, 298-306.

Robie, R.A., Hemingway, B.S. and Fisher, J.R. 1979. *Thermodynamic properties of minerals and related substances at 298.15 K and 1 bar (10^5 Pascals) pressure and at high temperatures*. U.S. Geol. Surv. Bull. 1452.

Robie, R.A., Hemingway, B.S. and Takei, H. 1982. Heat capacities and entropies of Mg_2SiO_4, Mn_2SiO_4, and Co_2SiO_4 between 5 and 380 K. Am. Miner. 67, 470-482.

Robin, P.-Y.F. 1974. Stress and strain in cryptoperthite lamellae and the coherent solvus of alkali feldspars. Am. Miner. 59, 1286-1298.

Robin, P.-Y.F. 1979. Theory of metamorphic segregation and related processes. Geochim. Cosmochim. Acta 43, 1587-1600.

Robinson, A.L. 1986. Consensus on silicon surface structure near. Science 232, 451-453.

Robinson, K., Gibbs, G.V. and Ribbe, P. 1973. The crystal structure of the humite minerals. IV Clinohumite and titanoclinohumite. Am. Miner. 58, 43-49.

Roedder, E. 1984. Fluid inclusions. Rev. Miner. 12, 1-644.

Roedder, E. 1990. Fluid inclusion analysis — prologue and epilogue. Geochim. Cosmochim. Acta 54, 495-507.

Roeder, P.L., Campbell, I.H. and Jamieson, H.E. 1979. A re-evaluation of the olivine-spinel geothermometer. Contr. Miner. Petrol. 68, 325-334.

Rogers, J.J.W. and Bogy, D.B. 1958. A study of grain contacts in granitic rocks. Science 127, 470-471.

Roozeboom, H.W.B. 1891. Über die Löslichkeit von Mischkrystallen, speziell zweier isomorpher Körper. Zeit. Phys. Chemie 8, 504-530.

Rosenberg, P.E. 1967. Susolidus relations in the system $CaCO_3$-$MgCO_3$-$FeCO_3$ between 350 and 550 °C. Am. Miner. 52, 787-796.

Rosenfeld, J.L. 1955. Interfacial tensions and measurement of stress differences in metamorphic rocks. Am. J. Sci. 253, 574-579.

Rothman, S.J. 1984. The measurement of tracer diffusion coefficients in solids. In *Diffusion in crystalline solids*, ed. by G.E. Murch and A.S. Nowick. Academic Press, Orlando.

Rubie, D.C. 1986. The catalysis of mineral reactions by water and restrictions on the presence of aqueous fluid during metamorphism. Miner. Mag. 50, 399-415.

Rubie, D.C. and Thompson, A.B. 1985. Kinetics of metamorphic reactions at elevated temperatures and pressures: an appraisal of available experimental data. Adv. Phys. Geochem. 4, 27-79.

Rumble, D. III. 1970. Thermodynamic analysis of phase equilibria in the system Fe_2TiO_4-Fe_3O_4-TiO_2. Carnegie Inst. Washington, Year Book 69, 198-207.

Rumble, D. III 1973. Fe-Ti oxide minerals from regionally metamorphosed quartzites of western New Hampshire. Contr. Miner. Petrol. 42, 181-195.

Russ, J.C. 1990. *Practical stereology*. Plenum, New York.

Sack, R.O. 1982. Reaction skarns between quartz-bearing and olivine-bearing rocks. Am. J. Sci. 282, 970-1011.

Saikumar, V. and Goldstein, J.I. 1988. An evaluation of the methods to determine the cooling rates of iron meteorites. Geochim. Cosmochim. Acta 52, 715-726.

Sanford, R.F. 1982. Growth of ultramafic reaction zones in greenschist to amphibolite facies metamorphism. Am. J. Sci. 282, 543-616.

Saxena, S.K. 1968. Chemical study of phase equilibria in charnockites, Varberg, Sweden. Am. Miner. 53, 1674-1695.

Saxena, S.K. 1971. Mg^{2+}-Fe^{2+} order-disorder in orthopyroxene and the Mg^{2+}-Fe^{2+} distribution between coexisting minerals. Lithos 4, 345-356.

Saxena, S.K. 1973. *Thermodynamics of rock-forming crystalline solutions*. Springer-Verlag, Berlin.

Saxena, S.K. 1974. Order-disorder in sillimanite. Contr. Miner. Petrol. 45, 161-167.

Saxena, S.K. and Ghose, S. 1971. Mg^{2+}-Fe^{2+} order-disorder and the thermodynamics of orthopyroxene crystalline solutions. Am. Miner. 56, 532-559.

Saxena, S.K., Sykes, J. and Eriksson, G. 1986. Phase equilibria in the pyroxene quadrilateral. J. Petrology 27, 843-852.

Schindler, P. and Ghose, S. 1970. Electron paramagnetic resonance of Mn^{2+} in dolomite and magnesite, and Mn^{2+} distribution in dolomites. Am. Miner. 55, 1889-1896.

Schmid, R. and Wood, B.J. 1976. Phase relationships in granulitic metapelites from the Ivrea-Verbano Zone (northern Italy). Contr. Miner. Petrol. 54, 255-279.

Schmidt, W. 1932. *Tektonik und Verformungslehre*. Borntraeger, Berlin.

Schock, H.H. and Puchelt, H. 1971. Rubidium and cesium distribution in salt minerals-I. Experimental investigations. Geochim. Cosmochim. Acta 35, 307-317.

Schreurs, J. 1984. The amphibolite-granulite facies transition in west Uusimaa, S.W. Finland. A fluid inclusion study. J. Metamorphic Geol. 2, 327-341.

Schreurs, J. 1985. Prograde metamorphism of metapelites, garnet-biotite thermometry and prograde changes of biotite chemistry in high-grade rocks of west Uusimaa, southwest Finland. Lithos. 18, 69-80.

Schreyer, W. 1988a. Experimental studies on metamorphism of crustal rocks under mantle pressures. Miner. Mag. 52, 1-26.

Schreyer, W. 1988b. Subduction of continental crust to mantle depths: petrological evidence. Episodes 11, 97-104.

Schrijver, K. 1973. Bimetasomatic plagioclase-pyroxene reaction zones in granulite facies. N. Jb. Miner. Abh. 119, 1-19.

Schubert, G. and Anderson, O.L. 1974. The Earth's thermal gradient. Physics Today, March issue, 28-34.

Schulien, S. 1980. Mg-Fe partitioning between biotite and a supercritical chloride solution. Contr. Miner. Petrol. 74, 85-93.

Schwarcz, H.P. 1967. The effect of crystal field stabilization on the distribution of transition metals between metamorphic minerals. Geochim. Cosmochim. Acta 31, 503-517.

Scotford, D.M. 1973. Strontium partitioning between coexisting K-feldspar and plagioclase as an indicator of metamorphic grade in southwestern New Hampshire. Geol. Soc. Am. Bull. 84, 3985-3995.

Seck, H.A. 1971a. Koexistierende Alkalifeldspäte und Plagioklase im System $NaAlSi_3O_8$-$KAlSi_3O_8$-$CaAl_2Si_2O_8$-H_2O bei Temperaturen von 650 °C bis 900 °C. N. Jb. Miner. Abh. 115, 315-345.

Seck, H.A. 1971b. Der Einfluss des Drucks auf die Zusammensetzung koexistierender Alkalifeldspäte und Plagioklase im System $NaAlSi_3O_8$-$KAlSi_3O_8$-$CaAl_2Si_2O_8$-H_2O. Contr. Miner. Petrol. 31, 67-86.

Seck, H.A. 1972. The influence of pressure on the alkali feldspar solvus from peraluminous and persilicic materials. Fortschritte Miner. 49, 31-49.

Seifert, F. 1970. Low-temperature compatibility relations of cordierite in haplopelites of the system K_2O-MgO-Al_2O_3-SiO_2-H_2O. J. Petrology 11, 73-99.

Sella, C. and Deicha, G. 1962. Études au microscope électronique des pores intergranulaires des gangues et des roches. Comp. Rend. Acad. Sci. (Paris) 254, 2796-2798.

Seltz, H. 1935. Perfect ternary solid solutions. J. Chem. Phys. 3, 503-505.

Sempels, J.-M. 1978. *Mathematical simulation of crystal nucleation, growth, and the formation of textures.* Unpublished Ph.D. Thesis, The University of Ottawa, Ottawa, Canada.

Sempels, J.-M. and Raymond, J. 1980. Mathematical simulation of the growth of single crystals. Computers and Geosciences 6, 211-226.

Sen, S.K. and Chakraborty, K.R. 1968. Magnesium-iron exchange equilibrium in garnet-biotite and metamorphic grade. N. Jb. Miner. Abh. 108, 181-207

Seng, H. 1936. Die Migmatitfrage und der Mechanismus parakristalliner Prägung. Geol. Rund. 27, 471-492.

Shackelford, J.F. 1983. A canonical hole model of grain boundaries and interfaces in oxides. In *Character of grain boundaries*, ed. by M.F.Yan and A.H. Heuer. Advances in Ceramics 8, 96-101.

Shaw, D.M., Moxham, R.L., Filby, R.H. and Lapkowsky, W.W. 1963. The petrology and geochemistry of some Grenville skarns. Can. Miner. 7, 578-616.

Shearer, C.K., Papike, J.J., Simon, S.B., Laul, J.C. and Christian, R.P. 1984. Pegmatite/wall rock interactions, Black Hills, South Dakota: progressive boron metasomatism adjacent to the Tip Top pegmatite. Geochim. Cosmochim. Acta 48, 2563-2579.

Shelley, D. 1974. Mechanical production of metamorphic banding -a critical appraisal. Geol. Mag. 111, 287-292.

Shewmon, P.G. 1963. *Diffusion in solids*. McGraw-Hill, New York.

Shewmon, P.G. 1969. *Transformations in metals*. McGraw-Hill, New York.

Sipling, P.J. and Yund, R.A. 1976. Experimental determination of the coherent solvus for sanidine-high albite. Am. Miner. 61, 897-906.

Skinner, B.J. 1966. Thermal expansion. In *Handbook of physical constants*, ed. by S.P. Clark Jr., Geol. Soc. Am., Mem. 97.

Skinner, B.J., Clark, S.P. Jr. and Appleman, D.E. 1961. Molar volumes and thermal expansion of andalusite, kyanite, and sillimanite. Am. J. Sci. 259, 651-668.

Skippen, G.B. 1971. Experimental data for reactions in siliceous marbles, J. Geol. 79, 457-481.

Skippen, G. 1974. An experimental model for low pressure metamorphism of siliceous dolomitic marble. Am. J. Sci. 274, 487-509.

Skippen, G.B. 1977. Dehydration and decarbonation equilibria. In *Short course in applications of thermodynamics*, ed. by H.J. Greenwood. Miner. Assoc. Canada.

Skippen, G. and Trommsdorff, V. 1986. The influence of NaCl and KCl on phase relations in metamorphosed carbonate rocks. Am. J. Sci. 286, 81-104.

Skogby, H., Bell, D.R. and Rossman, G.R. 1990. Hydroxide in pyroxene: variations in the natural environment. Am. Miner. 75, 764-774.

Slater, J.C. 1939. *Introduction to chemical physics*. McGraw-Hill, New York.

Slaughter, J., Kerrick, D.M. and Wall, V.J. 1975. Experimental and thermodynamic study of equilibria in the system $CaO-MgO-SiO_2-H_2O-CO_2$. Am. J. Sci. 275, 143-162.

Smith, C.S. 1948. Grains, phases, and interfaces: an interpretation of microstructure. Trans. Am. Inst. Min. Metal. Engrs. 175, 15-51.

Smith, C.S. 1952. Interphase interfaces. In *Imperfections in nearly perfect crystals*, ed. by W. Shockley. J. Wiley & Sons, New York.

Smith, C.S. 1954. The microstructure of polycrystalline materials. Trans. Chalmers Univ. Tech., Nr. 152, 1-49.

Smith, D.G.W. 1965. The chemistry and mineralogy of some emery-like rocks from Sithean Slaigh Strachur, Argyllshire. Am. Miner. 50, 1982-2022.

Smith, J.V. 1961. Explanation of strain and orientation effects in perthites. Am. Miner. 46, 1489-1493.

Smith, J.V. 1962. Crystal structures with a chabazite framework. I Dehydrated Ca chabazite. Acta Crystal. 15, 835-845.

Smith, J.V. 1974. *Feldspar minerals*, 2. *Chemical and textural properties*. Springer-Verlag, New York.

Smith, J.V. and Brown, W.L. 1988. *Feldspar minerals*. Springer-Verlag, Berlin.

Smith, P.P.K. 1977. An electron microscope study of amphibole lamellae in augite. Contr. Miner. Petrol. 59, 317-322.

Smithson, S.B. 1962. Symmetry relations in alkali feldspars of some amphibolite facies rocks from the southern Norwegian Precambrian. Norsk Geol. Tidsskr. 42, 586-599.

Sobolev, N.V. and Shatsky, V.S. 1990. Diamond inclusions in garnets from metamorphic rocks: a new environment for diamond formation. Nature 343, 742-746.

Somorjai, G.A. 1978. Surface science. Science 201, 489-496.

Somorjai, G.A. 1985. Surface science and catalysis. Science 227, 902-908.

Sorby, H.C. 1858. On the microscopical structure of crystals, indicating the origin of minerals and rocks. Quart. J. Geol. Soc. 14, 453-500.

Sørensen, K. 1979. Are coronas cooling products? Rapp. Grønlands geol. Unders. 89, 113-114.

Soula, J.-C. and Debat, P. 1976. Dévelopment et caractères des litages tectoniques. Bull Soc. géol. France XVIII, 1515-1537.

Spear, F.S. 1993. Metamorphic phase equilibria and pressure-temperature-time paths. Mineralogical Society of America, Monograph.

Spear, F.S., Selverstone, J., Hickmott, D., Crowley, P. and Hodges, K.V. 1984. $P-T$ paths from garnet zoning: a new technique for deciphering tectonic processes in crystalline terranes. Geology 12, 87-90.

Spencer, K.J. and Lindsley, D.H. 1981. A solution model for coexisting iron-titanium oxides. Am. Miner. 66, 1189-1201.

Spry, A. 1969. *Metamorphic textures*. Pergamon Press, Oxford.

Spycher, N.F. and Reed, M.H. 1988. Fugacity coefficients of H_2, CO_2, CH_4, H_2O, and of H_2O-CO_2-CH_4 mixtures. A viral equation treatment for moderate pressures and temperatures applicable to calculations of hydrothermal boiling. Geochim. Cosmochim. Acta 52, 739-749.

Stanton, R.L. and Gorman, H. 1968. A phenomenological study of grain boundary migration in some common sulfides. Econ. Geol. 63, 907-923.

Steiger, R.H. and Hart, R. 1967. The microcline-orthoclase transition within a contact aureole. Am. Miner. 52, 87-116.

Steinfeld, J.I., Francisco, J.S. and Hase, W.L. 1989. *Chemical kinetics and dynamics.* Prentice-Hall, Englewood Cliffs, NJ.

Stephansson, O. 1974. Stress-induced diffusion during folding. Tectonophysics 22, 233-251.

Stephenson, N.C.N. 1979. Coexisting garnets and biotites from Precambrian gneisses of the south coast of Western Australia. Lithos 12, 73-87.

Stillwell, F.L. 1918. The metamorphic rocks of Adelie Land. Australasian Antarctic Expedition, 1911-1914. Sci. Rept. Ser. A., Vol. III, Pt 1.

St-Onge, M.R. 1987. Zoned poikiloblastic garnets: $P-T$ paths and syn-metamorphic uplift through 30 km of structural depth, Wopmay Orogen, Canada. J. Petrology 28, 1-21.

Stormer, J.C. Jr. 1975. A practical two-feldspar geothermometer. Am. Miner. 60, 667-674.

Strens, R.G.J. 1968. Stability of Al_2SiO_5 solid solutions. Miner. Mag. 36, 839-849.

Strickland-Constable, R.F. 1968. *Kinetics and mechanisms of crystallization.* Academic Press, London.

Strömgärd, K.-E. 1973. Stress distribution during formation of boudinage and pressure shadows. Tectonophysics 16, 215-248.

Sunagawa, I. 1964. Growth spirals on phlogopite crystals. Am. Miner. 49, 1427-1434.

Sunagawa, I. 1984. Growth of crystals in nature. In *Materials science of the Earth's interior*, ed. by I. Sunagawa. Terra Scientifica Pub. Co., Tokyo.

Sunagawa, I. and Koshino, Y. 1975. Growth spirals on kaolinite group minerals. Am. Miner. 60, 407-412.

Suzuki, K. 1977. Local equilibrium during the contact metamorphism of siliceous dolomites in Kasuga-mura, Gifu-ken, Japan. Contr. Miner. Petrol. 61, 79-89.

Swalin, R.A. 1962. *Thermodynamics of solids.* J. Wiley & Sons, New York.

Talbot, J.L. and Hobbs, B.E. 1968. The relation of metamorphic differentiation to other structural features at three localities. J. Geol. 76, 581-587.

Tanner, S.B., Kerrick, D.M. and Lasaga, A.C. 1985. Experimental kinetic study of the reaction: calcite + quartz $\rightleftarrows$ wollastonite + carbon dioxide from 1 to 3 kilobars and 500 to 850 °C. Am. J. Sci. 285, 577-620.

Thierry, P., Chatillon-Colinet, C., Mathieu, J.C., Regnard, J.R. and Amossé, J. 1981. Thermodynamic properties of forsterite-fayalite (Mg_2SiO_4-Fe_2SiO_4) solid solutions. Determination of heat of formation. Phys. Chem. Miner. 7, 43-46.

Thompson, A.B. 1970. A note on the kaolinite-pyrophyllite equilibrium. Am. J. Sci., 268, 457-458.

Thompson, A.B. 1974. Calculation of muscovite-paragonite-alkali feldspar phase relations. Contr. Miner. Petrol. 44, 173-194.

Thompson, A.B. 1975. Calc-silicate diffusion zones between marble and pelitic schist. J. Petrology 16, 314-346.

Thompson, A.B. 1976. Mineral reactions in pelitic rocks: II. Calculation of some $P-T-X$ (Fe-Mg) phase relations. Am. J. Sci. 276, 425-454.

Thompson, A.B. 1983. Fluid-absent metamorphism. J. geol. Soc. (London) 140, 533-547.

Thompson, J.B. 1957. The graphical analysis of mineral assemblages in pelitic schists. Am. Miner. 42, 842-857.

Thompson, J.B. 1959. Local equilibrium in metasomatic processes. In *Researches in geochemistry*, ed. by P.H. Abelson. J. Wiley & Sons, New York.

Thompson, J.B. 1967. Thermodynamic properties of simple solutions. In *Researches in geochemistry*, vol. 2, ed. by P.H. Abelson. J. Wiley & Sons, New York.

Thompson, J.B. 1969. Chemical reactions in crystals. Am. Miner. 54, 341-375.

Thompson, J.B. 1982. Composition space: an algebraic and geometric approach. Rev. Miner. 10, 1-31.

Thompson, J.B. and Hovis, G.L. 1979. Entropy of mixing in sanidine. Am. Miner. 64, 57-65.

Thompson, J.B. and Norton, S.A. 1968. Paleozoic regional metamorphism in New England and adjacent areas. In *Studies of Appalachian geology*, ed. by E-An Zen, W.S. White, J.B. Hadley and J.B. Thompson Jr. J. Wiley & Sons, New York.

Thompson, J.B. and Waldbaum, D.R. 1968. Mixing properties of sanidine crystalline solutions: I Calculations based on ion-exchange data. Am. Miner. 53, 1965-1999.

Thompson, J.B. and Waldbaum, D.R. 1969. Mixing properties of sanidine crystalline solutions: III Calculations based on two-phase data. Am. Miner. 54, 811-838.

Thompson, P.H. 1976. Isograd patterns and pressure-temperature distributions during regional metamorphism. Contr. Miner. Petrol. 57, 277-295.

Thomson, W. 1887. On the division of space with minimum partitional area. Phil. Mag. 24, 503-514.

Topor, L., Kleppa, O.J., Newton, R.C. and Kerrick, D.M. 1989. Molten salt calorimetry of Al_2SiO_5 polymorphs at 1000 K. EOS 70, 493.

Torgeson, D.R. and Sahama, Th.G. 1948. A hydrofluoric acid solution calorimeter and the determination of the heats of formation of Mg_2SiO_4, $MgSiO_3$, and $CaSiO_3$. J. Am. Chem. Soc. 70, 2156-2160.

Törnebohm, A.E., 1877. Om Sveriges viktigare Diabas- och Gabbro- Arter. Kgl. Svesta Vetenskaps Akademiens Handlingar, B.14, No. 13.

Touret, J. 1971. Le facies granulite en Norvége Méridionale. II. Les inclusions fluides. Lithos 4, 423-436.

Trommsdorff, V. 1966. Progressive Metamorphose kieseliger Karbonatgesteine in den Zentralalpen zwischen Bernina und Simplon. Schweizer. Miner. Petrograph. Mitt. 46, 421-460.

Trommsdorff, V. 1972. Change in $T-X$ during metamorphism of siliceous dolomite rocks of the Central Alps. Schweizer. Miner. Petrograph. Mitt. 52, 567-571.

Truskinovskiy, L.M., Sigalovskaya, Yu.I., Senderov, E.E. and Urusov, V.S. 1988. A general quasichemical model of mineral disordering. Geochem. Int. 25, No. 6, 4-19.

Tubandt, C. and Reinhold, H. 1929. Chemische Gleichgewichte zwischen festen Salzen. Z. Phys. Chemie 140, 291-308.

Tullis, J. and Yund, R.A. 1982. Grain growth kinetics of quartz and calcite aggregates. J. Geol. 90, 301-318.

Turekian, K.K. and Phinney, W.C. 1962. The distribution of Ni, Co, Cr, Cu, Ba and Sr between biotite-garnet pairs in metamorphic sequences. Am. Miner. 47, 1434-1441.

Turner, F.J. 1958. Concept of metamorphic facies. In *Metamorphic reactions and metamorphic facies*. Geol. Soc. Am. Mem. 73.

Turner, F.J. 1967. Thermodynamic appraisal of steps in progressive metamorphism of siliceous dolomitic limestone. N. Jb. Miner. Mh. 1967, 1-22.

Turner, F.J. 1981. *Metamorphic petrology*. McGraw-Hill, New York.

Turner, F.J. and Verhoogen, J. 1960. *Igneous and metamorphic petrology*. McGraw-Hill, New York.

Tuttle, O.F. and Smith, J.V. 1958. The nepheline-kalsilite system II. Phase relations. Am. J. Sci. 256, 571-589.

Tyler, I.M. and Ashworth, J.R. 1982. Sillimanite-potash feldspar assemblages in graphitic pelites, Strontian area, Scotland. Contr. Miner. Petrol. 81, 18-29.

Ulbrich, H.H. and Waldbaum, D.R. 1976. Structural and other contributions to the third-law entropies of silicates. Geochim. Cosmochim. Acta 40, 1-24.

Valiev, R.Z., Mulyukov, R.R. and Ovchinnikov, V.V. 1990. Direction of a grain-boundary phase in submicrometre-grained iron. Phil. Mag. Letters 62, 253-256.

Valley, J.W. and O'Neil, J.R. 1984. Fluid heterogeneity during granulite facies metamorphism in the Adirondacks: stable isotope evidence. Contr. Miner. Petrol. 85, 158-173.

Vance, J.A. 1968. Metamorphic aragonite in the prehnite-pumpellyite facies, Northwest Washington. Am. J. Sci. 266, 299-315.

van Laar, J.J. 1908. Die Schmelz-oder Erstarrungskurven bei beinären Systemen, wenn die feste Phase ein Gemisch (amorphe feste Lösung oder Mischkristalle) der beiden Komponenten ist. Z. Phys. Chem. 64, 257-297.

Veblen, D.R. 1985a. Direct TEM imaging of complex structures and defects in silicates. Ann. Rev. Earth Planet. Sci. 13, 119-146.

Veblen, D.R. 1985b. Extended defects and vacancy non-stoichiometry in rock-forming minerals. In *Point defects in minerals*, ed. by R.N. Schock. Geophysical Monographs 31, American Geophysical Union, Washington, DC.

Veblen, D.R. and Buseck, P.R. 1981. Hydrous pyriboles and sheet silicates in pyroxenes and uralites: intergrowth microstructures and reaction mechanisms. Am. Miner. 66, 1107-1134.

Verhoogen, J. 1948. Geological significance of surface tension. J. Geol. 56, 210-217.

Vernon, R.H. 1970. Comparative grain-boundary studies of some basic and ultrabasic granulites, nodules, and cumulates. Scottish J. Geol. 6, 337-351.

Vernon, R.H. 1974. Controls of mylonitic compositional layering during non-cataclastic ductile deformation. Geol. Mag. 11, 121-123.

Vernon, R.H. 1976. *Metamorphic processes*. George Allen & Unwin, London.

Vernon, R.H. and Flood, R.H. 1977. Interpretation of metamorphic assemblages containing fibrolitic sillimanite. Contr. Miner. Petrol. 59, 227-235.

Vidale, R.J. 1974. Vein assemblages and metamorphism in Dutchess County, New York. Geol. Soc. Am. Bull. 85, 303-306.

Villiers, J.P.R. De 1971. Crystal structure of aragonite, strontianite, and witherite. Am. Miner. 56, 758-767.

Vinokurov, V.M., Zairpov, M.M. and Stepanov, V.G. 1961. A study of some Mn-containing carbonates by the method of electron paramagnetic resonance. Sov. Phys. Crystallogr. 6, 83-86.

Virgo, D. 1968. Partition of strontium between coexisting K-feldspar and plagioclase in some metamorphic rocks. J. Geol. 76, 331-346.

Virgo, D. and Hafner, S.S. 1970. Fe^{2+}-Mg order-disorder in natural orthopyroxenes. Am. Miner. 55, 532-559.

Vistelius, A.B. 1966. Genesis of the Mt. Belaya granodiorite, Kamchatka (an experiment in stochastic modeling). Dok. Akad. Nauk SSSR 167, 48-50.

Vistelius, A.B. 1972. Ideal granite and its properties. I. The stochastic model. Math. Geol. 4, 89-102.

Vistelius, A.B. 1992. *Principles of mathematical geology*. Kluwer Academic, Boston.

Volfinger, M. and Robert, J.-L. 1980. Structural control of the distribution of trace elements between silicates and hydrothermal solutions. Geochim. Cosmochim. Acta 44, 1455-1461.

Volkova, N.I. and Sheplev, V.S. 1980. A model of diffusion metasomatism in quasi-stationary approximation. Dok. Akad. Nauk SSSR 252, 1224-1227.

Voll, G. 1960. New work on petrofabrics. Lpool & Manchr. geol. J. 2, 503-567.

Volmer, M. 1939. *Kinetik der phasenbildung*. Steinkopff, Dresden.

Volmer, M. and Weber, A. 1926. Keimbildung in übersättigten Gebilden. Z. Phys. Chem. 119, 277-301.

Vrána, S. and Barr, M.W.C. 1972. Talc-kyanite-quartz schists and other high-pressure assemblages from Zambia. Miner. Mag. 38, 837-846.

Vry, J.K., Brown, P.E. and Valley, J.W. 1990. Cordierite volatile content and the role of CO_2 in high-grade metamorphism. Am. Miner. 75, 71-88.

Wager, L.R. 1959. Differing powers of crystal nucleation as a factor producing diversity in layered igneous intrusions. Geol. Mag. 96, 75-80.

Wagner, C. 1938. Über den Mechanismus von doppelten Umsetzungen durch Reaction im festen Zustand. Z. anorg. allg. Chem. 236, 320-338.

Wagner, C., 1975. Equations for transport in solid oxides and sulfides of transition metals. Progress in solid state chemistry, 10, 3-16.

Walther, J.V. and Orville, P.M. 1982. Volatile production and transport in regional metamorphism. Contr. Miner. Petrol. 79, 252-257.

Walther, J.V. and Wood, B.J. 1984. Rate and mechanism in prograde metamorphism. Contr. Miner. Petrol. 88, 246-259.

Wang, G.-F., Banno, S. and Takeuchi, K. 1986. Reactions to define the biotite isograd in the Ryoke metamorphic belt, Kii Peninsula, Japan. Contr. Miner. Petrol. 93, 9-17.

Warner, R.D. and Luth, W.C. 1974. The diopside-orthoenstatite two-phase region in the system $CaMgSi_2O_6$-$Mg_2Si_2O_6$. Am. Miner. 59, 98-109.

Weare, J.H., Stephens, J.R. and Eugster, H.P. 1976. Diffusion metasomatism and mineral reaction zones: general principles and application to feldspar alteration. Am. J. Sci. 276, 767-816.

Wegner, W.W. and Ernst, W.G. 1983. Experimentally determined hydration and dehydration reaction rates in the system MgO-SiO_2-H_2O. Am. J. Sci. 283-A, 151-180.

Wehmiller, J. 1972. Strontium: element and geochemistry. In *The encyclopedia of geochemistry and environmental sciences*, ed. by R.W. Fairbridge. Van Nostrand Reinhold, New York.

Weill, D.F. 1966. Stability relations in the Al_2O_3-SiO_2 system calculated from solubilities in the Al_2O_3-SiO_2-Na_3AlF_6 system. Geochim. Cosmochim. Acta 30, 223-237.

Wenk, H.-R. 1974. Howieite, a new type of chain silicate. Am. Miner. 59, 86-97.

Wenk, H.-R. 1985. (ed.) *Preferred orientation in deformed metals and rocks: an introduction to modern texture analysis.* Academic Press, Orlando, Fla.

Whalen, J.W. 1961. Heats of immersion and silica-water systems. In *Solid surfaces and the gas-solid interface.* Advances in Chemistry Series, No. 33, pp. 281-290. Am. Chem. Soc., Washington.

White, J.C. and White, S.H. 1981. On the structure of grain boundaries in tectonites. Tectonophysics 78, 613-628.

Whittaker, E.J.W. and Muntus, R. 1970. Ionic radii for use in geochemistry. Geochim. Cosmochim. Acta 34, 945-956.

Whitten, E.H.T., Dacey, M.F. and Thompson, K.D. 1975. Markov grain relationships of a Grenville granulite. Am. J. Sci. 275, 1164-1182.

Wickham, S.M. and Taylor, H.P. Jr. 1985. Stable isotope evidence for large-scale seawater infiltration in a regional metamorphic terrane; the Trois Seigneurs Massif, Pyrenees, France. Contr. Miner. Petrol. 91, 122-137.

Wildeman, T.R. 1970. The distribution of Mn^{2+} in some carbonates by electron paramagnetic resonance. Chem. Geol. 5, 167-177.

Wilkins, R.W.T. and Sabine, W. 1973. Water content of some nominally anhydrous silicates. Am. Miner. 58, 508-516.

Wilkinson, J.F.G. 1971. The petrology of some vitrophyric calc-alkaline volcanics from the Carboniferous of New South Wales. J. Petrology 12, 587-619.

Willaime, C. and Brown, W.L. 1974. A coherent elastic model for the determination of the orientation of exsolution boundaries: application to the feldspars. Acta Crystal. A30, 316-331.

Williams, M.L. and Grambling, J.A. 1990. Manganese, ferric iron, and the equilibrium between garnet and biotite. Am. Miner. 75, 886-908.

Williams, P.F. 1972. Development of metamorphic layering and cleavage in low grade metamorphic rocks at Bermagui, Australia. Am. J. Sci. 272, 1-47.

Williams, P.F. 1990. Differentiated layering in metamorphic rocks. Earth-Sci. Rev. 29, 267-281.

Williams, R.E. 1968. Space-filling polyhedron: its relation to aggregates of soap bubbles, plant cells, and metal crystallites. Science 161, 276-277.

Winkler, H. 1974. *Petrogenesis of metamorphic rocks*. Springer-Verlag, New York.

Winter, J.K. and Ghose, S. 1979. Thermal expansion and high-temperature crystal chemistry of the Al_2SiO_5 polymorphs. Am. Miner. 64, 573-586.

Wiser, N.M. and Wood, B.J. 1989. Experimental determination of activities of Fe-Mg olivines at 1400 K. EOS, p. 1401, Abstract V32C-2.

Wohl, K. 1946. Thermodynamic evaluation of binary and ternary liquid systems. Trans. Am. Inst. Chem. Eng. 42, 215-249.

Wohl, K. 1953. Thermodynamic evaluation of binary and ternary liquid systems. Chem. Eng. Progress 49, 218-219.

Wones, D.R. and Eugster, H.P. 1965. Stability of biotite: experiment, theory, and application. Am. Miner. 50, 1228-1272.

Wood, B.J. 1988. Activity measurements and excess entropy-volume relationships for pyrope-grossular garnets. J. Geol. 96, 721-729.

Wood, B.J. and Kleppa, O.J. 1981. Thermochemistry of forsterite-fayalite olivine solutions. Geochim. Cosmochim. Acta 45, 529-534.

Wood, B.J. and Walther, J.V. 1986. Fluid flow during metamorphism and its implication for fluid-rock ratios. Adv. Phys. Geochem. 5, 89-108.

Wood, R.M. 1979. The iron-rich blueschist-facies minerals. Miner. Mag. 43, 251-259, 363-370, 605-614.

Wright, T.L. 1967. The microcline-orthoclase transformation in the contact aureole of the Eldora Stock, Colorado. Am. Miner. 52, 117-136.

Wulff, G. 1901. Zur Frage der Geschwindigkeit des Wachsthums und der Auflösung der Kristall-flächen. Z. Kristal. Miner. 34, 449-530.

Wyllie, P.J. 1962. The effect of 'impure' pore fluids on metamorphic dissociation reactions. Miner. Mag. 33, 9-25.

Yamaguchi, Y., Akai, J. and Tomita, K. 1978. Clinoamphibole lamellae in diopside of garnet lherzolite from Alpe Arami, Bellinzona, Switzerland. Contr. Miner. Petrol. 66, 263-270.

Yardley, B.W.D. 1978. Genesis of the Skagit gneiss migmatites, Washington, and the distinction between possible mechanisms of migmatization. Geol. Soc. Am. Bull. 89, 941-951.

Yardley, B.W.D. 1989. *An introduction to metamorphic petrology*. Longman, London.

Yardley, B.W.D., MacKenzie, W.S. and Guilford, C. 1990. *Atlas of metamorphic rocks and their textures*. Longman Sci. & Tech., and J. Wiley & Sons, New York.

Yoder, H.S. 1950. The jadeite problem. Am. J. Sci. 248, 225-248, 312-334.

Yoder, H.S. 1955. Role of water in metamorphism. Geol. Soc. Am. Special Paper 62, 505-523.

Yokoi, K. 1983. Fe_2O_3 content of co-existing andalusite and sillimanite in the Ryoke metamorphic rocks occurring in the Hiraoka-Kadoya area, central Japan. Soc. Petrol., Miner. and Ore Deposits J. 78, 246-254.

Youden, W.J. 1959. Accuracy and precision; evaluation and interpretation of analytical data. In *Treatise on analytical chemistry*, ed. by I.M. Kolthoff and P.J. Elving. Part I, Vol. 1. Interscience Encyclopedia, New York.

Young, D.A. 1966. Decomposition of solids. *International encyclopedia of physical chemistry and chemical physics*. Pergamon Press, Oxford.

Yund, R.A. 1983a. Alkali feldspar exsolution: kinetics and dependence on alkali interdiffusion. In *Feldspars and feldspathoids*, ed. by W.L. Brown. D. Reidel, Dordrecht.

Yund, R.A. 1983b. Diffusion in feldspar. Rev. Miner. 2 (2nd edition), 203-222.

Yund, R.A. and Davidson, P. 1978. Kinetics of lamellar coarsening in cryptoperthites. Am. Miner. 63, 470-477.

Yund, R.A. and McCallister, R.H. 1970. Kinetics and mechanisms of exsolution. Chem. Geol. 6, 5-30.

Yund, R.A., McCallister, R.H. and Savin, M. 1972. An experimental study of nepheline-kalsilite exsolution. J. Petrology 13, 255-272.

Yutzy, H.C. and Kolthoff, I.M. 1937. Aging of fresh precipitates. XII. The equilibrium between mixed crystals of silver chloride and bromide and solutions. J. Am. Chem. Soc. 59, 916-921.

Zangwill, A. 1988. *Physics at surfaces*. Cambridge University Press, Cambridge.

Zen, E-An. 1966. Construction of pressure-temperature diagrams for multicomponent systems after the method of Schreinemakers—a geometric approach. U.S. Geol. Surv. Bull. 1225.

Zen, E-An. and Roseboom, E.H. 1972. Some topological relationships in multisystems of n+3 phases III. Ternary Systems. Am. J. Sci. 272, 677-710.

Zharikov, V.A. and Zaraiski, G.P. 1991. Experimental modelling of wall-rock metasomatism. In *Progress in metamorphic and magmatic petrology*, ed. by L.L. Perchuk. Cambridge University Press, Cambridge.

Zwart, H.J. 1960. Relation between folding and metamorphism in the central Pyrenees, and their chronological succession. Geol. en Mijnb. 22, 163-180.

Index